ULTRAHIGH-PRESSURE METAMORPHISM AND GEODYNAMICS IN COLLISION-TYPE OROGENIC BELTS

ULTRAHIGH-PRESSURE METAMORPHISM AND GEODYNAMICS IN COLLISION-TYPE OROGENIC BELTS

Final Report of the Task Group III-6 (1994–1998) of the International Lithosphere Project International Book Series, Vol. 4

W. G. Ernst and J. G. Liou, editors

International Lithosphere Program

Contribution No. 344

Bellwether Publishing, Ltd. **for the**
Geological Society of America

Published by

Bellwether Publishing, Ltd.
8640 Guilford Road, Suite 200
Columbia, MD 21046-2655
http://www.bellpub.com
for the Geological Society of America

Exclusively marketed throughout the world by:

Geological Society of America
P. O. Box 9140
Boulder, CO 80301-9140
email: pubs@geosociety.org
or order from the GSA Bookstore on the World Wide Web at
http://www.geosociety.org/bookstore.

Library of Congress Catalog Card Number 00-102362
ISBN 0-9665869-3-X

Printed by The Sheridan Press, Hanover, Pennsylvania.

ULTRAHIGH-PRESSURE METAMORPHISM AND GEODYNAMICS IN COLLISION-TYPE OROGENIC BELTS: FINAL REPORT OF THE TASK GROUP III-6 OF THE INTERNATIONAL LITHOSPHERE PROJECT

W. G. Ernst and J. G. Liou, editors

Contents

Overview of Ultrahigh-Pressure Metamorphism

Petrologic Evolution of Important Ultrahigh-Pressure Metamorphic Rock Types

Experimental Ultrahigh-Pressure Recrystallization

Classic Areas of Ultrahigh-Pressure Metamorphism

OVERVIEW OF ULTRAHIGH-PRESSURE METAMORPHISM

Preface[1]

WERNER SCHREYER

Research Group for High-Pressure Metamorphism, Institut für Mineralogie, Ruhr-Universität, D-44780 Bochum, Germany

AND CHENG YUQI

Chinese Academy of Geological Sciences, Baiwanzhuang Road 26, Beijing 100037, People's Republic of China

IN THE TITLE of his epoch-making 1984 paper, in which he described coesite, the high-pressure polymorph of silica from a metamorphic rock of the continental crust, Christian Chopin added the following specification: "A first record and some consequences" (Chopin, 1984). Since then the geoscientific world is experiencing the discovery of more and more such records and continues to debate all the consequences in what may well be regarded as another revolution in the understanding of the internal dynamics of our globe after the acceptance of plate tectonics.

Chopin's discovery was achieved using the methods of mineralogy and petrology, but the consequences affect nearly all other branches of solid Earth sciences as well. Thus, the concerted efforts of researchers in all of these fields are definitely necessary for a more complete understanding. Nevertheless, Chopin's publication, as well as the history of subsequent research on this topic, shows the fundamental importance of the Earth material sciences of mineralogy and petrology, which make it possible to decipher the record of rock burial and exhumation with the aid of experimental phase equilibria. It is fair to say that we would still not know about "ultradeep metamorphism" without the work of researchers in these fields as "mineral detectives."

After Chopin's first and some following records, a group of American scientists—in the spirit of the sensational—coined the term "ultrahigh-pressure metamorphism" (UHPM) for all pressure-temperature conditions endured by metamorphic rocks that lie within the stability field of coesite. Others, however, feared that this superlative was premature and might—in the long term—even be unjustified. What if the lower PT limits of the coesite field were drastically surpassed in rock metamorphism? Would we then use terms like "super-ultrahigh pressures"? Recent discoveries of mineral inclusions in diamonds suggest precisely this possibility, that the 100 to 150 km deep burial of UHPM rocks is not the "end of the line" in terms of preserved mineralogical evidence of deep-earth processes in rocks that can be collected at the Earth's surface. Consequently, at least one group of investigators already has retreated from the new terminology and now refers to typical UHPM mineral assemblages only as "very high pressure metamorphic" (VHPM).

Notwithstanding such semantic nuances, the ensuing discoveries of coesite elsewhere in the world and the identification of the—thus far—largest region of outcrop of UHPM rocks in eastern China proved that the time was ripe for a concerted international effort. Therefore, during 1993 the Inter-Union Commission on the Lithosphere operating the International Lithosphere Program (ILP) established a Task Group (III-6) with the mandate to foster research on "Ultrahigh-Pressure Metamorphism and Geodynamics in Collision-Type Orogenic Belts."

It thus was "officially" recognized that UHPM cannot be considered as simply a new branch of mineralogy-petrology, but that other fields of geosciences must necessarily be involved as well. Problems such as the mechanics of ultradeep subduction of low-density crustal materials into the denser upper mantle and the variable kinds and pathways of exhumation of UHPM rocks, together with possible velocities of these processes required input from specialists in structural geology, rock mechanics, geochronology, and many other fields.

Much has been achieved by the Task Group over the past five years. The present volume of its Final Report summarizes the major results, which obviously reflect not just the work of the relatively few (28) Task Group Members and their local groups. However, it became clear during the years of its existence that the Task Group, and especially its workshops, represented focal points that attracted

[1] This volume is contribution No. 344 of the International Lithosphere Program, representing the Final Report of Task Group III-6 of the International Lithospheric Project.

many other scientists interested in the field of UHPM. Needless to say, this Final Report cannot be considered the "end of the road" for UHPM studies. On the contrary, the field is still wide open for more, and especially for more detailed, efforts.

REFERENCES

Chopin, C., 1984, Coesite and pure pyrope in high-grade blueschists of the Western Alps: A first record and some consequences: Contrib. Mineral. Petrol., v. 86, p. 107–118.

Overview of UHP Metamorphism and Tectonics in Well-Studied Collisional Orogens

W. G. ERNST AND J. G. LIOU

Department of Geological and Environmental Sciences, Stanford University, Stanford, California 94305-2115

Abstract

In Eurasia, the Qinling-Dabie-Sulu belt of eastern China, the Kokchetav Complex of northern Kazakhstan, the Maksyutov Complex of the southern Urals, the Dora-Maira massif of the Western Alps, and the Western Gneiss Region of Norway mark profound intracontinental collisional sutures. Adjacent regions exhibit scant evidence of contemporaneous calc-alkaline volcanism/plutonism. Each ultrahigh-pressure (UHP) metamorphic complex contains mineralogic and textural relics of coesite ± diamond as well as other very high P, moderate-T phases such as K-rich clinopyroxene, Mg-rich garnet, ellenbergerite, lawsonite, Al-rutile, glaucophane, high-Si phengite, and the phase assemblages coesite + dolomite, magnesite + diopside, and talc + kyanite, diopside, jadeite, or phengite. In each of these well-studied Eurasian complexes, maximum pressures approached or exceeded 2.8 GPa. Deep-seated recrystallization of old, cool continental crust took place during Phanerozoic time. Subduction zones constitute the only known plate-tectonic environment where such high-P, low-T conditions exist. Disaggregated, exhumed ultrahigh-pressure terranes consist of relatively thin sialic sheets 5 ± 3 km thick. After cessation of UHP recrystallization, tectonic slices ascended largely because of buoyancy to shallow depths along stress guides provided by the subduction zones themselves. Collisional sheets that retain UHP relics (micro-inclusions enclosed in strong, impermeable, unreactive mineralogic host grains) lost heat by conduction across both upper, normal-fault and lower, reverse-fault contacts. These sheets rose to mid-crustal levels rapidly at exhumation rates approaching 10 mm/yr. Backreaction attending decompression in all cases was nearly complete; where UHP relics survive, retrogression evidently was limited by the coarse grain size and relative impermeability of the rocks, as well as by declining temperature and lack of aqueous fluids.

Introduction

SEVERAL EURASIAN intracratonic collisional orogens exhibit rare, scattered effects of subsolidus UHP recrystallization (Liou et al., 1994; Coleman and Wang, 1995; Harley and Carswell, 1995; Schreyer, 1995; Maruyama et al., 1996). Well-investigated tracts include the Qinling-Dabie-Sulu belt of eastern China, the Kokchetav Complex of northern Kazakhstan, the Maksyutov Complex of the southern Urals, the Dora-Maira massif of the Western Alps, and the Western Gneiss Region, southwestern Norway. These five terranes occupy tectonic suture zones juxtaposing markedly dissimilar continental entities. Slices of these sialic complexes contain traces of phases that attest to recrystallization at mantle depths of, or exceeding, 100 km (Wang et al., 1995; Schreyer, 1988). Mineralogic evidence includes the very rare preservation of minerals and assemblages such as coesite, diamond, K-rich clinopyroxene, pyropic garnet, Al-rutile, and volatile-bearing assemblages such as magnesite + diopside, coesite + dolomite, talc + kyanite and/or jadeite and/or diopside and/or high-Si phengite, ellenbergerite, lawsonite, and glaucophane. Phase-equilibrium data and thermobarometric computations document attending temperatures of ~700 to 900°C and pressures approaching 2.8 to 4.0 GPa. The discovery of Fe-Ti oxide inclusions in olivine grains in a peridotite from Alpe Arami, southern Switzerland in rocks that have been studied for years, and the provocative interpretation of recovered mantle crystallization depths of ~300 km (Dobrzhinetskaya et al., 1996) raises the possibility that some UHP complexes ascended from depths far greater than currently recognized. Even more recently, the report of majorite component in garnet from the Western Gneiss Region (van Roermund et al., 1998), and allegedly very high pressure clinoenstatite in garnet lherzolites from the Sulu terrane (Zhang et al., 1998) and Alpe Arami (Green et al., 1998), are additional examples of potentially very high pressure phase configurations.

As with relatively high pressure (HP) metamorphic terranes, specialized UHP conditions appear to have been generated through subduction, reflecting the fact that geologic materials are poor thermal conductors (Ernst, 1971; Peacock, 1995). No other terrestrial setting provides the appropriate P-T conditions indicated for this type of solid-state recrystallization. In order to explain the generation of ultrahigh-pressure complexes, the deep underflow of coherent tracts of ancient, cool, sialic-crust-capped lithosphere seems to be required (Ernst and Peacock, 1996). Thus, UHP orogenic belts attest to lithospheric descent prior to final suturing of the colliding continental crust–capped plates. The subsequent decoupling and ascent of such collisional complexes is also understandable, based on the buoyancy of the transformed sialic crust (Cloos, 1993; Ernst and Liou, 1995).

Late-stage adiabatic decompression would result in the passage of rising UHP complexes through regimes characterized by high T and relatively low P—700 to 800°C and 0.2 to 1.0 GPa. Complete annealing—even partial fusion—would be expected, and the very rare occurrence of UHP relics suggests that phase assemblages have indeed backreacted thoroughly on return toward the Earth's surface. However, because of the short duration of individual UHP events, it appears that during prograde, subduction-zone metamorphism, metastable low-P mineral assemblages in some cases failed to produce the stable UHP configuration (Erambert and Austrheim, 1993; Harley and Carswell, 1995; Zhang and Liou, 1997); similarly, on later decompression, retrogression may not run to completion either (Rubie, 1986, 1990; Mosenfelder and Bohlen, 1997). The persistance of metastable phases, even under extreme P-T conditions, evidently is a reflection of the dominant kinetic effect of an aqueous fluid: transformations run rapidly when H_2O is present as a free phase, but rates are orders of magnitude slower when it is absent.

Brief Sketches of Eurasian Ultrahigh-Pressure Complexes

Qinling-Dabie-Sulu belt

The 2000 km long Qinling-Dabie-Sulu orogen in eastern China defines the E-W–trending junction between the Sino-Korean craton to the north and the Yangtze craton to the south (Huang, 1978; Dong et al., 1986; Dong, 1993). The central part of this tectonic zone consists of a series of fault-bounded, penetratively deformed blocks that expose, from south to north: (1) Proterozoic, uppermost Proterozoic (Sinian), and Lower Paleozoic metavolcanogenic, metaclastic, and metacarbonate strata, overlain by lesser amounts of bimodal metavolcanic + metasedimentary rocks of Late Proterozoic age, all resting on latest Archean(?) to Early + Middle Proterozoic, dominantly quartzofeldspathic gneisses of the Yangtze basement; (2) orthogneisses and migmatites of uncertain cratonal affinities but chiefly Early Cretaceous ages of origin and/or metamorphism; and (3) feebly recrystallized, chiefly metaclastic Foziling and spatially associated greenschists. The first two groups of lithologies represent northerly portions of the Yangtze craton, whereas Foziling rocks are accreted to the Sino-Korean shield (Ernst et al., 1991; Liou et al., 1995; Hacker et al., 1996). Except in the Dabie and Hong'an blocks, schistosites dip steeply to the north, and folds verge to the south (Huang and Wu, 1992; Hacker et al., 1995); in these latter areas, late-stage deformation may be analogous to the backfolding characteristic of the Western Alps (e.g., Dal Piaz et al., 1972).

Within the southern part of the Hong'an block, HP recrystallized rocks contain, from south to north, weakly reconstituted chlorite-actinolite schists, grading to transitional blueschist/greenschist-facies assemblages; farther north, greenschist, epidote amphibolite, and eclogite assemblages progressively replace lower-grade phase compatibilities (Zhang and Liou, 1994; Eide, 1995). Actinolite, crossite, and Mg-riebeckite associated with chlorite on the south give way northward to winchite and yet farther north to barroisite or omphacite + garnet. Late, high-rank amphibolite-facies metamorphism in the northern part of the terrane has thoroughly overprinted precursor higher-pressure assemblages. Phase-equilibrium data, mineral chemistry, and thermobarometry indicate physical conditions for the earlier HP metamorphism (Ernst et al., 1991; Liou et al., 1996) as follows: blueschist-greenschist = 325 to 425°C, 0.5 to 0.7 GPa; eclogites = 500 to 520°C, 1.0 to 1.2 GPa. Similar but lower-grade parageneses are displayed to the WNW in the Qinling Mountains.

Coesite inclusions in garnet, zoisite, zircon, kyanite, dolomite, and omphacite occur in the central part of the belt, confined to UHP eclogitic rocks of the Hong'an, Dabie Shan, and Sulu blocks (Okay et al., 1989; Wang et al., 1989; Wang and Liou, 1991; Zhang et al., 1994, 1995); associated paragneisses, schists, and calc-silicate rocks contain coesite in zircon (Sobolev et al., 1994) and in dolomite (Schertl and Okay, 1994; Zhang and Liou, 1996), and coesite pseudomorph inclusions in garnet (Wang and Liou, 1991). Rare microdiamonds have also been reported from garnetiferous UHP rocks (Xu et al., 1992; Okay, 1993). Phase-equilibrium data and mineralogic exchange reactions indicate that Dabie ultrahigh-pressure coesite-bearing eclogites and enclosing country-rock gneisses ± marbles formed at pressures >2.8 GPa and diamond-bearing lithologies at pressures >3.5 GPa, both at temperatures of 750°C or more. Based on U-Pb, Sm-Nd, and $^{40}Ar/^{39}Ar$ systems, recrystallization ages for the HP/UHP mineral assemblages are Triassic, 240 to 245 Ma, with exhumation to deep crustal levels by ~230 Ma (Okay et al., 1993; Ames et al., 1995; Rowley et al., 1997; Hacker et al., 1998; Webb et al., 1999). This assemblage has been intensely overprinted by younger, high-rank amphibolite-facies associations, and UHP phases are present only as rare relics; mid-crustal annealing apparently took place by 200 ± 5 Ma. Variably deformed plutons, orthogneisses, and migmatites, principally of Cretaceous age, lie directly north of, and intrude, the ultrahigh-pressure belt (Hacker et al., 1996, 1998). Eastward from the Dabie Shan, across the NNE-trending Tan-Lu fault, similar HP/UHP mafic and quartzofeldspathic assemblages crop out in the Sulu area (Hirajima et al., 1990; Wang et al., 1993; Zhang et al., 1994, 1995; Ye et al., 1996; Zhang and Liou, 1998).

Blueschist + eclogite mineral parageneses, bulk-rock + phase chemistries, and regional structural-lithologic relations of the Qinling-Dabie-Sulu orogenic belt suggest formation attending northward underflow of the Yangtze continental margin beneath the Sino-Korean block after the subduction of an intercratonic ocean basin. Fragments of this consumed ophiolitic basin are not preserved in the rock record, but remnant magnetic studies demonstrate the postulated latitudinal closure (Yang et al., 1992; Nie and Rowley, 1994). Accretion of the Yangtze shield to the enlarging Asiatic assembly evidently took place near the end of Triassic time (Liou et al., 1996). Especially profound subduction, now recovered, that produced the UHP Qinling-Dabie-Sulu metamorphic complex apparently occurred as a consequence of plate descent involving a narrow, cool microcontinental salient (Yin and Nie, 1993) carried along with negatively buoyant ocean crust–capped lithosphere. Decoupling of the subducted continental crust at great depth, followed by differential return toward the surface of deformed slices, ensued. Delayed separation from the sinking oceanic lithosphere may account for the extremely high pressures of some of the recovered terranes; episodic ascent of individual sialic slabs after different storage intervals may explain the contrasting metamorphic facies and later overprinting in different segments of the orogenic belt (Ernst and Liou, 1995). Maruyama et al. (1994) and Maruyama (1997) hypothesized the subhorizontal extrusion of thin, fault-bounded slices of high-pressure and ultrahigh-pressure rocks to account for their spatial distribution in the Dabie Shan. Hacker et al. (1995) advanced a tectonic model involving the northward emplacement of a UHP nappe on the order of 5 to 10 km maximum thickness, which arrived at mid-crustal levels by Early Jurassic time. In any case, all Dabie-Sulu tectonic models invoke relatively thin slab geometries for the UHP and HP lithostratigraphic units, in order to account for the observed structural relationships and P-T histories (Hacker and Peacock, 1995; Ernst and Peacock, 1996).

Kokchetav Complex

The Kokchetav Complex is an isolated massif within the Central Asiatic accretionary zone. It lies between the Siberian platform on the northeast, the Russian platform + bounding Uralian suture zone on the west, and the Kazakhstan–North Tienshan microcontinent on the south (Dobretsov and Sobolev, 1970, 1984; Hamilton, 1970; Zonenshain et al., 1990; Sengör et al., 1993; Zhang et al., 1997). The complex comprises dislocated, reassembled fragments of the pre–Late Riphean sialic crust, and constitutes a penetratively deformed Precambrian basement high surrounded by a Phanerozoic fold belt. The amalgamated assembly of differentially exhumed blocks lies between two major WNW-trending transcurrent shear zones. It is comprised of gneissic basement processing 2.2 to 2.3 Ga Sm-Nd model ages (Dobretsov et al., 1995), overlain by Upper Proterozoic–Vendian platform strata. REE patterns for the metasedimentary units suggest a passive-margin setting (Dobrzhinetskaya, 1994). The complex is overlapped on all sides by Middle and Upper Paleozoic multicycle platform and arc/backarc strata, reflecting the Devonian initiation of a postorogenic supracratonal basin. Syntectonic granitoids of chiefly Late Ordovician age, but including Late Cambrian(?) gabbroic precursors, and Silurian–Devonian + younger anorogenic granitic plutons have invaded the Kokchetav Complex, especially along its southern margin (Dobretsov et al., 1995).

A tectonic mega-mélange of seven pre-Ordovician units (I–VII) of constrasting lithologies and P-T conditions of metamorphism, overlain and/or intruded by four post-recrystallization entities, constitutes the complex. The outcrop area of diamondiferous unit I is approximately 15 × 50 km. Detailed studies of the mineralogy and bulk-rock geochemistry of these units have been carried out by numerous investigators (e.g., Sobolev et al., 1986, 1991; Shatsky et al., 1989, 1991, 1995; Sobolev and Shatsky, 1990; Shatsky and Sobolev, 1993; Dobrzhinetskaya et al., 1994). Paraschists and paragneisses, which contain biogenically produced carbon, were laid down near the surface of the Earth (Sobolev and Sobolev, 1980; Chopin and Sobolev, 1995). Microdiamonds occur principally as inclusions in strong, refractory, neoblastic garnet, zircon, clinopyroxene, and kyanite in metasedimentary unit I. Rare coesite micro-inclusions and pseudomorphs have been discovered in zircon and garnet of the interlayered biotite gneiss and within eclogitic zircon (Sobolev et al., 1994; Dobretsov et al., 1995; Zhang et al., 1997). Systematic mineral parageneses and textural relationships support the conclusion that the metamorphic assemblages represented a close approach to chemical equilibrium during their formation.

Metamorphism of diamond-bearing paragneisses and schists of unit I and associated tectonic entities took place in Early Cambrian time, as indicated by Sm-Nd and U-Pb ages between 530 and 540 Ma for the UHP eclogitic and diamondiferous metasedimentary rocks (Jagoutz et al., 1991; Claoue-Long et al., 1991; Shatsky et al., 1993). P-T conditions involved minimum pressures of 4.0 GPa and temperatures >900°C (Sobolev et al., 1986, 1991; Shatsky et al., 1989, 1991, 1993, 1995, in press; Sobolev and Shatsky, 1990). Adjacent units II, III, and IV(?) contain mineralogic evidence of slightly to considerably lower pressures and temperatures. Magnesite + diopside pairs, coesite, pyropic garnet, potassic clinopyroxene, Si-rich phengite, barroisite and/or crossite, and Al-rutile, as well as the assemblage talc + kyanite + pyropic garnet, all testify to elevated pressures of crystallization. The metamorphosed lithotectonic units represent relict stages in what may initially have been a continuous, synchronous P-T series, but intense post-metamorphic tectonic disruption has resulted in a chaotic mixture. Only units I-III, and probably part of VI, represent unambiguous portions of a dismembered HP/UHP, subduction-zone paragenesis.

Late Vendian–Early Cambrian subduction of the Kazakhstan–North Tienshan microcontinental salient to depths exceeding 125 km, followed by decoupling from the descending oceanic crust–capped lithosphere, is hypothesized to account for the ultrahigh-pressure metamorphism of the Kokchetav Complex (Dobretsov et al., 1995). The dimensions of UHP unit I are poorly constrained as a result of limited exposures, but judging from the outcrop pattern (Dobretsov, 1998), it represents a tectonic slice that cannot be thicker than 5 to 10 km, and may be much thinner. New mapping by a Japanese research group has provided additional mineralogic, structural, and petrologic information (Ishikawa et al., 1998; Kaneko et al., 1998). Low-angle tectonic contacts have been recognized for the first time. Provisional conclusions from the new field and laboratory studies are as follows (Maruyama, 1998): The HP/UHP terrane is structurally overlain by a weakly recrystallized unit, and is underlain by the low-pressure Daulet suite. These two major tectonic boundaries are subhorizontal, sharp, and locally offset by high-angle normal faults. NE-SW– and NW-SE–trending, vertical strike-slip faults suggest post-orogenic transpression. Fabric analyses confirm the presence of systematic normal faulting along the top and reverse faulting along the bottom of the HP/UHP slice. The structurally intermediate zone is occupied by the highest-grade diamondiferous assemblages. The HP/UHP complex roots to the south. Exhumation to mid-crustal levels and cooling of the mega-mélange evidently took place by ~515 to 517 Ma, according to $^{40}Ar/^{39}Ar$ ages on white mica and biotite from retrogressed garnetiferous paragneiss (Troesch and Jagoutz, 1993). Thus, the presently exposed complex was exhumed rapidly, and by Late Cambrian time had been stabilized as part of the Kazakhstan–North Tienshan sialic collage.

Maksyutov Complex

The relatively high pressure metamorphic Maksyutov Complex, southern Ural Mountains, Russia, is part of the Paleozoic Ural-Mongolian fold system (Sengör et al., 1993; Berzin et al., 1996). It forms an elongate N-S belt along the junction between the Siberian + Kazakhstan–North Tienshan craton and the Russian platform (Zonenshain et al., 1984, 1990; Zacharov, 1991; Puchkov, 1993; Matte, 1995). The metamorphic complex is bounded to the east by the Main Ural thrust, juxtaposing it against feebly recrystallized Ordovician–Silurian ophiolites + associated flysch. The Maksyutov Complex (Lennykh, 1977; Moscovchenko, 1982; Valizer and Lennykh, 1988; Dobretsov and Dobretsova, 1988) consists of two superimposed tectonic entities that together constitute a composite slab ~6 to 12 km thick and ~150 km in length. The lithotectonic entities are characterized by contrasting mineral parageneses: the polymetamorphic eclogitic micaschist, unit #1, underwent HP to UHP metamorphism, overprinted by later, lower-P assemblages, whereas the meta-ophiolite, unit #2, was subjected to blueschist-, then greenschist-facies metamorphism (Beane et al., 1995). Unit #1 probably represents part of the Russian platform of Late Proterozoic age, 1150 to 1200 Ma, based on U-Pb zircon ages of metavolcanics, according to Dobretsov and Sobolev (1984); some of the protolith may be younger, 589 ± 69 Ma, judging from the Sm-Nd garnet-omphacite bulk-rock isochron reported by Dobretsov et al. (1996). Unit #1 consists of supracrustal lithologies, dominantly pelitic, graphitic schists + subequal volumes of quartzofeldspathic paragneisses, with less abundant interlayered boudins and tectonic blocks of mafic + ultramafic composition. The eclogitic phase assemblages crystallized at ~375 Ma, as indicated by the $^{40}Ar/^{39}Ar$ phengite and U/Pb rutile data of Beane et al. (1996). Unit #2 is comprised of Lower Paleozoic oceanic crust and associated sediments, formed during Ordovician–Silurian time (Sm-Nd isochron on oceanic greenstones) (Edwards and Wasserburg, 1985).

The Maksyutov Complex is the initial locality where quartz intergrowth aggregates, encapsulated within garnet, were first interpreted as coesite pseudomorphs (Chesnokov and Popov, 1965). They are confined to unit #1, regarded as the structurally lower of two lithotectonic units by Lennykh (1977). Relict coesite was later reported from quartz + almandine + jadeite rocks by Dobretsov and Dobretsova (1988). More recent investigations have failed to identify coesite. If present, this phase would indicate that tectonic unit #1 underwent UHP metamorphism at subduction-zone depths exceeding 90 km. The mineral parageneses for HP/UHP metamorphosed rocks of unit #1 described by Dobretsov et al. (1996) include: (a) coesite (pseudomorphs) + garnet + omphacite + rutile + zoisite; (b) jadeite + quartz + garnet + kyanite + paragonite; (c) garnet + omphacite + barroisite; and (d) garnet + glaucophane + lawsonite. Culminating physical conditions attending formation of the eclogitic phase assemblages were 1.5 to 2.7 GPa at ~600 to 650°C. Blocky graphite cuboids were recently described by Lennykh et al. (1995) and Leech and Ernst (1998). If these cuboids are pseudomorphs after coarsely crystalline, neoblastic diamond, temperatures and pressures would have been even more extreme. Retrogression of unit #1 involving production of garnet-glaucophane schists, then greenschists, indicates a clockwise P-T trajectory (Beane et al., 1995). Ophiolitic unit #2 is characterized by sheets of serpentinite containing lawsonite-bearing metarodingite and gabbroic amphibolite. Tabular lawsonite porphyroblasts have been replaced by clinozoisite. Meta-mélanges contain blocks of ultramafic lithology, metabasaltic blueschists, and metasedimentary quartz-jadeite rocks.

The generation of associated Maksyutov unit #2 Ordovician–Silurian ophiolites, the suturing of units #1 and #2, and the suturing of the amalgam against the Devonian Magnitogorsk island arc were Middle Paleozoic events. The UHP recrystallization in both the Maksyutov and Kokchetav terranes may reflect an Early Paleozoic collision of the Russian platform along its eastern edge with a prong of the Kazakhstan–North Tienshan microcontinental assembly on the south (Shatsky et al., 1995). The last stages of penetrative deformation and blueschist-facies metamorphism of units #1 and #2 at ~355 Ma ($^{40}Ar/^{39}Ar$ mica data of Matte et al., 1993; Beane et al., 1996) were contemporaneous with subduction and accretion of intervening intracontinental ocean crust as young as the Late Silurian (Edwards and Wasserburg, 1985, Sm-Nd isochron). This date provides only an upper limit for the time of initial exhumation of the HP/UHP complex. Late-stage greenschist-facies metasomatism took place at ~335 Ma, when both Maksyutov units #1 and #2 were sequestered at relatively shallow crustal levels. The Mugodzhar and Ilmen microcontinents of Vendian–Ordovician ages consist of migmatites and granulites that were thrust beneath the eastern edge of the Devonian Magnitogorsk island arc (Zonenshain et al., 1990). Continuing deformation of the Maksyutov Complex was related to mid-Paleozoic sinistral strike slip along the Main Uralian fault (Dobretsov et al., 1996). This large-scale transcurrent motion indicates a change in Middle Paleozoic lithospheric plate kinematics that resulted in surface exposure of the subducted but buoyant Maksyutov HP/UHP quartzofeldspathic rocks.

Western Alps

The Alps have long been recognized as a contractional orogen reflecting the consumption of an intervening narrow seaway, Mesozoic Tethys, and subduction of the European foreland beneath the stable, nonsubducted African plate. Metamorphic assemblages reflect increasing depths of descent of now-exhumed segments of the European continental margin and Tethyan oceanic crust progressing toward and beneath the Po Plain. The nonsynchronous mineral-facies sequence ranges from anchimetamorphic to eclogitic, reflecting the polarity of subduction (Ernst, 1971). Throughout the Alps, effects of the earlier HP metamorphism have been partly obliterated by later greenschist-facies overprinting. Although UHP recrystallization has been recognized at isolated localities near Alpe Arami in the southernmost central Alps (Ernst, 1977; Evans et al., 1979; Becker, 1993), and the Zermatt area (Reinecke, 1991), the largest and most thoroughly documented tract crops out in the environs of the Dora-Maira massif (Chopin, 1984; Chopin et al., 1991; Schertl et al., 1991).

Similar to other Alpine internal complexes, the lowermost structural unit of the Dora-Maira massif consists of Hercynian (~300 Ma) and older European granodioritic gneisses + metagranites (Monie and Chopin, 1991; Tilton et al., 1997), and an autochthonous cover series of Carboniferous(?) graphitic and conglomeratic schists, Permo-Triassic metaclastics, and Middle Triassic carbonate strata (Michard, 1965, 1967; Vialon, 1966; Borghi et al., 1985). This central massif is surrounded and overlain by allochthonous Penninic nappes, consisting of Mesozoic Tethyan oceanic crust and associated deep-water limestones + overlying calcareous graywackes; slices of orthogneiss and polymetamorphic schist constituting portions of the pre-Alpine basement also are associated. Structurally higher Austroalpine nappes and the Sesia Lanzo zone represent portions of the more southerly continent, but the lowermost levels of both thrust sheets exhibit the effects of HP metamorphism; accordingly, at least tectonically lower slices of the African lithospheric plate must have decoupled from the parental slab and were entrained with the more northerly subducted terrane (Dal Piaz et al., 1972; Ernst, 1973).

The Dora-Maira massif consists of a number of fault-bounded, subhorizontal lithotectonic packages (Henry, 1990; Chopin et al., 1991; Avigad, 1992). The basal Pinerolo nappe consists of Carboniferous orthogneisses, as well as Permo-Triassic and Triassic metasedimentary rocks + Permian diorites, incipiently recrystallized to blueschist-facies assemblages (Borghi et al., 1985). The Venasca complex has been thrust over the basal nappe, and consists of several tectonically juxtaposed slices of Late Paleozoic, polymetamorphic schist and orthogneiss. The lower part of this unit includes boudinaged quartzite layers that contain coesite inclusions in pyropic garnets, but rare UHP relict assemblages also are scattered throughout fine-grained orthogneisses, and in pelitic schists + eclogitic pods (Michard et al., 1995). This sheetlike unit, approximately 1 to 2 km thick, has been juxtaposed against a tectonically higher portion of the Venasca nappe which, while carrying HP eclogitic rocks, apparently is devoid of UHP indicators. The overlying polymetamorphosed Dronero unit is similar to the Venasca complex in terms of protolith compositions, but contains traces of pre-Alpine amphibolite-facies recrystallization (Compagnoni et al., 1995), and subsequently was recrystallized under HP metamorphic conditions. Major P-T discontinuities, therefore, characterize tectonic contacts of the Venasca with the underlying and overlying nappes. Faulting has placed a thin slice of Penninic ophiolitic material of the Monte Viso complex + younger Mesozoic metasediments over the Venasca unit (Phillipot, 1988; Henry, 1990; Blake et al., 1995). In turn, the allochthonous oceanic nappe has been overthrust by a sialic crustal sheet consisting of polymetamorphic schists and orthogneisses, associated with metaclastics of roughly mid-Mesozoic depositional age; both nappes exhibit HP mineral parageneses.

Bulk-rock compositions of the quartzites that retain coesite and coesite pseudomorphs are highly magnesian. The UHP assemblage consists of pyrope + kyanite + talc + phengite + coesite (now mostly quartz). The quartzite boudins occur within fine-grained gneisses consisting of the assemblage albite + microcline + phengite + biotite + quartz + clinozoisite, along with armored, back-reacted grossular-rich garnet and rutile. Metabasaltic eclogite assemblages present within the Venasca nappe record the Alpine HP event, but do not contain surviving coesite relics. Other phases compatible with the UHP assemblages in the quartzites, however, suggest coherence of the terrane during isofacial ultrahigh-pressure Alpine recrystallization (Chopin, 1981, 1984; Kienast et al., 1991; Reinecke, 1991). The peak metamorphism of the Dora-Maira massif is estimated at 2.8 to 3.5 GPa and 700 to 750°C for the coesite-bearing quartzites (Schertl et al., 1991; Compagnoni et al., 1995). Thermobarometric measurements (Michard et al., 1995) suggest that eclogites of the other Venasca units were attended by physical conditions on the order of 1.7 to 2.0 GPa and 550 to 580°C. Eclogitic assemblages of the Monte Viso nappe experienced approximately 1.2 GPa at 500 to 550°C. Fine-grained quartzofeldspathic gneisses enclosing the UHP layers of pyropic quartzite carry mineral associations reflecting overprinting at ~0.3 to 0.6 GPa and 350 to 500°C. The earlier sets of conditions attest to tectonic assembly of various lithostratigraphic units within an active subduction zone, whereas the much lower pressures of retrograde metamorphism undoubtedly reflect cooling accompanying exhumation of the imbricated complex to mid-crustal levels.

At least two distinct stages of relatively high pressure metamorphism seem to be represented in the Alps overall. Underflow and continental collision were initiated in the Cretaceous (90 to 125 Ma), and may account for some of the HP parageneses, especially in the Western Alps (Paquette et al., 1989; Monie and Chopin, 1991). Oligocene–Miocene gneissic and granitoid domal uplift record a widespread Alpine thermal event, but surviving UHP mineral assemblages also appear to have been produced at ~35 to 40 Ma (Tilton et al., 1991; Gebauer et al., 1993, 1997). Perhaps a brief episode of intracontinental convergence at ~35 to 40 Ma caused renewed recrystallization and the rapid uplift of deeply buried rocks (Gebauer, 1996).

Western Gneiss Region

Along the southwestern coast of Norway, the suture zone occupied by rocks of the Western Gneiss Region reflects collision of the eclogite-bearing Greenland sialic crust–capped lithospheric plate with Fennoscandia during the Caledonian orogeny (Gilotti, 1993; Krogh and Carswell, 1995; Brueckner et al., 1998). The gneiss complex constitutes the basal tectonic member in a sequence of nappes that were thrust eastward over the Fennoscandian continental margin. These allochthonous sheets occupy an area 350 km in length and up to 150 km in width. Basement protoliths have Middle Proterozoic or older ages of formation (Kullerud et al., 1986), but polymetamorphic nappes as young as Vendian and/or Early Paleozoic also are part of the structural stack. Two principal lithotectonic entities are recognized in the Western Gneiss Region (Krogh, 1977; Cuthbert et al., 1983; Griffin et al., 1985; Mörk and Krogh, 1987)—(1) the HP/UHP metamorphosed Fjordane complex, a mixed assembly of supracrustals, anorthosites, and augen gneisses on the north and west; and (2) the Josuedl complex, a more coherent migmatitic orthogneiss unit recrystallized under lower-pressure conditions, on the east.

Eclogite phase assemblages are predominantly omphacite + garnet, with minor quartz, rutile, phengite, and/or kyanite; coesite has been documented as rare micro-inclusions within garnet and clinopyroxene in eclogite from westernmost exposures of the Fjordane complex (Smith, 1984; Smith and Lappin, 1989; Wain, 1997). The discovery of microdiamond in kyanite + garnet + biotite gneiss cropping out to the north on Fjörtoft Island (Dobrzhinetskaya et al., 1995; Smith, 1995) further supports the UHP nature of these intensely recrystallized rocks. Peak metamorphic conditions for the parautochthonous coesite ± diamond-bearing rocks of the Western Gneiss Region are estimated as 700 to 850°C at 3.0 to 3.5 GPa (Carswell et al., 1985). Associated garnet peridotites possess comparable temperatures and pressures of formation (Krogh and Carswell, 1995). Structural concordance between the HP and UHP lithologies in gneisses, metaperidotites, meta-anorthosites, and augen gneisses supports the hypothesis of a regionally extensive metamorphic event within the Fjordane and Josuedl complexes. Estimated physical conditions attending the relatively high pressure event range from 500°C and 1.0 to 1.2 GPa in the east, to 775 ± 75°C at 3.0 to 3.5 GPa in the extreme west (Griffin et al., 1985; Medaris and Wang, 1986; Smith and Lappin, 1989; Krogh, 1997). Resulting from the consumptive collision between the Greenland and Fennoscandian cratons, decoupling and ascent of the sialic crustal slices apparently were driven by buoyancy to mid-crustal levels (Dewey et al., 1993). Overprinted mineral assemblages attest to a retrograde P-T trajectory indicating nearly isothermal decompression from eclogite to granulite- or high-rank amphibolite-facies conditions; these relationships suggest a moderately rapid, virtually adiabatic rise of the gneiss complex. Recent mapping in the north-central segment of the Western Gneiss Region by Robinson and Terry (1998) indicates the presence of a stack of nappes, the individual units being no more than 1 to 2 km thick.

The UHP subduction event can be no older than the Early Paleozoic polymetamorphic gneiss complexes, for these are the youngest protoliths involved in the Caledonian continental collision. Ummetamorphosed Upper Devonian sediments that fill fault-bounded basins along the Norwegian coast provide a younger limit to the UHP metamorphism (Cuthbert et al., 1983). Based on geochronologic data employing various isotopic systems, Krogh and Carswell (1995) concluded that the metamorphic culmination in the Western Gneiss Region took place at approximately 420 to 430 Ma, with cooling and isotopic readjustment continuing until ~395 Ma. These data suggest that ~30 m. y. elapsed between the peak metamorphism and final unroofing of the eclogitic complex. Exhumation may have reflected slap rise, involving extensional ascent beneath the overlying, non-UHP Caledonian nappes, and low-angle reverse thrusting above the Fennoscandian basement (Dewey et al., 1998).

Petrotectonic commonalities among UHP terranes

Geologic/mineralogic, radiometric, maximum P-T, and tectonic constraints for the five Eurasian ultra high-pressure occurrences described above allow the following scenario to be proposed. Phanerozoic metamorphism of chiefly old sialic crust took place at depths of 90 to 125 km. Attending plate convergence, subduction of continental material carried along as an integral part of dominantly oceanic lithosphere appears to have been responsible for the profound descent of low-density sialic materials (Ernst and Liou, 1995). Times and rates of exhumation to mid-crustal levels range from 10 to 25 m.y., 3 to 12 mm/yr in the Qinling-Dabie-Sulu, Kokchetav, and Dora-Maira terranes, to 30 to 55 m.y., 2 to 4 mm/yr for the Maksyutov Complex and the Western Gneiss Region (Ernst et al., 1997). Metamorphic overprinting presumably reflects the achievement of neutral buoyancy of the ascending UHP complexes, and storage at mid-crustal levels. Exposed areas of the described UHP terranes vary considerably, but third dimensions are characteristically thin—approximately 5 ± 3 km. Although few firm constraints are available, it is clear from earlier descriptions that decompression of the most profoundly subducted lithotectonic sections was geologically rapid, that the exhumed UHP assemblies are chiefly old sialic crust, and that their dimensions are modest and sheetlike. All supracrustal rocks subjected to ultrahigh-pressure metamorphism appear to have been portions of pre-existing, relatively cool cratons, rather than segments of young, relatively warm oceanic crust or calc-alkaline arcs. Lower-T terranes such as the Maksyutov, Dora-Maira, and Dabie Shan include blueschists, whereas more elevated thermal conditions attending UHP metamorphism of the Western Gneiss Region and the Kokchetav Complex evidently caused destruction of such assemblages on decompression. Also, because Kokchetav unit I rocks achieved the highest temperatures of all, it seems conceivable that the rarity of coesite there is a consequence of elevated reaction rates in the system SiO_2 during unloading at 900°C, compared to kinetics of the diamond → graphite transformation. Among the terranes considered, the Maksyutov Complex and the Western Gneiss Region, which evidently rose relatively more slowly to mid-crustal levels, seem to have retained the fewest UHP relics.

Pressure-Temperature-Time Paths Followed by Ultrahigh-Pressure Complexes

The quantitative UHP significance of individual phases and mineral assemblages described in previous sections can be readily assessed employing experimental phase-equilibrium data and thermobarometric constraints for appropriate bulk-rock compositions. Extremely low P-T gradients—approximately 8 ± 3°C/km—are required for deeply buried rocks to enter the stability fields of some of the observed minerals and phase associations. Appropriate terrestrial HP and UHP environments are generated only in the vicinity of downgoing lithospheric plates, as shown by numerical computations (Oxburgh and Turcotte, 1971; Peacock, 1992, 1995). Moreover, the systematic progression in subduction-zone phase compatibilities supports the conclusion that relict assemblages represented close approaches to chemical equilibrium during formation. It is nevertheless true that various investigations of natural, coarse-grained, dry UHP assemblages have documented the metastable retention of low-pressure mineral assemblages at surprisingly great depths (Austrheim, 1990; Jamveit et al., 1990; Hirajima et al., 1993; Austrheim et al., 1997; Wallis et al., 1997; Zhang and Liou, 1997), as well as the metastable persistence of HP and UHP phase assemblages and mineral inclusions during later exhumation (Chopin, 1984; Smith, 1984; Sobolev and Shatsky, 1990; Liou and Zhang, 1996). Cases in which equilibrium was not achieved under extreme P-T conditions of recrystallization undoubtedly are due to the absence of a rate-enhancing aqueous fluid attending the metamorphism.

In all five Eurasian cases described earlier, a range of temperatures and pressures is preserved in the lithic assemblages, reflecting stages of partial re-equilibration on ascent toward mid-crustal levels. The exhumation process must have involved a continuum of changing conditions. Incompleteness of backreaction apparently attests, at least in part, to the rapidity of exhumation and/or the absence of a rate-enhancing aqueous fluid phase, as noted above. Rock and mineral oxygen-isotope anomalies in UHP eclogitic rocks from the Sulu block (Yui et al., 1995; Baker et al., 1997; Rumble, 1998) also seem to reflect failure to achieve equilibrium, indicating the lack of an aqueous fluid during metamorphism. The lack of H_2O apparently is responsible for the absence of coeval calc-alkaline magmatism and volcanic-plutonic arcs associated with these UHP terranes (Liou et al., 1995; Ernst et al., 1998).

Even so, for sections undergoing nearly adiabatic decompression at 800 ± 100°C, residence times outside the stability fields of UHP minerals for tens of millions of years ought to have caused the total obliteration of such phases. It therefore seems plausible that during initial stages of exhumation, some deeply subducted sections that retain traces of UHP precursors may have faithfully retraced the prograde P-T trajectory; this retrograde P-T path could be achieved by migrating back up the subduction channel—the zone between the overlying hanging wall and underlying footwall lithospheric plates—during continued (refrigerated) plate descent. For instance, in the Dora-Maira UHP massif, talc + phengite remained stable during decompression from peak metamorphic conditions of 725 ± 50°C at depths of 90 to 120 km. This assemblage dehydrates to phlogopite + kyanite + quartz + H_2O at ~600°C and 1.0 GPa (assuming $P_{aqueous}$ fluid = $P_{lithostatic}$), and if it is similar to other devolatilization reactions, proceeds relatively rapidly at such temperatures. The fact that this reaction did not take place constrains the maximum T of the Dora-Maira terrane during decompression. Moreover, if the activity of H_2O were low, the maximum temperature for stability of the talc + phengite assemblage would be correspondingly reduced. Temperatures must have declined from a maximum of ~725°C at 2.8 to 3.5 GPa to less than 500 to 600°C at 0.7 to 0.9 GPa accompanying the low-rank amphibolite/greenschist facies overprint (Schertl et al., 1991; Chopin and Sobolev, 1995). The Dora-Maira massif evidently did not pass through conditions of the high-rank amphibolite or granulite facies during exhumation; other UHP terranes such as Kokchetav unit I (Dobretsov et al., 1995) and the Sulu block evidently did (Wang et al., 1993).

Ultradeep Subduction and Resurrection of Continental Crust

Previous sections have shown that remarkably deep subduction of continental crust accounts for the generation of ultrahigh-pressure (UHP) terranes. An outstanding petrotectonic problem consists of explaining why these complexes were conducted so far down into the mantle, and then returned to shallow crustal levels while preserving intact relics of the UHP phase assemblages. The "two-way street" nature of subduction zones was recognized long ago (Ernst, 1970; Suppe, 1972). Specifically for UHP terranes, salients or peninsulas of old continental crust firmly attached to chiefly cool, oceanic crust–capped lithospheric plates, are thought to have descended rapidly, generating the characteristic UHP mineralogy (Peacock, 1995; Ernst et al., 1997). This evidently was the situation for the five Eurasian complexes described earlier: all represent tectonic regimes in which old continental crust constituted an integral part of the sinking lithospheric plate. These UHP slabs apparently were carried to depths where the buoyancy tending to drive them upward exactly balanced the dynamic forces tending to subduct them further.

The entrance of increasing amounts of strongly bonded crust + mantle into the subduction zone would intensify the braking effect of buoyancy. This deceleration would promote loss of the high-density lithospheric anchor pulling down the plate at intermediate upper-mantle depths—a plate regime where the sinking lithosphere is in extension (Isacks et al., 1968). Slab breakoff (Sacks and Seacor, 1990; von Blanckenburg and Davies, 1995) should enhance the buoyancy propulsion of the sialic promontory—or at least a slice thereof—allowing it to decouple from the descending lithosphere and move back up the subduction channel. Exhumation would be enhanced by progressive shallowing of the ruptured, buoyant, rebounding continental crust–capped section, and especially by reduction of shear forces acting along the slab base because of its more ductile behavior as it gradually warmed in the deep upper mantle. Because of continued subduction-induced refrigeration tectonically beneath the rising UHP complex, and observed extensional faulting against the overlying, cooler hanging-wall plate, relatively thin slices of UHP terranes should effectively lose heat by conduction along both upper and lower surfaces during ascent; thus, some such complexes may nearly retrace the subduction-zone P-T trajectory during decompression (Rubie, 1984; Ernst and Peacock, 1996).

Even in favorable circumstances, only packets of rock sufficiently thin to be relatively efficiently cooled along both upper and lower tectonic contacts would be able to preserve evidence of their earlier UHP history. The proposed mechanism involving ascent of thin tectonic slices is similar to the general process advanced by Dobretsov (1991) and Dobretsov and Kirdyashkin (1994), as well as the tectonic-wedge-extrusion scenario proposed for the Dabie-Sulu UHP belt by Maruyama et al. (1994), and the contrasting Dabie Shan nappe models advanced by Okay et al. (1993) and Hacker and Peacock (1995). In any event, bulk-rock density contrasts in the various subducted units are more than adequate to promote the buoyancy-driven ascent of ultrahigh-pressure metamorphosed sialic crust (Ernst et al., 1997).

Tentative Conclusions

In several Eurasian intracontinental collisional belts, promontories of sialic crust are postulated as having been carried down along subduction zones at least 100 km or more, attending the consumption of oceanic crust–capped lithosphere. Such rocks are transformed to UHP phase assemblages at pressures exceeding 2.8 GPa and temperatures on the order of 750°C. Geologic/structural relations, combined with mineral parageneses and experimental phase-equilibrium data, allow additional conclusions to be proposed regarding the generation and exhumation of portions of these UHP complexes from great depths.

Subduction of oceanic and/or sialic massifs at shallow depths results in the partial hydration of relatively dry units and in the evolution of volatiles from more hydrous lithologies. Attending this process, and the deeper prograde evolution of volatiles, some H_2O diffuses into the overlying, nonsubducted lithospheric hanging wall (mantle wedge), whereas the rest of the aqueous fluid migrates back up the subduction channel, depending on the relative permeabilities of the lithologic units involved. Where volatiles are not present, earlier assemblages may be preserved as scattered relics, on both prograde and retrograde P-T paths. Presence or absence of an aqueous fluid is therefore critical in determining the rates of prograde and retrograde reaction.

Deeply subducted salients of chiefly continental crust are disaggregated, and may decouple from the downgoing oceanic lithosphere in response to slab breakoff. The intrinsic buoyancy of sialic slices tends to drive them back toward the surface. Where these tectonically bounded entities are neither too thin to overcome frictional resistance to differential ascent, nor too thick to cool efficiently along the faulted slab margins, UHP relics may survive the exhumation processes of decompression and annealing under more normal sets of crustal P-T conditions. Suitable geometric configurations appear to involve slices 5 ± 3 km thick. Their ascent velocities over 20 to 40 m.y. approach an average exhumation rate of ~10 mm/yr.

Acknowledgments

This study was supported by Stanford University, and by the U.S. National Science Foundation through grant no. EAR97-25347. An early version of the paper was reviewed by S. Banno, R. G. Coleman, and S. Maruyama. To the above researchers and institutions, we express our sincere thanks.

REFERENCES

Ames, L., Zhou, G., and Xiong, B., 1995, Geochronology and geochemistry of ultrahigh-pressure metamorphism with implications for collision of the Sino-Korean and Yangtze cratons, central China: Tectonics, v. 15, p 472–489.

Austrheim, H., 1990, The granulite-eclogite facies transition: A comparison of experimental work and a natural occurrence in the Bergen Arcs, western Norway: Lithos, v. 25, p. 163–169.

Austrheim, H., Erambert, M., and Engvik, A. K., 1997, Processing of crust in the root of the Caledonian continental collision zone: The role of eclogitization: Tectonophys., v. 273, p. 129–153.

Avigad, D., 1992, Exhumation of coesite-bearing rocks in the Dora Maira massif (western Alps, Italy): Geology, v. 20, p. 947–950.

Baker, J., Matthews, A., Mattey, D., Rowley, D. B., and Xue, F., 1997, Fluid-rock interactions during ultrahigh pressure metamorphism, Dabie Shan, China: Geochim. et Cosmochim. Acta, v. 61, p. 1685–1696.

Beane, R. J., Connelly, J. N., and Liou, J. G. 1996, Evidence for Devonian eclogite-facies metamorphism in the Maksyutov Complex, southern Ural Mountains, Russia [abs.]: Geol. Soc. Amer. Abs. Prog., v. 28, p. 170.

Beane, R. J., Liou, J. G., Coleman, R. G., and Leech, M. L., 1995, Petrology and retrograde P-T path for eclogite of the Maksyutov Complex, Southern Ural Mountains, Russia: The Island Arc, v. 4, p. 254–266.

Becker, H., 1993, Garnet peridotite and eclogite Sm-Nd mineral ages from the Lepontine dome (Swiss Alps): New evidence for Eocene high-pressure metamorphism in the central Alps: Geology, v. 21, p. 599–602.

Berzin, R., Oncken, O., Knapp, J. H., Pérez-Estaún., A., Histmatulin, T., Yunusov, N., and Lipilin, A., 1996, Orogenic evolution of the Ural Mountains: Results from an integrated seismic experiment: Science, v. 274, p. 220–221.

Blake, M. C., Jr., Moore, D. E., and Jayko, A. S., 1995, The role of serpentinite melanges in the unroofing of ultrahigh-pressure metamorphic rocks: An example from the Western Alps, *in* Coleman, R. G., and Wang, X., eds., Ultrahigh pressure metamorphism: New York, Cambridge University Press, p. 182–205.

Borghi, A., Cadoppi, P., Porro, A., and Sacchi, R., 1985, Metamorphism in the northern part of the Dora-Maira massif (Cottian Alps): Mus. Reg. Sci. Natural. di Torino, Bolletino, v. 3, p. 369–380.

Breuckner, H. K., Gilotti, J. A., and Nutman, A., 1998, Caledonian eclogite-facies metamorphism of Early Proterozoic protoliths from the North-East Greenland Eclogite Province: Contrib. Mineral. Petrol., v. 130, p. 103–120.

Carswell, D. A., Krogh, E. J., and Griffin, W. L., 1985, Norwegian orthopyroxene eclogites: Calculated equilibration conditions and petrogenetic implications, *in* The Caledonide Orogen: New York, John Wiley and Sons, p. 823–841.

Chesnokov, B. V., and Popov, V. A., 1965, Increasing volume of quartz grains in eclogites of the South Urals: Doklady Akad. Nauk, v. 62, p. 909–910 (in Russian).

Chopin, C., 1981, Talc-phengite: A widespread assemblage in high-grade pelitic blueschists of the western Alps: Jour. Petrol., v. 22, p. 628–650.

———, 1984, Coesite and pure pyrope in high-grade blueschists of the Western Alps. A first record and some consequences: Contrib. Mineral. Petrol., v. 86, p. 107–118.

Chopin, C., Henry, C., and Michard, A., 1991, Geology and petrology of the coesite-bearing terrain, Dora Maira massif, Western Alps: Eur. Jour. Mineral., v. 3, p. 263–291.

Chopin, C., and Sobolev, N. V., 1995, Principal mineralogic indicators of ultrahigh-pressure in crustal rocks, *in* Coleman, R. G., and Wang, X., eds., Ultrahigh pressure metamorphism: New York, Cambridge University Press, p. 96–131.

Claoue-Long, J. C., Sobolev, N. V., Shatsky, V. S., and Sobolev, A. V., 1991, Zircon response to diamond-pressure metamorphism in Kokchetav Massif: Geology, v. 19, p. 710–713.

Cloos, M., 1993, Lithospheric buoyancy and collisional orogenesis: Subduction of oceanic plateaus, continental margins, island arcs, spreading ridges, and seamounts: Geol. Soc. Amer. Bull., v. 105, p. 715–737.

Coleman, R. G., and Wang, X., eds., 1995, Ultrahigh pressure metamorphism: New York, Cambridge University Press, 528 p.

Compagnoni, R., Hirajima, T,. and Chopin, C., 1995, UHPM rocks in the Western Alps, *in* Coleman, R. G., and Wang, X., eds., 1995, Ultrahigh pressure metamorphism: New York, Cambridge University Press, p. 206–243.

Cuthbert, S. J., Harvey, M. A., and Carswell, D. A., 1983, A tectonic model for the metamorphic evolution of the Basal Gneiss Complex, Western South Norway: Jour. Metamor. Geol., v. 1, p. 63–90.

Dal Piaz, G. V., Hunziker, J. C., and Martinotti, G., 1972, La zona Sesia-Lanzo e l'evoluzione tettonico-metamorfica delle Alpi nordoccidentali interne: Memoir Soc. Geol. Italy, v. 11, p. 433–460.

Dewey, J. F., Krabbendam, M., and Wain, A., 1998, Transtensional exhumation of the UHP rocks of the Western Gneiss Region of southwest Norway [abs.], *in* International Workshop on Ultrahigh-Pressure Metamorphism and Exhumation: Stanford, CA, Stanford University, Abs. vol., p. A84–A85.

Dewey, J. F., Ryan, P. D., and Andersen, T. B., 1993, Orogenic uplift and collapse, crustal thickness, fabrics and metamorphic phase changes: The role of eclogites, *in* Prichard, H. M., Alabaster, T., Harris, N. B. W., Neary, C. R., eds., Magmatic processes and plate tectonics: Geol. Soc. Lond., Spec. Publ. No. 76, p. 325–343.

Dobretsov, N. L., 1991, Blueschists and eclogites: A possible plate tectonic mechanism for the emplacement from the upper mantle: Tectonophys., v. 186, p. 253–268.

———, ed., 1998, Geological map of Kokchetav region (northern Kazakhstan): Novosibirsk, United Inst. Geol., Geophys., Mineral.

Dobretsov, N. L., and Dobretsova, L. V., 1988, New mineralogic data on the Maksyutov eclogite-glaucophane schist complex, Southern Urals: Trans. (Doklady) U.S.S.R. Acad. Sci., Earth Sci. Sect., v. 300, p. 111–116.

Dobretsov, N. L., and Kirdyashkin, A. G., 1994, Blueschists of North Asia and models of subduction-accretion wedge, *in* Coleman, R. G., ed., Reconstruction of the Paleo-Asian ocean: Proc. 29th Int. Geol. Cong., Kyoto, Japan, Part B, p. 99–114.

Dobretsov, N. L., Shatsky, V. S., Coleman, R. G., Lennykh, V. I., Valizer, P. M., Liou, J. G., Zhang, R., and Beane, R. J., 1996, Tectonic setting of ultrahigh-pressure metamorphic rocks in the Maksutov complex, Ural Mountains, Russia: Int. Geol. Rev., v. 38, p. 136–160.

Dobretsov, N. L., and Sobolev, N. V., 1970, Eclogites from metamorphic complexes of the USSR: Phys. Earth Planet. Inter., v. 3, p. 401–424.

———, 1984, Glaucophane schists and eclogites in the folded systems of northern Asia: Ofioliti, v. 9, p. 401–424.

Dobretsov, N. L., Sobolev, N. V., Shatsky, V. S., Coleman, R. G., and Ernst, W. G., 1995, Geotectonic evolution of diamondiferous paragneisses, Kokchetav Complex, northern Kazakhstan: The geologic enigma of ultrahigh-pressure crustal rocks within a Paleozoic foldbelt: The Island Arc, v. 4, p. 267–279.

Dobrzhinetskaya, L. F., 1994, REE-patterns of metamorphic rocks from the Kokchetav microdiamond province of Kazakhstan and their implication for tectonics: EOS (Trans. Amer. Geophys. Union), v. 75, p. 701.

Dobrzhinetskaya, L. F., Braun, T. V., Sheshkel, G. G., and Podkuiko, Y. A., 1994, Geology and structure of diamond-bearing rocks of the Kokchetav massif, Kazakhstan: Tectonophys., v. 233, p. 293–313.

Dobrzhinetskaya, L. F., Eide, E. A., Larsen, R. B., Sturt, B. A., Tronnes, R. G., Smith, D. C., Taylor, W. R., and Posukhova, T. V., 1995, Microdiamond in high-grade metamorphic rocks of the Western Gneiss region, Norway: Geology, v. 23, p. 597–600.

Dobrzhinetskaya, L., Green, H. W., and Wang, S., 1996, Alpe Arami: A peridotite massif from depths of more than 300 kilometers: Science, v. 271, p. 1841–1845.

Dong, S. B., 1993, Metamorphic and tectonic domains of China: Jour. Metamor. Geol., v. 11, p. 465–481.

Dong, S. B., Shen, Q. H., Sun, D. Z., and Lu, L. Z., 1986, Metamorphic map of China (1:4,000,000) with explanatory text: Beijing, Geol. Publ. House.

Edwards, R. L., and Wasserburg, G. J., 1985, The age and emplacement of obducted oceanic crust in the Urals from Sm-Nd and Rb-Sr systematics: Earth Planet. Sci. Lett., v. 72, p. 389–404.

Eide, E. A., 1995, A model for the tectonic history of high- and ultrahigh-pressure metamorphic regions in east central China, *in* Coleman, R. G., and Wang, X., eds., 1995, Ultrahigh pressure metamorphism: New York, Cambridge University Press, p. 391–426.

Erambert, M., and Austrheim, H., 1993, The effect of fluid and deformation on zoning and inclusion patterns in polymetamorphic garnets: Contrib. Mineral. Petrol., v. 115, p. 204–214.

Ernst, W. G., 1970, Tectonic contact between the Franciscan melange and the Great Valley sequence, crustal expression of a Late Mesozoic Benioff zone: Jour. Geophys. Res., v. 75, p. 886–901.

———, 1971, Metamorphic zonations on presumably subducted lithospheric plates from Japan, California, and the Alps: Contrib. Mineral. Petrol., v. 34, p. 43–59.

———, 1973. Interpretive synthesis of metamorphism in the Alps: Geol. Soc. Amer. Bull., v. 84, p. 2053–2078.

———, 1977, Mineralogic study of eclogitic rocks from Alpe Arami, Lepontine Alps, southern Switzerland: Jour. Petrol., v. 18, p. 371–398.

Ernst, W. G., and Liou, J. G., 1995, Contrasting plate-tectonic styles of the Qinling-Dabie-Sulu and Franciscan metamorphic belts: Geology, v. 23, p. 353–356.

Ernst, W. G., and Peacock, S. M., 1996, A thermotectonic model for preservation of ultrahigh-pressure phases in metamorphosed continental crust: Geophys. Monogr. 96, Amer. Geophys. Union, p. 171–178.

Ernst, W. G., Maruyama, S., and Wallis, S., 1997, Buoyancy-driven, rapid exhumation of ultrahigh-pressure metamorphosed continental crust: Proc. Nat. Acad. Sci., v. 94, p. 9532–9537.

Ernst, W. G., Mosenfelder, J. L., Leech, M. L., and Liu, J., 1998, H_2O recycling during continental collision: Phase-equilibrium and kinetic considerations, *in* Hacker, B. R., and Liou, J. G., eds., When continents collide: Geodynamics and geochemistry of ultrahigh-pressure rocks: Dordrecht, Kluwer Acad. Publ., p. 275–295.

Ernst, W. G., Zhou, J. G., Liou, J. G., Eide, E., and Wang, X., 1991, High-pressure and superhigh-pressure metamorphic terranes in the Qinling-Dabie mountain belt, central China: Early- to mid-Phanerozoic accretion of the western paleo-Pacific Rim: Pacific Sci. Assoc. Informat. Bull., v. 43, p. 6–15.

Evans, B. W., Trommsdorff, V., and Richter, W., 1979, Petrology of an eclogite-metarodingite suite at Cima de Gagnone, Ticino, Switzerland: Amer. Mineral., v. 64, p. 15–31.

Gebauer, D., 1996, A P-T-t path for an (ultra?-) high-pressure ultramafic/mafic rock association and its felsic country rocks based on SHRIMP-dating of magmatic and metamorphic zircon domains. Example: Alpe Arami (central Swiss Alps): Amer. Geophys. Union, Geophys. Monog. 95, p. 307–329.

Gebauer, D., Schertl, H. P., Briz, M., and Schreyer, W., 1997, 35 Ma old ultrahigh-pressure metamorphism and evidence for very rapid exhumation in the Dora Maira Massif, Western Alps: Lithos, v. 41, p. 5–24.

Gebauer, D., Tilton, G. R., Schertl, H. P., and Schreyer, W., 1993, Eocene/Oligocene ultrahigh-pressure (UHP)-metamorphism in the Dora-Maira massif (western Alps) and its geodynamic implications [abs.]: Fourth Int. Eclogite Conf. (Calabria, Italy), Terra Abs., v. 4, p. 10–11.

Gilotti, J. A., 1993, Discovery of a medium-pressure eclogite province in the Caledonides of North-East Greenland: Geology, v. 21, p. 523–526.

Green, H. W., Bozhilov, K. N., and Dobrzhinetskaya, L., 1998, Exsolution of high-pressure clinoenstatite in diopside of the Alpe Arami peridotite: Further evidence of a very high pressure origin [abs.], *in* International Workshop on Ultrahigh-Pressure Metamorphism and Exhumation: Stanford, CA, Abs. vol., p. A8.

Griffin, W. L., Austrheim, H., Brastad, K., Bryhni, I., Krill, A. G., Krogh, E. A., Mörk, M. B. E., Qvale, H., and Torudbakken, B., 1985, High-pressure metamorphism in the Scandinavian Caledonides, *in* Gee, D. G., and Sturt, B. A., eds., The Caledonide Orogen: New York, John Wiley and Sons, p. 783–801.

Hacker, B. R., and Peacock, S. M., 1995, Creation, preservation, and exhumation of UHPM rocks, *in* Coleman, R. G., and Wang, X., eds., 1995, Ultrahigh pressure metamorphism: New York, Cambridge University Press, p.159–181.

Hacker, B. R., Ratschbacher, L., Webb, L., and Dong, S., 1995, What brought them up? Exhuming the Dabie Shan ultrahigh-pressure rocks: Geology, v. 23, p. 743–746.

Hacker, B. R., Ratschbacher, L., Webb, L., Ireland, T., Walker, D., and Dong, S., 1998, U/Pb zircon ages constrain the architecture of the ultrahigh-pressure Qinling-Dabie orogen, China: Earth Planet. Sci. Lett., v. 161, p. 215–230.

Hacker, B. R., Wang, X., Eide, E. A., and Ratschbacher, L., 1996, The Qinling-Dabie ultrahigh-pressure collisional orogen, *in* Harrison, T. M., and Yin, A., eds., The tectonic development of Asia: New York, Cambridge University Press, p. 345–370.

Hamilton, W., 1970, The Uralides and the motion of the Russian and Siberian platforms: Geol. Soc. Amer. Bull., v. 81, p. 2553–2576.

Harley, S. L., and Carswell, D. A., 1995, Ultradeep crustal metamorphism: A prospective review: Jour. Geophys. Res., v. 100, p. 8367–8380.

Henry, C., 1990, L'unite a coesite du massif Dora-Maira dans son cadre petrologique et structural (Alpes occidentales, Italie): Unpubl. doctoral thesis, Université de Paris VI, 453 p.

Hirajima, T., Ishiwatari, A., Cong., B., Zhang, R. Y., Banno, S., and Nozaka, T., 1990, Coesite from Mengzhong eclogite at Dhonghai county, northeastern Jiangsu province, China: Mineral. Mag., v. 54, p. 579–583.

Hirajima, T., Wallis, S., Zhai, R., and Ye, K., 1993, Eclogitized metagranitoid from the Su-Lu ultrahigh-pressure (UHP) province, eastern China: Proc. Japan Acad., v. 69, sect. B, p. 249–254.

Huang, T. C., 1978, An outline of the tectonic characteristics of China: Eclogae Geol. Helvetiae, v. 71, p. 611–635.

Huang, W., and Wu, Z. W., 1992, Evolution of the Qinling orogenic belt: Tectonics, v. 11, p. 371–380.

Isacks, B., Oliver, J., and Sykes, L. R., 1968, Seismology and the new global tectonics: Jour. Geophys. Res., v. 73, p. 5855–5899.

Ishikawa, M., et al., 1998, Structures of ultra-high pressure terrain: Sulu-Tube, Kokchetav, Kazakhstan [abs.]: EOS (Trans. Amer. Geophys. Union), v. 79, p. F983.

Jagoutz, E., Shatsky, V. S., and Sobolev, N. V., 1991, The origin and history of ultrahigh PT rocks from Kokchetav Massif [abs.]: Terra Abs., v. 3, p. 83.

Jamtveit, B., Bucher-Nurminen, K., and Austrheim, H., 1990, Fluid controlled eclogitization of granulites in deep crustal shear zones, Bergen arcs, Western Norway: Contrib. Mineral. Petrol., v. 104, p. 184–193.

Kienast, J. R., Lombardo, B., Biino, G., and Pinardon, J. L., 1991, Petrology of very-high-pressure eclogitic rocks from the Brossasco-Isasca Complex, Dora-Maira Massif, Italian Western Alps: Jour. Metamor. Geol., v. 9, p. 19–34.

Kaneko, Y., et al., 1998, Geology of the Kokchetav UHP-HP Unit, northern Kazakhstan [abs.]: EOS (Trans. Amer. Geophys. Union), v. 79, p. F983.

Krogh, E. J., 1977, Evidence for a Precambrian continent-continent collision in western Norway: Nature, v. 267, p. 17–19.

Krogh, E. J., and Carswell, D. A., 1995, HP and UHP eclogites and garnet peridotites in the Scandinavian Caledonides, *in* Coleman, R. G., and Wang, X., eds., 1995, Ultrahigh pressure metamorphism: New York, Cambridge University Press, p. 244–298.

Kullerud, L., Tirudbakken, B. O., and Ilebekk, S., 1986. A compilation of radiometric age determinations from the Western Gneiss Region, South Norway: Norges Geol. Undersök. Bull., v. 406, p. 17–42.

Leech, M. L., and Ernst, W. G., 1998, Graphite pseudomorphs after diamond? A carbon isotope and spectroscopic study of graphite cuboids from the Maksyutov Complex, south Ural Mountains, Russia: Geochim. et Cosmochim. Acta, v. 62, p. 2143–2154.

Lennykh, V. I., 1997, The eclogite-glaucophane belt of the southern Urals: Moscow, Nauka Press, 158 p. (in Russian).

Lennykh, V. I., Valizer, P. M., Beane, R., Leech, M., and Ernst, W. G., 1995, Petrotectonic evolution of the Maksyutov Complex, southern Urals, Russia: Implications for ultrahigh-pressure metamorphism: INT. GEOL. REV., v. 37, 584–600.

Liou, J. G., and Zhang, R. Y., 1996, Occurrences of intergranular coesite in ultrahigh-P rocks from the Sulu region, eastern China: Implications for lack of fluid during exhumation: Amer. Mineral., v. 81, p. 1217–1221.

Liou, J. G., Zhang, R., and Ernst, W. G., 1994, An introduction to ultrahigh-pressure metamorphism: The Island Arc, v. 3, p. 1–24.

———, 1995, Occurrence of hydrous and carbonate phases in ultrahigh-pressure rocks from east-central China—implications for the roles of volatiles deep in cold subduction zones: The Island Arc, v. 4, p. 362–375.

Liou, J. G., Zhang, R. Y., Wang, X., Eide, E. A., Ernst, W. G., and Maruyama, S., 1996, Metamorphism and tectonics of high-pressure and ultrahigh-pressure belts in the Dabie-Sulu region, China, *in* Yin, A., and Harrison, T. M., eds., The tectonic evolution of Asia: New York, Cambridge University Press, p. 300–344.

Maruyama, S., 1997, Pacific-type orogeny revisited: Miyashiro-type orogeny proposed: The Island Arc, v. 6, p. 91–120.

———, 1998, The Kokchetav UHP-HP metamorphism: Current state of our research (1998) [abs.], *in* International Workshop of UHP Metamorphism and Exhumation: Stanford, CA, Stanford University, Abs. vol., p. A58–A59.

Maruyama, S., Liou, J. G., and Terabayashi, M., 1996, Blueschists and eclogites of the world, and their exhumation: INT. GEOL. REV., v. 38, p. 485–594.

Maruyama, S., Liou, J. G., and Zhang, R., 1994, Tectonic evolution of the ultrahigh-pressure (UHP) and high-pressure (HP) metamorphic belts from central China: The Island Arc, v. 3, p. 112–121.

Matte, P., 1995, Southern Uralides and Variscides: Comparison of their anatomies and evolutions: Geol. en Mijnbouw, v. 74, p. 151–166.

Matte, P., Maluski, H., Caby, R., Nicholas, A., Kepezhinskas, P., and Sobolev, S., 1993, Geodynamic model and $^{39}Ar/^{40}Ar$ dating for the generation and emplacement of the high pressure metamorphic rocks in SW Urals: Comptes Rendus Acad. Sci. Paris, v. 317, serie II, p. 1667–1674.

Medaris, L. G. J., and Wang, H. F., 1986, A thermal-tectonic model for high-pressure rocks in the Basal Gneiss Complex of Western Norway: Lithos, Second Int. Eclogite Conf., p. 323–332.

Michard, A., 1965, Une nappe de socle dans les Alpes cottiennes internes? Implications paleogeographiques et role eventuel des movements cretaces: Comptes Rendus Acad. Sci. Paris, v, D-260, p. 4012–4015.

———, 1967, Etudes geologiques dans les zones internes des Alpes cottiennes: Paris, C. N. R. S., 447 p.

Michard, A., Henry, C., and Chopin, C., 1995, Structures in UHPM rocks: A case study from the Alps, *in* Coleman, R. G., and Wang, X., eds., 1995, Ultrahigh pressure metamorphism: New York, Cambridge University Press, p. 132–158.

Monie, P., and Chopin, C., 1991, $^{40}Ar/^{39}Ar$ dating in coesite-bearing and associated units of the Dora-Maira massif, Western Alps: Eur. Jour. Mineral., v. 3, p. 239–262.

Mörk, M. B. E., and Krogh, E. J., 1987, Excursion guide for the eclogite field symposium in western Norway: Tromso, University of Tromso, Spec. Publ., 115 p.

Mosenfelder, J. L., and Bohlen, S. R., 1997, Kinetics of the coesite to quartz transformation: Earth Planet. Sci. Lett., v. 153, p. 133–147.

Moskovchenko, N. I., 1982, Precambrian high-pressure complexes in Phanerozoic folded belts: Moscow, Nauka, 161 p. (in Russian).

Nie, S., and Rowley, D. B., 1994, Comment on "Paleomagnetic constraints on the geodynamic history of the major blocks of China from the Permian to the Present" by R. J. Enkin et al.: Jour. Geophys. Res., v. 99, p. 18,035–18,042.

Okay, A. I., 1993, Petrology of a diamond and coesite-bearing metamorphic terrane, Dabie Shan, China: Eur. Jour. Mineral., v. 5, p. 659–673.

Okay, A. I., Sengör, A. M. C., and Satir, M., 1993, Tectonics of an ultrahigh-pressure metamorphic terrane: The Dabie Shan/Tongbai Shan orogen, China: Tectonics, v. 12, p. 1320–1334.

Okay, A. I., Shutong, X., and Sengör, A. M. C., 1989, Coesite from the Dabie Shan eclogites, central China: Eur. Jour. Mineral., v. 1, p. 595–598.

Oxburgh, E. R., and Turcotte, D. L., 1971, Origin of paired metamorphic belts and crustal dilation in island arc regions: Jour. Geophys. Res., v. 76, p. 1315–1327.

Paquette, J. R., Chopin, C., and Peucat, J. J., 1989, U-Pb zircon, Rb-Sr and Sm-Nd geochronology of high- to very-high-pressure meta-acidic rocks from the western Alps: Contrib. Mineral. Petrol., v. 101, p. 280–289.

Peacock, S. M., 1992, Blueschist-facies metamorphism, shear heating, and P-T-t paths in subduction shear zones: Jour. Geophys. Res., v. 97, p. 17,693–17,707.

———, 1995, Ultrahigh-pressure metamorphic rocks and the thermal evolution of continent collision belts: The Island Arc, v. 4, p. 376–383.

Philippot, P., 1988, Deformation et eclogitization progressives d'une croute oceanique subductee: Le Monviso, Alpes occidentales. Contraintes cinematiques durant la collision alpine: Travaux Centre Geol. Geophys. Montpelier, v. 19, 269 p.

Puchkov, V. N., 1993, Paleoceanic structures of the Ural Mountains: Geotectonics, v. 27, p. 184–196 (English translation).

Reinecke, T., 1991, Very-high-pressure metamorphism and uplift of coesite-bearing metasediments from the Zermatt-Saas zone, Western Alps: Eur. Jour. Mineral., v. 3, p. 7–17.

Robinson, P., and Terry, M. P., 1998, Evolution of amphibolite-facies structural features during transition from convergence to extensional collapse, Western Gneiss Region, Norway [abs.], *in* International Workshop on UHP Metamorphism and Exhumation: Stanford, CA, Stanford University, Abs. vol., p. A79–A83.

Rowley, D. B., Xue, F., Tucker, R. D., Peng, Z., Baker, J., and Davis, A., 1997, Ages of ultrahigh pressure metamorphism and protolith orthogneisses form the eastern Dabie Shan: U/Pb zircon geochronology: Earth Planet. Sci. Lett., v. 151, p. 191–204.

Rubie, D. C., 1984, A thermal-tectonic model for high-pressure metamorphism in the Sesia zone, Western Alps: Jour. Geol., v. 92, p. 21–36.

———, 1986. The catalysis of mineral reactions by water and restrictions on the presence of aqueous fluid during metamorphism: Mineral. Mag., v. 50, p. 399–415.

———, 1990, Role of kinetics in the formation and preservation of eclogites, *in* Carswell, D.A., ed., Eclogite facies rocks: New York, Chapman and Hall, p. 111–140.

Rumble, D., 1998, Stable isotope geochemistry of ultrahigh-pressure rocks, *in* Hacker, B. R., and Liou, J. G., eds., When continents collide: Geodynamics and geochemistry of ultrahigh-pressure rocks: Dordrecht, Kluwer Acad. Publ., p. 241–259.

Sacks, P. E., and Secor, D. T., Jr., 1990, Delaminations in collisional orogens: Geology, v. 18, p. 999–1002.

Schertl, H. P., Schreyer, W., and Chopin, C., 1991, The pyrope-coesite rocks and their country rocks at Parigi, Dora-Maira Massif, Western Alps: Detailed petrography, mineral chemistry and PT-path: Contrib. Mineral. Petrol., v. 108, p. 1–21.

Schertl, H. P., and Okay, A. I., 1994, Coesite inclusion in dolomite of Dabieshan, China: Petrological and rheological significance: Eur. Jour. Mineral., v. 6, p. 995–1000.

Schreyer, W., 1988, Experimental studies on metamorphism of crustal rocks under mantle pressures: Mineral. Mag., v. 53, p. 1–26.

———, 1995, Ultradeep metamorphic rocks: The retrospective viewpoint: Jour. Geophys. Res., v. 100, p. 8353–8366.

Sengör, A. M. C., Natal'in, B. A., and Burtman, V. S., 1993, Evolution of the Altaid tectonic collage and Palaeozoic crustal growth in Eurasia: Nature, v. 364, p. 299–307.

Shatsky, V. S., Jagoutz, E., Kozmenko, O. A., Blinchik, T. M., and Sobolev, N. V., 1993, The age and origin of eclogites from the Kokchetav massif, north Kazakhstan: Geol. i Geofiz., v. 34 no. 12, p. 47–58 (in Russian).

Shatsky, V. S., Jagoutz, E., Kozmenko, O. A., Parkhomenko, V. S., Troesch, M., and Sobolev, N. V., in press, The geochemistry of ultra high pressure rocks from Kokchetav massif (northern Kazakhstan): Geochim. et Cosmochim. Acta.

Shatsky, V. S., and Sobolev, N. V., 1993, Some specific features of the origin diamonds in metamorphic rocks: Doklady Akad. Nauk SSSR, v. 331, p. 217–219 (in Russian).

Shatsky, V. S., Sobolev, N. V., and Gilbert, A. E., 1989, Eclogites of the Kokchetav massif, *in* Sobolev, N. V., ed., Eclogites and glaucophane schists in folded belts: Novosibirsk, Nauka Press, p. 54–82 (in Russian).

Shatksy, V. S., Sobolev, N. V., and Vavilov, M. A., 1995, Diamond-bearing metamorphic rocks of the Kokchetav massif (northern Kazakhstan), *in* Coleman, R. G., and Wang, X., eds., Ultrahigh pressure metamorphism: New York, Cambridge University Press, p. 427–455.

Shatsky, V. S., Sobolev, N. V., Zayachkovsky, A. A., Zorin, Yu, M., and Vavilov, M. A., 1991, A new find of microdiamond occurrence as evidence of the regional ultra-high-pressure metamorphism at the Kokchetav Massif: Doklady Akad. Nauk SSSR, v. 321, p. 189–193 (in Russian).

Smith, D. C., 1984, Coesite in clinopyroxene in the Caledonides and its implications for geodynamics: Nature, v. 310, p. 641–644.

———, 1995, Microcoesites and microdiamonds in Norway: An overview: *in* Coleman, R. G., and Wang, X., eds., 1995, Ultrahigh pressure metamorphism: New York, Cambridge University Press, p. 299–355.

Smith, D. C., and Lappin, M. A., 1989, Coesite in the Straumen kyanite-eclogite pod, Norway: Terra Res., v. 1, p. 47–56.

Sobolev, N. V., Dobretsov, N. L., Bakirov, A. B., and Shatsky, V. S., 1986, Eclogites from various types of metamorphic complexes in the USSR and problems of their origin, *in* Evans, B. W., and Brown, E. H., eds., Blueschists and eclogites: Geol. Soc. Amer. Memoir, v. 164, p. 349–363.

Sobolev, N. V., and Shatsky, V. S., 1990, Diamond inclusions in garnets from metamorphic rocks: A new environment for diamond formation: Nature, v. 343, p. 742–746.

Sobolev, N. V., Shatsky, V. S., and Vavilov, M. A., 1991, Inclusions of microdiamonds and coexisting minerals in garnets and zircons from metamorphic rocks of Kokchetav Massif, USSR [abs.]: Terra Abs., v. 3, p. 83.

Sobolev, N. V., Shatsky, V. S., Vavilov, M. A., and Goryainov, S. V., 1994, Zircon from ultrahigh-pressure metamorphic rocks of folded regions as an unique container of inclusions of diamond, coesite, and coexisting minerals: Doklady Akad. Nauk, v. 334, p. 488–492 (in Russian).

Sobolev, V. S., and Sobolev, N. V., 1980, New proof of very deep subsidence of eclogitized crustal rocks: Doklady Akad. Nauk SSSR, v. 250, p. 683–685 (in Russian).

Suppe, J., 1972, Interrelationships of high-pressure metamorphism, deformation, and sedimentation in Franciscan tectonics, U.S.A.: 24th Int. Geol. Cong., Montreal, Reports, sect. 3, p. 552–559.

Tilton, G. R., Ames, L., Schertl, H. P., and Schreyer, W., 1997, Reconnaissance isotopic investigations on rocks of an undeformed granite contact within the coesite-bearing unit of the Dora Maira Massif: Lithos, v. 41, p. 25–36.

Tilton, G. R., Schreyer, W., and Schertl, H. P., 1991, Pb-Sr-Nd isotopic behavior of deeply subducted crustal rocks from the Dora-Maira Massif, Western Alps-II: What is the age of the ultrahigh-pressure metamorphism?: Contrib. Mineral. Petrol., v. 108, p. 22–33.

Troesch, M., and Jagoutz, E., 1993, Mica cooling ages of a diamond-bearing gneiss from the Kokchetav Massif, Kazakhstan [abs.]: Terra Abs., v. 7, p. 396.

Valizer, P. M., and Lennykh, V. I., 1988, Amphiboles of blueschists of the Urals: Nauka, Moscow, 202 p. (in Russian).

van Roermund, H. L. M., Drury, M. R., Barnhoon, A., and de Ronde, A., 1998, Ultra-high pressure (P > 6 GPa) garnet peridotites in western Norway: Implications for exhumation processes of mantle rocks [abs.], *in* International Workshop on Ultrahigh-Pressure Metamorphism and Exhumation: Stanford, CA, Stanford University, Abs. vol., p. A9–A12.

Vialon, P., 1966, Etude geologique du massif cristallin Dora-Maira, Alpes cottiennes internes, Italie: Thesis d'Etat, Université de Grenoble, 293 p.

von Blanckenburg, F., and Davies, J. H., 1995, Slab breakoff: A model for syncollision magmatism and tectonics in the Alps: Tectonics, v. 14, p. 120–131.

Wain, A., 1997, New evidence for coesite in eclogite and gneisses: Defining an ultrahigh-pressure province in the Western Gneiss region of Norway: Geology, v. 25, p. 927–930.

Wallis, S. R., Ishiwatari, A., Hirajima, T., Ye, K., Guo, J., Nakamura, D., Kato, T., Zhai, M., Enami, M., Cong, B., and Banno, S., 1997, Occurrence and field relationships of UHP meta-granitoid and coesite eclogite in the Su-Lu terrane, Eastern China: Jour. Geol. Soc. Lond., v. 154, p. 45–55.

Wang, Q., Ishiwatari, A., Zhao, Z., Hirajima, T., Enami, M., Zhai, M., Li, J., and Cong, B. L., 1993, Coesite-bearing granulite retrograded from eclotite in Wehai, eastern China: A preliminary study: Eur. Jour. Mineral., v. 5, p. 141–152.

Wang, X., and Liou, J. G., 1991, Regional ultrahigh-pressure metamorphic terrane in central China: Evidence from coesite-bearing eclogite, marble, and metapelite: Geology, v. 19, p. 933–936.

Wang, X., Liou, J. G., and Mao, H. K., 1989, Coesite-bearing eclogites from the Dabie Mountains in central China: Geology, v. 17, p. 1085–1088.

Wang, X., Zhang, R., and Liou, J. G., 1995, UHPM terrane in east central China, *in* Coleman, R. G., and Wang, X., eds., 1995, Ultrahigh pressure metamorphism: New York, Cambridge University Press, p. 356–390.

Webb, L. E., Hacker, B. R., Ratschbacher, L., McWilliams, M. O., and Dong, S., 1999, $^{40}Ar/^{39}Ar$ thermochronologic constraints on deformation and cooling history of high and ultrahigh-pressure rocks in the Qinling-Dabie orogen: Tectonics, v. 18, in press.

Xu, S., Okay, A. I., Ji, S., Sengör, A. M. C., Su, W., Liu, Y., and Jiang, L., 1992. Diamond from Dabie Shan eclogite and its implication for tectonic setting: Science, v. 256, p. 80–82.

Yang, Z., Courtillot, V., and Besse, J., 1992, Jurassic paleomagnetic constraints in the collision of the North and South China blocks: Geophys. Res. Lett., v. 19, p. 577–580.

Ye, K., Hirajima, T., Ishiwatari, A., Guo, J., and Zhai, M., 1996, Finding of coesite in eclogite in Qingdao and its implications: Chinese Sci. Bull., v. 11, p. 1407–1408 (in Chinese).

Yin, A., and Nie, S., 1993, An indentation model for the north and south China collision and the development of the Tan-lu and Honam fault systems, eastern Asia: Tectonics, v. 12, p. 801–813.

Yui, T. F., Rumble, D., and Lo, C. H., 1995, Unusually low ^{18}O ultrahigh-pressure metamorphic rocks from the Sulu terrain, eastern China: Geochim. et Cosmochim. Acta, v. 59, p. 2859–2864.

Zakharov, O. A., 1991, New data on the geological structure of the Maksyutov metamorphosed complex in the upper part of the Sakmara River, *in* Overthrust-nappe tectonics and its role in the formation of natural deposits: Ufa, Inst. Geol., Urals Section, USSR Acad. Sci. (in Russian).

Zhang, R. Y., and Liou, J. G., 1994, Coesite-bearing eclogite in Henan Province, central China: Detailed petrography, glaucophane stability and PT path: Eur. Jour. Mineral., v. 6, p. 217–233.

———, 1996, Coesite inclusions in dolomite from eclogite in the southern Dabie Mountains, China: The significance of carbonate minerals in UHPM rocks: Amer. Mineral., v. 81, p. 181–186.

———, 1997, Partial transformation of gabbro to coesite-bearing eclogite from Yangkou, the Sulu terrane, eastern China: Jour. Metamor. Geol., v. 15, p. 183–202.

Zhang, R. Y., and Liou, J. G., 1998, Petrochemical constraints for dual origin of garnet peridotites of the Dabie-Sulu UHP terrane, China, *in* International Workshop on Ultrahigh-Pressure Metamorphism and Exhumation: Stanford, CA, Stanford University, Abs. vol., p. A134–A137.

Zhang, R. Y., Liou, J. G., and Cong, B., 1994, Petrogenesis of garnet-bearing ultramafic rocks and associated eclogites in the Su-Lu ultrahigh-pressure metamorphic terrane, China: Jour. Metamor. Geol., v. 12, p. 169–186.

———, 1995, Talc-, magnesite-, and Ti-clinohumite-bearing ultrahigh-pressure metamafic and ultramafic complex in the Dabie Mountains, China: Jour. Petrol., v. 36, p. 1011–1037.

Zhang, R. Y., Liou, J. G., Ernst, W. G., Coleman, R. G., Sobolev, N. V., and Shatsky, V. S., 1997, Metamorphic evolution of diamond-bearing and associated rocks from the Kokchetav Massif, northern Kazakhstan: Jour. Metamor. Geol., v. 15, p. 479–496.

Zhang, R. Y., Shau, Y. H., Liou, J. G., and Lo, C. H., 1998, Discovery of clinoenstatite in the Sulu garnet pyroxenite and its implication for continental subduction, *in* International Workshop on Ultrahigh-Pressure Metamorphism and Exhumation: Stanford, CA, Stanford University, Abs. vol., p. A16–A19.

Zonenshain, L. P., Korinevsky, V. G., Kazmin, V. G., Pecherskiy, D. M., Khain, V. V., and Matveyenkov, V. V., 1984, Plate tectonic model of the South Urals development: Tectonophys., v. 109, p. 95–135.

Zonenshain, L. P., Kuzmin, M. I., and Natapov, L. M., 1990, Geology of the USSR: A plate-tectonic synthesis: Amer. Geophys. Union Geodynamics Series, v. 21, 242 p. (Page, B. M., ed.).

Petrotectonic Summary of Less Intensively Studied UHP Regions

J. G. Liou

Department of Geological and Environmental Science, Stanford University, Stanford, California 94305-2115

Abstract

This paper summaries tectonic settings and mineral parageneses of several recognized ultrahigh-pressure (UHP) terranes other than the well-studied Kokchetav Massif of northern Kazakhstan, the Western Gneiss Region of Norway, the Dora-Maira massif of the Western Alps, and the Dabie-Sulu terrane of east-central China. Diamond-bearing terranes include the Beni Bousera/Ronda peridotite massif, the Erzgebirge Crystalline Complex, mantle peridotite from a Tibetan ophiolite, and possibly the Maksyutov Complex. Coesite-bearing UHP terranes are the Zermatt-Saas area, Western Alps; the Mali eclogites from West Africa; the Makabal complex of western Tien Shan; the Bohemian massif; and the newly reported Central Indonesia terrane and Himalayan eclogites from the upper Kaghan Valley of Pakistan. Except for the diamond-bearing Tibetan ophiolite, most UHP terranes share similar petrotectonic assemblages and lie within major continental collision belts in Eurasia and Africa. Using new approaches (including new geobarometers) and technologies, additional UHP terranes are expected to be recognized in other Phanenorozic orogenic sutures.

Introduction

More than a dozen Eurasian UHP terranes have been documented (see Fig. 1); well-investigated UHP terranes including the Dabie-Sulu belt of east-central China, the Kokchetav massif of northern Kazakhstan, the Dora-Maira massif of the Western Alps, and the Western Gneiss Region of Norway (Liou et al., 1994; Coleman and Wang, 1995; Ernst et al., 1995) have been described by many workers. These four classic and several other UHP terranes described below share common structural and lithological characteristics: (1) They are A-type subduction zones, constituting major continental collision belts in Eurasia and Africa, and extend several hundred kilometers or more; many are associated with contemporaneous high-pressure (HP) blueschist and eclogite belts. (2) The UHP records occur mainly in eclogites and garnet peridotites included as pods and slabs within quartzofeldspathic gneissic units, a few of which contain microinclusions of coesite in garnet and omphacite, and contain microdiamonds in garnet, kyanite, and zircon. (3) These lithologies have continental and subcontinental geochemical and petrological characteristics. (4) The exhumed UHP units are now present in the upper crust as thin subhorizontal slabs, bounded by normal faults on the top, reverse faults on the bottom, and sandwiched in among HP or lower-grade units. (5) Coeval calc-alkalic volcanic and plutonic rocks do not occur, whereas post-collisional or late-stage granitic plutons are common in some occurrences. Supracrustal rocks of these UHP regions experienced subduction-zone metamorphism at mantle depths, followed by a retrograde amphibolite-granulite facies overprint during exhumation, and finally thermal recrystallization and deformation during post-collisional granitic intrusion and orogeny.

Only trace UHP minerals are partially preserved in strong containers such as zircon or garnet in eclogite and garnet peridotite. Such preservation is being increasingly recognized. The latest reported occurrences include: (1) coesite in a few Himalayan eclogites from the Upper Kaghan Valley, Pakistan (O'Brien et al., 1999); (2) coesite in UHP rocks from Sulawesi, Indonesia, where minute coesite inclusions recently have been confirmed in zircons from jadeite-bearing quartzite (Parkinson et al., 1998; Parkinson, in press); and (3) microdiamond inclusions in garnet, kyanite, and zircon from gneisses of the Saxonian Erzgebirge, Germany (Massonne, in press). The geotectonic settings and general petrologic features of other smaller and less thoroughly investigated UHP terranes are described below.

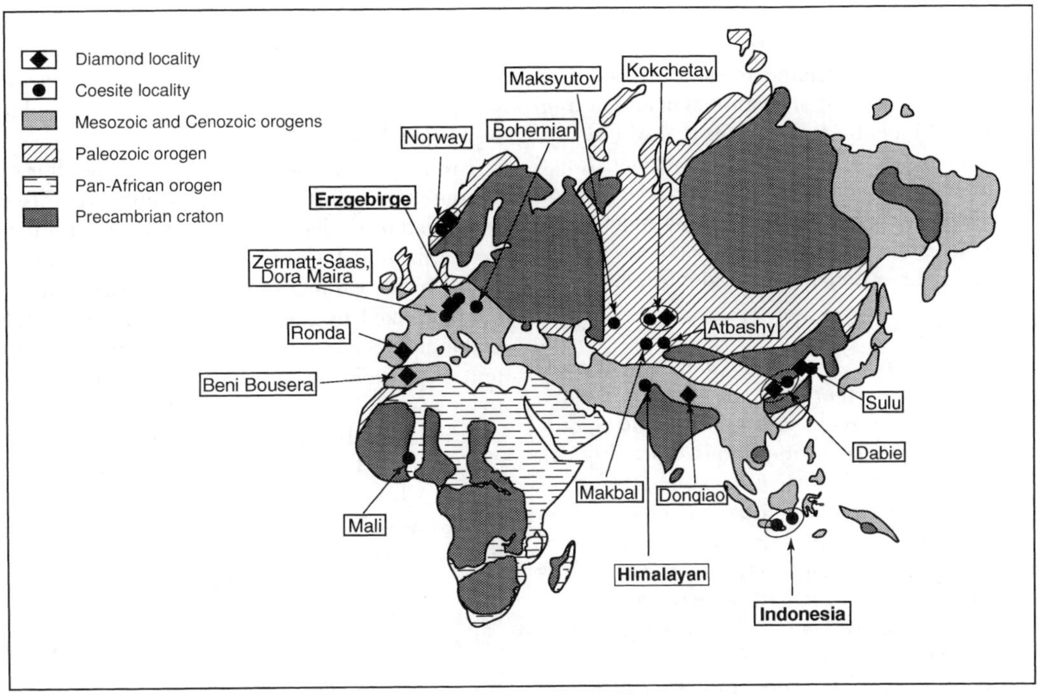

FIG. 1. Distribution of coesite- and diamond-bearing UHP terranes in the Eurasian and African continents. New occurrences from the Himalayas, Erzgebirge, and Indonesia are highlighted in bold-face type.

Diamond-bearing UHP Terranes

Beni Bousera, Morocco

The Beni Bousera peridotite massif of northern Morocco forms part of the arcuate Betico-Rifean orogenic belt and is the southern extension of the Ronda massif in Spain. The peridotite body is a 70 km^2, SE-plunging antiformal dome, faulted along its northeastern margin, and largely overlain by migmatitic, graphite-garnet-sillmanite gneiss (Pearson et al., 1989, 1993). Pyroxenite layers of the massif contain multicrystalline aggregates of graphite as octahedral and cubic forms that are interpreted as graphitized diamonds. $\delta^{13}C$ values of graphite range from −16 to −27.6‰ and are anomalously low, indicating an origin from subducted kerogenous carbon (Pearson et al., 1991). Subsolidus re-equilibration has destroyed most evidence of UHP minerals in these ultramafics. Calculated equilibration temperatures from the bulk compositions of exolved pyroxenes indicate derivation of the massif from the mantle at ~1400°C at 5 GPa. The age of diamond-forming metamorphism is not known. The surrounding gneiss is metamorphosed to granulite facies at P < 1.0 GPa, and has a plateau $^{39}Ar/^{40}Ar$ age of 21.5 Ma for K-feldspar (Pearson et al., 1993). This age is identical to the Sm-Nd isochron age of 21.5 Ma for a pyroxenite layer in the Ronda massif (Zindler et al., 1983).

Ronda, Spain

The Ronda peridotite massif, surrounded by granulite-facies gneiss, is the largest and possibly the best-studied peridotite massif of the world (Obata, 1980; Suen and Frey, 1987). Garnet-, spinel-, and plagioclase/lherzolite-facies peridotites occur within an outcrop of 300 km^2 (Obata, 1980). Graphite occurs predominantly as coated octahedra in a quartz-bearing garnet clinopyroxenite within the garnet peridotite mass. The graphite aggregates range in size from 1 to over 20 mm (Davies et al.,1992) and have $\delta^{13}C$ values (+1‰) plotting outside the typical mantle values exemplified by MORB (−4 to −8‰); they may have been derived from sea-floor carbonate subducted to mantle depths and transformed to diamond within its stability field at pressures exceeding 5 GPa (Pearson et al., 1995). The emplacement of such mantle rocks into an orogenic belt may be related to mantle upwelling (Pearson et al., 1995) or mantle convection from the lower (?) mantle to 100–150 km, where subducted continental materials trapped and transported them to the surface by wedge extrusion (Maruayama et al., 1996). The exact mechanism of emplacement for both Beni Bousera and Ronda ultramafic slices remains to be identified.

Saxonian Erzgebirge, Germany

The Erzgebirge Crystalline Complex (ECC) is situated at the northern margin of the Bohemian Massif, and is in fault contact with a low-grade Paleozoic sequence at its margins (see Fig. 2). The Variscan ECC consists of abundant ortho- and paragneisses that include numerous lenses of eclogites. Based on petrological criteria, three eclogite-bearing HP units have been mapped, as shown in Figure 2. Broadly from east to west, the eclogites record a systematic decrease in P-T conditions of recrystallization, from 830 to 910°C and >2.9 GPA in the UHP unit, to 650–750°C and 2.4–2.6 GPa in HP unit 1, and finally to 600–650°C and 2.0–2.4 GPa in HP unit 2. P-T conditions in the HP units increase progressively toward the east; such polarity indicates a WSW-directed tectonic transport direction during exhumation.

The "Gneiss-Eclogite Unit" (HP unit 1 of Schmadicke et al., 1995) at the core of this massif contains some UHP rocks typified by inclusions of coesite pseudomorph in garnets from eclogite (Schmadicke, 1991) and inclusions of microdiamonds in garnet, kyanite, and zircon from gneisses that contain abundant quartz, plagioclase, phengite, and minor rutile and graphite (Massonne, in press). Thus far, the diamondiferous gneisses seem to occur only in a 1 km long strip near the eastern shore of the Saidenbach Reservoir. Diamond microinclusions occur exclusively in garnet, kyanite, and zircon of several gneissic rocks. Two examples of microdiamonds with characteristic striation patterns are shown in Figure 3. In addition to completely preserved diamonds with grain sizes between 1 and 25 μm, partially graphitized diamonds and graphite pseudomorphs after diamond occur in the host minerals. Such microdiamond inclusions in garnet and zircon are similar to those in the diamondiferous biotite gneiss of the Kokchetav Massif, documented by Sobolev and Shatsky (1990); many photomicrographs of such microdiamond inclusions in garnet with similar striation patterns have been published by Massonne (in press); an example is reproduced in Figure 4.

From textural observations and geothermobarometric calculations, Massonne (in press) identified three distinct stages of metamorphic recrystallization for such diamondiferous gneissic rocks, as shown in Figure 5. The earliest stage is characterized by pressure conditions of ~2 GPa and temperatures of ~700°C, based on the Si content of phengite inclusions in garnet and phengite-garnet geothermometry. The peak stage is characterized by formation of diamond at depths of at least 130 km and temperatures of 900 to 1000°C, based on high Ti content in garnet and breakdown of phengite to kyanite + an unidentified K-bearing phase, which might have been a siliceous fluid. During exhumation accompanied by cooling, surviving microdiamonds were enclosed in garnet, kyanite, or zircon before they were completely graphitized. In a third stage, phengite was newly formed at ~750°C and pressures between 2 and 1.5 GPa. These UHP gneisses possess similar P-T paths to diamondiferous gneisses from the Kokchetav massif.

In contrast to the widespread distribution of metabasaltic eclogite in felsic gneiss of the entire ECC, garnet peridotites/pyroxenites are restricted only to UHP unit 1 (Schmadicke and Evans, 1997) and do not enclose eclogites as do most garnet peridotites of other UHP terranes. Garnet-peridotites occur as elongate bodies (up to 1–2 km in length) in leucocratic granulites; garnet-pyroxenites form sporadic layers within the peridotites, with a thickness ranging from ~1 to 10 cm. Geochemical data (ε_{Nd} ~ 1.9) suggest that isotopically enriched garnet pyroxenite may represent subcontinental mantle wedge materials, whereas eclogites with ε_{Nd} of +4.4 to +6.9 may have protoliths derived from a long-term depleted mantle, probably a MORB source.

Schmadicke et al. (1995) reported Sm-Nd isochrons for garnet-Cpx-WR at 360 ± 7 Ma for eclogite and 353 ± 6 Ma for garnet pyroxenites from the Erzgebirge UHP rocks. $^{40}Ar/^{39}Ar$ spectra of phengite from two eclogite samples give plateau ages of 348 ± 2 and 355 ± 2 Ma. Such similarity in ages for garnet peridotites and eclogites enclosed in gneiss suggests a coeval

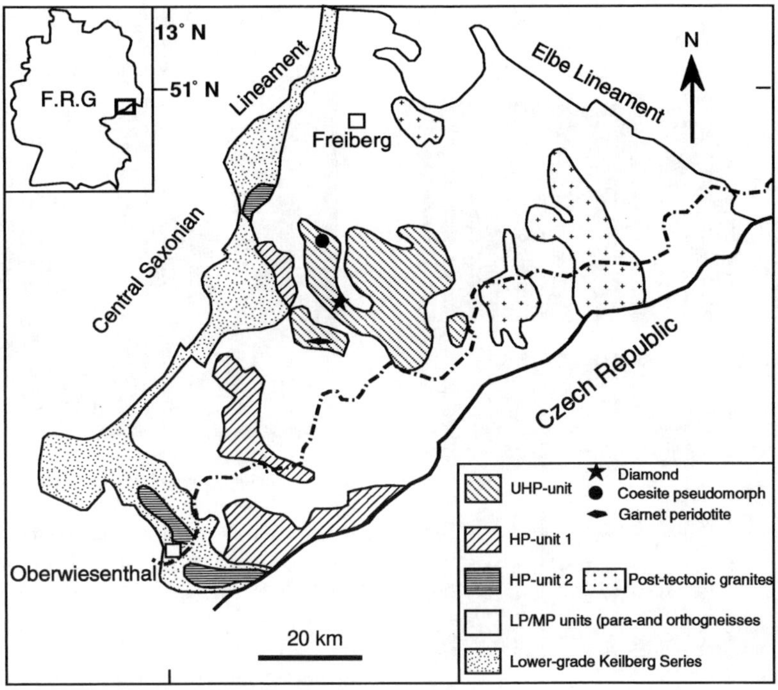

FIG. 2. Geologic sketch map of the preliminary subdivision of HP and UHP units in the Erzgebirge Crystalline Complex (modified after Schmadicke et al., 1995). Location of diamond, coesite pseudomorph, and garnet peridotite of the UHP unit was provided by Hans-J. Massonne.

Variscan UHP in-situ metamorphism of the ECC at ~350 Ma.

Tibetan Ophiolite, China

The Luobsa ophiolite of the Yarlung-Zangbo (southern Tibet) suture zone and the Donqiao ophiolite of the Bangong-Nujiang suture zone (northern Tibet) are allochthonous bodies that contain diamond-bearing mantle peridotites and podiform chromitites (Bai et al., 1993a, 1993b). The mantle sections in both massifs consist chiefly of harzburgite and diopside-bearing harzburgite, with abundant lenses of dunite and chromitite. These ultramafic rocks are more strongly depleted than typical abyssal peridotites; host rocks and mineral chemistries suggest their formation above subduction zones.

Three independent studies of mineral separates from more than one ton of serpentinized peridotite samples have yielded consistent ultrahigh-pressure minerals, including diamond and SiC (moissanite) (Yan et al., 1986; Bai et al., 1993a, 1993b; Robinson et al., 1998). The diamonds occur as colorless, octahedral, dodecahedral, and cubo-octahedral crystals, typically 0.2 to 0.4 mm across; the largest diamond grain is 1.0 mm across. SiC crystals are typically euhedral, up to 1 mm in diameter, and vary in color from blue-green or pale green to nearly colorless, yellow-green or blue-black. They are accompanied by an unusual mineral association of graphite, native chromium, Ni-Fe alloy, chromite, CrC, Cr-diopside, Cr-chlorite, olivine, various sulfides, PGE minerals, Ni-Fe alloys, Au-Ag alloys, native iron, and many others (e.g., Robinson et al., 1998). Olivine with an octahedral crystal form was suggested to be a possible relic of β-polymorph (Bai, pers. commun., 1998). This association suggests its occurrence at extremely high pressure and a reducing environment.

There is no evidence that the podiform chromitites or the host ophiolite have experienced diamond-facies UHP metamorphism. In fact, many of the ophiolitic rocks retain their original igneous textures and minerals. It is not clear whether the diamonds were formed by UHP metamorphism of

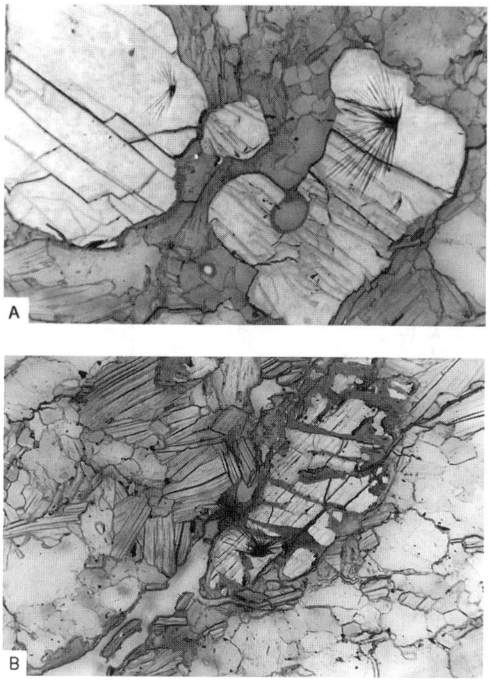

FIG. 3. Photomicrograph of Erzgebirge diamondiferous felsic gneiss (sample E97-3) taken under reflected light. It shows characteristic striation patterns around microdiamond inclusions in garnet (A) and in kyanite (B). The kyanite in (B) is partially replaced by white mica (length of the image = 3 mm; reflected light, open polarizer) (photos provided by Hans-J. Massonne).

the oceanic crust or by crystallization in the mantle wedge above the subduction zone; their occurrence in chromities and harzburgites suggests a metamorphic origin. Bai et al., 1993a, 1993b) and Robinson et al. (1998) speculated that the diamonds and SiC grains could be either xenocrysts incorporated into the magmas from which the chromities crystallized, or that they formed metastably shortly after formation of the chromitites. The origin of these ultradeep minerals in an ophiolitic terrane remains enigmatic.

Coesite-Bearing UHP Terranes

Zermatt-Saas area, Western Alps

In contrast to most other UHP terranes where the protolith is continental crust, the eclogites and metasediments at Lago di Cignana of the Western Alps have been derived from oceanic crust and constitute an ophiolite sequence of the

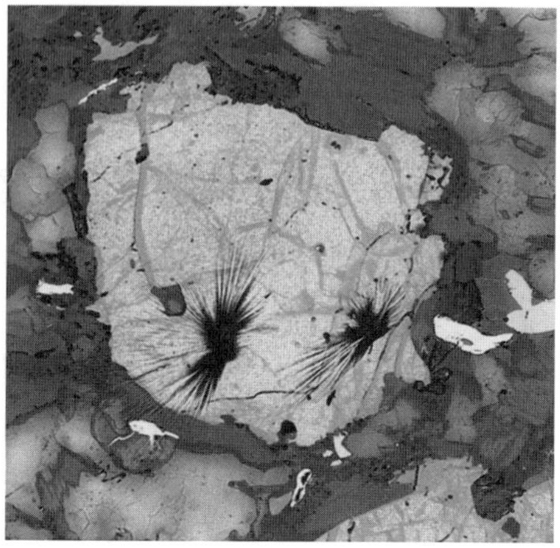

FIG. 4. Photomicrograph of Kokchetav diamondiferous felsic gneiss, showing similar characteristic striation patterns around microdiamond inclusions in garnet (length of the image = 3 mm; reflected light, open polarizer) (photo provided by Hans-J. Massonne).

Zermatt-Saas zone. Inclusions of coesite and coesite pseudomorphs occur in tourmaline and garnet of manganiferous quartzite and in omphacite and garnet of the underlying eclogite (Reinecke, 1991). The metasediments are mainly garnet-phengite-quartz schists, with variable amounts of garnet-clinopyroxene quartzites, and piemontite-phengite-quartz schist. The metabasic rocks include eclogites and retrograded eclogite in various stages of greenschist-facies overprint (Van Klauw et al., 1997). The eclogite with perfectly preserved UHP mineral parageneses consists of zoned garnet, omphacite, glaucophane, zoisite, phengite, and dolomite. Zoned garnets preserve prograde quartz single crystals in their cores and coesite inclusions in mantles and rims. Coesite inclusions, partly or completely pseudomorphed by quartz, are much more common in omphacite than in garnet. Based on mineral parageneses of various groups of eclogite and metasediment combined with thermobarometric calculations, a detailed P-T path was established; it is reproduced in Figure 8 of Carswell and Zhang (1999). UHP conditions were estimated at 2.6 to 3.0 GPa and 590 to 630°C. Sm-Nd isotopic analyses of an essentially unretrogressed eclogitic metabasalt suggest that eclogite-facies metamorphism occurred at 52 ± 18 Ma (Bowtell et al., 1994), consistent with a Tertiary age of eclogitiza-

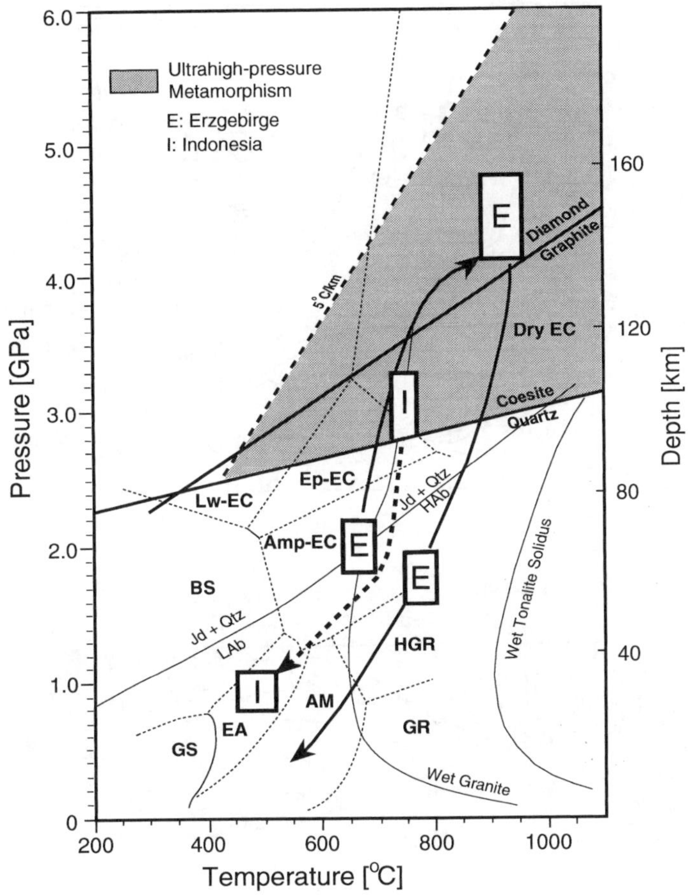

FIG. 5. P-T evolution of the diamondiferous gneisses from the Erzgebirge (after Fig. 10 of Massonne, 1999) and coesite-bearing jadeite quartzite from the Batimala Complex, Indonesia (Parkinson, in press).

tion recently established for Alpine metamorphism in the Western Alps (Gebauer et al., 1997).

Thus far, the recognized areal extent of the coesite-bearing rocks is ~2 km^2. The apparent thickness of the UHP slice does not exceed a few hundred meters. This area is the best example of subduction of oceanic lithosphere to mantle depth, and later exhumation as a fragmented tectonic slab. Similarly, the Allalin metagabbro, part of the oceanic Piemonte zone, contains magnesite + talc + kyanite + garnet recrystallized at ~600°C and 3.5 GPa, with an estimated geothermal gradient of 5 to 6°C/km (Barnicoat, 1996); however, inclusions of coesite or coesite pseudomorphs have not been reported for the Allalin metagabbro. It should be noted this metagabbro preserves both primary textures and minerals similar to the coesite-bearing metagabbro from the Sulu terrane of eastern China (Zhang and Liou, 1997).

Mali, Africa

Major parts of Africa were stabilized through the Pan-African orogeny (~800 to 550 Ma), when the West African, Congo, and Kalahari cratons, as well as several island arcs and microcontinents, all collided and amalgamated to form a Gondwana supercontinent during latest Proterozoic to Early Cambrian time (e.g., Gass, 1981; Caby et al., 1989). Blueschist and eclogite terranes in a suture zone between the West African and Congo cratons (Caby et al., 1981) form a nappe overlying the leading edge of the West African craton (Fig. 6). This zone continues from Accra (Ghana), through Togo and the Gourma Basin to the Hoggar

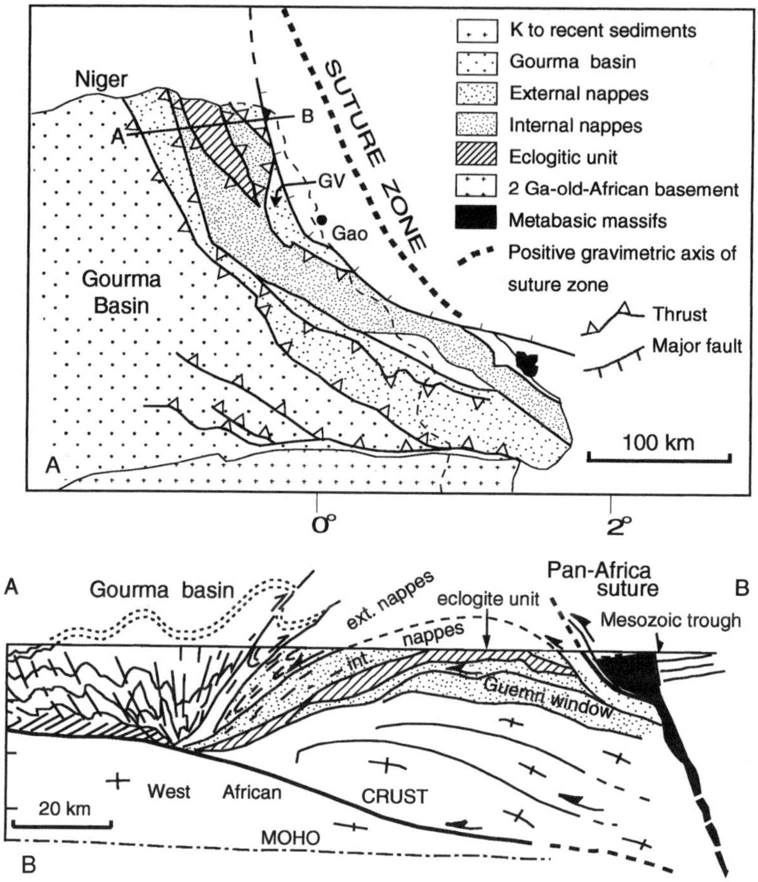

FIG. 6. A. Geological sketch map of the coesite-bearing eclogitic unit. B. Interpretative crustal section in Mali, western Africa (modified after Figs. 2 and 3 of Caby, 1994). The cross section shows the coesite-bearing eclogite unit as a subhorizontal layer sandwiched between lower-grade rocks and the doming structure.

Mountains (central Sahara) and bends abruptly in Mali because of Cambrian indentation of the West African craton during continent-continent collision (Ball, 1980). Sodic amphibole and jadeite occur in Gourma (Caby et al., 1981). Quartz-bearing eclogites in western Hoggar, central Hoggar, and Lato Hill, Togo (Affaton et al., 1980; Ménot and Seddoh, 1985) have P-T estimates >1.4 GPa and 650 to 750°C (Sautter, 1985), and >1.7 GPa and 700 to 750°C (Ménot and Seddoh, 1985).

Rocks in the Pan-African collision zone of the trans-Saharan segment cropping out in northern Mali contain inclusions of coesite in omphacite from mafic nodules within a calc-silicate layer enclosed in impure marble, and coesite pseudomorphs in garnets from the surrounding eclogitic metasediments (Caby, 1994). These eclogitic mica schists contain UHP assemblages including kyanite, pyrope-rich garnet, jadeite-rich omphacite, phengite ($Si_{3.35-3.40}$), and rutile, and have P-T estimates of 700 to 750°C and >2.7 GPa. A well-defined plateau $^{39}Ar/^{40}Ar$ age of 1045 ± 9 Ma has been obtained on the coarse-grained phengite from an omphacite-kyanite mica schist (Caby, 1994).

This coesite-bearing UHP eclogitic unit ~3 km thick, as shown in the cross section of Figure 6, lies within an internal nappe as a flat, thin slice bounded on the top by a thick shaly formation of the inner part of the Gourma passive margin, and on the bottom by low-grade greenschist-facies phyllites. Such a subhorizontal nappe structure is similar to many other UHP and HP terranes documented by Maruyama et al. (1996). Caby (1994) suggested that a large portion of the terrigenous metasediments of passive-continental-margin affinity of the West African plate was subducted eastward to mantle depths (>90 km) beneath a Late Proterozoic oceanic domain. Collision of the West African plate with an island arc resulted in a low-angle, westward subhorizontal extrusion of this subducted slice of sialic materials to their present position. The coesite-bearing eclogite unit may also exist 1500 km to the south in Togo, where kyanite-bearing eclogites occur as part of the passive-margin assemblage. Zircon ages (upper intercept) of ~820 Ma (Bernard-Griffiths et al., 1991) for kyanite eclogite may represent the magmatic age of the eclogitic protolith. Therefore, the timing of UHP metamorphism for the Mali coesite-bearing eclogitic unit remains to be investigated.

Makbal (480 Ma) and Atbashy (270 Ma), Kazakhstan

The Tien Shan is divided into three mountain ranges—northern, middle, and southern. Within the Kyrgyzstan Tien Shan, between the Kazakhstan and Tarim continents, several eclogite localities have been described (e.g., Sobolev et al., 1986). UHP eclogite localities with coesite pseudomorph inclusions in garnet and omphacite have been described from the northern and southern Tien Shan, respectively (Tagiri and Bakirov, 1990; Tagiri et al., 1995). These northern and southern localities represent Caledonian and Hercynian orogenic belts, respectively. Petrochemical studies and P-T determinations of selected eclogitic samples by Tagiri and his associates are summarized in Carswell and Zhang (1999). However, the extent of eclogite distribution, the contact with other high P-T units, precise timing for eclogite-facies metamorphism, and tectonic interpretation of these UHP rocks remain to be studied.

The Makabal complex in the western Kyrgyz Ridge of the northern Tien Shan consists of two lithologic units, the Neldy and Makbal formations. The Makbal Formation is characterized by quartzose schist alternating with pelitic schists intercalated with thin marbles and calciphyres. Eclogites and amphibolitized eclogites crop out as layers and lenticular bodies one to several tens of m thick. Tagiri and Bakirov (1990) found inclusions of quartz pseudomorphs after coesite in garnet from a garnet-chloritoid-talc schist of the Makbal Formation. The mineral assemblage for the peak metamorphism is coesite + almandine + chloritoid + talc + phengite (Si = 3.42 p.f.u.) + phlogopite + rutile. The Makbal eclogites were subjected to later retrograde recrystallization at 2 GPa, 560°C. The Makbal coesite eclogite has a paragonite K-Ar age of 480 Ma.

The eclogite-bearing complex of Atbashy Ridge from the southern Tien Shan consists of pelitic and siliceous schists alternating with thin UHP eclogitic layers, and is unconformably overlain by Upper Paleozoic molasse and silicic volcanics. Coesite-bearing eclogites consist of euhedral garnet, omphacite, and phengite with retrograde albite, actinolite, and chlorite of greenschist-facies overprint. Inclusions of quartz pseudomorphs after coesite were identified in omphacite cores (together with glaucophane, which has been replaced by talc + albite intergrowths) and in garnet. The garnet shows three distinct and essentially euhedral zones—core, mantle, and rim. The core has low pyrope (Pyr = 5.8 to 8.6) but high grossular content (Grs = 32 to 38) and includes omphacite, titanite, and quartz inclusions. The mantle is rich in pyrope and poor in grossular, and contains inclusions of omphacite and quartz pesudomorphs after coesite. The rim is free of inclusions. Eclogites from the Atbashy area preserve prograde and retrograde paths with a peak metamorphic condition at 660°C, 2.5 GPa; their P-T path is shown in Figure 8 of Carswell and Zhang (1999). The Atbashy coesite-eclogites yielded an Rb-Sr mineral isochron age of 270 Ma (Tagiri et al., 1995).

Maksyutov Complex, southern Urals, 375 ± 10 Ma

The Maksyutov Complex is an elongate (10 × 120 km), N-S belt between the Russian and Siberian cratons in the southern Ural Mountains (Coleman and Wang, 1995; Ernst et al., 1995; Dobretsov et al., 1996). The complex is the locality from which quartz aggregates within garnet exhibiting radial cracks were first recognized as coesite-pseudomorphs by Chesnokov and Popov (1965), nearly 20 years prior to the independent finding of coesite inclusions by Chopin (1984) from the Dora-Maira massif of the Western Alps and Smith (1984) from the Western Gneiss Region of Norway. This complex is composed of two tectonic units—an overlying meta-ophiolite unit and an underlying eclogite-bearing schist unit (Beane et al., 1995; Dobretsov et al., 1996). The eclogite-bearing unit contains boudins of eclogite, layers of eclogitic gneiss, and rare ultramafic bodies within host metasedimentary mica schist and quartzite. The meta-ophiolite unit consists of lenses of serpentinite mélange and blocks of metasomatic rock, metabasalt, and marble within mica schist and graphite quartzite host rock. The HP/UHP event is characterized by recumbent folds and shear zones. Later, large-scale left-lateral strike-slip movements deformed both tectonic units.

Garnet-clinopyroxene thermobarometry on Maksyutov eclogites yields equilibrium T of 594 to 637°C. Minimum pressure estimates using the jadeite component of clinopyroxene range from 1.5 to 1.7 GPa (Beane et al., 1995; Lennykh et al., 1995; Leech and Ernst, 1998) or up to 2.7 GPa if coesite pseudomorphs (Chesnokov and Popov, 1965; Dobretsov and Dobretsova, 1988) are indeed present. However, subsequent study has not revealed the occurrence of inclusions of coesite pseudomorph in garnet and zircon (Beane et al., 1995; Shatsky et al., 1997). Leach and Ernst (1998) described unusual cuboid graphite aggregates (up to 13 mm edge length) that may be diamond pseudomorphs. This suggestion was based on the blocky morphology of the cuboid graphite aggregates' cross-cutting rock foliation, pressure shadows, biogenic isotopic signatures, and other spectroscopic characteristics similar to those graphitized diamonds from the Beni Bousera massif. This study has yielded the suggestive, though unproven, possibility that the eclogitic unit of the Maksyutov Complex experienced diamond-grade UHP metamorphism.

Radiometric data of mica and rutile from eclogite and metasediments indicate that the Maksyutov Complex underwent HP metamorphism during the Devonian. Mica separates from five samples yielded $^{40}Ar/^{39}Ar$ plateau ages of 372 to 374 Ma. Rutile separates from two samples yielded concordant U/Pb ages of 378 ± 4 Ma and 390 ± 5 Ma. These new data challenge previous interpretations that the eclogite-facies metamorphism occurred in Early Paleozoic or even Middle Proterozoic time (Dobretsov, 1984; Coleman and Wang, 1995; Dobretsov et al., 1996). Recent radiometric dating yielded peak metamorphic ages of 370 to 374 Ma (Shatsky et al., 1997; Beane, 1997). The UHP/HP metamorphism of the Maksyutov Complex may be related to the collision of the Russian platform with a fragment of the Siberian craton during Devonian time, but was affected later by the Triassic collision of the Kazakhstan block with the Russian platform (e.g., Matte, 1998).

Bohemian massif, 324 to 374 Ma

The Bohemian massif, 300 × 600 km^2 in area, is composed of Proterozoic continental basement, Hercynian metamorphic rocks and granites, and a post-Paleozoic cover sequence. The Hercynian metamorphic rocks consist mainly of orthogneiss, eclogite, spinel- and garnet peridotite, and an amphibolite/granulite–facies metabasite-metafelsite complex. Inclusions of coesite pseudomorphs in garnet and occurrence of quartz lamellae in omphacite as supersilicic clinopyroxene were first described in eclogite-granulite rock series of the Zlote Mountains in southwestern Poland by Bakun-Czubarow (1991). Two different eclogites occur: (1) prograde eclogite lenses and boudins in gneiss; and (2) retrograde eclogite lenses or layers in garnet peridotites. The estimated P-T conditions of the former are 615 to 730°C, 1.3 to 1.6 GPa, whereas those of the latter are much higher at 765 to 1190°C, 1.2 to 2.0 GPa (Medaris et al., 1995). Eclogites contain coesite and kyanite in the Münchberg body, reaching a maximum P of 2.5 to 2.8 GPa at 600°C, followed by a Barrovian overprint (O'Brien, 1993). Garnet peridotites and associated eclogites from the Gföhl unit were formed at 3.5 to 4.7 GPa, 900 to 1100°C (Medaris et al., 1995). The Nd-Sm ages for Gt-Cpx pairs from eclogites range from 324 to 374 Ma; those from their equivalents in Germany are 423 to 460 Ma. Hence, two distinct subduction-zone metamorphic events may have occurred (see Maruyama et al., 1996).

Central Indonesian UHP terrane, 110 to 120 Ma

The pre-Tertiary basement of the Indonesian region comprises a variety of imbricated terranes, mélange, ophiolite, and variably metamorphosed accretionary complexes; some of these are high- to ultrahigh-P metamorphic rocks due to the collision of an Australia-derived continent with

Eurasia (Parkinson et al., 1998). High-P metamorphic rocks including eclogites and garnet peridotites are widely distributed in Cretaceous accretionary complexes throughout Java, Sulawesi (formerly Celebes), and southeast Kalimantan (Indoesia Borneo), as shown in Figure 7. Many of these rocks occur as imbricate slices of carbonate, quartzite, and pelitic schist of shallow-marine or continental-margin parentage, interthrust with subordinate mafic schist and serpentinite. They are predominantly of low-to-intermediate metamorphic grade and yield mica K-Ar ages of 110 to 120 Ma.

Metamorphic rocks that exhibit evidence of exhumation from much greater depths (>60 km) are sporadically exposed as tectonic blocks throughout the Cretaceous accretionary complexes. They include eclogite, garnet-glaucophane rock (P = 1.8 to 2.4 GPa, T = 580 to 620°C), and jadeite-garnet quartzite (P > 2.7 GPa, T = 720 to 760°C) in Bantimala, southwestern Sulawesi; eclogite and garnet granulite in west-central Sulawesi; eclogite and jadeite-glaucophane-quartz rock (P ~ 2.2 GPa, T ~ 530°C) in Luk Ulo, central Java; and Mg-chloritoid-bearing whiteschists (P ~ 1.8 GPa) in the Meratus Mountains, southeastern Kalimantan. Garnet lherzolites from depths of >60 km also are associated with HP schists in east-central Sulawesi (P = 2.2 to 2.8 GPa, T = 1000 to 1100°C) and west-central Sulawesi (1.6 to 2.0 GPa, T = 1050 to 1100°C); garnet pyroxenite (P ~2 GPa, T ~ 850°C) occurs as blocks with pyrope-kyanite amphibolite, eclogite, and blueschist within Miocene conglomerate in Sabah, northeastern Borneo. Evidence for UHP rocks include: (1) inclusions of coesite in zircons and coesite pseudomorphs in jadeite from jadeite quartzites and eclogites of the Bantimala Complex of southern Sulawesi (Fig. 8) (Parkinson, in press); and (2) P-T estimates of peak-stage recrystallization at 2.7 to 3.5 GPa and 1000 to 1100°C for the majority of Palu-Koro garnet peridotites and 3.3 to 3.5 GPa and 1150°C for Bongka River lherzolites (Kardarusman and Parkinson, in review). The P-T path for coesite-bearing jadeite quartzite and eclogites of the Bantimala Complex is shown in Figure 5. Many of these rocks probably were recrystallized in a N-dipping subduction zone at the margin of the Sundaland craton in the Early Createceous. Exhumation may have been facilitated by the collision of a Gondwana continental fragment with the Sundaland margin at ~120 to 115 Ma.

Himalayan eclogites (~49 Ma)

Coesite was first discovered in omphacite in several Himalayan eclogites from the Upper Kaghan valley, Pakistan by O'Brien et al. (1999); this finding represents the first record of UHP metamorphism in the Himalaya. The coesite-bearing eclogite occurs in the Loihaloi Nala, north of Gittidas, opposite the settlement Faqir Shah di Baihk. This location lies only a few km from the Indus suture. The eclogite is massive and contains fine-grained (<1 mm) garnet and omphacite together with aggregates of quartz, chains of rutile rimmed by titanite, and mm-sized phengitic micas overgrown by randomly oriented amphiboles up to 1 cm in length. Coesite and coesite pseudomorphs occur as inclusions in omphacite (the largest grain is 60 microns) and show the characteristic palisade-quartz aggregates and the typical radiating fractures in the host pyroxene.

Garnet shows distinct compositional zoning, with a relatively Ca-rich core ($Alm_{50}Prp_{18}Grs_{29}Sps_{3}$), surrounded by a zone with lower Ca ($Alm_{56}Prp_{21}Grs_{21}Sps_{1}$) and followed by a narrow (but distinct) overgrowth where Ca is again higher than Mg. Omphacite ranges in Jd content from 25 to 33 mol% with 4 to 13 mol% aegirine content. Phengite has a maximum Si per formula unit of 7.2 (Mg/Mg + Fe = 0.74) in the cores; both values decrease toward the margins. Applying the conventional geobarometer with the Jd isopleth and Fe-Mg distribution between garnet and omphacite, O'Brien et al. (1999) obtained equilibrium conditions of ~650°C and P > 1.4 GPa, which is similar to the estimates by Spencer et al. (1995) at 600 to 650°C and P > 1.4 GPa. On the other hand, the new garnet-omphacite-phengite geobarometer proposed by Carswell et al. (1997) yielded 2.7 GPa and 680°C. In fact, this new finding resulted from the application of this geobarometer to the data of Spencer (1993) from Kaghan Valley eclogites, indicating pressures well above 2.0 GPa and prompting new field work in this area. Some eclogite samples described from the Tso Morari area of Ladakh by de Sigoyer et al. (1997) also yielded pressures in the coesite field. Further research may confirm wider occurrences of this newly documented UHP terrane along the Indus suture.

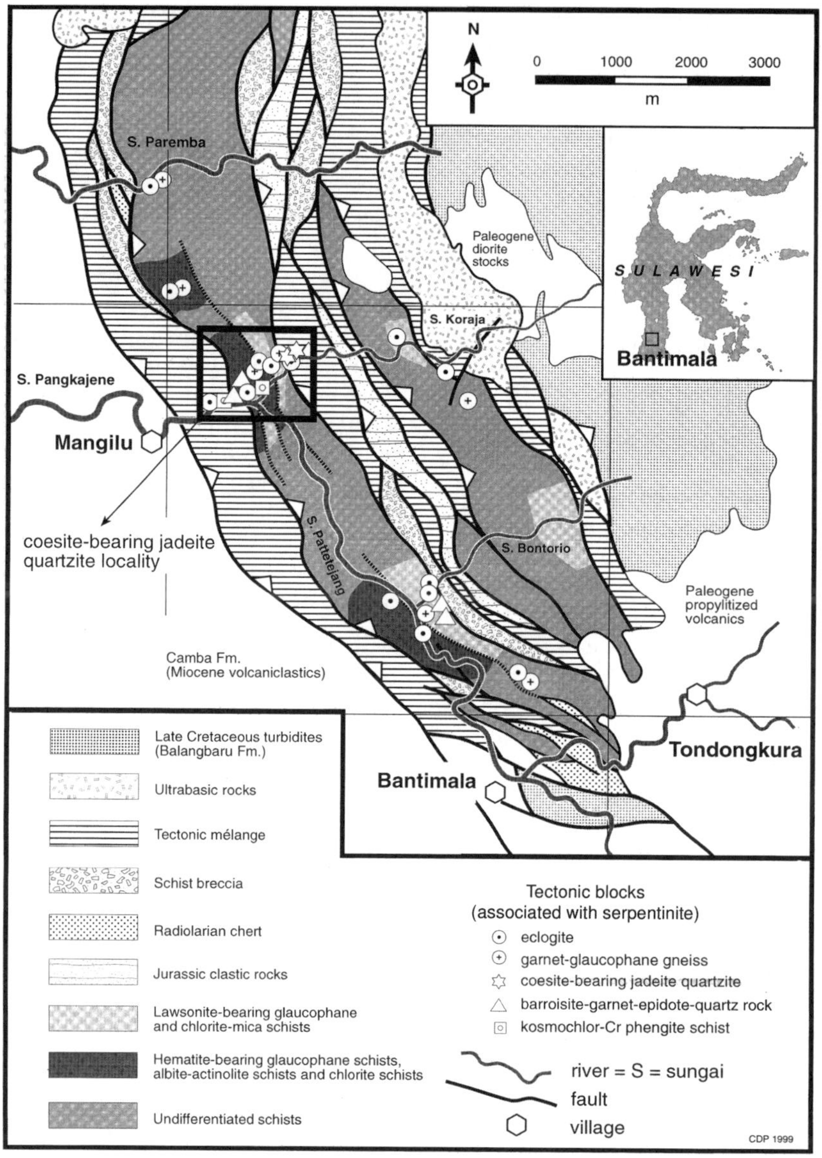

FIG. 7. Geologic sketch map of the newly discovered UHP terrane in central Indonesia (map provided by Chris Parkinson).

This first discovery of coesite (combined with the new pressure estimates) indicates that the leading edge of the Indian continent has been subducted to mantle depths of >75 km, and that these rocks must have been exhumed at least twice as fast as previously estimated. As stated by O'Brien et al. (1999), "the exhumation of rocks from coesite-forming depths, and more importantly the metastable preservation of coesite in such rocks, may well be entirely due to the bouyancy of the relatively low density (compared to the mantle) subducted continental crust slice hosting the eclogites. Tectonic models promoting this aspect, such as that proposed by Chemenda et al. (1995), may therefore be directly appplicable to the Himalaya."

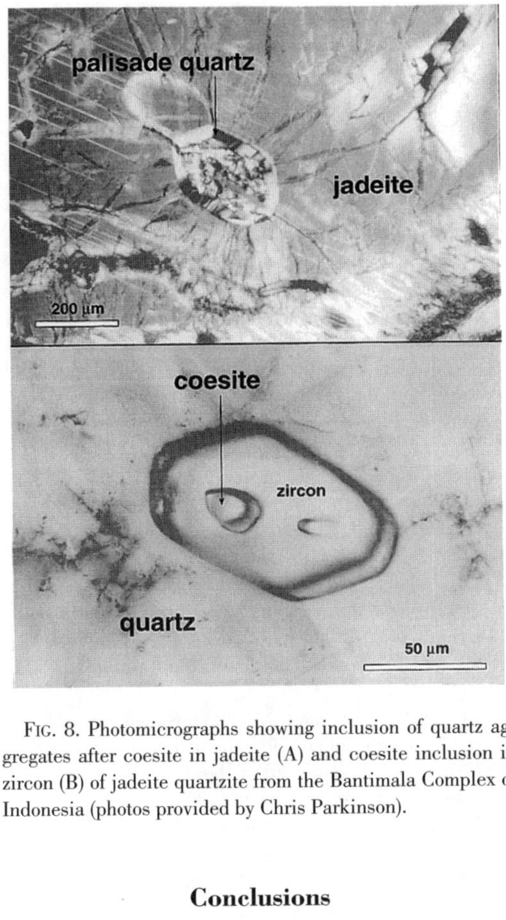

FIG. 8. Photomicrographs showing inclusion of quartz aggregates after coesite in jadeite (A) and coesite inclusion in zircon (B) of jadeite quartzite from the Bantimala Complex of Indonesia (photos provided by Chris Parkinson).

Conclusions

Among the recognized global UHP terranes described in this and previous reports, most contain both eclogites and garnet peridotites with conclusive index UHP minerals. UHP rocks are enclosed within felsic gneisses intercalated with pelitic schists, marbles, and quartzites. These country rocks also have been subjected to coeval UHP metamorphism; however, subsequent amphibolite- to greenschist-facies overprinting has entirely re-equilibrated their mineral assemblages. The suggestions of foreign origin for UHP blocks in central Indonesia by Parkinson et al. (1998) and xenocrysts for ultradeep minerals in Tibetan ophiolite by Robinson et al. (1998) remain to be substantiated.

Also among the UHP terranes, the Kokchetav Massif may be the oldest at 540 to 530 Ma. In order of descending geologic age, the UHP rocks are 480 Ma at Makbal, 425 to 408 Ma in the Western Gneiss Region, 375 to 380 Ma in the Maksyutov Complex, 374 to 324 Ma in the Bohemian massif, 270 Ma in the Atbashy Complex, 240 to 210 Ma in the Dabie-Sulu regions, 48 to 41 Ma at Sifnos, and 35 Ma in the Dora-Maira massif and Zermatt-Saas areas. The UHP rocks from Mali have been dated at 1045 to 820 Ma. However, this age is suspect, inasmuch as the UHP terrane lies within the 750 to 550 Ma Pan-African collision zone between the West African and East Saharan cratons (Caby, 1994). It is apparent that UHP metamorphism occurred throughout the Phanerozoic; no clear indications of UHP metamorphism have survived from Archean or Proterozoic time (Maruyama and Liou, 1998).

UHP rocks of intracontinental suture zones have experienced complex, multi-stage metamorphic recrystallization histories; hence, most UHP minerals have been obliterated during their return to the surface. Strong, rigid, unreactive crystals—including garnet, zircon, kyanite, and omphacite—evidently function as pressure vessels, preserving some early-formed UHP mineral inclusions. Examination of zircons and garnets for included primary UHP phases such as coesite and diamond using Raman spectroscopy has proven to be a powerful technique, especially for highly retrograded metamorphic rocks (e.g., Tabata et al., 1998; Katayama et al., in review) and has revealed new UHP localities, such as the one in Indonesia described by Parkinson (in press).

A new method for simple identification and quantification of microdiamonds in rock thin sections was described by Massonne et al. (1998). This method reveals microdiamonds as small as 1 micron by a distinctive striation pattern on the polished surface of the thin sections from diamondiferous gneiss and marble of the Kokchetav massif and Erzgebirge gneiss, as shown in Figures 4 and 5. Using this method, Massonne (in press) discovered up to 140 exposed microdiamond inclusions in garnet, kyanite, and zircon on a 6 cm^2 surface area from Erzgebirge quartzofeldspathic rocks described in the previous sections. It is possible that microdiamonds may be much more common than thus recognized in other areas where metamorphic quartzofeldspathic rocks dominate.

The discovery of coesite in Himalayan eclogites is significant, as the collision between the Indian and Eurasian continents has been continuous since ~50 Ma. Applying new technologies, including new geobarometers, and approaches, more UHP terranes will be identified in other Phanerozoic orogenic sutures; new interpretations of tectonic evolution involving continental subduction and collision remain to be proposed. It is appropriate to quote the statement of O'Brien et al. (1999) as a concluding paragraph in this review.

> The first definite eclogites in the Himalayas were first reported only in the early 1990s and there are still only very few areas where eclogites (or their retrograded or overprinted equivalents) have been identified. In comparison to European continental collision belts such as the Alps, Variscides, or Caledonides, eclogites are extraordinarily scarce in the Himalayas, and other characteristic high-pressure rocks, such as garnet peridotite, are completely unknown. Is this just a product of the lack of investigation, or is the lack of high-pressure rocks a function of the pervasive amphibolite-facies event? One thing is certain: the eclogites were encased in continental crust at these mantle depths and it is only a question of time before a high-pressure evolution is confirmed in these rocks as well. It should not be forgotten that easily accessible rocks preserving relics of ultrahigh-pressure metamorphism in the Dora-Maira massif, Western Alps, had been completely overlooked by generations of geologists before their true worth was discovered!

Acknowedgments

Preparation of this manuscript for the Task Group III-6 Final Report was supported by both the NSF Earth Science Division and International Program for a U.S.-Japan-Russia-Kazakhstan cooperative project (NSF EAR 9725347) and the Stanford China Industrial Affiliate Program. This manuscript has been critically reviewed by W. G. Ernst. Photomicrographs of microdiamonds from Erzgebirge and Kokchetav, and the tectonic map of Indonesia, were provided by Hans Massonne and Chris Parkinson respectively. I thank Paddy O'Brien for sending me his unpublished extended abstract announcing the new discovery of coesite in the Himalayas. The author appreciates the aforementioned individuals and institute for their help and support.

REFERENCES

Affaton, P., Sougy, J., and Trompette, R., 1980, The tectono-stratigraphic relationships between the upper Precambrian and lower Paleozoic Volta basin and the Pan-African Dahomeyide orogenic belt (west Africa): Amer. Jour. Sci., v. 280, p. 224–248.

Bai, W., Robinson, P. T., and Zhou, M., 1993a, Diamond-bearing peridotites from Tibetan ophiolites: Implications for a subduction-related origin of diamond: Edmundston, NB, Geol. Soc. Canada/Min. Assoc. Canada, Proc. Mid-Continent Diamond Symp., p. 77–92.

Bai, W. J., Zhou, M. F., and Robinson, P. T., 1993b, Diamond-bearing mantle peridotites and podiform chromitites in the Luobusa and Donqiao ophiolite massifs. Tibet: Can. Jour. Earth Sci., v. 30, p. 1650–1659.

Bakun-Czubarow, N., 1991, On the possibility of occurrence of quartz pseudomorphs after coesite in the eclogite-granulite rock series of the Zlote Mountains in the Sudetes, SW Poland: Archiwum Mineral., v. XLVII, p. 5–18.

Ball, E., 1980, An example of very consistent brittle deformation over a wide intracontinental area: The late Pan-African fracture system of the Tuareg and Nigerian Shield: Tectonophys., v. 61, p. 363–379.

Barnicoat, A. C., 1996, Dolomite breakdown under ultrahigh-pressure conditions in the Allalin gabbro of SW Switzerland [abs]: EOS (Trans. Amer. Geophys. Union), v. 77, p. F762.

Beane, R., 1997, Ar/Ar evidence for Paleozoic high-pressure metamorphism in the south Urals [abs]: Geol. Soc. Amer., Abs. Prog., v. 29, p. A-399.

Beane, R., Liou, J. G., Coleman, R. G., and Leech, M. L., 1995, Petrology and retrograde P-T path for eclogites of the Maksyutov Complex, southern Ural Mountains, Russia: The Island Arc, v. 4, p. 254–266.

Bernard-Griffiths, J., Peucat, J. J., and Ménot, R. P., 1991, Isotopic (Rb-Sr, U-Pb, and Sm-Nd) and trace element geochemistry of eclogites from the Pan-African belt: A case study of REE fractionation during high-grade metamorphism: Lithos, v. 27, p. 43–57.

Bowtell, S. A., Cliff, R. A., and Barnicoat, A. C., 1994, Sm-Nd isotopic evidence on the age of eclogitization in the Zermatt-Saas ophiolite: Jour. Metamor. Geol., v. 12, p. 187–196.

Caby, R., 1994, Precambrian coesite from northern Mali: First record and implications for plate tectonics in the trans-Saharan segment of the Pan-African belt: Eur. Jour. Mineral., v. 6, p. 235–244.

Caby, R., Andreopoulos-Renaud, U., and Pin, C., 1989, Late Proterozoic arc-continent and continent-

continent collision in the Pan-African trans-Saharan belt of Mali: Can. Jour. Earth Sci., v. 26, p. 1136–1146.

Caby, R., Bertrand, J. M. L., and Black, R., 1981, Pan-African ocean closure and continental collision in the Hoggar-Iforas segment, central Sahara, *in* Kröner, A., ed., Precambrian plate tectonics: Amsterdam, Elsevier, p. 407–434.

Carswell, D. A., O'Brien, P. J., Wilson, R. N., and Zhai, M., 1997, Thermobarometry of phengite-bearing eclogites in the Dabie Mountains of central China: Jour. Metamor. Geol., v. 15, p. 239–252.

Carswell, D. A., and Zhang, R. Y., 1999, Petrographic characteristics and metamorphic evolution of ultrahigh-pressure eclogites in plate-collision belts: INT. GEOL. REV., v. 41, in press.

Chemenda, A. I., Mattauer, M., Malavieille, J., and Bokum, A. N., 1995, A mechanism for syncollisional rock exhumation and associated normal faulting: Results from physical modeling: Earth Planet. Sci. Lett., v. 132, p. 225–232.

Chesnokov, B. V., and Popov, V. A., 1965, Increasing volume of quartz grains in eclogites of the south Urals: Doklady Akad. Nauk, v. 162, p. 176–178 (in Russian).

Chopin, C., 1984, Coesite and pure pyrope in high grade pelitic blueschists of the Western Alps: A first record and some consequences: Contrib. Mineral. Petrol., v. 86, p. 107–118.

Coleman, R. G., and Wang, X., 1995, Overview of the geology and tectonics of UHPM, *in* Coleman, R. G., and Wang, X., eds., Ultrahigh-pressure metamorphism: New York, Cambridge University Press, p. 1–32.

Davies, G. R., Nixon, P. H., Pearson, D. G., and Obata, M., 1992, The tectonic implications of graphitized diamonds from the Ronda peridotite massif, S. Spain: Geology, v. 21, p. 471–474.

de Sigoyer, J., Guillot, S., Lardeaux, J.-M., and Mascle, G., 1997, Glaucophane-bearing eclogites in the Tso Morari dome (eastern Ladakh, NW Himalaya): Eur. Jour. Mineral., v. 9, p. 1073–1083.

Dobretsov, N. L., 1984, Problem of the jadeite rocks, associated with peridotites: Mineralia Slov., v. 16, p. 3–12.

Dobretsov, N. L., and Dobretsova, L. V., 1988, New mineralogic data on the Maksyutov eclogite-glaucophane schist complex, southern Urals: Doklady Akad. Nauk SSSR, v. 300, p. 111–116.

Dobretsov, N. L., Shatsky, V. S., Coleman, R. G., Lennykh, V. I., Valizer, P. M., Liou, J. G., Zhang, R. Y., and Beane, R. J., 1996, Tectonic setting and petrology of ultrahigh-pressure metamorphic rocks in the Maksyutov Complex, Ural Mountains, Russia: INT. GEOL. REV., v. 38, p. 136–160.

Ernst, W. G., Liou, J. G., and Coleman, R. G., 1995, Comparative petrotectonic study of five Eurasian ultrahigh-pressure metamorphic complexes: INT. GEOL. REV., v. 37, p. 191–211.

Gass, I. G., 1981, Pan-African (Upper Proterozoic) plate tectonics of the Arabian-Nubian Shield, *in* Kröner, A., ed., Precambrian plate tectonics: Amsterdam, Elsevier, p. 388–405.

Gebauer, D., Schertl, H.-P., Brix, M., and Schreyer, W., 1997, 35 Ma old ultrahigh-pressure metamorphism and evidence for very rapid exhumation in the Dora Maira Massif, Western Alps: Lithos, v. 41, p. 5–24.

Katayama, I., Zayachkovsky, A., and Maruyama, S., in review, Progressive P-T records in zircon from the Kokchetav UHP-HP rocks, northern Kazakhstan: The Island Arc.

Kadarusman, A., and Parkinson, C. D., in review, Petrology and P-T evolution of garnet peridotites from central Sulawesi, Indonesia: Jour. Metamor. Geol.

Leech, M. L., and Ernst, W. G., 1998, Graphite pseudomorphs after diamond? A carbon isotope and spectroscopic study of graphite cuboids from the Maksyutov Complex, south Ural Mountains, Russia: Geochim. et Cosmochim. Acta, v. 62, p. 2143–2154.

Lennykh, V., Valizer, P. M., Beane, R., Leech, M., and Ernst, W. G., 1995, Petrotectonic evolution of the Maksyutov Complex, south Urals, Russia: Implications for ultrahigh-pressure metamorphism: INT. GEOL. REV., v. 37, p. 584–600.

Liou, J. G., Zhang, R., and Ernst, W. G., 1994, An introduction to ultrahigh-P metamorphism: The Island Arc, v. 3, p. 1–24.

Maruyama, S., and Liou, J. G., 1998, Initiation of UHP metamorphism and its significance on the Proterozoic/Phanerozoic boundary: The Island Arc, v. 7, p. 6–35.

Maruyama, S., Liou, J. G., and Terabayashi, M., 1996, Bleuschists/eclogites of the world and their exhumation: INT. GEOL. REV., v. 38, p. 490–596.

Massonne, H., in press, A new occurrence of microdiamonds in quartzofeldspathic rocks of the Saxonian Erzgebirge, Germany, and their metamorphic evolution: Proc. 7th Int. Kimb. Conf., Cape Town.

Massonne, H., Bernhardt, H., Dettmar, D., Kessler, E., Medenbach, O., and Westphal, T., 1998, Simple identification and quantification of microdiamonds in rock thin-sections: Eur. Jour. Mineral., v. 10, p. 497–504.

Matte, P., 1998, Continental subduction and exhumation of HP rocks in Paleozoic orogenic belts: Uralides and Variscides: Geologiska Foreningen i Stockholm Forhandlingar, v. 120, p. 209–222.

Medaris, L. G., Jr., Beard, B. L., Johnson, C. M., Valley, J. W., Spicuzza, M. J., Jelinek, E., and Misar, Z., 1995, Garnet pyroxenite and eclogite in the Bohemian Massif: Geochemical evidence for Variscan recycling of subducted lithosphere: Geol. Rundsch., v. 84, p. 489–505.

Medaris, L. G., Jr., Jelinek, E., and Misar, Z., 1995, Czech eclogites: Terrane setting and implications for Variscan tectonic evolution of the Bohemian Massif: Eur. Jour. Mineral., v. 7, p. 7–28.

Ménot, R. P., and Seddoh, K. F., 1985, The eclogites of the Lato Hills, south Togo, west Africa: Relics from the early tectonometamorphic evolution of the Pan-African orogeny: Chem. Geol., v. 50, p. 313–330.

Obata, M., 1980, The Ronda peridotite: Garnet-, spinel-, and plagioclase-lherzolite and the P-T trajectories of a high-temperature mantle intrusion: Jour. Petrol., v. 21, p. 533–572.

O'Brien, J., 1993, Partially retrograded eclogites of the Muenchberg Massif, Germany: Records of a multi-stage Variscan uplift history in the Bohemian Massif: Jour. Metamor. Geol., v. 11, p. 241–260.

O'Brien, P. J., Zotov, N., Law, R., Khan, M. A., and Jan, M. Q., 1999, Coesite in eclogite from the Upper Kaghan Valley, Pakistan: A first record and implications [abs.], *in* 14th Himalaya-Karakorum-Tibet Workshop, Kloster Ettal, Germany: Terra Nostra, v. 99, No. 2, p. 109–111.

Parkinson, C. D., in press, Coesite-bearing quartzites from Sulawesi, Indonesia: A first record of UHP metamorphism in SE Asia, *in* Maruyama, S., and Parkinson, D. D., eds., Current concepts of orogeny and future perspectives: Tectonophys., spec. vol.

Parkinson, C. D., Miyazaki, K., Wakita, K., Barber, A. J., and Carswell, D. A., 1998, An overview and tectonic synthesis of the pre-Tertiary very-high-pressure metamorphic and associated rocks of Java, Sulawesi and Kalimantan, Indonesia: The Island Arc, v. 7, p. 184–200.

Pearson, D. G., Davies, G. R., and Nixon, P. H., 1993, Geochemical constraints on the petrogenesis of diamond facies pyroxenites from the Beni Bousear peridotite massif, N. Morocco: Jour. Petrol., v. 34, p. 125–172.

———, 1995, Orogenic ultramafic rocks of UHP diamond facies origin, *in* Coleman, R. G., and Wang, X., eds., Ultrahigh pressure metamorphophism: New York, Cambridge University Press, p. 456–510.

Pearson, D. G., Davies, G. R., Nixon, P. H., and Mattey, D. P., 1991, A carbon isotope study of diamond facies pyroxenites and associated rocks from the Beni Bousera peridotite massif, North Morocco: Jour. Petrol., v. 32, p. 175–189.

Pearson, D. G., Davies, G. R., Nixon, P. H., and Milledge, H. J., 1989, Graphitized diamonds from a peridotite massif in Morocco and implications for anomalous diamond occurrences: Nature, v. 338, p. 60–62.

Reinecke, T., 1991, Very-high-pressure metamorphism and uplift of coesite-bearing metasediments from the Zermatt-Saas zone, Western Alps: Eur. Jour. Mineral, v. 3, p. 7–17.

Robinson, P. T., Bai, W., Fang, Q., Hu, X., Davies, G., Grieselon, M., Staudigel, H., Cameron, T., Malpas, J., and Zhou, M., 1998, Ultra-high pressure minerals in chromitites of the Luobusa ophiolite, Tibet [abs.], *in* Int. Workshop on UHP Metamorphism and Exhumation: Stanford, CA, Stanford University, p. A21.

Sautter, V., 1985, An eclogite paragenesis from the Aleksod basement, central Hoggar, south Algeria: Chem. Geol., v. 50, p. 331–347.

Schmadicke, E., 1991, Quartz pseudomorphs after coesite in eclogites from the Saxonian Erzgebirge: Eur. Jour. Mineral., v. 1, p. 231–238.

Schmadicke, E., and Evans, B. W., 1997, Garnet-bearing ultramafic rocks from the Erzgebirge, and their relation to other settings in the Bohemian Massif: Contrib. Mineral. Petrol., v. 127, p. 57–74.

Schmadicke, E., Mezger, K., Cosca, M. A., and Okrusch, M., 1995, Variscan Sm-Nd and Ar-Ar ages of eclogite facies rocks from the Erzgebirge, Bohemian Massif: Jour. Metamor. Geol., v. 13, p. 537–552.

Shatsky, V. S., Jagoutz, E., and Koz'menko, O. A., 1997, Sm-Nd dating of the high-pressure metamorphism of the Maksyutov Complex, Southern Urals: Trans. Russian Acad. Sci., Earth Sci. Sect. v. 353, p. 285–287.

Spencer, D. A., 1993, Tectonics of the Higher and Tethyan Himalaya, Pakistan: Implications of an early collisional, high pressure (eclogite facies) metamorphism to the Himalayan belt: Unpubl. Ph.D. dissertation, ETH-Zürich, 953 p.

Spencer, D. A., Tonarini, S., and Pognante, U., 1995, Geochemical and Sr-Nd isotopic characterization of Higher Himalayan eclogites (and associated metabasites): Eur. Jour. Mineral., v. 7, p. 89–102.

Smith, D. C., 1984, Coesite in clinopyroxene in the Caledonides and its implication for geodyanmics: Nature, v. 310, p. 641–644.

Sobolev, N. V., and Shatsky, V. S., 1990, Diamond inclusions in garnets from metamorphic rocks: A new environment for diamond formation: Nature, v. 343, p. 742–746.

Sobolev, N. V., Dobretsov, N. L., Bakirov, A. B., and Shatsky, V. S., 1986, Eclogites from various types of metamorphic complexes in the USSR and the problems of their origin, *in* Evans, B. W., and Brown, E. H., eds., Blueschists and eclogites: Geol. Soc. Amer., Memoir, v. 164, p. 349–363.

Suen, C. J., and Frey, F. A., 1987, Origins of the mafic and ultramafic rocks in the Ronda peridotite: Earth Planet. Sci. Lett., v. 85, p. 183–202.

Tabata, H., Yamauchi, K., Maruyama, S., and Liou, J. G., 1998, Tracing the extent of a UHP metamorphic terrane: Mineral-inclusion study of zircons in gneisses from the Dabie Shan, *in* Hacker, B., and Liou, J. G., eds., When continents collide: Geodynamics and geochemistry of ultrahigh-pressure rocks: Dordrecht, Kluwer Acad. Publ., p. 261–274.

Tagiri, M., and Bakirov, A., 1990, Quartz pseudomorph after coesite in garnet from a garnet-chloritoid-talc schist, northern Tien-Shan, Kirghiz, USSR: Proc. Japan Acad., v. 66, p. 135–139.

Tagiri, M., Yano, T., Bakirov, A., Nakajima, T., and Uchiumi, S., 1995, Mineral parageneses and metamorphic P-T paths of ultrahigh-pressure eclogites from Kyrgyzstan Tien-Shan: The Island Arc, v. 4, p. 280–292.

Van der Klauw, S. N. G. C., Reinecke, T., and Stöckhert, B., 1997, Exhumation of ultrahigh-pressure metamorphic oceanic crust from Lago di Cignana, Piemontese zone, western Alps: The structural record in metabasites: Lithos, v. 41, p. 79–102.

Yan, B., Liang, R., Fan, Q., Yang, F., and Wan, C., 1986, Characteristics of diamond and diamond-bearing ultramafic rocks in Qiaoxi and Hongqu, Xizang: Bull. Inst. Geol. Chinese Acad. Geol. Sci., v. 14, p. 61–124 (in Chinese with English abstract).

Zhang, R. Y., and Liou, J. G., 1997, Partial transformation of gabbro to coesite-bearing eclogites from Yangkou, the Sulu Terrane, eastern China: Jour. Metamor. Geol., v. 15, p. 183–202.

Zindler, A., Staudigel, H., Hart, S. R., Endres, R., and Golstein, S., 1983, Nd and Sr isotopic study of a mafic layer from Ronda ultramafic complex: Nature, v. 304, p. 226–240.

PETROLOGIC EVOLUTION OF IMPORTANT ULTRAHIGH-PRESSURE METAMORPHIC ROCK TYPES

Petrographic Characteristics and Metamorphic Evolution of Ultrahigh-Pressure Eclogites in Plate-Collision Belts

D. A. CARSWELL

Department of Earth Sciences, University of Sheffield, Dainton Building Brookhill, Sheffield S3 7HF, United Kingdom

AND R. Y. ZHANG

Department of Geological and Environmental Sciences, Stanford University, Stanford, California 94305

Abstract

Ultrahigh-pressure eclogites are metabasic rocks that have equilibrated at pressures within the coesite P-T stability field. This paper reviews their occurrences and petrographic characteristics in exhumed ultrahigh-pressure metamorphic terranes within certain major Phanerozoic continental plate–collision belts.

Unequivocal identification of ultrahigh-pressure eclogites depends on the presence of relict coesite or of polycrystalline quartz pseudomorphs after earlier coesite. However, preservation of such distinctive petrographic evidence is at best sporadic, leading to serious problems over whether or not particular eclogite samples have experienced "peak" pressures consistent with coesite stability. Accordingly, we also review here other petrographic, mineral-assemblage, and mineral-chemistry features that may aid in the identification of ultrahigh-pressure eclogites.

In addition, we discuss the quantitative evaluation of P-T conditions for formation based on various reaction equilibria, including the dominant garnet and omphacite phases. Various aspects of the mineral compositions are taken to indicate that the "peak" metamorphic pressures were most likely attained at lower than "peak" temperatures. Hence it is concluded that tight, hairpin-style P-T-time paths for the prograde and retrograde stages in the metamorphic evolution of ultra-high pressure eclogites that assume synchronous achievement of P_{max} and T_{max} conditions, as deduced in many recently published studies, may be invalid. A rather more open style, clockwise P-T-time path, as deduced for the metamorphic evolution of a typical ultrahigh-pressure phengite-bearing eclogite sample from the Dabie Shan in central China and characterized by a significant further temperature rise during the initial phase of relatively rapid exhumation, is considered to be more realistic.

Definition of Ultrahigh-Pressure Eclogites

WITHIN THE CONTEXT of the IGCP Task Group III-6 project on "Ultrahigh-Pressure Metamorphism and Geodynamics in Collision-Type Orogenic Belts" and the book *Ultrahigh Pressure Metamorphism* edited by Coleman and Wang (1995), it has become established that ultrahigh-pressure metamorphism is synonymous with eclogite-facies metamorphism that has occurred within the stability field of coesite. Ultrahigh-pressure (UHP) metamorphism was not specifically defined and characterized in the review texts on eclogites and eclogite-facies rocks edited by Smith (1988) and Carswell (1990), respectively, and published somewhat earlier. It is therefore appropriate that this review article focuses specifically on UHP eclogites—in other words, on metabasic rocks with a dominant plagioclase-free, eclogite-facies mineral paragenesis of garnet and omphacitic clinopyroxene, but with additional petrographic or mineral-chemistry features that indicate formation under UHP, coesite-stable conditions.

A generalized P-T field for the formation of UHP eclogites is indicated in Figure 1, with minimum pressures defined by those required for coesite stability. This P-T field encompasses the higher-pressure parts of the lawsonite eclogite, epidote eclogite, and dry eclogite stability fields as experimentally defined by Okamoto and Maruyama (1999). The stability of a zoisite/clinozoisite-epidote phase in some UHP eclogites has been confirmed by observations of coesite or polycrystalline quartz pseudomorphs within zoisite/clinozoisite-epidote in some Chinese samples (Liou et al., 1995, 1998). On the other hand, lawsonite has not yet been reported in any UHP eclogite samples, perhaps as might be expected given the exceptionally low geothermal gradient (ca. 6°C/km maximum) required for its sta-

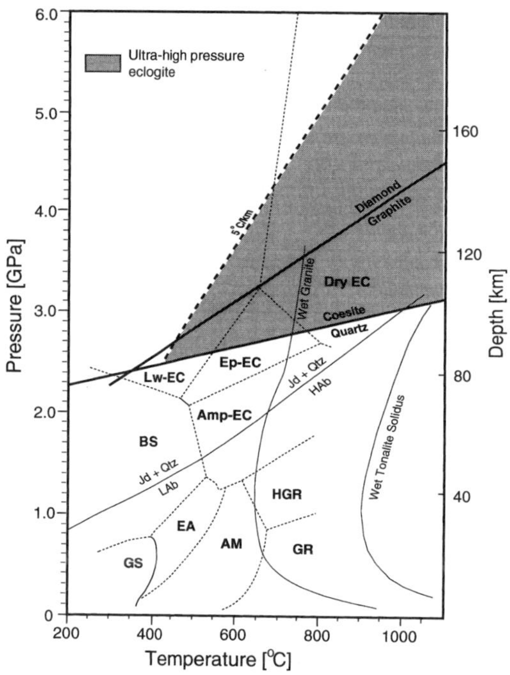

FIG. 1. Simplified P-T facies diagram highlighting the P-T stability field for ultrahigh-pressure eclogites. Subdivision of the eclogite facies field into amphibole eclogite (Amp-EC), epidote eclogite (Ep-EC), lawsonite eclogite (Lw-EC), and dry eclogite (Dry-EC) is based on Okamoto and Maruyama (1999). Also shown are the stability fields for diamond (Bundy, 1980), coesite (Hemingway et al., 1998), and jadeite + quartz (Holland, 1980).

bility. The stability of primary calcic clinoamphibole-bearing eclogite is, however, anticipated to be restricted to pressures less than those required for coesite stability (e.g., Liu et al., 1996), consistent with its absence in UHP eclogites. On the other hand, sodic clinoamphiboles such as nyböite (Smith, 1988, 1995; Hirajima et al., 1992) and glaucophane (Zhang and Liou, 1994a) have been reported to occur as "peak" pressure phases in certain UHP eclogites. Moreover, the P-T field labeled by Okamoto and Maruyama (1999) as "dry eclogite," implying stability only of an entirely anhydrous mineral assemblage, is somewhat misleading. Certainly in K_2O-bearing UHP eclogites, primary phengitic white mica has been observed to be commonly present (e.g., Carswell et al., 1997) and has been demonstrated by experiments to be stable to high temperatures (>900°C) and to pressures in excess of 3.5 to 4.0 GPa (e.g., Massonne and Szpurka, 1997; Skjerlie and Patino Douce, 1998). In addition, talc has been identified to be stable at "peak" pressures in certain more magnesian UHP eclogites (e.g., Liou et al., 1995; Zhang et al., 1995a, 1995c).

Occurrences and Lithological Associations

Smith (1984) merits credit as the first to describe (and to confirm by Raman spectroscopy) the occurrence of relict coesite within a crustal eclogite sample and then, in his expansive 1988 review article, to recognize the existence within the Western Gneiss Region of the Norwegian Caledonides of a specific coesite-eclogite province containing rocks that had experienced UHP metamorphic conditions.

Spectacular, relatively unretrograded eclogites are a conspicuous feature of the Western Gneiss Region in the coastal areas of Nordfjord, Sunnmore, and Romsdal counties in Norway, yet by the time of Smith's (1988) eclogite review, coesite had only been positively confirmed at two localities (Grytting and Straumen) located some 14 km apart. Textural evidence in the form of polycrystalline or at least multicrystalline quartz inclusions within garnet or omphacite, thought to be indicative of the previous presence of coesite, was recorded from five additional, more widely scattered localities, but did not enable Smith (1988) to establish any boundaries to the west Norway UHP coesite-eclogite province. Moreover, Smith (1988) emphasized that the gneisses intervening between the various eclogite localities lacked mineralogical evidence that they had experienced high-pressure, HP (quartz-eclogite), let alone UHP (coesite-eclogite stable) conditions. Hence he proposed the existence of a highly imbricated tectonic mélange of UHP and HP eclogites with lower-pressure lithologies rather than a regionally extensive, structurally coherent UHP province/terrane.

More recent studies by Wain (1997) and Cuthbert et al. (1998) have greatly increased the number of west Norway eclogite localities with recognized relict coesite or quartz pseudomorphs after coesite, but the regional extent of any UHP eclogite terrane much beyond the Stadlandet Peninsula area (and the nature of its boundary in the Outer Nordfjord district with a different metamorphic unit that appears to contain only HP, quartz-stable eclogites) remains unclear. Importantly, Wain (1997) and Krabbendam and Wain (1997) have also reported that some felsic gneisses intimately interlayered with UHP eclogites also contain quartz pseudomorphs after coesite, demonstrating that they also experienced UHP metamorphic conditions.

At around the same time that Smith (1984) published evidence of relict coesite in a west Norway eclogite, Chopin (1984) reported a startling occurrence of remarkedly well preserved coesite in a pyrope-bearing quartzite from the Dora-Maira massif in the Western Alps. Relict coesite, or polycrystalline quartz pseudomorphs after coesite, are reported to be more rarely preserved in associated eclogites within the Brossasco-Isasca UHP Metamorphic Unit of the Dora-Maira massif (Chopin et al., 1991; Kienast et al., 1991; Compagnoni et al., 1995, Nowlan et al., 1998). Importantly, the intimate field association with the coesite-bearing quartzite and other UHP paraschists, as well as marbles, emphasized the point that rocks of the upper continental crust have witnessed UHP metamorphism following subduction down to depths of 100 km or more, in a P-T metamorphic regime once thought to be experienced only by mantle rocks (Schreyer, 1985).

Triggered by the coesite discoveries of Chopin (1984) and Smith (1984), the last 10 years has seen a remarkable increase in the number of eclogite-bearing metamorphic terranes now recognized to have experienced UHP metamorphism in various major continental plate–collision zones. The global distribution of the currently recognized UHP eclogite occurrences is shown in Figure 2, but based on recent experience can be anticipated to further increase in number with additional new discoveries.

A notable feature is that all recognized UHP eclogite-bearing terranes have relatively young metamorphic ages, the oldest reliable ages being ~530 Ma for the UHP metamorphism in the Kokchetav Massif (Claoué-Long et al., 1991). This may well indicate that the lithospheric plates have only been sufficiently cool, strong, and/or thick enough during the Phanerozoic era (Ernst, 1990) for the transient subduction of continental-crust lithologies to sufficient depths for the stability of UHP mineralogies and equally important for their at least partial survival during exhumation (Maruyama and Liou, 1998).

Relict coesite inclusions within garnet or omphacite in eclogites were reported in the Dabie Shan of central China first by Okay et al. (1989) and Wang et al. (1989), but later by many other researchers, some of whom have extended the reported occurrences of coesite within this major Triassic plate-collision belt both west into Hongan and Tongbaishan (e.g., Zhang and Liou, 1994a) and eastward across the Tanlu fault into its displaced continuation in the Sulu region (e.g., Hirajima et al., 1990; Zhang et al., 1995c; Zhang and Liou, 1998). Even more startling have been the published reports of microdiamond inclusions in garnets within certain of the Dabie Shan UHP eclogites (Xu et al., 1992; Okay, 1993). This discovery somewhat disquietingly has not yet been confirmed by other research groups, but importantly, signals that the confining pressures for formation of UHP eclogites in Dabie Shan may extend into the diamond stability field.

This certainly appears to have been the case in the UHP metamorphic terrane within the Kokchetav massif of Kazakhstan, which contains abundant microdiamonds (especially as armored inclusions in garnets and zircons) within garnet-mica schists and carbonate-bearing calc-silicate gneisses (Sobolev and Shatsky, 1990; Shatsky et al., 1995). Perhaps somewhat surprisingly, in situ microdiamonds have not been reported from associated eclogites at Kokchetav and only very recently has relict coesite preserved within zircon been reported from a Kokchetav eclogite sample (Korsakov et al., 1998) to confirm formation under UHP conditions.

In the Western Gneiss Region of Norway, scarce microdiamonds have been recovered by dissolution of a garnet-kyanite-biotite gneiss with common matrix graphite from the island of Fjortoft (Dobrzhinetskaya et al., 1995). This locality is ~ 100 km distant from the coesite-bearing UHP eclogites identified by Smith (1984, 1988), Wain (1997), and Cuthbert et al. (1998), but a nearby kyanite-bearing eclogite on Fjortoft has been observed (Cuthbert et al., 1999) to contain multi-crystalline quartz inclusions within garnet with a distinctive elongate tabular form that could well be distinctive of original coesite.

As yet no microdiamonds have been reported from either eclogites or associated metasedimentary rocks in the Brossasco-Isasca UHP Unit of the Dora-Maira massif, despite some eclogites containing large graphite flakes (Compagnoni et al., 1995). By contrast, in the Variscan Saxonian Erzgebirge of Germany, UHP metamorphic conditions are convincingly signaled by both polycrystalline quartz inclusions after coesite in eclogites (Schmadicke, 1991) and in situ microdiamonds in associated garnet-kyanite-phengite gneisses (Massonne, 1998). Elsewhere in the central European Variscides, polycrystalline quartz pseudomorphs after coesite have to date only been recorded in certain eclogite and

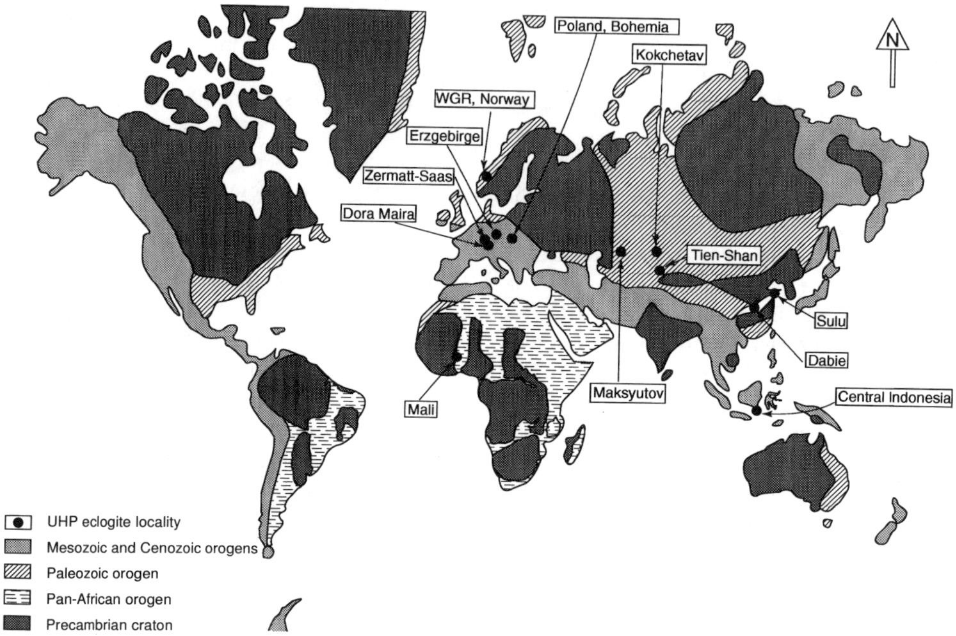

FIG. 2. Global map of the recognized occurrences of ultrahigh-pressure eclogites (modified after Maruyama and Liou, 1998).

omphacite-bearing granulite samples of the Zlote Mountains in the Polish Sudetes (Bakun-Czubarow, 1991, 1992).

Caby (1994) has reported relict coesite enclosed in omphacite in a garnet-omphacite-carbonate-quartz-phengite-rutile assemblage mafic nodule (essentially a carbonate-bearing eclogite) within impure marble from northern Mali, west Africa. In addition, associated eclogitic mica schists were reported to contain quartz pseudomorphs after coesite within garnets. The precise metamorphic age of these UHP rocks is uncertain but most likely relates to Pan-African plate collision at ~ 550 to 600 Ma.

Tagiri et al. (1995) have documented polycrystalline quartz pseudomorphs after coesite within both garnet and omphacite in eclogite from Atbashy in the southern Tien Shan of Kyrgyzstan. Again the timing of UHP metamorphism is somewhat uncertain, but it may relate to Hercynian collision between the Kazakhstan and Tarim continental plates. Two further eclogite localities at Aktyuz and Makbal in the northern Tien Shan also were described by Tagiri et al. (1995). The studied eclogite sample from Aktyuk was heavily retrograded and lacked any mineralogical evidence for formation under UHP conditions. The better-preserved sample from Makbal was reported to contain primary paragonite and to have equilibrated at ~ 2.0 GPa, well outside of the coesite stability field, although rather enigmatically, an associated garnet-chloritoid-talc schist was previously reported by Tagiri and Bakirov (1990) to contain polycrystalline quartz pseudomorphs after coesite within garnet.

The Maksyutov Complex in the southern Urals is the location where an eclogite sample was first reported (Chesnokov and Popov, 1965) to contain polycrystalline quartz aggregates within garnets showing adjacent radial fractures, interpreted as pseudomorphs after earlier coesite. Unfortunately to date this observation of quartz pseudomorphs after coesite has not been confirmed in further eclogite samples from the Maksyutov Complex (e.g., Beane et al., 1995), although intriguingly, an associated graphite-bearing mica schist has been described to contain large graphite cuboids (up to 1.3 × 1 cm), interpreted as possible pseudomorphs after diamond (Leech and Ernst, 1998).

A comprehensive evaluation of the geochemical characteristics of UHP eclogites enclosed in presumed continental crust metasediments and orthogneisses has been provided by Jahn (1998), for a large sample suite from the Dabie Shan and Sulu sectors of the Triassic UHP terrane in central China. It was reported that the eclogites were mainly of quartz-normative basaltic composition but showed a wide range of major- and trace-element abundances suggesting derivation from heterogeneous sources complicated by the effects of fractional crystallization, metamorphic segregation, and fluid metasomatism. The observed variation in immobile element ratios (Ti/V, La/Nb) exceeds the entire range of basaltic rocks from different tectonic settings. However, these rocks were reported not to resemble typical MORB-type ocean-floor basalts or ophiolites. Instead their characteristic LREE enrichment and negative Nb, Zr, and Ti anomalies on primitive-mantle normalized spidergrams were taken to signal protoliths of continental affinity comparable to the mafic components of many Precambrian gneiss terranes. Comparable in-depth geochemical studies to identify the protoliths of UHP eclogites are lacking from other terranes, but similar lithological associations with granitic orthogneisses and continental platform metasediments (including limestones) in most other recognized UHP terranes point to an origin from comparable continental crust protoliths.

An apparent exception to this appears to be the occurrence of coesite-bearing UHP ophiolitic rocks reported by Reineke (1991) and van der Klauw et al. (1997) at Lago di Cignana in the Zermatt-Saas Zone of the Western Alps. There, both coesite relicts and polycrystalline quartz pseudomorphs after coesite were reported as inclusions within tourmaline or garnet in a manganiferous micaceous metaquartzite by Reineke (1991). Associated UHP eclogites have a MORB affinity and are reported to contain quartz pseudomorphs after coesite within coarse omphacite (van der Klauw et al., 1997).

Another example of UHP eclogite formation from subducted oceanic crust would appear to be the reported occurrence by Parkinson et al. (1998) of polycrystalline quartz pseudomorphs after coesite hosted in jadeites in jadeite-garnet quartzite blocks associated with phengite-bearing eclogites in an ophiolitic mélange sequence of the Bantimala Complex in central Sulawesi, Indonesia—thought to have been formed in a N-dipping B-type subduction zone at the margin of the Sundaland craton in the Early Cretaceous, then exhumed in response to collision of a Gondwanan continental fragment with the Sundaland margin at ~ 115 to 120 Ma.

Perhaps these Western Alps and Sulawesi occurrences represent exceptional instances where UHP eclogites formed from deeply subducted oceanic crust in Benioff zones have been returned to the surface because of unusual tectonic circumstances. By contrast, the successful exhumation of recognizable UHP eclogites in deeply subducted continental-crust slabs may be more commonplace as a result of the inherent buoyancy of the associated dominant acidic orthogneisses, psammitic metasediments, and marbles.

Petrographic Characteristics and Identification Problems

In the case of coesite grains partially replaced by quartz, the coesite may be provisionally identified microscopically as irregularly bounded, slightly higher relief, lower birefringence, and grain core fragments (Fig. 3A) encased in a fine-grained aggregate of replacement quartz. As coesite is monoclinic, larger grains also may be distinguished from quartz as biaxial positive.

With an electron microprobe, both phases can be confirmed as essentially pure SiO_2, but it often is possible to distinguish between them by blue fluorescence for coesite under the electron beam, in contrast to orange fluorescence for quartz. With sensitive backscattered electron imaging, the relict denser coesite will appear slightly lighter (higher reflectance) than the surrounding quartz (Fig. 3B). Conclusive identification of relict coesite is most easily achieved by micro-Raman spectroscopy with an especially diagnostic strong spectral peak at 521 cm^{-1} (Boyer et al., 1985).

In the absence of an actual coesite relict, a strongly indicative textural feature of polycrystalline quartz aggregates pseudomorphing coesite (Smyth, 1977; Gillet et al., 1984) is that the initial replacive quartz usually grows with a distinctive palisade intergrowth texture (Fig. 3C), with the long axes roughly perpendicular to the original grain boundary. However, with further fluid-assisted recrystallization, a less diagnostic multicrystalline quartz-grain aggregate often forms, and ultimately just a monocrystalline quartz grain. The unusually large palisade-textured quartz pseudomorph after coesite in Figure 3C has been fortuitously preserved in a large, isolated, relatively unfractured garnet within an almost entirely quartz matrix into which most of the tectonic strain has been partitioned. More commonly, inclusions of petrographically distinct polycrystalline quartz pseudomorphs after coesite are of <200 microns size, whereas larger inclusions, even within the same garnet, are usually

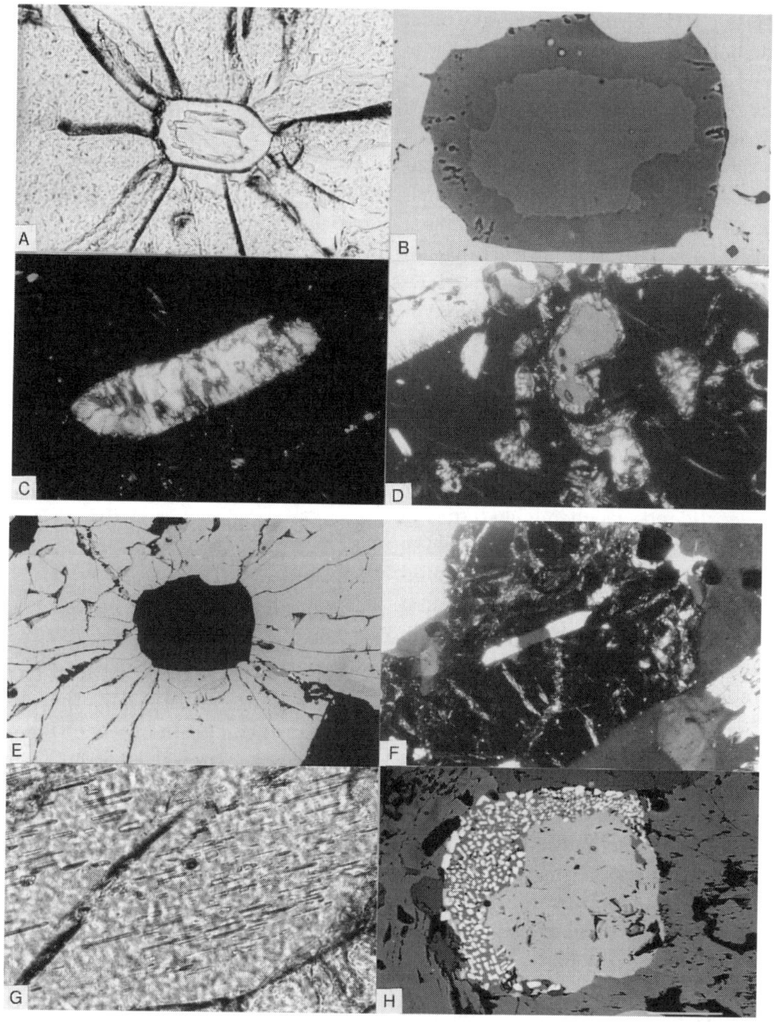

FIG. 3. Photomicrographs of key petrographic features of UHP eclogites. A. Plane polarized light image of coesite relict (~200 microns in length), with partial replacement at margins by lower-relief quartz, enclosed within garnet. Sample 94-18E from Xinyan, Dabie Shan, central China. B. Backscattered electron image of ~180 micron-sized inclusion within garnet, showing a relict core of denser and brighter coesite enclosed within darker replacement quartz. Sample CD97 from Shima, Dabie Shan, central China. C. Cross-polarized light image of a ~700 micron-length polycrystalline quartz pseudomorph after coesite with a distinctive palisade texture (enclosed within garnet). Sample A306 from Salta, Western Gneiss Region, Norway. D. Cross-polarized light image, ~1.2 mm across, of part of a large garnet enclosing a mix of polycrystalline and monocrystalline quartz inclusions. Note that both the largest monocrystalline quartz inclusion and the polycrystalline quartz inclusion below it are separated from garnet by a corona of secondary aegirine augite. Sample A226 from Liset, Western Gneiss Region, Norway. E. Lower-magnification backscattered electron image of the same ~180 micron-sized coesite inclusion within garnet as in Figure 3B, showing distinctive radial expansion cracks in the brighter garnet around the darker coesite inclusion, now partly replaced by polycrystalline quartz. Sample CD97 from Shima, Dabie Shan, central China. F. Cross-polarized light image of a ~350 micron-length elongate tabular multicrystalline quartz inclusion after coesite enclosed within garnet, in a heavily retrograded eclogite sample (CD61 from south of Yuexi, Dabie Shan, central China). G. Plane-polarized light image, ~0.33 mm across, of quartz exsolution lamellae within a supersilicic omphacite. Sample from Rongcheng, Sulu, east-central China. H. Backscattered electron image of a ~300 micron-sized high-alumina UHP titanite, showing distinctive partial replacement by a secondary symplectite of mostly bright ilmenite and darker amphibole. Sample CD29 from Shuanghe, Dabie Shan, central China.

monocrystalline quartz (Fig. 3D). The most likely explanation is that the larger inclusions have been connected to the rock matrix along fractures that acted as pathways for fluid ingress, which has promoted annealing to a minimum-surface-energy, single-grain quartz inclusion, whereas the smaller inclusions remained isolated and protected, thus retaining the crucial textural evidence that coesite previously was stable.

A further distinctive petrographic feature of partly or completely replaced coesite inclusions, within garnet in particular, is the development of radial expansion cracks (Figure 3E), extending from the inclusion boundary into the enclosing mineral. This reflects the roughly 10% volume increase from coesite to alpha-quartz.

Mostly the generally small inclusions of partly or completely pseudomorphed coesite preserved in garnets or other robust minerals have an ovoid shape with indistinct crystal faces, but a few larger inclusions are more elongate, with a pair of parallel faces or more rarely of distinctly tabular form with pseudo-hexagonal prismatic and cross sections, consistent with the fact that the symmetry of coesite is monoclinic pseudohexagonal (Yang et al., 1998). Coesites also have been shown to grow as lath-shaped and tabular crystals in synthesis experiments (e.g., Sclar et al., 1962; Renner et al., 1997), so the presence of quartz inclusions with these distinctive shapes may provide a tenuous indication of quartz that has replaced coesite, even when a distinctive polycrystalline aggregate texture is absent (e.g., Fig. 3F).

The fact that even in now well-authenticated UHP metamorphic terranes, such as the Western Gneiss Region of Norway or the Kokchetav massif in Kazakhstan, the preservation of relict coesite or even just of distinctive polycrystalline quartz pseudomorphs after coesite is sporadic and rare means that easy microscopic identification of many UHP eclogites is not possible. It is exceedingly likely that many UHP eclogite samples have not been recognized as such simply because of the lack of obvious petrographic evidence that coesite was previously stable. In addition, not all eclogite samples contain a free-silica phase. This begs the question of whether such eclogites might contain any other identifiable petrographic features indicative of formation under UHP conditions.

One possibility may be evidence of exsolved needles of quartz in omphacite (Fig. 3G), as identified in samples of eclogites from several UHP terranes (Smith, 1984, 1988; Bakun-Czubarow, 1992; Liou et al., 1998; Zhang and Liou, 1998). The quartz lamellae are in a single crystallographic orientation and appear to have a topotaxial relation to the host omphacite. Hence they have been interpreted to have exsolved from a supersilicic clinopyroxene that is stable at high pressures. Evidence for the stability of supersilicic clinopyroxenes at high pressures has been provided by several experimental studies (e.g., Mao, 1976; Wood and Henderson, 1978; Angel et al., 1988). However, the P-T dependence of uptake of excess SiO_2 by omphacites has not been quantified and at this stage this feature cannot be taken to provide unequivocal proof of eclogite formation under UHP conditions, inasmuch as similar exsolved rods of quartz have been reported by Smith and Cheeney (1980) and Gayk et al. (1995) from eclogite and granulite samples from terranes currently not recognized to have experienced UHP conditions.

Several experimental studies have demonstrated the strongly pressure-dependent uptake of K in clinopyroxenes (e.g., Luth, 1995; Harlow, 1997; Okamoto and Maruyama, 1998), such that the K_2O content of Ca-Na clinopyroxenes may well be an important barometer in UHP rocks. Shatsky et al. (1995) have reported K_2O contents in clinopyroxenes of up to 1.2 wt% in diamond-bearing metamorphic rocks in the Kokchetav massif and noted that during exhumation, the K_2O has exsolved as K-feldspar exsolution lamellae within the clinopyroxenes. However, these are diopsidic clinopyroxenes in carbonate-bearing rocks and to date, exsolution of K-feldspar lamellae within omphacites in eclogites has not been recorded in this or other UHP terranes, presumably because of the low K_2O contents and high Na/K ratios of metabasaltic rocks.

The potential for the recognition in eclogites of the stability of specific mineral parageneses that might be restricted to UHP (coesite-stable) P-T conditions is very limited given the limited range of bulk-rock compositions and the fact that the dominant garnet + clinopyroxene assemblage can accommodate considerable variation in P, T, and X merely by changes in mineral chemistry.

Perhaps the only mineral paragenesis identified to date that may be limited (and hence restricted) to UHP eclogites is magnesite + diopside/omphacite. Kushiro et al. (1975) have shown that the pure end member assemblage magnesite + diopside is only stable at pressures well within the coesite stability field, consistent with the observed stability of magnesite + omphacite in a Dabie Shan coesite eclogite and of magnesite + diopsidic clinopyroxene in an enstatite-bearing garnet websterite from a nearby, large layered coesite-bearing complex of mafic-ultramafic rocks (Zhang and Liou, 1994b). However, natural magnesite + omphacite compositions in eclogites are far removed from the end-member magnesite and diopside compositions and this may well substantially lower the pressure required for stability. This may explain the stability of magnesite in a kyanite eclogite from the Tauern window, Austria (Holland, 1978), which is unlikely to have formed in the coesite stability field.

Mineral Chemistry Indicators of UHP Conditions

Given the difficulties encountered in recognition of UHP eclogites from preserved petrographic features alone, it also is worth exploring whether any other particular aspects of the mineral chemistry of the mineral phases encountered in eclogites may provide a potential means of signaling that the constituent minerals formed under UHP conditions, over and above the possible unusual Si and/or K contents in omphacites mentioned above.

The jadeite content of clinopyroxenes in eclogites is pressure dependent but also is strongly dependent on rock chemistry. On its own, the jadeite content of clinopyroxene only indicates minimum pressures for formation of most eclogites and does not enable discrimination between HP (quartz-stable) and UHP (coesite-stable) eclogites. Moreover, pure jadeite is stable at pressures significantly below those required for coesite (Holland, 1980), and is even well known from the relatively low T and low P eclogites in blueschist-facies terranes.

Similarly, the pyrope content of the garnets, the other discriminatory composition parameter in the early classification scheme for eclogites of Coleman et al. (1965), also is too strongly dependent on rock chemistry to enable reliable discrimination between HP and UHP eclogites. The grossular content of garnet in metabasic rocks also can be expected to be markedly pressure dependent, but again strong dependence also on rock chemistry means that by itself it is an unreliable indicator of formation pressures for eclogites.

On the other hand, the low yet significant solubility of Na in garnets coexisting with omphacites in eclogites may be an important mineralogical indicator of UHP conditions. In a recent experimental study on a natural eclogite composition, Okamoto and Maruyama (1998) demonstrated that the Na content of garnet at 1000°C increased from 0.05 wt% at 5 GPa to 0.08 wt% at 7 GPa. It therefore is apparent that the critical content levels of Na in garnets of UHP eclogites can be anticipated to be barely above detection levels with routine microprobe analysis procedures, but there could be potential in precise determinations of Na contents in the garnets using long counting times or perhaps better by laser ablation ICP-MS analysis.

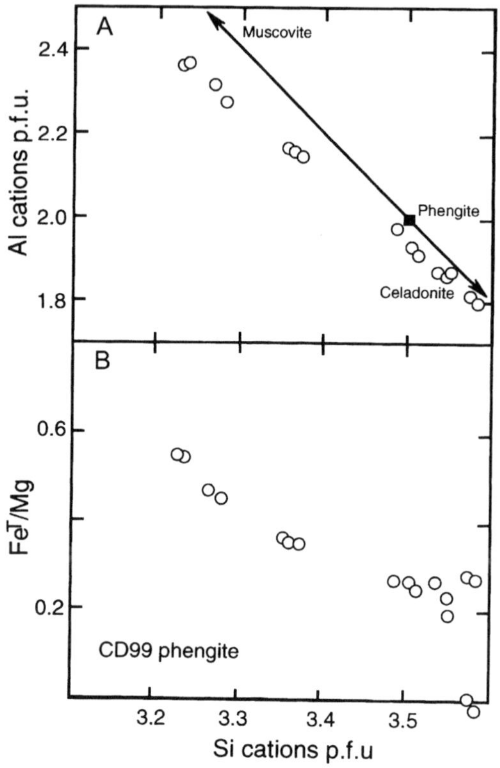

FIG. 4. Plots of Al contents and Fe^{total}/Mg ratio against Si contents for phengites in UHP eclogite sample CD99 from the Shima locality in Dabie Shan, central China.

It is well known that Si contents of phengitic white micas are strongly pressure dependent (Massonne and Schreyer, 1987; Carswell and Harley, 1990), increasing from >3.00 to values <4.00 in a 12 (O,OH) formula unit in response to $Mg^{[6]} Si^{[4]} Al^{[6]}_{-1} Al^{[4]}_{-1}$ substitution, since $Al^{[4]}$ is not favored at high pressures. Phengitic white micas are minor phases in some metabasic eclogites but their Si contents often vary considerably between 3.2 and 3.6 per formula unit even in the same sample, as illus-

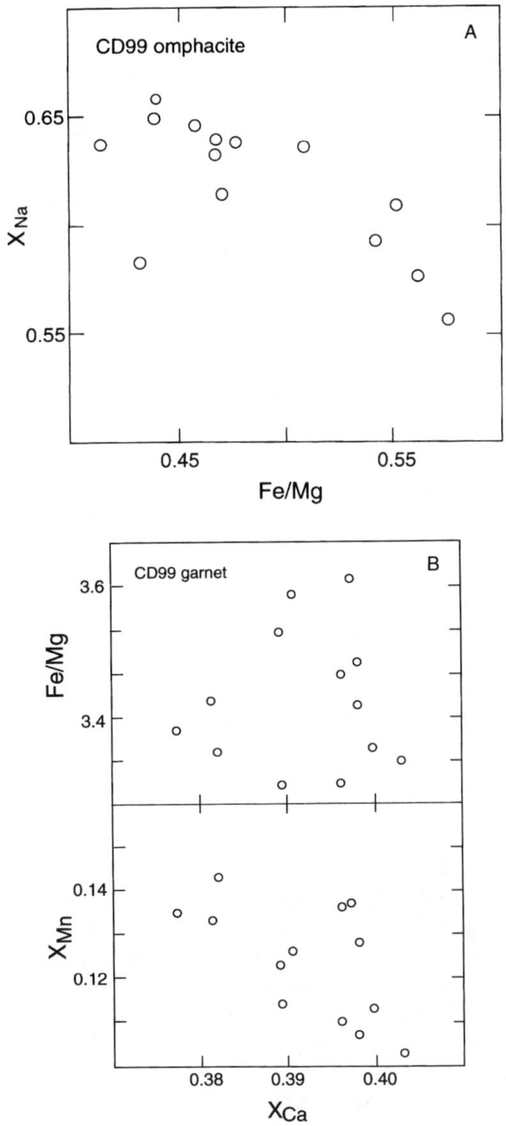

FIG. 5. Plots of X_{Na} versus Fe/Mg for omphacites (A) and of X_{Ca} and X_{Mn} versus Fe/Mg (B) for garnets in UHP eclogite sample CD99 from the Shima locality in Dabie Shan, central China.

rated in Figure 4A for phengite spot analyses in UHP eclogite sample CD99 from Dabie Shan in China. Plots of important composition parameters for microprobe analyses of the coexisting omphacite and garnet in this UHP eclogite sample are shown in Figure 5 and will be discussed in more detail later.

It is apparent from studies of phengite-bearing eclogites from Dabie Shan (Carswell et al., 1997) that the highest Si contents in phengites usually are observed in relict highest-pressure phengites preserved as inclusions within robust grains, especially in large garnets. Conversely, much lower Si contents are recorded in retrograded matrix grains. Phengites with Si contents greater than about 3.55 may be restricted to UHP eclogites, but this remains to be properly confirmed. Unfortunately, at this time there is no quantitative experimental calibration available for the P-T dependence of the Si content of phengite in metabasic rock compositions, unlike for phengites in the limiting assemblage with K feldspar, phlogopite, and quartz in pelites with compositions close to the experimentally studied KMASH system (Massonne and Scheyer, 1987).

Some carbonate-bearing UHP eclogites from western Norway (Smith, 1977, 1988), Dabie Shan (Carswell et al., 1996), and Sulu, China (Hirajima et al., 1992; Zhang et al., 1995c) have been shown to contain primary titanites with substantial aluminium and fluorine contents as a result of $AlFTi_{-1}O_{-1}$ substitution, which may be a valuable indication of formation under UHP conditions. An experimental study by Troitzsch and Ellis (1998) has demonstrated that a complete solid solution between titanite and $CaAlFSiO_4$ exists at high P and T, although to date reported Al contents in natural titanites do not significantly exceed $X_{Al} = 0.5$. The unit cell volume of $CaAlFSiO_4$ is 7% smaller than that of ideal titanite, consistent with the observed high Al and F contents of the primary titanites found in the UHP rocks and endorsing the significance of this compositional feature as a potentially important geobarometer. Petrographic indications that texturally early titanites are of UHP type with significant substitution toward $CaAlFSiO_4$ are provided by evidence of decompression-induced breakdown to symplectites of ilmenite + amphibole + plagioclase (Fig. 3H) or to a secondary lower Al titanite with plagioclase (see Carswell et al., 1996).

Thermobarometry and Deduced P-T-Time Paths

Demonstration through quantitative evaluation of appropriate mineral reaction equilibria that particular eclogite samples have formed under P-T conditions consistent with coesite stability is an important goal, especially for samples that for one reason or another provide no direct evidence of coesite stability. As already discussed, in the absence of relict coesite or distinctive quartz pseudomorphs after coesite, the common essentially bimineralic eclogites present a real problem unless it can be demonstrated with certainty that there is something uniquely distinctive about the chemistry of their constituent garnets or omphacites to indicate formation under UHP conditions. On the other hand, for phengite-bearing eclogites, a barometric evaluation methodology proposed by Waters and Martin (1993) has been demonstrated by Carswell et al. (1997) and Wain (1997) to successfully discriminate between the P-T formation conditions for HP (quartz-stable) and UHP (coesite-stable) eclogites. Importantly, this methodology combines key pressure-dependent composition features of coexisting garnet-omphacite-phengite, as expressed in the KMASH-system exchange reaction equilibrium:

pyrope + grossular + 3 ideal celadonite =
6 diopside + 3 ideal muscovite.

Application of this methodology has demonstrated—both in the Dabie Shan of China (Carswell et al., 1997) and in the Western Gneiss Region of Norway (Wain, 1997; Cuthbert et al., 1999)—that different eclogite-bearing units can be recognized that have respectively experienced HP (quartz-stable) and UHP (coesite-stable) conditions. Hence at least for phengite-bearing eclogites, there is now a reliable alternative means of positively recognizing UHP eclogites in the absence of coesite or its distinctive quartz pseudomorphs.

In the absence of phengite, or even of surviving omphacite, it is possible to alternatively consider the strongly pressure dependent reaction equilibrium (see Manning and Bohlen, 1991):

grossular + rutile + quartz/coesite + water =
zoisite/clinozoisite + titanite.

However, this methodology is not so robust inasmuch as, in effect, the calculated pressure is virtually entirely dependent on a single composition parameter in just one peak pressure phase (the grossular content in garnet). In addition, the calculated pressure is dependent on the assumed water activity. Moreover, clinozoisite and titanite are usually seen petrographically as secondary retrograde minerals in eclogites, so all that can be established is that this reaction equilibrium has been crossed on the exhumation path of the eclogite sample. On the other hand, for kyanite-bearing, excess-SiO_2 eclogites, the strongly pressure dependent reaction equilibrium 3 diopside + 2 kyanite = grossular + pyrope + 2 quartz/coesite may well provide a powerful and more reliable means for determination of whether the equilibration pressures were within the coesite or quartz stability fields (see Nakamura and Banno, 1997; Cuthbert et al., 1999).

For some 25 years now, since the publication of the first experimental calibration by Raheim and Green (1974), the partitioning of Fe^{2+} and Mg between coexisting garnet and omphacite has been the mainstay for the temperature evaluation of eclogites. This strongly temperature-dependent cation exchange equilibrium has been the basis for the recognition of regional temperature gradients across certain eclogite-bearing terranes, such as the Western Gneiss Region, Norway (Krogh, 1977; Griffin et al., 1985) and the Dabie Shan, China (Wang and Liou, 1991). Over the years, many different calibrations of this geothermometer have been proposed—see Krogh Ravna (1999) for the most recent comprehensive re-evaluation and proposed calibration based on experimental data assessed as most reliable. However, for the most part, the differences between temperatures derived from most recent proposed calibrations are quite small, especially relative to the much larger uncertainties that often arise because of inadequate knowledge of the amount of Fe specifically present as Fe^{2+} in the omphacites of eclogites. For relatively Fe-rich eclogite samples it should be possible to calculate reasonably accurate Fe^{3+} contents in omphacites by stoichiometric charge balance from high-quality wavelength-dispersive electron-microprobe analyses. However, even with the best-quality analyses, the customary levels of precision in determined Si contents are such that for low Fe eclogites, an unrealistically high proportion of the total Fe in omphacites may be calculated as Fe^{3+} on the basis of stoichiometric considerations.

The seriousness of this problem can be illustrated for eclogites in the central Dabie Shan UHP terrane, where assessed temperatures for individual samples published by Wang et al. (1992), Okay (1993), Zhang et al. (1995a), Carswell et al. (1997), and Tabata et al. (1998) range widely from 610 to 870°C. Wang et al. (1992) proposed that there was a regional gradient of decreasing temperature southward across this UHP terrane. However, with the enlarged data set now available, it is no longer possible to discern a consistent regional temperature gradient. Instead, Figure 6 illustrates how for the recent analyses data sets for Dabie Shan UHP eclo-

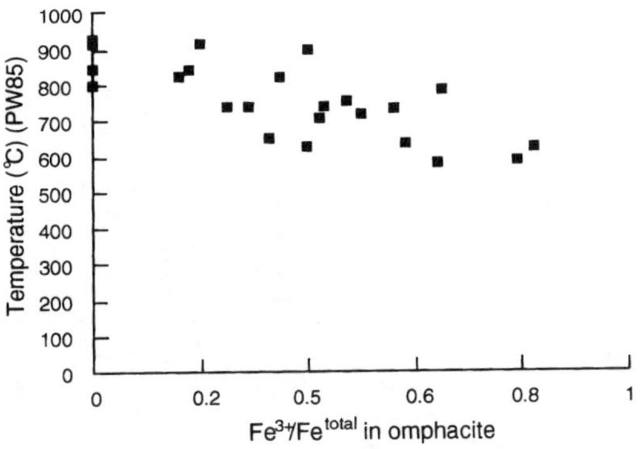

FIG. 6. Plot of temperature estimates for Dabie Shan UHP eclogites based on Fe^{2+}/Mg partitioning between garnet and omphacite pairs (Powell, 1985), at an assumed pressure of 3.0 GPa, against the calculated Fe^{3+}/Fe^{total} ratio for the omphacite.

gites of Carswell et al. (1997) and Tabata et al. (1998), there is a strong negative correlation between the calculated ratio of Fe^{3+}/Fe^{total} in the omphacite and the temperature calculated from Fe^{2+}/Mg partitioning between the garnet and omphacite. Hence, if there is any systematic regional temperature gradient across this UHP terrane, it is effectively masked by the large error brackets that should be attached to individual temperature estimates (at least ± 100°C) simply because of uncertainties regarding the proportions of Fe^{3+} and Fe^{2+} present in the omphacites. Determination by Mössbauer spectroscopy (P. O'Brien, pers. commun.) of the Fe^{3+}/Fe^{total} ratio in omphacite separates from two Dabie Shan UHP eclogite samples indicates that roughly 50% of the Fe is present as Fe^{3+}. On the other hand, in the coexisting garnets only around 6% of the Fe was determined to be present as Fe^{3+}.

Unfortunately, the extensive replacement of omphacite by secondary symplectites means that it is only possible to obtain pure separates of primary omphacites from a few UHP eclogite samples. However, applying an assumed ratio in omphacite of $Fe^{3+}/Fe^{total} = 0.5$ (and of 0.06 in coexisting garnet) to other Dabie Shan UHP eclogite samples mostly yields garnet-omphacite temperatures just over 700°C (e.g., Fig. 7), values that are encouragingly close to independent estimates based on Fe^{2+}/Mg partitioning between garnet and phengite, using the Green and Hellman (1982) calibration.

Alternatively, if it is assumed that all Fe is present as Fe^{2+} in both the coexisting omphacite and garnet, then it is likely that the equilibration tem-

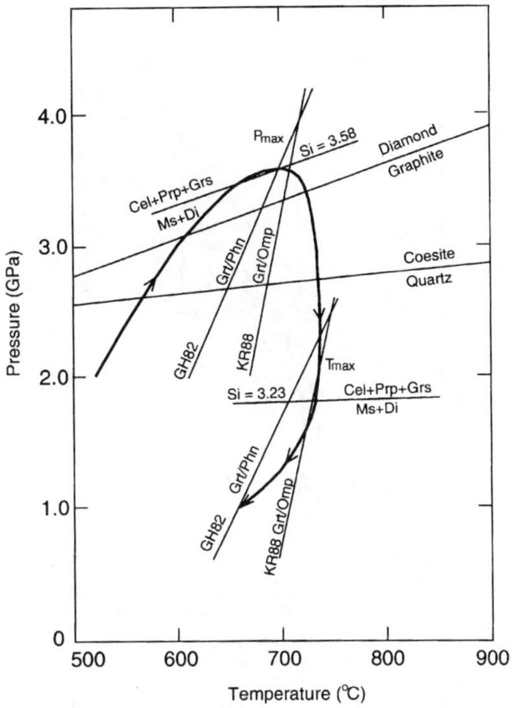

FIG. 7. Calculated reaction equilibria and deduced P-T path for the metamorphic evolution of UHP eclogite sample CD99 from the Shima locality, Dabie Shan, central China. The calculated reaction equilibria for pyrope + grossular + 3 ideal celadonite = 6 diopside + 3 ideal muscovite are based on the calibration of Waters and Martin (1993), and those for Fe/Mg partitioning between garnet-omphacite and garnet-phengite pairs on the calibrations of Krogh (1988) and Green and Hellman (1982), respectively.

peratures will be significantly overestimated. This may perhaps explain the suspiciously high temperatures of 800 to 1000°C indicated for the UHP eclogites in the Kokchetav massif of Kazakhstan (Shatsky et al., 1995; Zhang et al., 1997; Maruyama, 1998).

Published P-T paths deduced for the metamorphic evolution of UHP eclogites in a number of UHP terranes are shown in Figure 8. In some instances, certain constraints controlling the P-T location and shape of the P-T path have been derived from the mineral assemblages in other associated lithologies, such as metapelites. For several localities, only an exhumation P-T path from the peak pressure has been deduced, as no mineralogical constraints are available to define the prograde, subduction-related segment of the path. However, in favorable circumstances, attempts have been made to establish the

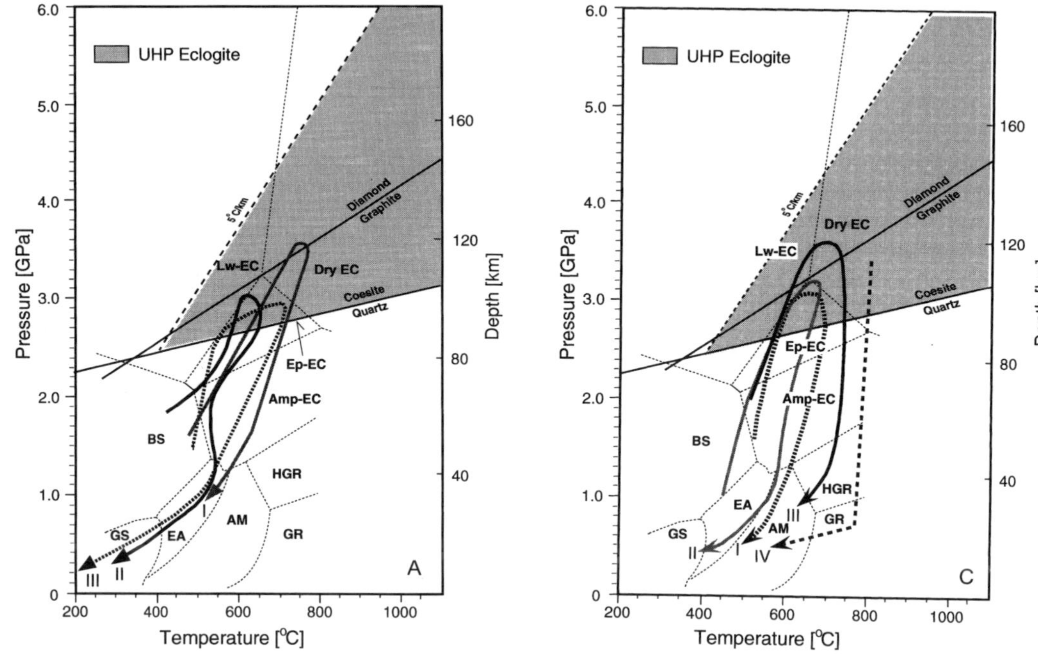

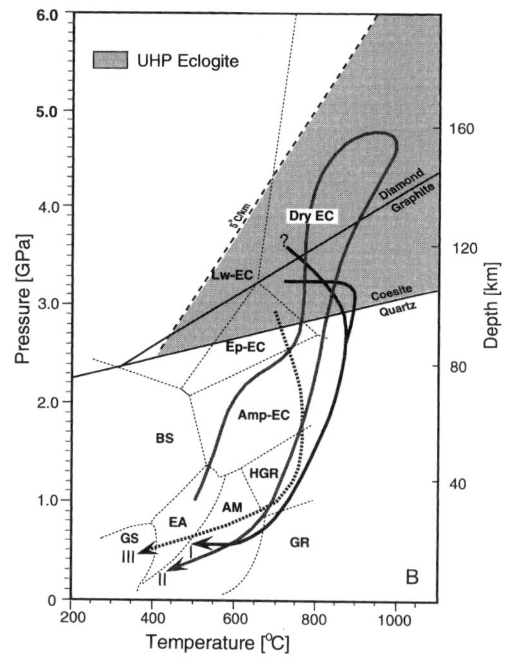

FIG. 8. Comparisons between previously published P-T metamorphic evolutionary paths for UHP eclogites from various locations and that proposed in this paper for a typical UHP eclogite from Dabie Shan, central China. The following P-T paths are shown. A. I = for Dora-Maira massif, Western Alps (Nowlan et al., 1998); II = for Zermatt-Saas Zone, Western Alps (van der Klauw et al., 1997); III = for Tien Shan, Kyrgzstan (Tagiri et al., 1995). B. I = for Saxonian Erzgebirge, Germany (Schmadicke et al., 1992); II = for Kokchetav massif, Kazakhstan (Zhang et al., 1997); III = for WGR, Norway (Krogh and Carswell, 1995). C. I = for Hongan, Henan Province, China (Zhang and Liou, 1994); II = for north Hubei, China (Cui and Wang, 1995); III = for central Dabie Shan, China (this paper; see Fig. 7); IV = for Sulu region, Weihei locality, China (Zhang et al., 1995b).

prograde path from mineral inclusions within prograde growth zoned garnets.

In the absence of age data on the relative timing of growth of the inclusions and the enclosing garnet, an assumption is often made that the identified inclusions are representative of mineral parageneses stable at various stages during the prolonged prograde garnet growth. A potential pitfall with this approach is that it is well known that the poikiloblastic growth of minerals such as garnet may enclose and preserve inclusions of older mineral phases that were either in excess or not participating in the garnet-forming reaction. Even if a particular mineral inclusion was in equilibrium with the enclosing garnet at the time of its growth, there is a further problem that the currently observed immediately adjacent compositions of the inclusion and the garnet may have been modified by diffusion-controlled cation exchange at a later stage along the prograde or retrograde P-T path followed by the rock.

Such problems may explain why several of the best-constrained prograde segments of deduced P-T paths for UHP eclogites surprisingly tend to signal that the rocks concerned were effectively subducted down along a low (<10°C/km) steady-state, slightly concave-upward geothermal gradient. This would imply that, during geodynamically driven subduction, the rocks were able to rapidly re-equilibrate to the ambient temperature conditions down a geotherm. However, this runs contrary to expectations from the thermal modeling of subducted crustal slabs that deeply subducted rocks will follow a more open, markedly convex-upward, clockwise P-T path with time, because of the slow rates of thermal re-equilibration by conductive heat transfer relative to tectonic movement rates (England and Thompson, 1984). The precise form of the uplift trajectory will be controlled by the rate of buoyancy or tectonically driven exhumation relative to the rate of thermal relaxation and re-equilibration of the deeply subducted crustal slab. However, unless the uplift rate is exceedingly fast, deeply subducted rocks may well continue to experience heating during the initial stages of exhumation (England and Thompson, 1984). In that case, it may be anticipated that the maximum subduction depth (P_{max}) will be reached well before the rocks attain their T_{max} condition and the latter will only be achieved at much shallower depths along the exhumation path.

Contrary to this expectation, the majority of the deduced P-T-time paths for UHP eclogites shown in Figure 8 have been constructed with an underlying assumption that P_{max} and T_{max} conditions were reached synchronously. In particular, we suspect the several really tight, hairpin-style P-T-time paths showing the synchronous achievement of P_{max} and T_{max} conditions to be unrealistic representations of the actual paths followed by these rocks, although the deduced paths do emphasize the important point that for UHP eclogites to survive they almost certainly had to be exhumed back toward the surface relatively rapidly.

If the constraint is removed of maximum temperatures having been attained during a short-lived period at maximum subduction depth, then it is possible to construct alternative, much more open, clockwise P-T-time paths that would be more akin to the form of the path deduced by Zhang and Liou (1994a) for a UHP eclogite from Xinxian in the Henan province sector of the Tongbaishan–Dabie Shan collision belt. It is therefore appropriate to examine closely any mineral composition data that have a bearing on whether the P_{max} and T_{max} conditions are likely to have been experienced more or less synchronously by samples of UHP eclogites.

TABLE 1. Mineral Compositions in Phengite-Bearing Eclogites Selected for Evaluation of P-T Conditions During Metamorphic Evolution

For P-T conditions at P_{max}	For P-T conditions at T_{max}	For P-T conditions at P_{min}
Phengite with maximum Si content	Phengite with maximum Fe/Mg	Phengite with minimum Si content
Omphacite with maximum X_{Jd} content	Omphacite with maximum Fe/Mg	Omphacite with minimum X_{Jd} content
Garnet with maximum X_{Ca} content	Garnet with minimum Fe/Mg	Garnet with minimum X_{Ca} content

In many UHP eclogite samples, considerable variations in mineral compositions are apparent between individual spot electron microprobe analyses for each of the major "peak" metamorphic phases—garnet, omphacite, and, when present, phengite. In some, but not all, cases regular within-grain zoning patterns can be discerned that relate to either the prograde (up-temperature) or retrograde (down-temperature) evolution of the rock.

A commonly observed feature is for the cores of phengite grains to have higher Si contents than do rims, and for the cores of omphacite grains to have the highest jadeite contents. Loss of Si content from phengite grain margins and of jadeite at omphacite grain margins can be readily linked to decompression during exhumation, but importantly in terms of Fe/Mg partitioning with coexisting garnet, the grain margins have higher Fe/Mg ratios than do grain cores, signaling higher, rather than lower, equilibration temperatures.

For example, Figure 4B illustrates the strong negative correlation between Si content and Fe/Mg ratio for phengite analyses from a Dabie Shan UHP eclogite sample. In the same sample there is also a strong negative correlation between X_{Na} (jadeite content) and Fe/Mg ratio for omphacite analyses (Fig. 5A). Although in Figure 5B the correlation is less strong, it also appears that the most calcic (grossular-rich) garnet compositions have the lowest Fe/Mg ratios. All of these mineral-composition features are consistent with the peak-pressure mineral compositions having formed at significantly lower temperatures than the peak temperatures experienced by these rocks. It is therefore desirable to combine appropriately selected mineral compositions for these three minerals, in the manner outlined in Table 1, in order to calculate the respective conditions of pressure and temperature at P_{max}, T_{max}, and P_{min} during the metamorphic evolution of this and other UHP eclogite samples.

In reality, the calculated T_{max} and P_{min} conditions for stability of the garnet-omphacite-phengite assemblage usually are not greatly different in most samples, since the analyses selected on the different indicated composition criteria are frequently partly or even completely in common. However, very large differences are invariably indicated between the calculated P_{max} and T_{max} conditions as illustrated for the Dabie Shan UHP eclogite sample CD99 in Figure 7. This evaluation indicates that the maximum temperatures were probably attained at pressures well below those required for coesite stability and that a substantial part of the exhumation history probably was accomplished under near-isothermal conditions, contrary to the suspect, tight-hairpin-style P-T trajectories proposed in many other studies of UHP eclogites. If UHP eclogites customarily follow the style of P-T-time path that we propose, then there are important implications for the interpretation of radiometric age dates for UHP rocks. The closure of the U-Pb and Nd-Sm isotopic systems used to date the timing for formation of UHP mineralogies would be expected to occur some time after the attainment of maximum subduction depth, and consequently some reconsideration and significant reduction in deduced exhumation rates (e.g., Wallis, 1998) may be required.

Acknowledgments

This paper was written while the first author was on a research sabbatical in the Department of Geological and Environmental Sciences at Stanford University. The encouragement and stimulation provided during this visit by Professors Juhn Liou and Gary Ernst was much appreciated, and the financial support provided by a Cox Visiting Fellowship is most gratefully acknowledged. For the second author, this research was supported by NSF grant EAR 9814468.

REFERENCES

Angel, R. J., Gasparik, T., Ross, N. L., Finger, L. W., Prewitt, C. T., and Hazen, R. M., 1988, A silica-rich pyroxene phase with six-coordinated silicon: Nature, v. 335, p. 156–158.

Bakun-Czubarow, N., 1991, On the possibility of occurrence of quartz pseudomorphs after coesite in the eclogite-granulite rock series of the Zlote Mountains in the Sudetes (SW Poland): Archiwum Mineral., v. XLVII, p. 5–16.

______, N., 1992, Quartz pseudomorphs after coesite and quartz exsolutions in eclogitic omphacites of the Zlote Mountains in the Sudetes (SW Poland): Archiwum Mineral., v. XLVIII, p. 3–25.

Beane, R., Liou, J. G., Coleman, R. G., and Leech, M. L., 1995, Petrology and retrograde P-T path for eclogites of the Maksyutov Complex, Southern Ural Mountains, Russia: The Island Arc, v. 4, p. 254–266.

Boyer, H., Smith, D. C., Chopin, C., and Lasnier, B, 1985, Raman microprobe (RMP) determinations of natural and synthetic coesite: Phys. Chem. Minerals, v. 12, p. 45–48.

Bundy, F. R., 1980, The P, T phase, and reaction diagram for elemental carbon, 1979: Jour. Geophys. Res., v. 85, p. 6930–6936.

Caby, R., 1994, Precambrian coesite from northern Mali: First record and implications for plate tectonics in the trans-Saharan segment of the Pan-African belt: Eur. Jour. Mineral., v. 6, p. 235–244.

Carswell, D. A., ed., 1990, Eclogite facies rocks: Glasgow, Blackie, 396 p.

Carswell, D. A., and Harley, S. L., 1995, Mineral barometry and thermometry, *in* Carswell, D. A., ed., Eclogite facies rocks, Glasgow, Blackie, p. 83–110.

Carswell, D. A., O'Brien, P. J., Wilson, R. N., and Zhai, M., 1997, Thermobarometry of phengite-bearing eclogites in the Dabie Mountains of central China: Jour. Metamor. Geol., v. 15, p. 239–252.

Carswell, D. A., Wilson, R. N., and Zhai, M., 1996, Ultra-high pressure aluminous titanites in carbonate-bearing eclogites at Shuanghe in Dabieshan, central China: Mineral. Mag., v. 60, p. 461–471.

Chesnokov, B. V., and Popov, V. A., 1965, Increasing volume of quartz grains in eclogites of the South Urals: Doklady Akad. Nauk., v. 62, p. 909–910 (in Russian).

Chopin, C., 1984, Coesite and pure pyrope in high grade blueschists of the Western Alps: A first record and some consequences: Contrib. Mineral. Petrol., v. 86, p.107–118.

Chopin, C., Henry, C., and Michard, A., 1991, Geology and petrology of the coesite-bearing terrain, Dora Maira Massif, Western Alps: Eur. Jour. Mineral., v. 3, p. 263–291.

Claoué-Long, J. C., Sobolev, N. V., Shatsky, V. S., and Sobolev, A. V., 1991, Zircon response to diamond-pressure metamorphism in the Kokchetav massif, USSR: Geology, v. 19, p. 710–713.

Coleman, R. G., Lee, D. E., and Brannock, W. W., 1965, Eclogites and eclogites, their differences and similarities: Bull. Geol. Soc. Amer., v. 76, p. 483–508.

Coleman, R. G., and Wang, X., eds., 1995, Ultrahigh pressure metamorphism: New York, Cambridge University Press, 528 p.

Compagnoni, R., Hirajima, T., and Chopin, C., 1995, Ultrahigh-pressure metamorphic rocks in the Western Alps, *in* Coleman, R. G., and Wang, X., eds., Ultrahigh-pressure metamorphism: New York, Cambridge University Press, p. 206–243.

Cui, W., and Wang, X., 1995, Eclogites of southern Henan and northern Hubei Provinces, central China: The Island Arc, v. 4, p. 347–361.

Cuthbert, S. J., Carswell, D. A., and Krogh Ravna, E. J., 1998, Eclogites and eclogites in the Western Gneiss Region (WGR), Norwegian Caledonides [abs.]: EOS (Trans. Amer. Geophys. Union), v. 79, no. 45, V22D-12.

Cuthbert, S. J., Carswell, D. A., and Krogh Ravna, E. J., 1999, Eclogites and eclogites in the Western Gneiss Region (WGR), Norwegian Caledonides: Lithos (in press).

Dobrzhinetskaya, L., Eide, E. A., Larsen, R. B., Sturt, B. A., Tronnes, R. G., Smith, D. C., Taylor, W. R., and Posukhova, T. V., 1995, Microdiamond in high-grade metamorphic rocks of the Western Gneiss Region, Norway: Geology, v. 23, p. 597–600.

England, P. C., and Thompson, A. B., 1984, Pressure-temperature-time paths of regional metamorphism I. Heat transfer during the evolution of regions of thickened continental crust: Jour. Petrol., v. 25, p. 894–928.

Ernst, W. G., 1990, Thermobarometric and fluid-expulsion history of subduction zones: Jour. Geophys. Res. v. 95, p. 9047–9053.

Gayk, T., Kleinschrodt, R., Langosch, A., and Seidel, E., 1995, Quartz exsolution in clinopyroxene of high-pressure granulite from the Munchberg Massif: Eur. Jour. Mineral., v. 7, p. 1217–1220.

Gillet, P., Ingrin, J., and Chopin, C., 1984, Coesite in subducted continental crust: P-T history deduced from an elastic model: Earth Planet. Sci. Lett., v. 70, p. 426–436.

Green, D. H., and Hellman, P. L., 1982, Fe-Mg partitioning between coexisting garnet and phengite at high pressures, and comments on a garnet-phengite geothermometer: Lithos, v. 15, p. 253–266.

Griffin, W. L., Austrheim, H., Brastad, K., Bryhni, I., Krill, A. G., Krogh, E. J., Mork, M.-B. E., Qvale, H., and Torudbakken, B., 1985, High-pressure metamorphism in the Scandinavian Caledonides, *in* Sturt, B., and Gee. D., eds, The Caledonide orogen. New York, Wiley, p. 783–801.

Harlow, G. E., 1997, K in clinopyroxene at high pressure and temperature: An experimental study: Amer. Mineral., v. 82, p. 259–269.

Hemingway, B. S., Bohlen, S. R., Hankins, W. B., Westrum, E. J., and Kuskov, O. L., 1998, Heat capacity and thermodynamic properties for coesite and jadeite: Reexamination of the quartz-coesite equilibrium boundary: Amer. Mineral., v. 83, p. 419–433.

Hirajima, T., Ishiwatari, A., Cong, B., Zhang, R. Y., Banno, S., and Nozaka, T., 1990, Identification of coesite in Mengzhong eclogite from Donghai county, northeastern Jiangsu Province, China: Mineral. Mag., v. 54, p. 579–584.

Hirajima, T., Zhang, R. Y., Li, J. J., and Cong, B. L., 1992, Nyböite from Donghai area, Jiangsu province, eastern China: Mineral. Mag., v. 56, p. 37–46.

Holland, T. J. B., 1978, High water activities in the generation of high pressure kyanite eclogites of the Tauern window, Austria: Jour. Geol., v. 87, p. 1–27.

______, 1980, The reaction albite – jadeite + quartz determined experimentally in the range 600–1200°C: Amer. Mineral., v. 65, p. 129–134.

Jahn, B., 1998, Geochemical and isotopic characteristics of UHP eclogites of the Dabie orogen: Implications for continental subduction and collisional tectonics, *in* Hacker, B., and Liou, J. G., eds., When continents collide: Geodynamics and geochemistry of ultrahigh-pressure rocks: Dordrecht, Kluwer Acad. Publ., p. 203–240.

Kienast, J. R., Lombardo, B., Biino, G., and Pinardon, J. L., 1991, Petrology of very-high-pressure eclogitic rocks from the Brossasco-Isasca Complex, Dora-Maira Massif, Italian Western Alps: Jour. Metamor. Geol., v. 9, p. 19–34.

Korsakov, A. V., Shatsky, V. S., and Sobolev, N. V., 1998, First occurrence of coesite in eclogites from Kokchetav Massif: Doklady Akad. Sci., v. 359, p. 77–81.

Krabbendam, M., and Wain, A., 1997, Late-Caledonian structures, differential retrogression and structural position of (ultra)high-pressure rocks in the Nordfjord-Stadlandet area, Western Gneiss Region: Norges Geol. Undersok., v. 432, p. 127–139.

Krogh, E. J., 1977, Evidence for a continent-continent collision in western Norway: Nature, v. 267, p. 17–19.

______, 1988, The garnet-clinopyroxene Fe-Mg geothermometer—a reinterpretation of existing experimental data: Contrib. Mineral. Petrol., v. 99, p. 44–48.

Krogh, E. J., and Carswell, D. A., 1995, HP and UHP eclogites and garnet peridotites in the Scandinavian Caledonides, *in* Coleman, R. G., and Wang, X., eds., Ultrahigh pressure metamorphism: New York, Cambridge University Press, p. 244–298.

Krogh Ravna, E. J., 1999, The garnet-clinopyroxene geothermometer—an updated calibration: Jour. Metamor. Geol. (in press).

Kushiro, I., Satake, H., and Akimoto, S., 1975, Carbonate-silicate reactions at high pressures and possible presence of dolomite and magnesite in the upper mantle: Earth Planet. Sci. Lett. v. 28, p. 116–120.

Leech, M. L., and Ernst, W. G., 1998, Graphite pseudomorphs after diamond? A carbon isotope and spectroscopic study of graphite cuboids from the Maksyutov Complex, south Ural Mountains, Russia: Geochim. et Cosmochim. Acta, v. 62, p. 2143–2154.

Liou, J. C., Zhang, R. Y., and Ernst, W. G., 1995, Occurrences of hydrous and carbonate phases in ultrahigh-pressure rocks from east-central China: Implications for the role of volatiles deep in cold subduction zones: The Island Arc, v. 4, p. 362–375.

Liou, J. C., Zhang, R. Y., Ernst, W. G., Rumble, D., and Maruyama, S., 1998, High-pressure minerals from deeply subducted metamorphic rocks, *in* Hemley, R. J., ed., Ultrahigh-pressure mineralogy: Physics and chemistry of the earth's deep interior: Mineral. Soc. Amer., Rev. Mineral., v. 37, p. 33–96.

Liu, J., Bohlen, S. R., and Ernst, W. G., 1996, Stability of hydrous phases in subducting oceanic crust: Earth Planet. Sci. Lett., v. 143, p. 161–171.

Luth, R. W., 1995, Potassium in clinopyroxene at high pressure: Experimental constraints [abs.]: EOS (Trans. Amer. Geophys. Union), v. 76, p. F711.

Manning, C. E., and Bohlen, S. R., 1991, The reaction titanite + kyanite = anorthite + rutile and titanite-rutile barometry in eclogites: Contrib. Mineral. Petrol., v. 109, p. 1–9.

Mao, H. K., 1976, The system jadeite ($NaAlSi_2O_6$)-anorthite ($CaAl_2Si_2O_8$) at high pressures: Carnegie Inst. Year Book, v. 69, p. 163–168.

Maruyama, S., 1998, The Kokchetav UHP-HP metamorphism: Current state of our research (1998) [abs.]: International Workshop on UHP Metamorphism and Exhumation: Stanford, CA, Stanford University, Abs. vol., p. A58–A59.

Maruyama, S., and Liou, J. C., 1998, Initiation of ultrahigh-pressure metamorphism and its significance on the Proterozoic–Phanerozoic boundary: The Island Arc, v. 7, p. 6–35.

Massonne, H. J., 1998, Ultra-high pressure metamorphism of rocks from the Gneiss-Eclogite Unit of the Saxonian Erzegebirge, Germany [abs.]: EOS (Trans. Amer. Geophys. Union), v. 79, no. 24, V41B-1, p. W129.

Massonne, H. J., and Schreyer, W., 1987, Phengite barometry based on the limiting assemblage with K-feldspar, phlogopite, and quartz: Contrib. Mineral. Petrol., v. 96, p. 212–224.

Massonne, H. J., and Szpurka, Z., 1997, Thermodynamic properties of white micas on the basis of high-pressure experiments in the systems K_2O-MgO-Al_2O_3-SiO_2-H_2O and K_2O-FeO- Al_2O_3-SiO_2-H_2O: Lithos, v. 41, p. 229–250.

Nakamura, D., and Banno, S., 1997, Thermodynamic modeling of sodic pyroxene solid solution and its application in a garnet-omphacite-kyanite-coesite geothermobarometer for UHP metamorphic rocks: Contrib. Mineral. Pertol., v. 130, p. 93–102.

Nowlan, U., Schertl, H.-P., and Schreyer, W., 1998, P-T evolution from the coesite-bearing unit of the southern Dora Maira massif, Western Alps [abs.], *in* International Workshop on UHP Metamorphism and Exhumation: Stanford, CA, Stanford University, Abs. vol., p. A202–205.

Okamoto, K., and Maruyama, S., 1998, Multi-anvil re-equilibration experiments of a Dabie Shan ultrahigh-pressure eclogite within the diamond-stability fields: The Island Arc, v. 7, p. 52–69.

______, 1999, The high pressure stability limits of lawsonite in the MORB + H_2O system: Amer. Mineral. (in press).

Okay, A. I., Xu, S., and Sengör, A. M. C., 1989, Coesite from the Dabie Shan eclogites, central China: Eur. Jour. Mineral., v. 1, p. 595–598.

Okay, A. I., 1993, Petrology of a diamond and coesite-bearing metamorphic terrain: Dabie Shan, China: Eur. Jour. Mineral., v. 5, p. 659–675.

Parkinson, C. D., Miyazaki, K., Wakita, K., Barber, A. J., and Carswell, D. A., 1998, An overview and tectonic synthesis of the pre-Tertiary very-high-pressure metamorphic and associated rocks of Java, Sulawesi and Kalimantan, Indonesia: The Island Arc, v. 7, p. 184–200.

Powell, R., 1995, Regression diagnostics and robust regression in geothermometer/geobarometer calibration: The garnet-clinopyroxene geothermometer revisited: Jour. Metamor. Geol., v. 3, p. 327–342.

Raheim, A., and Green, D. H., 1974, Experimental determination of the temperature and pressure dependence of Fe-Mg partition coefficient for coexisting clinopyroxene and garnet: Contrib. Mineral. Petrol., v. 48, p. 179–203.

Reinecke, T., 1991, Very-high-pressure metamorphism and uplift of coesite bearing metasediments from the Zermatt-Saas zone, Western Alps: Eur. Jour. Mineral, v. 3, p. 7–17.

Renner, J., Zerbian, A., and Stockhert, B., 1997, Microstructures of synthetic polycrystalline coesite aggregates: The effect of pressure, temperature, and time: Lithos, v. 41, p. 169–184.

Schmadicke, E., 1991, Quartz pseudomorphs after coesite in eclogites from the Saxonian Erzgebirge: Eur. Jour. Mineral., v. 3, p. 231–238.

Schmadicke, E., Okrusch, M., and Schmidt, W., 1992, Eclogite-facies rocks in the Saxonian Erzegebirge, Germany: High pressure metamorphism under contrasting P-T conditions: Contrib. Mineral. Petrol., v. 110, p. 226–241.

Schreyer, W., 1985, Metamorphism of crustal rocks at mantle depths: High-pressure minerals and mineral assemblages in metapelites: Fortschr. Mineral., v. 63, p. 227–261.

Sclar, C. B., Carrison, L. C., and Schwartz, C. M., 1962, Optical crystallography of coesite: Amer. Mineral., v. 47, p. 1292–1302.

Shatsky, V. S., Sobolev, N. V., and Vavilov, M. A., 1995, Diamond-bearing metamorphic rocks of the Kokchetav massif (Northern Kazakhstan), *in* Coleman, R. G., and Wang, X., eds., Ultrahigh pressure metamorphism: New York, Cambridge University Press, p. 427–455.

Skjerlie, K. P., and Patino Douce, E., 1998, Fluid-absent melting of a zoisite-bearing eclogite from the western Gneiss Region of Norway: Preliminary results with implications for fluid-absent melting in the deep crust and in subduction zones [abs.]: EOS (Trans. Amer. Geophys. Unoin), v. 79, no. 45, V32A-03, p. F994.

Smith, D. C., 1977, Aluminium-bearing sphene in eclogites from Sunnmore (Norway): Geolognytt, v. 10, p. 32–33.

______, 1984, Coesite in clinopyroxene in the Caledonides and its implications for geodynamics: Nature, v. 310, p. 641–644.

______, 1988, Eclogites and eclogite facies rocks: Amsterdam, Elsevier, 524 p.

______, 1995, Microcoesites and microdiamonds in Norway: An overview, *in* Coleman, R. G., and Wang, X., eds., Ultrahigh pressure metamorphism: New York, Cambridge University Press, p. 299–355.

Smith, D. C., and Cheeney, R. F., 1980, Orientated needles of quartz in clinopyroxene: Evidence for exsolution of SiO_2 from a non-stoichiometric super-silicic "clinopyroxene" [abs.]: 26th Int. Geol. Cong. Paris, Abs. sect. 02.3.1, 145.

Smyth, J. R., 1977, Quartz pseudomorphs after coesite: Amer. Mineral., v. 62, p. 828–830.

Sobolev, N. V., and Shatsky, V. S., 1990, Diamond inclusions in garnets from metamorphic rocks: A new environment for diamond formation: Nature, v. 343, p. 742–746.

Tabata, H., Maruyama, S., and Shi, Z., 1998, Metamorphic zoning and thermal structure of the Dabie ultra-high-pressure/high-pressure terrane, central China: The Island Arc, v. 7, p. 142–158.

Tagiri, M., and Bakirov, A., 1990, Quartz pseudomorph after coesite in garnet from a garnet-chloritoid-talc schist, northern Tien-Shan, Kirghiz, USSR: Proc. Japan Acad., v. 66, p. 135–139.

Tagiri, M., Yano, T., Bakirov, A. Nakajima, T., and Uchiumi, S., 1995, Mineral parageneses and metamorphic P-T paths of ultrahigh-pressure eclogites from Kyrghyzstan Tien-Shan: The Island Arc, v. 4, p. 280–292.

Troitzsch, U., and Ellis, D. J., 1998, The first synthesis of Al-rich titanite (X_{Al} >>0.5) including the new end member $CaAlFSiO_4$ [abs.]: EOS (Trans., Amer. Geophys. Union), V21E-11, p. F972.

van der Klauw, S. N. G. C., Reinecke, T., and Stöckhert, B., 1997, Exhumation of ultrahigh-pressure metamorphic oceanic crust from Lago di Cignana, Piemontese zone, western Alps: The structural record in metabasites: Lithos, v. 41, p. 79–102.

Wain A., 1997, New evidence for coesite in eclogite and gneisses: Defining an ultrahigh-pressure province in the Western Gneiss Region of Norway: Geology, v. 25, p. 927–930.

Wallis, S. R., 1998, Rapid exhumation of the Su Lu UHP terrain [abs.]: EOS (Trans. Amer. Geophys. Union), V22D-04, p. F983.

Wang, X., and Liou, J. G., 1991, Regional ultrahigh-pressure coesite-bearing eclogitic terrane in central China: Evidence from country rocks, gneiss, marble, and metapelite: Geology, v. 19, p. 933–936.

Wang, X., Liou, J. G., and Mao, H. K., 1989, Coesite-bearing eclogite from the Dabie Mountains, central China: Geology, v. 17, p. 1085–1088.

Wang, X., Liou, J. G., and Maruyama, S., 1992, Coesite-bearing eclogites from the Dabie Mountains, Central China: Petrogenesis, P-T paths, and implications for regional tectonics: Jour. Geol., v. 100, p. 231–250.

Waters, D. J., and Martin, H. N., 1993, Geobarometry of phengite-bearing eclogites [abs.]. Terra Abs., v. 5, p. 410–411.

Wood, B. J., and Henderson, C. M. B., 1978, Compositions and unit-cell parameters of synthetic non-stoichiometric tschermakitic clinopyroxene: Amer. Mineral., v. 63, p. 66–72.

Xu, S. T., Okay, A. I., Ji, S. Y., Sengör, A. M. C., Su, W., Lui, Y. C., and Jiang, L. L., 1992, Diamond from the Dabie Shan metamorphic rocks and its implication for tectonic setting: Science, v. 256, p. 80–82.

Yang, J., Godard, G., and Smith, D. C., 1998, K-feldspar-bearing coesite pseudomorphs in an eclogite from Lanshantou (Eastern China): Eur. Jour. Mineral., v. 10, p. 969–985.

Zhang, R. Y., and Liou, J. G., 1994a, Coesite-bearing eclogite in Henan Province, central China: Detailed petrography, glaucophane stability, and P-T path: Eur. Jour. Mineral., v. 6, p. 217–233.

______, 1994b, Significance of magnesite paragenesis in ultrahigh-P metamorphic rocks: Amer. Mineral., v. 79, p. 397–400.

______, 1998, Ultrahigh-pressure metamorphism of the Sulu terrane, eastern China: A prospective view: Continental Dynamics, v. 3, p. 32–53.

Zhang, R. Y., Liou, J. C., and Cong, B. L., 1995a, Talc-, magnesite- and Ti-clinohumite-bearing ultra-high-pressure meta-mafic and ultramafic complex in the Dabie Mountains, China: Jour. Petrol., v. 36, p. 1011–1037.

Zhang, R. Y., Liou, J. G., and Ernst, W. G., 1995b, Ultra-high-pressure metamorphism and decompressional P-T path of eclogites and country rocks from Weihai, eastern China: The Island Arc, v. 4, p. 293–309.

Zhang, R. Y., Hirajima, T., Banno, S., Cong, B., and Liou, J. C., 1995c, Petrology of ultrahigh-pressure rocks from the southern Su-Lu region, eastern China: Jour. Metamor. Geol., v. 13, p. 659–675.

Zhang, R. Y., Liou, J. C., Ernst, W. G., Coleman, R. G., Sobolev, N. V., and Shatsky, V. S., 1997, Metamorphic evolution of diamond-bearing and associated rocks from the Kokchetav Massif, northern Kazakhstan: Jour. Metamor. Geol., v. 15, p. 479–496.

Garnet Peridotites in Eurasian High-Pressure and Ultrahigh-Pressure Terranes: A Diversity of Origins and Thermal Histories

L. G. MEDARIS, JR.

Department of Geology and Geophysics, University of Wisconsin-Madison, Madison, Wisconsin 53706

Abstract

Garnet peridotites and pyroxenites have been reported from 11 of the 15 or so high-pressure/ultrahigh-pressure (HP/UHP) terranes in Eurasia. Most of these ultramafic rocks are Mg-Cr types, derived from depleted upper mantle, but some are more Fe-rich and originated by crystallization in ultramafic-mafic igneous complexes. The peridotites are polymetamorphic, with HP/UHP garnet-bearing assemblages being followed by a succession of retrograde assemblages related to exhumation and cooling; some also contain evidence for a pre-HP/UHP stage, such as spinel inclusions in garnet or the presence of Ti-clinohumite. Equilibration conditions have been calculated from all available analyses of garnetiferous assemblages, by application of the olivine-garnet Fe-Mg exchange thermometer and the Al-in-orthopyroxene barometer, resulting in two distinct P-T regimes for garnet peridotites—one at high P/T in the coesite and diamond fields, and another at low P/T in the vicinity of the spinel-to-garnet transition.

Garnet peridotites are thought to have evolved in at least four different tectonothermal settings, including: (1) emplacement of peridotites into oceanic or continental crust, followed by transport of peridotites and associated crust to UHP conditions by a subducting plate; (2) transfer of peridotites from a mantle wedge to the crust of an underlying, subducting plate; (3) origination from upwelling asthenosphere that passed through a high-temperature spinel peridotite stage, followed by cooling into the garnet peridotite field; and (4) extraction of garnet peridotites from ancient subcontinental lithosphere, perhaps by deep-seated faulting within a continental plate.

Introduction

PERIDOTITE MASSIFS (a.k.a. alpine peridotites or orogenic peridotites) are important, but volumetrically minor, features of orogenic belts worldwide and have long provided a wealth of information on the physical and chemical characteristics of the upper mantle. Spinel peridotite is predominant in such bodies, but garnet peridotite is common in those belts where continental collisions have taken place, particularly in Eurasian ultrahigh-pressure metamorphic terranes. The recognition that ultrahigh-pressure crustal rocks have been subjected to pressures above 25 kbar, and as high as 45 kbar, has evoked new interest in associated garnet peridotites, which provide an opportunity to evaluate the tectonothermal conditions related to the interaction of mantle and crust during ultrahigh-pressure metamorphism.

Because of the brevity of this report, a comprehensive summary of garnet peridotites is not practical, and emphasis will be placed on their geologic settings, critical high-pressure/ultrahigh-pressure (HP/UHP) mineral assemblages, and thermobarometry. Detailed descriptions are presented elsewhere of phase equilibria in ultramafic systems (Liou et al., 1998; Schmidt and Poli, 1998; Ernst, 1999), isotope geochemistry of garnet peridotites (Brueckner and Medaris, 1998; Jahn, 1998), and the recently recognized and controversial ilmenite rods and magnetite lamellae in olivine (Dobrzhinetskaya et al, 1996; Hacker et al., 1997; Zhang et al., in press). However, before focusing on the equilibration conditions of HP/UHP assemblages in garnet peridotites, it should be emphasized that these rocks are typically polymetamorphic and occur as relics within more thoroughly recrystallized, lower-grade ultramafic hosts. Retrograde assemblages in the peridotites formed in response to decompression during exhumation and commonly can be correlated with retrograde assemblages in associated crustal rocks. A generalized sequence of assemblages is illustrated in Figure 1, which is based on chemographic analysis of the system CMASH (Obata and Thompson, 1981) and observations of natural occurrences. Although such retrograde assemblages are important in establishing complete P-T-t histories

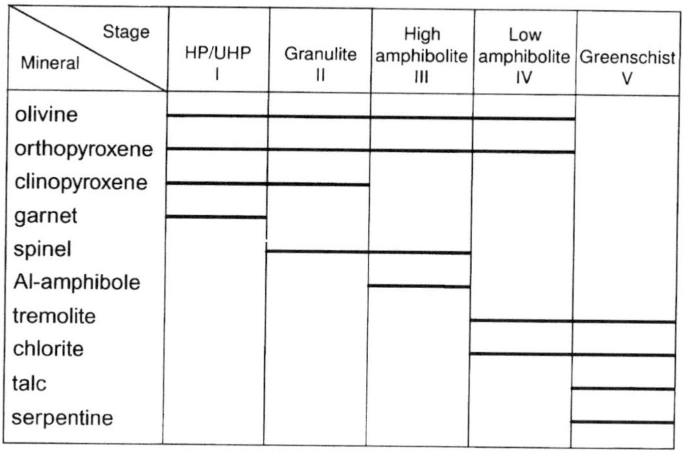

FIG. 1. Generalized sequence of mineral assemblages in peridotites, based on chemographic analysis of the system CMASH (Obata and Thompson, 1981) and on natural occurrences.

for peridotites in HP/UHP terranes, emphasis is placed here on P-T conditions achieved during the peak of UHP metamorphism and on any evidence for an earlier, lower-pressure, pre-garnet stage. Detailed descriptions of mineral parageneses in specific garnet peridotite occurrences are provided in the review by Medaris and Carswell (1990) and the following references herein.

A comparison of the petrology, mineralogy, and thermobarometry of Eurasian garnet peridotites, as summarized in this paper, demonstrates significant differences among peridotites in the various HP/UHP terranes, reflecting a diversity in origins and thermal histories. Such diversity results largely from the timing of juxtaposition of peridotites and crustal rocks and to the tectonothermal regime during HP/UHP metamorphism, as discussed below.

Summary of Garnet Peridotite Occurrences

Garnet peridotites have been reported from 11 of the 15 or so Eurasian HP/UHP terranes (Fig. 2). Occurrences of these garnet peridotites are summarized in the following section, beginning with the classic UHP terranes of Sulu-Dabie and Kokchetav, in which coesite and diamond have been thoroughly documented, followed by the Scandinavian Caledonides, where coesite is abundant in the Western Gneiss Region, Norway, and ending with the numerous and diverse terranes of the Variscan and Alpine fold belts.

Sulu-Dabie terrane, China

Garnet-bearing ultramafic rocks occur in the coesite- and diamond-bearing eclogite unit in the Qinling-Dabie-Sulu metamorphic complex, which is located at the Triassic suture between the Sino-Korean and Yangtze cratons (Liou et al., 1998). In the Sulu terrane, garnet peridotites and pyroxenites occur in four areas, including Rongcheng (locality Chijiadian), Rizhao (locality Hujaling), Junan (locality Zhoubian), and Donghai (localities Zhimafang, Mengzhong, and Jiangzhuang) (Yang et al., 1993; Zhang et al., 1994; Cong et al., 1996; Hacker et al., 1997). The ultramafic bodies are meters to kilometers in size and have concordant to discordant contacts with surrounding Proterozoic quartzofeldspathic gneisses. Chemically, the peridotites represent depleted mantle material, and lithologically, they consist of garnet lherzolite, garnet harzburgite, garnet wehrlite, and hornblende peridotite. Lenses and layers of garnet websterite and garnet clinopyroxenite occur in the peridotites, as do isolated nodules, lenses, and blocks of eclogite, which range in size from decimeters to meters.

The garnet-stage mineral assemblage in lherzolite is olivine + orthopyroxene (0.14 to 0.28 wt% Al_2O_3) + clinopyroxene + garnet ± phlogopite. Harzburgite from Zhimafang additionally contains magnesite, which is in textural equilibrium with minerals of the garnetiferous assemblage. Pressure-temperature estimates for the garnet ultramafites are 820 to 900°C/36 to 41 kbar. Evidence for a

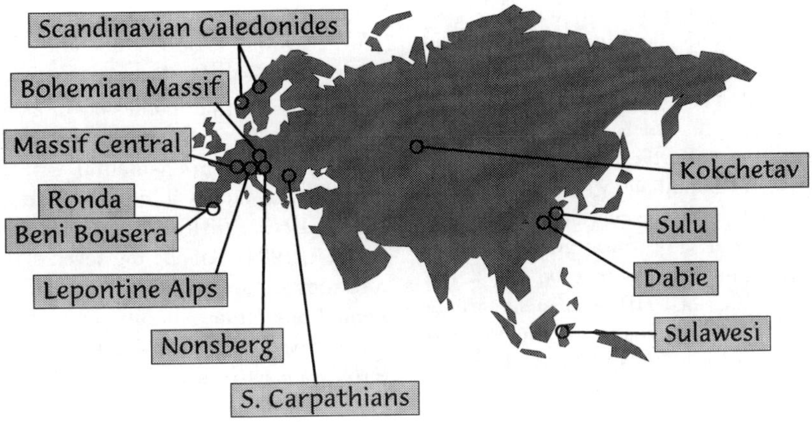

FIG. 2. Localities of garnet peridotites in Eurasian HP and UHP terranes.

pre-garnet stage in the Sulu ultramafic rocks is provided by chromite inclusions in garnet in harzburgite (Zhimafang) and spinel, and amphibole inclusions in garnet in clinopyroxenite (Hujaling), as well as rare Ti-clinohumite (~0.4 wt% F) inclusions in olivine and orthopyroxene in garnet lherzolite and harzburgite and in orthopyroxene in garnet websterite.

Garnet ultramafites in the Dabie terrane occur in two metamorphosed mafic-ultramafic complexes, Bixiling and Maowu, which were emplaced into Proterozoic quartzofeldspathic gneisses prior to Triassic UHP metamorphism (Okay, 1994; Zhang and Liou, 1994; Zhang et al., 1995; Zhang, Liou et al., 1995; Cong et al., 1996; Liou and Zhang, 1998). Field, mineralogical, and geochemical data indicate that the Bixiling and Maowu complexes originated as low-pressure crystal cumulates and thus contrast markedly in petrogenesis with the mantle-derived Sulu garnet ultramafites.

The Bixiling complex is ~1.5 km^2 in outcrop and consists mostly of layered eclogite, but includes about 20 lenses of ultramafic rocks, measuring 50 to 300 m in length and 5 to 50 m in width. The ultramafic rocks are garnet lherzolite, garnet wehrlite, and garnet websterite, which contain HP/UHP mineral assemblages of olivine + orthopyroxene (0.10 to 0.22 wt% Al_2O_3) + clinopyroxene + garnet ± olivine and ilmenite (after Ti-clinohumite), and orthopyroxene + clinopyroxene + garnet + magnesite + Ti-clinohumite (0.05 to 0.2 wt% F). UHP metamorphic conditions are estimated to be 700 to 800°C/47 to 67 kbar.

The Maowu complex is a lenticular body ~250 m × 100 m in size, consisting of layers, lenses, and pods of garnet orthopyroxenite, garnet clinopyroxenite, harzburgite, eclogite, and omphacitite, which are intercalated on a scale of 5 cm to 1.6 m. In contrast to Bixiling, olivine is uncommon at Maowu, occurring only in a few thin harzburgite layers, and the most abundant ultramafic rock type is orthopyroxenite. UP/UHP mineral assemblages are orthopyroxene (0.08 to 0.16 wt% Al_2O_3) + garnet + chlorite ± clinopyroxene ± Ti-clinohumite (0.8 to 1.3 wt% F) ± magnesite in orthopyroxenite, and olivine + orthopyroxene + garnet ± Ti-clinohumite ± magnesite ± chlorite in harzburgite. UHP conditions for Maowu ultramafic rocks are estimated to be 750 ± 50°C/35 to 50 kbar, with $X(CO_2) \leq 0.001$. Mineralogical evidence for a pre-UHP stage in Maowu orthopyroxenite is provided by inclusions of chlorite, orthopyroxene, sapphirine, gedrite, hornblende, talc, rutile, phlogopite, corundum, and zircon in garnet, which are interpreted to be the result of Precambrian(?) granulite-facies metamorphism (730 ± 30°C/4 ± 2 kbar), unrelated to Triassic UHP metamorphism.

Kokchetav massif, northern Kazakhstan

The Kokchetav massif is a fault-bounded Proterozoic metamorphic complex, surrounded by Caledonian rocks of the Ural-Mongolian fold belt, and subjected to UHP metamorphism in Middle Cambrian time (Liou et al., 1998). A single occurrence of garnet peridotite has been reported from the Kokchetav massif in the Kumdy Kol area (Zhang et al., 1997), where it is associated with diamond-bearing paragneiss, dolomitic marble, and eclogite in tectonic Unit I of the Proterozoic Zerenda Series. The garnet peridotite is unusual in that it consists of variable proportions of garnet, olivine (Fo 84.5), Ti-clinohumite (0.7 to 0.8 wt% F), pleonaste spinel, and minor pargasitic amphibole, with some samples being composed entirely of garnet and Ti-clinohumite. Locally, garnet, Ti-clinohumite, and olivine-ilmenite intergrowths appear to be in textural equilibrium. Estimates of UHP peak metamorphic conditions are ~900°C/>40 kbar for diamond-bearing paragneiss and 740–780°C/>28–35 kbar for eclogite and garnet peridotite.

Scandinavian Caledonides

The Caledonides are an Early to Middle Paleozoic mountain system that evolved during closure of the Iapetus ocean basin and ultimate Late Silurian collision of a western continent, Laurentia, with an eastern one, Baltica, resulting in a sequence of thrust sheets that were transported eastward onto the Baltoscandian platform in Middle to Late Silurian time. Caledonian tectonic units have been divided into five major complexes, including autochthon (parautochthon) and lower, middle, upper, and uppermost allochthons. Garnet peridotites occur in Proterozoic gneisses of the Western Gneiss Region in Norway (parautochthon) and in migmatitic gneisses in the Seve Nappe in northern Sweden (upper allochthon). Comprehensive reviews of the garnet peridotites have been given by Medaris and Carswell (1990), Krogh and Carswell (1995), and Brueckner and Medaris (1998).

Western Gneiss Region, Norway. Two compositionally distinct types of garnet peridotite (Mg-Cr and Fe-Ti) occur in the Western Gneiss Region (Carswell et al., 1983). The Mg-Cr type of garnet peridotite is magnesian (Mg# = 80 to 92), rich in Cr_2O_3 (~0.5 wt%), poor in TiO_2 (~0.25 wt%), and, together with subordinate garnet pyroxenite, is sporadically preserved in large peridotite bodies. In contrast, the Fe-Ti type is ferriferous (Mg# = 60 to 75), contains conspicuous amounts of Fe and Ti oxides, is poor in Cr_2O_3 (<0.05 wt%) and rich in TiO_2 (~1 wt%), and is associated with abundant garnet pyroxenite and eclogite. Mg-Cr garnet peridotites, which are the more abundant of the two types, represent tectonically emplaced fragments of depleted upper mantle, whereas Fe-Ti garnet peridotites were formed by early Caledonian high-pressure metamorphism of low-pressure crustal cumulates in layered gabbro-peridotite complexes of Sveconorwegian age (Mørk, 1985a, 1985b; Mørk and Mearns, 1986; Jamtveit, 1987a, 1987b; Jamtveit et al., 1991).

Mg-Cr garnet peridotites occur as high-pressure relics within large bodies of recrystallized chlorite peridotites that are isofacial with surrounding amphibolite-facies gneisses (Carswell, 1968a, 1968b; 1973, 1981, 1986; Medaris, 1980, 1984; Jamtveit, 1984). Among the seven stages of recrystallization in the peridotites, the earliest well-documented assemblage is olivine + orthopyroxene + clinopyroxene + garnet, which commonly occurs as porphyroclasts in a recrystallized matrix, followed by the five-phase assemblage olivine + orthopyroxene + clinopyroxene + garnet + spinel, which is developed in kelyphitic rims around earlier garnet (for discussion of a possible pre-garnet stage and other features, see p. 263–265 in Medaris and Carswell, 1990). Sm-Nd data demonstrate that the first garnet assemblage is Proterozoic in age, and that the second garnet assemblage is early Caledonian, ~440 Ma (Mearns, 1986; Jamtveit et al., 1991; Brueckner et al., 1996; Brueckner and Medaris, 1998). Equilibration conditions for the Proterozoic garnetiferous assemblage are in the range 725 ± 30°C/28 ± 3 kbar to 825 ± 30°C/32 ± 2 kbar and for the Caledonian garnetiferous assemblages are 680 ± 30°C/18 ± 2 kbar (Krogh and Carswell, 1995).

Recently, van Roermund and Drury (in press) have reported the presence of pyroxene lamellae in garnet from peridotite on the island of Otrøy, western Norway, and interpreted the lamellae to have formed by exsolution from majoritic garnet, initially containing 3.6 to 4.6 vol% pyroxene solid solution, corresponding to pressures of 60 to 70 kbar. This exciting discovery provides compelling mineralogical evidence that some Norwegian garnet peridotites experienced UHP conditions and raises the possibility that majoritic garnet, which until now was known to occur only in xenoliths, may be found in peridotite massifs in other UHP terranes, as well.

Seve Nappe, Sweden. In the Seve Nappe, Mg-Cr garnet peridotites occur in the Ertsekey and Tjeliken tectonic lenses (van Roermund, 1989). As in the Western Gneiss Region, the Seve peridotites contain an early four-phase garnet assemblage and a later five-phase garnet and spinel association. The five-phase assemblage equilibrated at 650°C/22 kbar and 680°C/20 kbar in the two tectonic lenses (Krogh and Carswell, 1995), but mineral composition data are incomplete for the four-phase assemblage, and neither assemblage has been dated isotopically.

European Variscides

The Middle to Late Paleozoic Variscan orogen evolved through the closure of one or more ocean basins and the Carboniferous collision of Laurasia and Gondwana (Rey et al., 1997). Continental collision was preceded by an amalgamation of microplates, volcanic arcs, and small ocean basins along the margins of Laurasia and Gondwana from Late Ordovician to Early Carboniferous time. The Variscan orogen presently is exposed in Europe in 20 or so massifs, which have long been divided into the Moldanubian, Saxothuringian, and Rhenohercynian zones (Kossmat, 1927), and in numerous basement blocks within the Alpine system (Neubauer and Raumer, 1993). Garnet peridotites occur in the Moldanubian and Saxothuringian zones in four of the massifs and in two basement blocks within the Alpine system.

Moldanubian zone. In the Moldanubian zone, which is the most internal, highly metamorphosed segment of the Variscan orogen, garnet peridotites occur only in the Gföhl Nappe in the Bohemian Massif and in its tectonostratigraphic equivalents in the Schwarzwald, Vosges, and Massif Central (*Bohemian Massif*: Fiala, 1965; Fiala and Padera, 1977; Bakun-Czubarow, 1983; Machart, 1984; Medaris et al., 1990; Carswell, 1991; Becker, 1996a, 1996b, 1997; Brueckner and Medaris, 1998; *Schwarzwald*: Kalt et al., 1995; Kalt and Altherr, 1996; *Vosges*: Altherr and Kalt, 1996; *Massif Central*: Lasnier, 1971; Gardien et al., 1990). Moldanubian garnet peridotites range in size from kilometers to meters (disrupted larger bodies) and can generally be divided into two types: those in which garnet is devoid of spinel inclusions, and those in which garnet contains spinel inclusions and formed from pre-existing spinel peridotite. In the Bohemian Massif, the garnet peridotites *sensu stricto* contain lenses and layers of garnet pyroxenite and eclogite and are associated with migmatitic quartzofeldspathic gneiss, and the spinel to garnet peridotites contain lenses of garnet pyroxenite, but lack eclogite, and are associated with felsic high-pressure granulite. Orthopyroxene in garnet peridotite *sensu stricto* typically contains <1 wt% Al_2O_3, and in spinel to garnet peridotite, 4 to 5 wt% Al_2O_3. Similar features and associations are found in Moldanubian peridotites in the Schwarzwald, Vosges, and Massif Central. Moldanubian garnet peridotites equilibrated over a wide range of P-T conditions, reflecting different thermal histories for the two different types (Medaris and Carswell, 1990; and see below). Sm/Nd mineral ages for the garnet peridotites and pyroxenites are broadly Variscan, ~370 to 330 Ma (Carswell and Jamtveit, 1990; Medaris et al., 1995; Brueckner et al., 1996; Becker, 1997).

Saxothuringian zone. In the Saxothuringian zone of the Bohemian massif, garnet peridotites occur in one of three HP units in the Erzgebirge Crystalline Complex, Germany and northwestern Czech Republic (Schmädicke and Evans, 1997). The garnet peridotites occur as elongate bodies (≤1 to 2 km in length) in high-grade gneisses and felsic granulites that contain crustal eclogites. The peridotites consist of olivine + orthopyroxene + clinopyroxene + garnet, with rare spinel inclusions in garnet, and locally contain thin lenses of garnet pyroxenite. Peak P-T conditions are estimated at 850 to 900°C/30 to 35 kbar, but the garnetiferous ultramafic assemblage has not yet been dated isotopically.

Nonsberg, Austridic Alps. In the Austrides of northern Italy, numerous small peridotite bodies (≤300 m × 100 m) occur in Austroalpine high-grade garnet-kyanite paragneiss, orthogneiss, and migmatite in the Nonsberg area, Ultental, between the Non and Ultimo valleys (Morten and Obata, 1983; Obata and Morten, 1987; Godard et al., 1996). The peridotites are divided into: (a) coarse-grained, protogranular spinel lherzolite, partly transformed to garnet lherzolite, and (b) fine-grained, porphyroclastic to equigranular garnet lherzolite, which contains amphibole and, locally, spinel inclusions in garnet. The fine-grained garnet lherzolite is thought to be the metamorphic derivative of coarse-grained spinel lherzolite. $^{207}Pb/^{206}Pb$ analyses of zircon from migmatitic gneiss yield two groups of ages—~470 and 365 Ma (Hauzenberger et al., 1993)—but the age of garnetiferous ultramafic assemblages remains undetermined.

Southern Carpathians, Romania. The Carpathian Mountains constitute a convex-eastward loop of the Alpine belt, extending through Slovakia, Hungary, and Romania, before rejoining the Dinaric Alps in Yugoslavia. The Southern Carpathians, Romania, are characterized by E- and S-verging thrust sheets and nappes of Alpine age, the uppermost of which is the Getic Nappe (Kräutner et al. 1988). The Getic Nappe consists of a Carboniferous to Early Cretaceous cover sequence and underlying Late Precambrian basement, which has been divided into four main tectonostratigraphic units, three of which contain eclogites and garnet peridotites (Iancu and Maruntiu, 1994). Mg-Cr peridotite lenses, meters to kilometers in size, occur in the uppermost of the four units, the Sebes, where they are associated with quartzofeldspathic gneiss and minor mica schist. Although the peridotites are largely recrystallized to chlorite-bearing assemblages, garnet lherzolites are locally preserved, which contain garnet with spinel inclusions and relatively aluminous orthopyroxene (3.5 to 4.0 wt% Al_2O_3). More Fe-rich ultramafic cumulate rocks occur in the two lowest tectonic units, the Lotru and Cumpana, where they have been partly recrystallized to garnet-bearing assemblages, such as orthopyroxene + garnet + kyanite + zoisite and orthopyroxene + garnet + spinel + corundum, as a result of reaction between magmatic olivine and plagioclase. P-T estimates for the garnet-stage assemblages are 1050 to 1090°C/23 to 24 kbar (Sebes), 640 to 665°C/18 to 19 kbar (Lotru), and 620 to 700°C/16 to 18 kbar (Cumpana) (Maruntiu et al., 1997). Although metamorphic ages of Southern Carpathian eclogites and garnet peridotites remain undetermined, $^{40}Ar/^{39}Ar$ hornblende and muscovite ages indicate that exhumation of basement units and cooling from ~500°C to ~375°C occurred between 320 and 295 Ma (Dallmeyer et al., 1998).

Alpine

Ronda, Spain and Beni Bousera, Morocco. The Ronda peridotite is a large (300 km^2), well-exposed, and thoroughly studied massif in the Internal Zone of the Betic Cordillera, southern Spain (Loomis, 1972a, 1972b; Lundeen, 1978; Obata, 1980; Frey et al., 1985; Suen and Frey, 1987). The peridotite, which is a slab approximately 1.5 km thick and sandwiched between an upper Casares unit and lower Blanca unit, is distinctive for having a high-temperature metamorphic aureole developed along its northwestern contact with the Casares unit, where P-T conditions of 800°C/4.3 kbar were attained. The tectonic evolution of the Ronda peridotite is complex and controversial in detail, but the general view is of two main stages, involving initial lithospheric extension, followed by thrusting (Loomis, 1972b; Lundeen, 1978; Tubia and Cuevas, 1986; Doblas and Oyarzun, 1989; Platt and Vissers, 1989; Van der Wal and Vissers, 1993). Emplacement of peridotite into the crust occurred at 22 Ma (Loomis, 1975; Priem et al., 1979), followed by rapid cooling, 150 to 350°C/m.y., and exhumation, 5 to 10 km/m.y., of associated mantle and crust (Zeck et al., 1992).

Internally, the Ronda peridotite has been divided into four zones of metamorphic facies, based on mineral assemblages in peridotite and mafic layers (Obata, 1980). From northwest to southeast, the zones are garnet lherzolite facies, ariégite subfacies (spinel lherzolite facies), seiland subfacies (spinel lherzolite facies), and plagioclase lherzolite facies. Structural, microstructural, and textural data, such as spinel inclusions in garnet in lherzolite, demonstrate that the sequence of recrystallization is ariégite subfacies, garnet lherzolite facies, seiland subfacies, and plagioclase lherzolite (Van der Wal and Vissers, 1996). P-T conditions for the ariégite subfacies, based on reconstructed compositions of exsolved pyroxenes, are 1100 to 1200°C/20 to 25 kbar, followed by recrystallization in the other three facies at more or less uniform temperatures, 800 to 900°C, but decreasing pressures, from ~15 to 5–7 kbar. Recrystallization to garnet-bearing assemblages was synchronous with crustal emplacement, as determined by garnet-clinopyroxene Sm-Nd ages of 21.5 ± 1.8 Ma for pyroxenite (Zindler et al., 1983) and 21.1 to 24.5 ± 7 Ma for peridotite (Reisberg et al., 1989).

A milestone in the investigation of peridotite massifs was reached with the discovery of graphite pseudomorphs after diamond in pyroxenites from Ronda (Davies et al., 1993) and from Beni Bousera (Pearson et al., 1993), a counterpart to Ronda in the Rif Cordillera of Morocco. Such graphite pseudomorphs point to an origin for the western Mediterranean peridotites from deep levels within the mantle, on the order of 200 km, and require a broadened consideration of possible tectonothermal models for the origin and evolution of UHP peridotites.

Lepontine Alps, Switzerland. The lower Pennine Adula/Cima Lunga Nappe in the Lepontine Alps contains numerous lenses of peridotite, several of which are garnetiferous, including Cima di Gagnone, Monte Duria, and Alpe Arami (Möckel, 1969; Ernst, 1978; Evans and Trommsdorff, 1978; Brenker and Brey, 1997). The peridotites are enclosed by semi-pelitic and quartzofeldspathic gneisses, commonly with amphibolite, eclogite, and impure carbonates. At Cima di Gagnone, garnet peridotite constitutes about 10% of a 40 × 50 m macroboudin of largely chlorite peridotite, which contains mafic boudins that are related to an eclogite-metarodingite suite in neighboring ultramafic rocks. The critical assemblage is olivine + orthopyroxene (0.54 wt% Al_2O_3) + clinopyroxene + pargasitic amphibole + garnet. Relict Ti-clinohumite was found in one sample and olivine-ilmenite intergrowths after Ti-clinohumite in several others. The presence of Ti-clinohumite in garnet peridotite and the association with metarodingites indicate that the Cima di Gagnone body was partly serpentinized at low pressure prior to HP/UHP metamorphism.

The Alpe Arami peridotite is a 1 × 0.5 km lensoidal body, bordered by discontinuous selvages of eclogite, hornblendite, and clinopyroxenite against surrounding quartzofeldspathic gneisses. The earliest ultramafic mineral assemblage consists of olivine + orthopyroxene (0.64 to 0.93 wt% Al_2O_3) + clinopyroxene + garnet, which is locally preserved in garnet lherzolite relicts within largely recrystallized chlorite peridotite. In contrast to Cima di Gagnone, Ti-clinohumite has not been found in situ at Alpe Arami, metarodingite is apparently absent, and there is no evidence for a pre-garnet, low-pressure stage (Medaris and Carswell, 1990, pp. 272–273).

Sm-Nd and U-Pb studies of Lepontine ultramafic and mafic rocks have given contrasting results and interpretations. Garnet-clinopyroxene Sm-Nd ages are ~40 Ma for garnet lherzolite, garnet clinopyroxenite, and eclogite at Alpe Arami and garnet lherzolite at Cima di Gagnone, and are thought to represent the time of eclogite-facies metamorphism (Becker, 1993). In contrast, concordant SHRIMP U-Pb ages from zoned zircons in Alpe Arami eclogite average 35 Ma, which is taken to be the eclogite protolith age, and preliminary SHRIMP data on well-equilibrated zircons from Cima di Gagnone garnet peridotite yield 28.5 Ma, which is interpreted as the age of HP metamorphism (Gebauer et al., 1992).

Sulawesi, Indonesia. Sulawesi is located in a tectonically active area at the junction of the Australian, Pacific, and Asia plates and is geologically complex, including elements of ophiolites, magmatic arcs, microcontinental fragments, and high-grade metamorphic terranes. Garnet peridotite occurs in two small outcrops on the floor of the left-lateral Palu-Koro fault valley, where it is associated with mica schist, kyanite eclogite, garnet granulite, and amphibolite of the Cretaceous Bantimala Complex (Helmers et al., 1990; Parkinson et al., 1998). The HP/UHP assemblage in peridotite is olivine + orthopyroxene (3.1 to 4.0 wt% Al_2O_3) + clinopyroxene + garnet, which was preceded by a spinel stage (spinel inclusions in garnet) and followed by a five-phase garnet + spinel stage. The polymetamorphic peridotite, kyanite eclogite, and granulite yield maximum P-T conditions of ~1100°C/15 to 20 kbar, ~700 to 750°C/~17 kbar, and ~750°C/12 kbar, respectively.

Pressure-Temperature Conditions

Thermobarometric methods

The defining assemblage for eclogite-facies peridotite is garnet + olivine (± orthopyroxene ± clinopyroxene), controlled by the reaction

orthopyroxene + clinopyroxene + spinel
= garnet + olivine,

which is largely pressure dependent, but also is a function of bulk composition. Fe and Cr have opposite effects on displacement of the reaction, with garnet being stabilized to lower pressures by higher Fe/(Fe+Mg) values, and spinel being stabilized to higher pressures by higher Cr/(Cr+Al) values (estimates are illustrated in Fig. 3). To illustrate the effects of bulk composition, experimental results from three different starting compositions are included in Figure 4—curve 1 for the system CFMAS (Gasparik, 1987), curve 2 for natural lherzolite minerals (O'Hara et al., 1971), and curve 3 for the system SMACCR + Fe, in which $X_{Mg}{}^{Ol} = 0.9$ and $X_{Cr}{}^{Spl} = 0.25$ (O'Neill and Wood, 1981). Thus, although the occurrence of garnet in peridotite indicates eclogite-facies conditions and pressures ≥15–20 kbar, its presence alone does not necessarily signify UHP conditions. Documentation of UHP conditions in garnet peridotite requires thermobarometric calculations or the occurrence of UHP minerals, such as majoritic garnet (van Roermund and Drury, in press) or diamond (Pearson et al., 1995).

Numerous geothermometers and calibrations of the Al-in-orthopyroxene geobarometer are available for garnet peridotites, but the determination of meaningful P-T values for the garnet stage in such rocks is not a trivial task, because of polymetamorphism, compositional zoning in minerals, and the inherent problem of utilizing a combination of exchange and net transfer reactions, which may have frozen in at different times under different P-T conditions. Medaris and Carswell (1990) used the core compositions of minerals to calculate P-T values for European garnet peridotites by combining the average of the olivine-garnet, orthopyroxene-garnet, and clinopyoxene-garnet geothermometers with the Al-in-orthopyroxene geobarometer. However, Brenker and Brey (1997) have pointed out that the best method for determining "peak" P-T

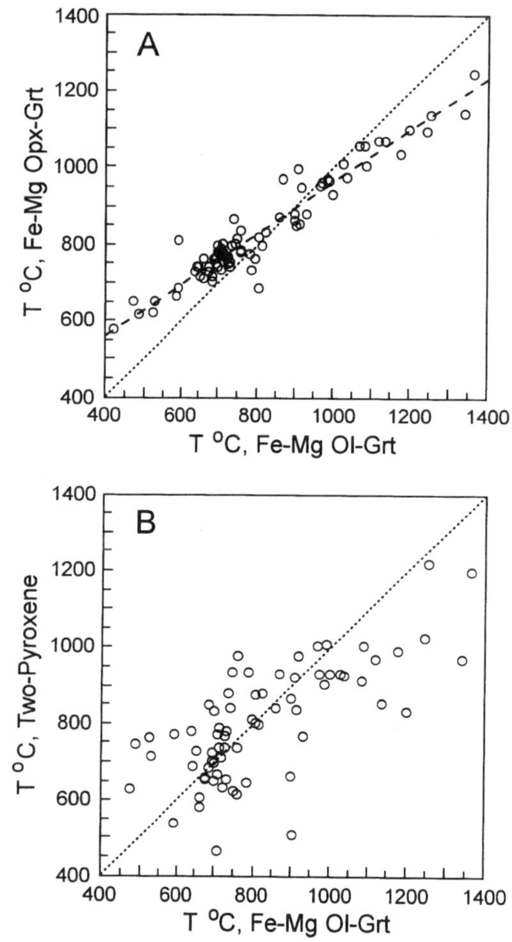

FIG. 3. Comparison of temperature estimates for garnet peridotites (at P = 15 kbar). A. Olivine-garnet Fe-Mg exchange (O'Neill and Wood, 1979) versus orthopyroxene-garnet Fe-Mg exchange (Harley, 1984). B. Olivine-garnet Fe-Mg exchange (O'Neill and Wood, 1979) versus two-pyroxene (Brey and Köhler, 1990).

values for garnet peridotites is to apply the olivine-garnet Fe-Mg exchange thermometer and the Al-in-orthopyroxene barometer to the core compositions of garnet and orthopyroxene and the composition of matrix olivine, because Fe and Mg diffusion are slow in garnet, Al diffusion is slow in orthopyroxene, and olivine, which is the predominant phase, will undergo the least compositional change during cooling. Accordingly, in this review P-T values have been calculated for Eurasian garnet peridotites from all published analyses of coexisting garnet (cores), olivine, and orthopyroxene (cores), using the olivine-garnet Fe-Mg exchange thermometer (O'Neill and Wood, 1979; O'Neill, 1980) and the Al-in-orthopyroxene barometer (Brey and Köhler, 1990). Results are summarized in Table 1 and illustrated in Figures 4–7.

For garnet websterites, in which olivine is absent or a minor phase, temperatures have been calculated by the orthopyroxene-garnet Fe-Mg exchange thermometer (Harley, 1984). A comparison of results from the olivine-garnet and orthopyroxene-garnet thermometers for garnet peridotites (Fig. 3A) shows that results from the two methods coincide at 885°C, but diverge at higher and lower temperatures. Relative to the olivine-garnet thermometer, the orthopyroxene-garnet thermometer gives lower values at higher temperatures (~1095°C at 1200°C) and higher values at lower temperatures (~695°C at 600°C). The scatter in the data probably represents the difficulty in selecting equilibrium mineral compositions in these polymetamorphic rocks. A significant feature of Figure 3A is the enormous range in temperature for Eurasian garnet peridotites, presumably reflecting very different tectonothermal regimes.

A comparison of results from the two-pyroxene (Brey and Köhler, 1990) and olivine-garnet thermometers (Fig. 3B) shows considerable scatter, probably because, as above, of the difficulty in selecting equilibrium compositions, compounded by the extensive degree of exsolution and re-equilibration in pyroxenes during cooling. Clearly, the two-pyroxene thermometer is unsuitable for determining "peak" conditions in these particular rocks.

Application

Among all reported Eurasian garnet ultramafites, those in the Sulu and Dabie terranes yield the highest P/T ratios (Fig. 4), plotting well within the diamond field at temperatures compatible with the presence of Ti-clinohumite and the stability of magnesite at low $X(CO_2)$. The exceptionally high pressures of the Sulu and Dabie suites reflect the consistently low Al_2O_3 contents of their orthopyroxenes. Sulu samples range from 670°C/41.2 kbar to 925°C/69.9 kbar, and Dabie samples from 705°C/41.9 kbar to 885°C/66.6 kbar.[1] P-T results for the Sulu and Dabie garnet ultramafites are consistent

[1]Note that temperature estimates for the Dabie samples are based on the orthopyroxene-garnet thermometer, because of a predominance of pyroxenites and unreasonably low olivine-garnet temperatures for associated garnet peridotites.

TABLE 1. P-T Estimates for Eurasian Garnet Peridotites and Pyroxenites

Rock type[1]	T, °C[2]	P, kbar[2]	Data[3]
Sulu			
Zhimafang, pd	925	69.9	1
Jiangzhuang, px	*782*	53.5	1
Chijiadian, pd	667	41.2	2
Chijiadian, pd	816	58.0	3
Chijiadian, pd	*747*	*41.4*	3
Dabie			
Bixiling, pd	*810*	*46.1*	4
Bixiling, pd	*847*	*62.6*	4
Bixiling, px	*850*	*58.9*	4
Maowu, pd	*706*	*41.9*	5
Maowu, px	*849*	*65.8*	5
Maowu, px	*885*	*66.6*	5
Maowu, px	*838*	*62.8*	5
Maowu, px	*772*	*46.7*	5
Kokchetav			
Kumdy Kol, pd	733	(30)	6
Norway, WGR, Proterozoic			
Kalskaret, pd	891	44.0	7
Kalskaret, pd	949	41.6	8
Aldalen, pd	898	49.4	7
Lien, pd	877	35.6	7
Lien, pd	895	39.1	9
Rødhaugen, pd	850	32.1	10
Rødhaugen, pd	737	27.3	10
Rødhaugen, pd	849	36.4	10
Rødhaugen, pd	752	23.4	10
Sandvika, pd	947	44.2	7
Sandvika, pd	934	50.2	11
Raudhaugene, pd	885	42.8	7
Raudhaugene, pd	744	23.2	12
Rødskar, pd	871	43.7	7
Ugelvik, pd	708	36.5	7
Ugelvik, pd	789	27.8	12
Norway, WGR, Caledonian			
Raudhaugene, pd	668	13.6	8
Fjørtoft, pd	771	26.0	8
Best estimate	680	18.0	13
Sweden, Seve Nappe			
Ertsekey, px	623	19.9	14
Tjeliken, pd	676	19.5	14
Tjeliken, px	*796*	*21.3*	14
Alpine, Indonesia			
Sulawesi, pd	994	18.2	27
Sulawesi, pd	1071	22.3	27
Sulawesi, pd	1027	20.8	27
Sulawesi, pd	976	16.2	27
Sulawesi, pd	1141	27.4	27

Rock type[1]	T, °C[2]	P, kbar[2]	Data[3]
Variscan, Moldanubian			
Gföhl, Becvary, pd	843	40.6	15
Gföhl, Hamry, pd	1252	46.4	15
Gföhl, Mohelno, pd	1280	28.8	16
Gföhl, Mohelno, pd	1332	29.2	16
Gföhl, Mohelno, pd	1120	23.5	16
Gföhl, Níhov, px	*935*	*35.7*	15
Gföhl, Níhov, px	*927*	*36.0*	15
Gföhl, Nové Dvory, pd	816	28.7	16
Gföhl, Nové Dvory, pd	994	38.2	16
Gföhl, Nové Dvory, pd	887	39.8	16
Gföhl, Nové Dvory, pd	899	39.2	16
Gföhl, Nové Dvory, pd	984	43.1	16
Gföhl, Úhrov, pd	1173	44.0	15
Poland, Gory Sowie, pd	1027	21.3	17
Poland, Gory Sowie, pd	1176	24.2	17
Schwarzwald, pd	646	15.4	18
Schwarzwald, pd	804	35.1	18
Schwarzwald, px	*779*	*33.1*	18
Schwarzwald, px	*841*	*39.9*	18
Vosges, pd	896	56.1	19
Variscan, Saxothuringian			
Erzgebirge, pd	805	30.1	20
Erzgebirge, pd	872	29.3	20
Erzgebirge, pd	904	31.9	20
Variscan, Austridic Alps			
Nonsberg, pd	679	15.4	21
Nonsberg, pd	679	13.9	21
Nonsberg, pd	707	16.0	21
Nonsberg, pd	696	13.4	22
Variscan, Romania			
S. Carpathians, pd	1297	32.2	15
S. Carpathians, pd	1145	25.1	15
Alpine, Betic Cordillera			
Ronda, pd	932	17.6	23
Ronda, pd	922	17.3	23
Ronda, pd	867	14.9	23
Ronda, pd	1076	21.2	23
Ronda, pd	1229	25.5	15
Ronda, pd	1165	26.0	15
Ronda, pd	1240	28.0	15
Alpine, Lepontine Alps			
Cima di Gagnone, pd	776	36.8	24
Monte Duria, pd	823	34.2	24
Alpe Arami, pd	837	34.4	24
Alpe Arami, pd	953	39.8	25
Alpe Arami, pd	1128	52.1	25
Alpe Arami, pd	990	45.3	25
Alpe Arami, pd	1091	48.2	26
Alpe Arami, pd	1095	51.5	15

[1]Rock types: pd = peridotite; px = pyroxenite.
[2]Numbers (regular font) = olivine-garnet Fe-Mg exchange; italicized numbers = orthopyroxene-garnet Fe-Mg exchange.
[3]Sources of mineral compositional data: 1 = Yang et al., 1993; 2 = Hacker et al., 1997; 3 = Zhang et al., 1994; 4 = Zhang, Liou et al., 1995; 5 = Liou and Zhang, 1998; 6 = Zhang et al., 1997; 7 = Medaris, 1984; 8 = Jamtveit et al., 1991; 9 = Medaris, 1980; 10 = Carswell, 1981; 11 = Jamtveit, 1984; 12 = Carswell, 1986; 13 = Krogh and Carswell, 1995; 14 = van Roermund, 1989; 15 = Medaris, unpubl.; 16 = Medaris et al., 1990; 17 = Bakun-Czubarow, 1983; 18 = Kalt et al, 1995; 19 = Altherr and Kalt, 1996; 20 = Schmädicke and Evans, 1997; 21 = Obata and Morton, 1987; 22 = Godard et al., 1996; 23 = Obata, 1980; 24 = Evans and Trommsdorff, 1978; 25 = Ernst, 1978; 26 = Brenker and Brey, 1997; 27 = Helmers et al., 1990.

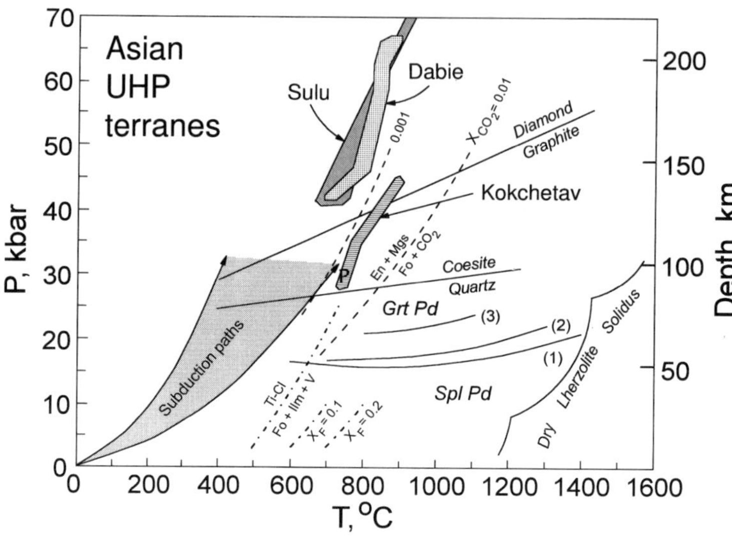

FIG. 4. P-T estimates for garnet peridotites and pyroxenites in the Sulu and Dabie terranes, and for gneiss, eclogite, whiteschist, and garnet peridotite (symbol P at 30 kbar) in the Kokchetav massif. Shown for reference are equilibria for diamond-graphite (Bundy, 1980), coesite-quartz (Bohlen and Boettcher, 1982), spinel-garnet peridotite (1 = Gasparik, 1987; 2 = O'Hara et al., 1971; 3 = O'Neill, 1981), magnesite (Ogasawara et al., 1995), and Ti-clinohumite (Engi and Lindsley, 1980). Also shown are a dry lherzolite solidus (Takahashi, 1986) and a field for steady-state subduction paths (Peacock, 1993, see text for discussion).

with steady-state P-T paths calculated by Peacock (1993) for the top of subducting oceanic crust at different shear stresses and convergence rates. The limiting curves to the field of subduction paths illustrated in Figure 4 are for τ = 20 Mpa and V = 10 cm/yr, on the left, and τ = 80 Mpa, V = 3 cm/yr, on the right.

Pressure estimates for the single outcrop of garnet peridotite near Kumdy Kol in the Kokchetav massif are problematic because of the absence of orthopyroxene. The temperature for garnet peridotite is 735°C at an assumed pressure of 30 kbar, and P-T estimates for associated diamond-bearing gneiss and marble, eclogite, and whiteschist are ~900°C/>40 kbar and 740 to 780°C/>28.0 to 35.0 kbar (Fig. 4). Such P-T conditions are compatible with the textural equilibrium of olivine, garnet, and Ti-clinohumite in peridotite, and with a subduction P-T path at higher shear stresses.

P-T estimates for Proterozoic assemblages from peridotites in the Western Gneiss Region, Norway (15 samples from 7 different bodies), plot in a tight array from 745°C/23.2 kbar to 935°C/50.2 kbar, well into the diamond stability field (Fig. 5). The composition of reconstructed majoritic garnet from the island of Oterøy indicates that some peridotites in the Western Gneiss Region were derived from depths on the order of 200 km (van Roermund and Drury, in press). A garnet pyroxenite layer, also on Otrøy, contains coarse-grained orthopyroxene with exsolution lamellae of garnet and Cr-diopside, whose reconstructed composition suggests initial crystallization at ~1380°C/27 kbar (domain P in Fig. 5) (Carswell et al., 1983; van Roermund and Drury, in press). It has been shown elsewhere that such garnet pyroxenite layers are the product of HP crystallization from melts passing through the upper mantle (Pearson et al., 1993; Medaris et al., 1995; Becker, 1996), and as such may record the temperature of transient melt rather than the ambient temperature of upper mantle. Pressure estimates for Caledonian assemblages in peridotites in the Western Gneiss Region and in the Seve Nappe (Fig. 5) are lower than those for the Proterozoic assemblages, ranging from 13.6 to 26.0 kbar, within the stability field of quartz, and broadly similar to those for many country-rock eclogites (Krogh and Carswell, 1995). The apparent similarity in P/T gradient for Caledonian and Proterozoic assemblages is probably fortuitous.

Variscan garnet peridotites equilibrated over a wide range of P-T conditions, reflecting occurrences

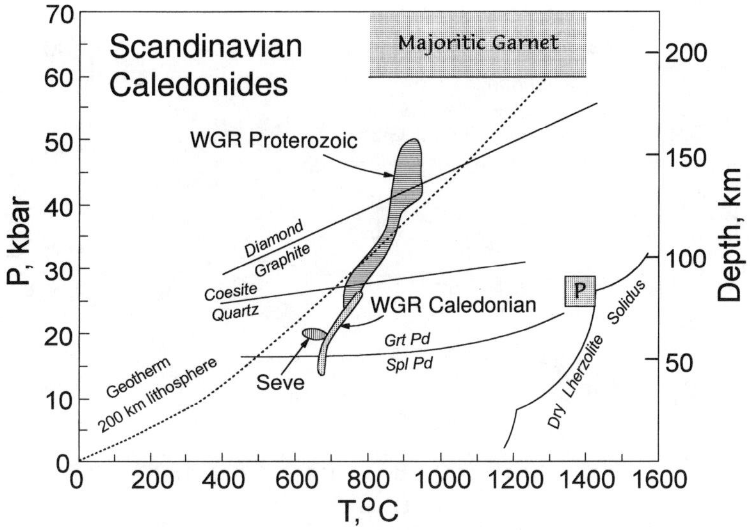

FIG. 5. P-T estimates for Proterozoic and Caledonian assemblages of garnet peridotites in the Western Gneiss Region, Norway, and for garnet peridotites and pyroxenites in the Seve Nappe, Sweden. Also indicated are a stability field for majoritic garnet containing 3 to 5 vol% pyroxene and reconstructed pre-garnet P-T conditions for Ugelvik pyroxenite (symbol P, see text for discussion). Shown for reference is a steady-state conductive geotherm for 200 km thick lithosphere.

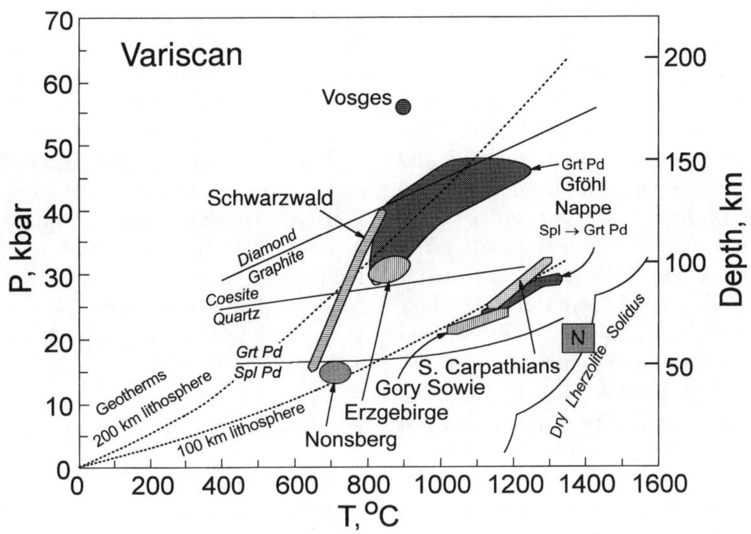

FIG. 6. P-T estimates for garnet peridotites and pyroxenites in European Variscan terranes. Note the distinction in the Gföhl Nappe between garnet peridotites and spinel→garnet peridotites. Symbol N indicates reconstructed conditions for Nonsberg spinel peridotites. Shown for reference are steady-state conductive geotherms for 200 km and 100 km thick lithosphere.

in different tectonstratigraphic units with different tectonothermal histories (Fig. 6). In the Moldanubian zone of the Bohemian massif, Schwarzwald, and Vosges, garnet peridotites generally fall into two groups with respect to P/T regimes. The low P/T group (Gföhl Nappe and Gory Sowie), is character-

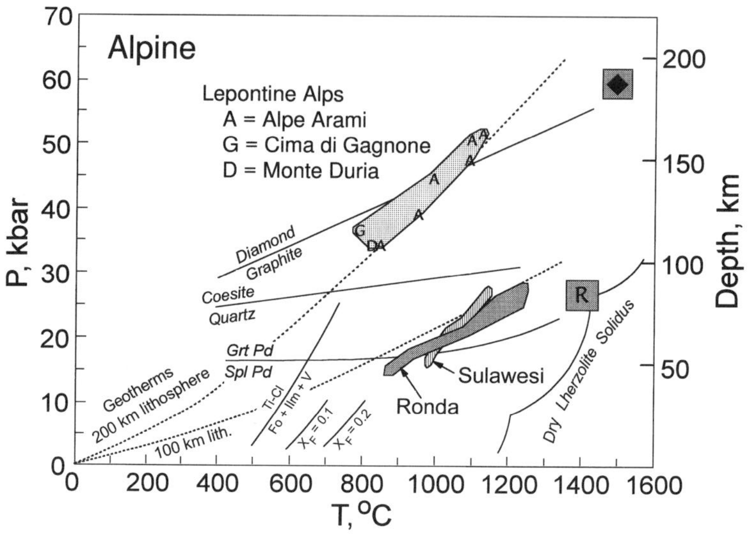

FIG. 7. P-T estimates for garnet peridotites in the Lepontine Alps (A = Alpe Arami; G = Cima di Gagnone; D = Monte Duria), Ronda, and Sulawesi. Symbol R shows reconstructed conditions for Ronda spinel peridotites, and diamond symbol indicates derivation of Ronda and Beni Bousera pyroxenites, which contain graphite pseudomorphs after diamond, from the diamond stability field.

ized by garnet peridotites that were derived from spinel peridotite protoliths, and the high P/T group, situated largely within the coesite field, is made up mostly, but not entirely, of garnet peridotites *sensu stricto* (Gföhl Nappe, Schwarzwald, and Vosges). Garnet peridotites in the Saxothuringian zone of the Bohemian Massif (Erzgebirge) plot at ~850°C/32 kbar within the coesite field (Fig. 6) and are related to the high P/T Moldanubian peridotites with respect to thermal regime. The spinel to garnet peridotites in the Southern Carpathians are similar to those in the Gföhl Nappe in terms of mineralogy and P-T conditions (Fig. 6). The Nonsberg spinel to garnet peridotites could be assigned to either P/T group, because of their position in P-T space, but the initial conditions for spinel peridotite protoliths, 1380°C/20 kbar, estimated from the reconstructed compositions of pyroxenes (box N in Fig. 6), and the P/T characteristics of associated country rocks suggest that the Nonsberg suite belongs to the low P/T group.

Garnet peridotites in the Lepontine Alps plot in a high P/T realm, with samples from Alpe Arami extending into the diamond field, at higher temperatures and pressures than those for Cima di Gagnone and Monte Duria (Fig. 7). The lower temperatures (~800°C) for the Cima di Gagnone and Monte Duria bodies are consistent with the presence of Ti-clinohumite in these meta-serpentinites. Garnet peridotites from the Ronda massif lie in a low P/T domain, plotting over a wide range of temperatures in the vicinity of the spinel to garnet peridotite transition (Fig. 7). Compositions of pyroxene porphyroclasts record a high-temperature, moderate-pressure stage for Ronda peridotite (symbol R, Fig. 7), and graphite pseudomorphs after diamond in Ronda and Beni Bousera pyroxenites signify initial derivation from depths ≥180 km (symbol ♦, Fig. 7) (Obata, 1980; Pearson et al., 1995). Garnet peridotites from Sulawesi, Indonesia also lie in a low P/T realm (Fig. 7), similar to Ronda and several of the Variscan bodies.

Tectonothermal Classification of Garnet Peridotites

Based on thermobarometric calculations, all presently known garnet peridotites in Eurasian HP and UHP terranes can be divided into two groups, one at high P/T and another at low P/T (Figs. 4-7). All but one of the peridotite occurrences in the high–P/T group can be related to subduction processes and are isofacial and approximately contemporaneous with associated UHP crustal rocks that contain coesite and diamond. The single exception in this high–P/T group is the Western Gneiss

Region, Norway, in which the garnet peridotites are relicts of old, cold lithosphere that formed earlier and under different conditions than the overprinting Caledonian metamorphism. Peridotites in the low–P/T group are associated with HP crustal rocks that are characterized by high-pressure granulites and strongly retrogressed eclogites, in which there is little evidence for coesite or diamond so far, although graphite pseudomorphs after diamond occur in the Ronda and Beni Bousera pyroxenites. The extremely high temperatures of the low–P/T group, for both garnet peridotite and reconstructed pyroxenes in spinel peridotite, would seem to require the influence of upwelling asthenosphere.

High–P/T subduction peridotites

This group of peridotites may include ultramafic rocks that were emplaced into the crust prior to subduction and others that were derived from the mantle wedge during subduction (see Fig. 4 in Liou et al., 1998). Pre-subduction examples include the Lepontine Cima di Gagnone and Monte Duria bodies, which were emplaced into oceanic crust and serpentinized before Alpine subduction, and the Dabie Bixiling and Maowu ultramafic-mafic igneous complexes, which intruded continental crust prior to Triassic subduction. These pre-subduction ultramafic rocks were subsequently transported to UHP conditions by a subducting plate. Within a subduction zone, it is likely that peridotites are transferred as slices or ductile masses from the overlying mantle wedge to the underlying, subducting plate, although the mechanism for this transfer is poorly understood (Brueckner, 1998). Subsequent to juxtaposition, the ultramafic and associated crustal rocks will follow the same P-T path along a high–P/T subduction gradient and should give the same recrystallization ages. Peridotites could be derived from both the spinel and garnet fields of the mantle wedge, but in either case, such peridotites may show a geochemical signature of indicating intrusion by melts or fluids emanating from underlying, subducting crust. Examples of mantle-wedge peridotites may be the Sulu, Erzgebirge, Vosges, Schwarzwald, and Gföhl (high–P/T) bodies.

High–P/T relict peridotites

Relict peridotites are represented by garnet peridotites of the Western Gneiss Region, Norway, which are fragments of ancient depleted mantle and whose origins are unrelated to Caledonian HP/UHP metamorphism in Scandinavia. The early garnet assemblage in the WGR peridotites is Proterozoic in age, and thus the P-T array for WGR peridotites (Fig. 5) records some cryptic Proterozoic thermal regime. Emplacement of such mantle into subducting continental crust poses an interesting tectonic problem. One possibility is that a major fault may develop within one of two colliding continental plates, away from the subduction zone. The fault must cut through continental crust into subcontinental lithosphere in order to allow continental crust in the footwall to be driven into the mantle, where tectonic mixing of crust and Proterozoic mantle could occur (Brueckner and Medaris, 1998). At present, the only recognized high–P/T relict peridotites are those of the Western Gneiss Region, but many Eurasian garnet peridotites remain undated, and with further isotopic investigations, others may be discovered.

Low–P/T peridotites

Numerous garnet peridotite bodies—including some in the Gföhl Nappe, Gory Sowie, Nonsberg, Southern Carpathians, Ronda, and Sulawesi—yield low P/T values in the vicinity of the spinel to garnet peridotite transition. These bodies evolved from spinel peridotite and contain abundant inclusions of spinel in garnet, in contrast to high–P/T subduction peridotites, in which inclusions of spinel in garnet are rare. Temperatures in the range from 1000 to 1300°C make it unlikely that the low–P/T peridotites were derived from a mantle wedge in a subduction zone, considering the existence of viscous coupling and temperature inversion at the lower boundary of a mantle wedge and the cooling effect of a subducting plate. Rather, such high temperatures seem to require the influence of upwelling asthenosphere, driven by some mechanism such as slab breakoff, lithospheric delamination, or diapirism. Many of the low–P/T bodies still contain substantial amounts of high-temperature spinel peridotite and have been recrystallized only partly to garnet peridotite. Several models involving kinetics and deformation-induced recrystallization have been invoked to account for such partial recrystallization and the P-T paths linking spinel to garnet peridotite (Obata, 1980; Obata and Morten, 1987; Medaris et al., 1990).

Conclusions

Field, mineralogical, thermobarometric, and geochemical data demonstrate that garnet peridotites in HP/UHP terranes have different origins and tectonothermal evolutions. Most garnet peridotites are depleted upper mantle material, variously modified by metasomatism, but some originated by low-pressure crystallization from ultramafic-mafic igneous complexes. Thermal histories of the peridotites depend on the timing of juxtaposition with crustal rocks and tectonic setting, and four scenarios are proposed: (1) emplacement of peridotites (serpentinites or igneous complexes) into the crust prior to UHP metamorphism, followed by transport of peridotites and associated crust to UHP conditions (high P/T) by a subducting plate; (2) transfer of peridotites from a mantle wedge to the crust of an underlying, subducting plate (high P/T); such peridotites are initially either spinel- or garnet-bearing, depending on the depth of derivation; (3) origination from upwelling asthenosphere that passed through a high-temperature spinel peridotite stage, followed by cooling into the garnet peridotite field (low P/T); and (4) extraction of garnet peridotites from ancient subcontinental lithosphere, perhaps by deep-seated faulting within a continental plate. The only known example of this type, the Western Gneiss Region, Norway, yields high P/T values. Future investigations, utilizing many of the sophisticated analytical techniques now available, will allow further testing of the tectonothermal models summarized here and will surely provide new perspectives on the evolution of garnet peridotites in HP/UHP terranes.

REFERENCES

Altherr, R., and Kalt, A., 1996, Metamorphic evolution of ultrahigh-pressure garnet peridotites from the Variscan Vosges Mts., France: Chem. Geol., v. 134, p. 27–47.

Bakun-Czubarow, N., 1983, Mineral chemistry of garnet lherzolites from the Sudetes, southwest Poland: Carnegie Inst. Yearbook, v. 82, p. 336–343.

Becker, H., 1993, Garnet peridotite and eclogite Sm-Nd mineral ages from the Lepontine dome, Swiss Alps: New evidence for Eocene high-pressure metamorphism in the central Alps: Geology, v. 21, p. 599–602.

______, 1996a, Crustal trace element and isotopic signatures in garnet pyroxenites from garnet peridotite massifs from Lower Austria: Jour. Petrol., v. 37, p. 785–810.

______, 1996b, Geochemistry of garnet peridotite massifs from lower Austria and the composition of deep lithosphere beneath a Paleozoic convergent plate margin: Chem. Geol., v. 134, p. 49–65.

______, 1997, Sm-Nd garnet ages and cooling history of high-temperature garnet peridotite massifs and high-pressure granulites from lower Austria: Contrib. Mineral. Petrol., v. 127, p. 224–236.

Bohlen, S. R., and Boettcher, A. L., 1982, The quartz-coesite transformation: A precise determination and effects of other components: Jour. Geophys. Res., v. 87, p. 7073–7078.

Brenker, F. E., and Brey, G. P., 1997, Reconstruction of the exhumation path of the Alpe Arami garnet-peridotite body from depths exceeding 160 km: Jour. Metamor. Geol., v. 15, p. 581–592.

Brey, G. P., and Kohler, T., 1990, Geothermobarometry in four-phase lherzolites II. New thermobarometers, and practical assessment of existing thermobarometers: Jour. Petrol., v. 31, p. 1352–1378.

Brueckner, H. K., 1998, Sinking intrusion model for the emplacement of garnet-bearing peridotites into continent collision orogens: Geology, v. 26, p. 631–634.

Brueckner, H. K., Blusztajn, J., and Bakun-Czubarow, N., 1996, Trace element and Sm-Nd "age" zoning in garnets from peridotites of the Caledonian and Variscan Mountains and tectonic implications: Jour. Metamor. Geol., v. 14, p. 61–73.

Brueckner, H. K., and Medaris, L. G., Jr., 1998, A tale of two orogens: The contrasting T-P-t history and geochemical evolution of mantle in high- and ultrahigh-pressure metamorphic terranes of the Norwegian Caledonides and the Czech Variscides: Schweiz. Mineral. Petrogr. Mitt., v. 78, p. 293–307.

Bundy, F. P., 1980, The P,T phase and reaction diagram for elemental carbon, 1979: Jour. Geophys. Res., v. 85, p. 6930–6936.

Carswell, D. A., 1973, Garnet pyroxenite lens within Ugelvik layered garnet peridotite: Earth Planet. Sci. Lett., v. 20, p. 347–352.

______, 1968a, Picritic magma–residual dunite relationships in garnet peridotite at Kalskaret near Tafjord, south Norway: Contrib. Mineral. Petrol., v. 19, p. 97–124.

______, 1968b, Possible primary upper mantle peridotite in Norwegian basal gneiss: Lithos, v. 1, p. 322–355.

______, 1981, Clarification of the petrology and occurrence of garnet lherzolites, garnet websterites and eclogite in the vicinity of Rodhaugen, Almklovdalen, West Norway: Norsk. Geologisk. Tidskrift., v. 61, p. 249–260.

______, 1986, The metamorphic evolution of Mg-Cr type Norwegian garnet peridotites: Lithos, v. 19, p. 279–297.

______, 1991, Variscan high P-T metamorphism and uplift history in the Moldanubian Zone of the Bohemian Massif in Lower Austria: Eur. Jour. Mineral., v. 3, p. 323–342.

Carswell, D. A., Harvey, M. A., and Al-Samman, A., 1983, The petrogenesis of contrasting Fe-Ti and Mg-Cr garnet peridotite types in the high-grade gneiss complex of Western Norway: Bull. Mineral., v. 106, p. 727–750.

Carswell, D. A., and Jamtveit, B., 1990, Variscan Sm-Nd ages for the high-pressure metamorphism in the Moldanubian Zone of the Bohemian Massif, Lower Austria: Neues Jahrb. Mineral. Abh., v. 162, p. 69–78.

Cong, B., Zhang, R., Liou, J. G., Ye, K., and Wang, Q., 1996, Metamorphic evolution of UHPM rocks, *in* Cong, B., ed., Ultrahigh-pressure metamorphic rocks in the Dabieshan-Sulu Region of China: Dordrecht, Kluwer Acad. Publ., p. 128–160

Dallmeyer, R. D., Neubauer, F., Fritz, H., and Mocanu, V., 1998, Variscan vs. Alpine tectonothermal evolution of the Southern Carpathian orogen: Constraints from $^{40}Ar/^{39}Ar$ ages: Tectonophys., v. 290, p. 111–135.

Davies, G. R., Nixon, P. H., Pearson, D. G., and Obata, M., 1993, Tectonic implications of graphitized diamonds from the Ronda peridotite massif, southern Spain: Geology, v. 21, p. 471–474.

Doblas, M., and Oyarzun, R., 1989, "Mantle core complexes" and Neogene extensional detachment tectonics in the western Betic Cordilleras, Spain: An alternative model for the emplacement of the Ronda peridotite: Earth Planet. Sci. Lett., v. 93, p. 76–84.

Dobrzhinetskaya, L., Green, H. W., III, and Wang, S., 1996, Alpe Arami: A peridotite massif from depths of more than 300 kilometers: Science, v. 271, p. 1841–1845.

Engi, M., and Lindsley, D. H., 1980, Stability of titanian clinohumite: Experiments and thermodynamic analysis: Contrib. Mineral. Petrol., v. 72, p. 415–424.

Ernst, W. G., 1978, Petrochemical study of lherzolitic rocks from the western Alps: Jour. Petrol., v. 19, p. 341–392.

Evans, B. W., and Trommsdorff, V., 1978, Petrogenesis of garnet lherzolite, Cima di Gagnone, Lepontine Alps: Earth Planet. Sci. Lett., v. 40, p. 333–348.

Fiala, J., 1965, Pyrope of some garnet peridotites of the Czech Massif: Krystalinikum, v. 3, p. 55–74.

Fiala, J., and Padera, K., 1977, The chemistry of the minerals of the pyrope dunite from borehole T-7 near Staré (Bohemia): Tschermaks Mineral. Petrogr. Mitt., v. 24, 205–219.

Frey, F. A., Suen, C. J., and Stockman, H. W., 1985, The Ronda high temperature peridotite: Geochemistry and petrogenesis: Geochim. et Cosmochim. Acta, v. 49, p. 2469–2491.

Gardien, V., Tegyet, M., Lardeaux, J. M., Misseri, M., and Dufour, E., 1990, Crust-mantle relationships in the French Variscan chain: The example of the Southern Monts du Lyonnais unit, eastern French Massif Central: Jour. Metamor. Geol., v. 8, p. 477–492.

Gasparik, T., 1987, Orthopyroxene thermobarometry in simple and complex systems: Contrib. Mineral. Petrol., v. 96, p. 357–370.

Gebauer, D., Grunenfelder, M., Tilton, G., Trommsdorff, V., and Schmid, S., 1992, The geodynamic evolution of garnet peridotites, garnet pyroxenites, and eclogites of Alpe Arami and Cima di Gagnone, Central Alps, from early Proterozoic to Oligocene: Schweiz.. Mineral. Petrog. Mitt., v. 72, p. 107–111.

Godard, G., Martin, S., Prosser, G., Kienast, J. R., and Morten, L., 1996, Variscan migmatites, eclogites and garnet peridotites of the Ulten zone, eastern Austroalpine system: Tectonophys., v. 259, p. 313–341.

Hacker, B. R., Sharp, T., Zhang, R. Y., Liou, J. G., and Hervig, R. L., 1997, Determining the origin of ultrahigh-pressure lherzolites: Science, v. 278, p. 702–704.

Harley, S. L., 1984, An experimental study of the partitioning of Fe and Mg between garnet and orthopyroxene: Contrib. Mineral. Petrol., v. 86, p. 359–373.

Hauzenberger, C., Höller, W., Hoinkes, G., Klözli, U., and Thöni, M., 1993, Metamorphic evolution of the Austroalpine basement in the Nonsberg area, Ultental (Val d'Ultimo), Southern Tyrol [abs.]: Terra Nova, v. 5, p. 13.

Helmers, H., Maaskant, P., and Hartel, T. H. D., 1990, Garnet peridotite and associated high-grade rocks from Sulawesi, Indonesia: Lithos, v. 25, p. 171–188.

Iancu, V., and Maruntiu, M., 1994, Pre-Alpine litho-tectonic units and related shear zones in the basement of the Getic-Supragetic nappes, South Carpathians: Rom. Jour. Tect. Reg. Geol., v. 75, Suppl. 2, p. 87–92.

Jahn, B.-M., 1998, Geochemical and isotopic characteristics of UHP eclogites and ultramafic rocks of the Dabie Orogen: Implications for continental subduction and collisional tectonics, *in* Hacker, B. R., and Liou, J. G., eds., When continents collide: Geodynamics and geochemistry of ultrahigh-pressure rocks: Dordrecht, Kluwer Acad. Publ., p. 203–239.

Jamtveit, B., 1984, High-P metamorphism and deformation of the Gurskebotn garnet peridotite, Sunnmore, western Norway: Norsk. Geol. Tidsskr., v. 64, p. 97–110.

______, 1987a, Metamorphic evolution of the Eiksunddal eclogite complex, Western Norway, and some tectonic implications: Contrib. Mineral. Petrol., v. 95, p. 82–99.

______, 1987b, Magmatic and metamorphic controls on chemical variations within the Eiksunddal eclogite complex, Sunnmore, western Norway: Lithos, v. 20, p. 369–389.

Jamtveit, B., Carswell, D. A., and Mearns, E. W., 1991, Chronology of the high-pressure metamorphism of Norwegian garnet peridotites/pyroxenites: Jour. Metamor. Geol., v. 9, p. 125–139.

Kalt, A., and Altherr, R., 1996, Metamorphic evolution of garnet-spinel peridotites from the Variscan Schwarzwald, F. R. G.: Geol. Rundsch., v. 85, p. 211–224.

Kalt, A., Altherr, R., and Hanel, M., 1995, Contrasting P-T conditions recorded in ultramafic high-pressure rocks from the Variscan Schwarzwald, F.R.G.: Contrib. Mineral. Petrol., v. 121, p. 45–60.

Kossmatt, F., 1927, Gliederung des varistischen Gebirgsbaus: Abhandl. des Sächsisch. geolog. landesamts, v. 1, p. 1–39.

Krautner, H. G., Berza, T., and Dimitrescu, R., 1988, South Carpathians, *in* Zoubek, V., ed., Precambrian in younger fold belts: Chichester, U.K., Wiley, p. 639–663.

Krogh, E. J., and Carswell, D. A., 1995, HP and UHP eclogites and garnet peridotites in the Scandinavian Caledonides, *in* Coleman, R. G. and Wang, X., eds., Ultrahigh pressure metamorphism: New York, Cambridge Univ. Press, p. 244–298.

Lasnier, B., 1971, Les peridotites et pyroxenolites a grenat du Bois des Feuilles, Monts du Lyonnais, France: Contrib. Mineral. Petrol., v. 34, p. 29–42.

Liou, J. G., and Zhang, R. Y., 1998, Petrogenesis of an ultrahigh-pressure garnet-bearing ultramafic body from Maowu, Dabie Mountains, east-central China: Island Arc, v. 7, p. 115–134.

Liou, J. G., Zhang, R. Y., Ernst, W. G., Rumble, D., III, and Maruyama, S., 1998, High-pressure minerals from deeply subducted metamorphic rocks: Mineral. Soc. Amer., Rev. Mineral., v. 37, p. 33–96.

Loomis, T. P., 1972a, Contact metamorphism of pelitic rock by the Ronda ultramafic intrusion, southern Spain: Geol. Soc. Amer. Bull., v. 83, p. 2449–2474.

______, 1972b, Diapiric emplacement of the Ronda high-temperature ultramafic intrusion, southern Spain: Geol. Soc. Amer. Bull., v. 83, p. 2475–2496.

______, 1975, Tertiary mantle diapirism, orogeny, and plate tectonics east of the Strait of Gibraltar: Amer. Jour. Sci., v. 275, p. 1–30.

Lundeen, M. T., 1978, Emplacement of the Ronda peridotite, Sierra Bermeja, Spain: Geol. Soc. Amer. Bull., v. 89, p. 172–180.

Machart, J., 1984, Ultramafic rocks in the Bohemian part of the Moldanubicum and central Bohemian islet zone, Bohemian massif: Krystalinikum, v. 17, p. 13–32.

Maruntiu, M., Iancu, V., Ledru, P., and Johan, V., 1997, High-pressure parageneses in ultramafic rocks of the Getic-Supragetic basement, South Carpathians, Romania: Rom. Jour. Mineral., v. 78, Supl. no. 2, p. 54–55.

Mearns, E. W., 1986, Sm-Nd ages for Norwegian garnet peridotite: Lithos, v. 19, p. 269–278.

Medaris, L. G., Jr., 1980, Petrogenesis of the Lien peridotite and associated eclogites, Almklovdalen, western Norway: Lithos, v. 13, p. 339–353.

______, 1984, A geothermobarometric investigation of garnet peridotites in the Western Gneiss Region of Norway: Contrib. Mineral. Petrol., v. 87, p. 72–86.

Medaris, L. G., Jr., Beard, B. L., Johnson, C. M., Valley, J. W., Spicuzza, M. J., Jelinek, E., and Misar, Z., 1995, Garnet pyroxenite and eclogite in the Bohemian Massif: Geochemical evidence for Variscan recycling of subducted lithosphere: Geol. Rundsch., v. 84, p. 489–505.

Medaris, L. G., Jr., and Carswell, D. A., 1990, The petrogenesis of Mg-Cr garnet peridotites in European metamorphic belts, *in* Carswell, D. A., ed., Eclogite facies rocks: Glasgow, Blackie, p. 260–290.

Medaris, L. G., Jr., Wang, H. F., Mísar, Z. and Jelínek, E., 1990, Thermobarometry, diffusion modelling and cooling rates of crustal garnet peridotites: Two examples from the Moldanubian zone of the Bohemian Massif: Lithos, v. 25, p. 189–202.

Möckel, J. R., 1969, Structural petrology of the garnet-peridotite of Alpe Arami, Ticino, Switzerland: Leidse Geol. Meded., v. 42, p. 61–130.

Mørk, M. B. E., 1985a, A gabbro to eclogite transition on Flemsøy, Sunmøre, Western Norway: Chem. Geol., v. 50, p. 283–310.

______, 1985b, Incomplete high P-T metamorphic transitions within the Kvansøy pyroxenite complex, west Norway: A case study of disequilibrium: Jour. Metamor. Geol., v. 3, p. 245–264.

Mørk, M. B. E., and Mearns, E. W., 1986, Sm-Nd isotopic systematics of a gabbro-eclogite transition: Lithos, v. 19, p. 255–267.

Morten, L., and Obata, M., 1983, Possible high-temperature origin of pyroxenite lenses within garnet peridotite, northern Italy: Bull. Mineral., v. 106, p. 775–780.

Neubauer, F., and von Raumer, J., 1993, The Alpine basement—linkage between variscides and east-Mediterranean mountain belts, *in* von Raumer, J. F., and Neubauer, F., eds., Pre-Mesozoic Geology in the Alps: Berlin, Springer-Verlag, p. 641–663.

Obata, M., 1980, The Ronda peridotite: Garnet-, spinel-, and plagioclase-lherzolite facies and the P-T trajectories of a high-temperature mantle intrusion: Jour. Petrol., v. 21, p. 533–572.

Obata, M., and Morten, L., 1987, Transformation of spinel lherzolite to garnet lherzolite in ultramafic lenses of the Austridic crystalline complex, northern Italy: Jour. Petrol., v. 28, p. 599–623.

Obata, M., and Thompson, A. B., 1981, Amphibole and chlorite in mafic and ultramafic rocks in the lower crust and upper mantle: Contrib. Mineral. Petrol., v. 77, p. 74–81.

Ogasawara, Y., Liou, J. G., and Zhang, R. Y., 1995, Petrogenetic grid for ultrahigh-pressure metamorphism in the model system CaO-MgO-SiO_2_CO_2_H_2O: The Island Arc, v. 4, p. 240–253.

O'Hara, M. J., Richardson, S. W., and Wilson, G., 1971, Garnet-peridotite stability and occurrence in crust and mantle: Contrib. Mineral. Petrol., v. 32, p. 48–68.

Okay, A. I., 1994, Sapphirine and Ti-clinohumite in ultra-high-pressure garnet-pyroxenite and eclogite from Dabie Shan, China: Contrib. Mineral. Petrol., v. 116, p. 145–155.

O'Neill, H. St.C., 1980, An experimental study of Fe-Mg partitioning between garnet and olivine and its calibration as a geothermometer: Corrections: Contrib. Mineral. Petrol., v. 72, p. 337.

______, 1981, The transition between spinel lherzolite and garnet lherzolite, and its use as a geobarometer: Contrib. Mineral. Petrol., v. 77, p. 185–194.

O'Neill, H. St.C., and Wood, B. J., 1979, An experimental study of Fe-Mg partitioning between garnet and olivine and its calibration as a geothermometer: Contrib. Mineral. Petrol., v. 70, p. 59–70.

Parkinson, C. D., Miyazaki, K., Wakita, K., Barber, A. J., and Carswell, D. A., 1998, An overview and tectonic synthesis of the very high pressure and associated rocks of Sulawesi, Java, and Kalimantan, Indonesia: Island Arc, v. 7, p. 184–200.

Peacock, S. M., 1993, The importance of blueschist to eclogite dehydration reactions in subducting oceanic crust: Geol. Soc. Amer. Bull., v. 105, p. 684–694.

Pearson, D. G., Davies, G. R., and Nixon, P. H., 1993, Geochemical constraints on the petrogenesis of diamond facies pyroxenites from the Beni Bousera peridotite massif, North Morocco: Jour. Petrol., v. 34, p. 125–172.

______, 1995, Orogenic ultramafic rocks of UHP, diamond facies, origin, *in* Coleman, R. G., and Wang, X., eds., Ultrahigh pressure metamorphism: New York, Cambridge Univ. Press, p. 456–510.

Platt, J. P., and Vissers, R. L. M., 1989, Extensional collapse of thickened continental lithosphere: A working hypothesis for the Alboran Sea and Gibraltar arc: Geology, v. 17, p. 540–543.

Priem, H. N., Hebeda, E. H., Boelrijk, N. A., Verdurmen, T. E. A., and Oen, I. S., 1979, Isotopic dating of the emplacment of the ultramafic masses in the Serrania de Ronda, Southern Spain: Contrib. Mineral. Petrol., v. 70, p. 103–109.

Reisberg, L., Zindler, A., and Jagoutz, E., 1989, Further Sr and Nd isotopic results from peridotites of the Ronda Ultramafic Complex: Earth Planet. Sci. Lett., v. 96, p. 161–180.

Rey, P., Burg, J.-P., and Casey, M., 1997, The Scandinavian Caledonides and their relationship to the Variscan belt, *in* Burg, J.-P., and Ford, M., eds., Orogeny through time: Geol. Soc. Lond. Spec. Publ., v. 121, p. 179–200.

Schmädicke, E., and Evans, B. W., 1997, Garnet-bearing ultramafic rocks from the Erzgebirge, and their relation to other settings in the Bohemian Massif: Contrib. Mineral. Petrol., v. 127, p. 57–74.

Schmidt, M. W., and Poli, S., 1998, Experimentally based water budgets for dehydrating slabs and consequences for arc magma generation: Earth Planet. Sci. Lett., v. 163, p. 361–379.

Suen, C. J., and Frey, F. A., 1987, Origins of the mafic and ultramafic rocks in the Ronda peridotite: Earth Planet. Sci. Lett., v. 85, p. 83–202.

Takahashi, E., 1986, Melting of a dry peridotite KLB-1 up to 14 GPa: Implications on the origin of peridotitic upper mantle: Jour. Geophys. Res., v. 91, p. 9367–9382.

Tubia, J. M., and Cuevas, J., 1986, High temperature emplacement of the Los Reales peridotite nappe (Betic Cordillera, Spain): Jour. Struct. Geol., v. 8, p. 473–482.

Van der Wal, D., and Vissers, R. L. M., 1993, Uplift and emplacement of upper mantle rocks in the western Mediterranean: Geology, v. 21, p. 1119–1122.

______, 1996, Structural petrology of the Ronda peridotite, SW Spain: Deformation history: Jour. Petrol., v. 37, p. 23–43.

van Roermund, H., 1989, High-pressure ultramafic rocks from the allochthonous nappes of the Swedish Caledonides, *in* Gayer, R. A., ed., The Caledonide geology of Scandinavia: London, Graham and Trotman, p. 205–219.

van Roermund, H., and Drury, M. R., in press, Ultra-high pressure, P > 6GPa, garnet peridotites in western Norway: Exhumation of mantle rocks from >185 km depth: Terra Nova.

Yang, J., Godard, G., Kienast, J.-R., Lu, Y., and Sun, J., 1993, Ultra-high pressure, 60 kbar, magnesite-bearing garnet peridotites from northeastern Jiangsu, China: Jour. Geol., v. 101, p. 541–554.

Zeck, H. P., Monie, P., Villa, I. M., and Hansen, B. T., 1992, Very high rates of cooling and uplift in the Alpine belt of the Betic Cordilleras, southern Spain: Geology, v. 20, p. 79–82.

Zhang, R.-Y. et al., 1995, Petrology of ultrahigh-pressure rocks from the southern Su-Lu region, eastern China: Jour. Metamor. Geol., v. 13, p. 659–675.

______, 1997, Metamorphic evolution of diamond-bearing and associated rocks from the Kokchetav massif, northern Kazakhstan: Jour. Metamor. Geol., v. 15, p. 479–496.

Zhang, R. Y., and Liou, J. G., 1994, Significance of magnesite paragenesis in ultrahigh-P metamorphic rocks: Amer. Mineral., v. 79, p. 397–400.

Zhang, R. Y., Liou, J. G., and Cong, B., 1994, Petrogenesis of garnet-bearing ultramafic rocks and associated eclogites in the Su-Lu ultrahigh-P metamorphic terrane, eastern China: Jour. Metamor. Geol., v. 12, p. 169–186.

______, 1995, Talc-, magnesite- and Ti-clinohumite-bearing ultrahigh-pressure meta-mafic and ultramafic complex in the Dabie Mountains, China: Jour. Petrol., v. 36, p. 1011–1037.

Zhang, R. Y., Shu, J. F., Mao, H. K., and Liou, J. G., in press, Magnetite lamellae in olivine and clinohumite from Dabie UHP ultramafic rocks, east-central China: Amer. Mineral.

Zindler, A., Staudigel, H., Hart, S. R., Endres, R., and Goldstein, S., 1983, Nd and Sr isotopic study of a mafic layer from Ronda ultramafic complex: Nature, v. 304, p. 226–230.

Characteristics of UHP Pelites, Gneisses, and Other Unusual Rocks

ROBERTO COMPAGNONI AND FRANCO ROLFO

Dipartimento di Scienze Mineralogiche e Petrologiche, Via Valperga Caluso 35, I-10125 Torino, Italy

Abstract

The main felsic lithologies experiencing ultrahigh-pressure metamorphism (UHPM) are described, with a focus on rock association, nature, and composition of the peak minerals, as well as nature of the protolith. Felsic rocks usually preserve poor evidence of UHPM, being pervasively deformed and retrogressed during exhumation. Locally, however, unstrained lithologies are preserved that show UHPM minerals and even microstructures inherited from the protolith. Among the number of localities where UHPM has been reported, the following have been considered in detail—the southern Dora-Maira massif and Lago di Cignana in the Western Alps, the Dabie Shan–Sulu region in northeastern China, the Kokchetav massif in northern Kazakhstan, and the Western Gneiss Region of the Scandinavian Caledonides.

At Lago di Cignana, metasediments of oceanic affinity are carbonate-bearing garnet-phengite-quartz schist and oxidized manganiferous quartz schists, all preserving rare coesite relics. In the Dora-Maira massif, the Brossasco-Isasca Unit includes coesite-bearing phengite-jadeite-almandine-quartz micaschist, pyrope whiteschist, jadeite-kyanite quartzite, sodic whiteschist, and jadeite-almandine orthogneiss. In the Dabie Shan–Sulu region, felsic rocks with clear evidence of UHPM include jadeite quartzite, talc-kyanite quartzite, and other kyanite-bearing quartzites. Moreover, other lithotypes like aegirine orthogneiss, paragneiss, and micaschist also may have been subjected to UHPM. In the Kokchetav massif, microdiamond-bearing rocks include garnet-biotite gneisses and schists, garnet-pyroxene and pyroxene-carbonate-garnet rocks, and possibly also whiteschists. Other localities where UHPM felsic rocks have been reported include the northern Tien Shan, northern Mali, the Maksyutov Complex in the southern Urals, and the Western Gneiss Region of the southwestern Caledonides.

In felsic rocks, the most "common" UHP minerals are coesite, clinopyroxenes, garnets, epidotes, kyanite, phengite, talc, aragonite, dolomite, and accessory rutile, titanite, apatite, tourmaline, and zircon. Exceptionally, rare or unique phases also may be found as inclusions armored in more competent minerals. In the highest-P areas, microdiamond occurs, its growth being strongly controlled by the composition of the fluid phase. Mineral chemistry is sensitive to both bulk-rock chemical composition and P-T conditions: garnets are solid solutions of the pyralspite series, the Mn-rich members being high in calderite; pyroxenes are jadeite-diopside solid solutions; epidotes include allanite, zoisite, clinozoisite-epidote solid solutions, and piemontite; and micas are highly celadonitic phengites. In rocks with microdiamond, extreme UHPM conditions are further suggested by the unusual composition of some minerals, such as garnet high in Si, Y, and Yb, or Na-pyroxenes rich in K.

Clear evidence of partial melting during UHMP is missing, although peak conditions are close to or above the minimum granite wet solidus. This may be due to either the $a_{H2O} < 1$, or to melt extraction.

Introduction

UNLIKE OTHER more competent lithologies such as metabasics, where the eclogite-facies minerals are usually in part at least preserved or still recognizable from their pseudomorphous breakdown products (e.g., diopside + plagioclase symplectite after omphacite), felsic rocks are much more pervasively retrogressed during exhumation to greenschist- to amphibolite-facies mineral assemblages, eliminating traces of the former UHPM. In fact, the most significant index minerals, such as coesite and jadeite, are very sensitive to decompression and are quickly converted to quartz and albite, respectively. This pervasive retrogression is further enhanced by the internal strain, induced during exhumation by the volume increase (~10%) connected to the inversion of coesite to quartz, a weak mineral under a wide range of metamorphic conditions.

Therefore, UHPM key minerals (especially coesite) normally are preserved only as tiny inclusions, armored in more resistant, but less significant, minerals, such as garnet or zircon. Consequently, the study of UHPM in felsic rocks requires sampling of a large number of specimens, which must be studied in detail with sophisticated techniques.

However, exceptionally unstrained lithologies may be found, which show well-preserved UHPM minerals and even microstructures inherited from an older protolith: the best examples are the metagranitoids and related intrusive contacts with hornfelses and xenoliths of the Brossasca-Isasca Unit of the southern Dora-Maira massif, Western Alps (Biino and Compagnoni, 1992), and the metagranitoids of the Sulu area, eastern China (Hirajima et al., 1993). This suggests that in the UHPM terranes, deformation may be strongly partitioned—i.e., low-strain rock portions are bounded by high-strain (shear) zones, which are repeatedly reactivated and recrystallized, destroying traces of the former UHPM.

Felsic rocks still preserving UHP peak assemblages are reported from a number of localities, but available gelogic, structural, petrographic, and mineralogical data differ significantly from place to place. This review reflects the state of the art of a body of research, however, that is evolving rapidly. The collisional orogenic belts considered here are two Penninic units of the southern Dora-Maira massif (continental crust) and Lago di Cignana, the Upper Valtournenche (oceanic crust) in the Western Alps, the Dabie-Sulu region in northeastern China, the Kokchetav massif in northern Kazakhstan, and the Western Gneiss Region of the Scandinavian Caledonides.

Lago di Cignana, Upper Valtournenche, and Val d'Aosta—Western Italian Alps

This tectonic slice, less than 300 m thick and exposed over an area of ~2 km^2, crops out within the metaophiolites and related metasedimentary cover of the Penninic Piemonte Zone close to the tectonic contact with the overlying Austroalpine Dent Blanche continental crust nappe (Reinecke, 1991, 1998; Reinecke et al., 1994). Thus far it is the only UHPM unit characterized by rocks of oceanic affinity. It includes MORB-type metabasalts (now coesite-eclogite) and metasediments with a compositional range between quartzite (locally Mn-rich) to calc-schist. This sequence, considered to have been derived from the Mesozoic Tethyan oceanic crust, has been metamorphosed under coesite/eclogite-facies conditions during the Alpine orogeny in 44 Ma (Amato et al., 1977; Rubatto et al., 1998). The peak metamorphic conditions have been estimated to be 615 ± 15°C and 2.8 ± 1.0 GPa (Reinecke, 1991, 1998; Reinecke et al., 1994).

The metasediments mainly consist of the following two lithotypes: (1) carbonate-bearing garnet-phengite-quartz schist; and (2) oxidized manganiferous quartz schists and quartzite. Rare relict coesite (or retrograde polycrystalline quartz aggregates) occurs as very fine grained inclusions in garnet, piemontite, tourmaline, and epidote of most lithologies. Preserved coesite also has been found as inclusions in matrix apatite in a garnet-phengite-quartz schist (Reinecke, 1998, p. 163).

The carbonate-bearing garnet-phengite-quartz schists, which derive from a carbonate-rich metapelite protolith, consist of carbonate-rich layers alternating with phengite-quartz-rich layers. The inferred UHP peak assemblage of one calc-schist analyzed in detail (sample Cig 90-25) (Reinecke, 1998, p. 163) is garnet + dolomite + coesite + (lawsonite) + phengite ($Si_{3.35\text{-}3.3.4}$) + (aragonite) + rutile. Lawsonite is inferred from the widespread occurrence of millimeter-sized, lozenge-shaped pseudomorphs consisting mainly of zoisite/clinozoisite + paragonite aggregates. Other samples of garnet-phengite-quartz schists at the UHP stage may have had clinozoisite/epidote (but no lawsonite) ± sodic amphibole (as shape relics, pseudomorphed by Ca-Na-amphibole, biotite, and albite), apatite, and tourmaline. The usual UHP garnet composition in the Mn-poor silicate domain of sample Cig 90-25 (Reinecke, 1998, Table 1) is ~$Prp_7Alm_{71}Sps_3Grs_{19}$.

The oxidized manganiferous quartz-schists and quartzite, in turn, include two lithotypes: (1) piedmontite-garnet–phengite-talc-quartz schists, and (2) aegirine-augite–garnet-phengite quartzites. The piemontite-garnet–phengite-talc-quartz schists were derived from a manganiferous (MnO typically in the range of 1 to 4 wt%), SiO_2-rich metapelite, and include scattered millimeter- to centimeter-sized hematite-braunite–rich aggregates, most likely derived from original nodules rich in Fe and Mn.

In the "Al-rich" domains, the inferred UHPM peak assemblage is coesite + kyanite + talc + phengite + garnet + hematite + rutile + paragonite + piemontite + zircon, with the garnet zoned from $Prp_{19}Sps_{67}Grs_{14}$ in the core through $Prp_{36}Sps_{55}Grs_9$ in the intermediate to $Prp_{21}Sps_{70}Grs_9$in the rim. In the "Al-poor" domains, the inferred UHPM peak assemblage is coesite + phlogopite + talc + phengite + garnet + piemontite + hematite + rutile + braunite + ardennite + dravite + apatite + paragonite + zircon, with the garnet zoned from $Prp_{21}Sps_{67}Grs_{12}$ (core) through $Prp_{35}Sps_{55}Grs_{10}$ (intermediate) to $Prp_{20}Sps_{17}Grs_9$ (rim).

The aegirine-augite–garnet-phengite quartzites are chemically similar to the previous schists, but are higher in Na and lower in Al. The inferred UHPM peak assemblage is garnet I + phengite ($Si_{3.38}$) + coesite + manganian epidote + aegirine-augite + hematite + rutile ± apatite ± dravite. The aegirine-augite composition is in the range $Jd_{26-36}Aeg_{42-51}Aug_{13-24}$ (Fig. 1). It is interesting to note that in the Al-poor and hematite-rich domains (the former Fe-Mn nodules) of the aegirine-augite-garnet-phengite quartzite, a calderitic ($Mn_3Fe_2Si_3O_{12}$) garnet with composition $Calderite_{44-83}Adr_{5-14}Sps_{47-7}Prp_1$ (Fig. 1) was found, which is peculiar to high-pressure conditions (Lattard and Schreyer, 1983).

All lithologies show a polyphase deformation (van der Klauw et al., 1997) and variable greenschist-facies retrogression. The decomposition products are poikiloblastic garnet II $[Sps_{60-84}Cald_{0-15}Prp_{0.5-2.0}(Grs + And)_{12-30}]$ ± manganocummingtonite (=tirodite), which overgrows garnet I and/or replaces piemontite, albite, calcite, and phlogopite. The aegirine-augite is partially replaced by very fine grained symplectites consisting either of microcline ($Or_{95}Ab_4Cls_1$) + albite ($Ab_{99}Or_1$) + manganoan phogopite + hematite + sodian tremolite or albite + microcline + hematite + sodian tremolite and a less jadeitic aegirine-augite ($Jd_{4-17}Aeg_{29-49}Aug_{42-67}$) (Fig. 1).

The Brossasco-Isasca Unit, Dora-Maira Massif—Western Italian Alps

This unit is of paramount importance for petrology of the UHPM, since the first occurrence of coesite in continental crust was reported by Chopin (1984) from a pyrope whiteschist (see below). The unit, referred to here as the "Brossasco-Isasca Unit" (BIU) (Compagnoni et al., 1995), is exposed in the southern Dora-Maira massif (DMM) Penninic domain of the Western Alps (see also Chopin and Schertl, 1999). It is a small tectonic sheet, approximately 10 × 4 × 1 km, sandwiched between two units and characterized by quartz-eclogite–facies recrystallization (~550°C and ~15 kbar) (Chopin et al., 1991; Schertl et al., 1991). Detailed petrologic descriptions of the various geological units of the southern DMM and inferred P-T paths are presented in a series of articles (e.g., Chopin, 1984; Chopin et al., 1991; Schertl et al., 1991; Michard et al., 1995). The early Alpine UHPM, formerly considered to have occurred at ~110 Ma (cf. Hunziker et al., 1992) on the basis of conventional radiometric methods, recently has been dated at ~35 Ma by Gebauer et al. (1997, with references therein), combining cathodoluminescence and SHRIMP techniques on zircon single crystals. The Alpine coesite-eclogite–facies peak conditions were estimated at ~750°C and ~35 kbar (Schertl et al., 1991; Compagnoni et al., 1995).

Geology

The BIU is a coherent slice of Penninic continental crust, originally composed of a Variscan amphibolite-facies metamorphic basement (Compagnoni et al., 1995) intruded 275 Ma by porphyritic granitoids (Gebauer et al., 1997), as indicated by local preservation of relict pre-Alpine igneous and metamorphic structures and mineralogies (Biino and Compagnoni, 1992; Compagnoni et al., 1994, 1995). During the Alpine orogeny, the BIU sustained an early Alpine coesite-eclogite– and a late Alpine greenschist-facies recrystallization, and the Variscan amphibolite-facies basement and late Variscan granitoids were converted into the present "Polymetamorphic Complex" and "Monometamorphic Complex," respectively (Compagnoni et al., 1995).

The Polymetamorphic Complex consists of a wide range of lithologies, including gneisses and micaschists with minor marbles and eclogites (Chopin et al., 1991; Compagnoni et al., 1994, 1995). The felsic lithotype that best preserves the UHP mineral assemblage is a phengite-jadeite-almandine-quartz/(coesite) micaschist.

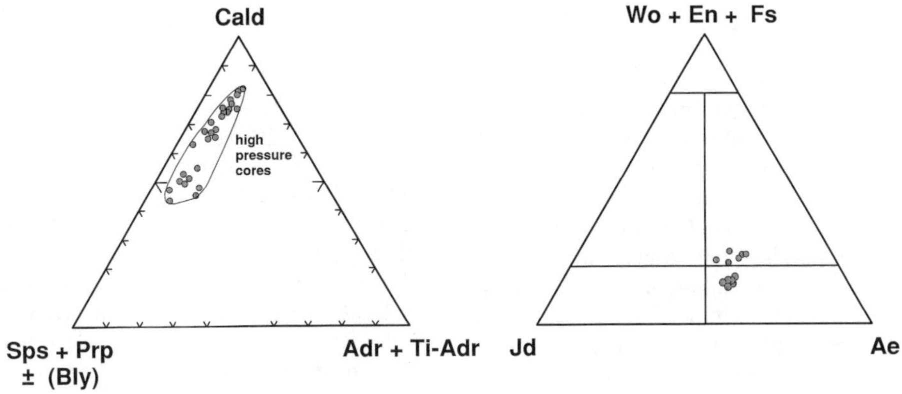

FIG. 1. Compositions of UHPM garnets and clinopyroxenes from the garnet-phengite-epidote-Na-pyroxene quartzite, Lago di Cignana. Mineral symbols according to Kretz (1983). Abbreviations: Cald = calderite; Bly = blythite.

The Monometamorphic Complex consists mainly of medium- to fine-grained greenschist facies, typically mylonitic orthogneiss, with local augen structure, where the only evidence of the early eclogite-facies recrystallization is porphyroclastic phengite, enveloped by upper-greenschist-facies foliation, and aggregates of titanite (after former rutile) + grossular-rich garnet. However, the orthogneiss, including slightly deformed relics of the late-Variscan granitoids with xenoliths of the country rocks, escaped polyphase Alpine pervasive deformation.

The most peculiar lithologies, best preserving the UHP mineral assemblages, are lens-like bodies of pyrope whiteschist with minor jadeite-kyanite quartzite, sodic whiteschist, and jadeite-almandine orthogneiss. Relict coesite or its polycrystalline quartz pseudomorphs are found in most lithologies as inclusions in garnet, kyanite, jadeite, omphacite, zoisite, and even in zircon (Sobolev et al., 1994).

Petrography

Quartz/(coesite)-jadeite-garnet-kyanite-phengite micaschist. This micaschist has the peak assemblage: quartz/(coesite) + phengite + porphyroblastic garnet and kyanite + jadeite + minor zoisite and accessory rutile, apatite, aggregates of yellow-greenish tourmaline, and local bearthite ($Ca_2Al[PO_4]_2OH$) (Chopin et al., 1993). Prograde compositional zoning of porphyroblastic garnet, up to 2 cm across, is especially interesting, from $Alm_{80}Prp_{20}$ in the core to $Alm_{60}Prp_{40}$ in the rim, accompanied by a change in the mineral inclusions, which are chlorite, chloritoid (up to ~50 mole% of the Mg end member), staurolite, kyanite, jadeite, paragonite, and quartz in the core, but only jadeite, kyanite, and coesite in the rim (Chopin et al., 1991). Both garnet zoning and distribution of mineral inclusions indicate a growth under prograde conditions from the quartz to the coesite stability field.

As a rule, jadeite (Jd_{90-95}), usually retrogressed to albite and paragonite in the matrix, is still preserved within the garnet. Almandine is partially altered to biotite and/or chlorite, and kyanite is replaced by paragonite and locally by an earlier margarite. The retrogression product is a greenschist-facies micaschist consisting mainly of the assemblage albite + muscovite + biotite + chlorite + clinozoisite/epidote.

Pyrope whiteschist. Pyrope whiteschist, really a kyanite-phengite-pyrope-quartz/(coesite) granofels, occurs within the orthogneiss of the Monometamorphic Complex as layers from a few centimeters to ~20 meters thick and from a few meters to hundreds of meters long (Chopin, 1984). The whiteschist mainly consists of pyrope embedded in a matrix of quartz (inverted from former coesite), 3T-polytype phengite, kyanite, minor talc, rare jadeite, and accessory rutile, zircon, and monazite. Pyrope grain size is variable, from a few millimeters to ~20 cm across; the largest crystals exhibit a trapezohedron habit, usually rounded by the polyphase post-crystalline deformation. Pyrope is poorly zoned with a reddish core and pale pinkish rim, with composition up to Prp_{96-98}. Pyrope megablasts are systematically crowded with minerals, including prograde phases such as Mg-chlorite, talc, and paragonite; phases stable at peak conditions, such as coesite; and retrograde phases, most likely formed during early decompression, such as corundum, enstatite, and gedrite. In addition to these more or less common minerals, pyrope megablasts also preserve inclusions of rare minerals, such as magnesiostaurolite, or new minerals, such as bearthite, magnesiodumortierite, and ellenbergerite (for a review, see Compagnoni et al., 1995; Chopin and Schertl, 1999). The whiteschist retrogression is controlled mainly by pyrope alteration, which occurred principally in two steps—an initial one producing the assemblage phlogopite + kyanite + quartz, and a second one characterized by the assemblage chlorite + kyanite + muscovite + quartz (Schertl et al., 1991).

The whiteschist, initially considered as the metamorphic product of original Mg-rich pelite in an evaporitic environment (Schertl et al., 1991; Chopin, 1987), is now believed to have been derived from former late-Variscan granitoids through a metasomatic process that occurred along ductile shear zones in the presence of a hydrous fluid phase either during Permo-Triassic time (Gebauer et al., 1997) or during the prograde Alpine tectonometamorphic evolution (Compagnoni and Hirajima, 1992, 1993). More detailed structural, mineralogic, and chemical descriptions of the whiteschist are provided in the review papers by Chopin and Sobolev (1995), Compagnoni et al. (1995), Schertl et al. (1991), Michard et al. (1995), and Chopin and Schertl (1999).

Jadeite-kyanite quartzite. This peculiar lithotype, originally described as a "weathered bluish rock devoid of mica" by Chopin (1984), was later reported as a "jadeite-kyanite quartzite" by Chopin (1987) and as a "garnet-jadeite-quartzite or garnet-jadeite-quartz band" by Schertl et al. (1991). For the sake of brevity, the short definition jadeite-kyanite quartzite is preferred here. The jadeite-kyanite quartzite, really a pyrope-jadeite-kyanite-quartz/coesite granofels, occurs within a few pyrope whiteschists as folded layers from 10 to 20 cm thick and several meters long. The UHPM peak assemblage consists of coesite + jadeite + kyanite + garnet + minor phengite and local talc, with accessory rutile, apatite, and zircon. Coesite is preserved as inclusions in garnet, jadeite, and kyanite. Possibly prograde inclusions in garnet other than the peak minerals are paragonite, colorless glaucophane, and wagnerite ($Mg_2[PO_4](F, OH)$). Very small crystalline inclusions of intergrown white mica and chlorite, filling cavities with the negative shape of the host-garnet trapezohedron, most likely correspond to former UHP fluid inclusions that later were decrepitated. Garnet is a pyrope with composition $Prp_{70\text{-}80}Alm_{20\text{-}30}Grs_{1\text{-}5}$ (Fig. 2); jadeite is very pure with average composition $Jd_{95}Di_4Ae_1$ (Fig. 2); and phengite is a 3T polytype with Si from 3.41 atoms per formula unit (a.p.f.u.) in the core to 3.25 in the rim (Fig. 2). The jadeite-kyanite quartzite is retrogressed to a fine-grained gneiss consisting of quartz, muscovite, and paragonite, albite/oligoclase, and minor clinozoisite.

This unusual rock type has been considered by Schreyer et al. (1987) and Sharp et al. (1993) as the possible product of partial melting of the host whiteschist protolith, but convincing evidence for this interpretation is still lacking.

Sodic whiteschist. A peculiar variety of whiteschist, named sodic whiteschist, is a Mg-Al-Na–rich lithotype, collected in the debris close to Case Ramello near Martiniana by R. Compagnoni and studied by Kienast et al. (1991). It is a coarse-grained rock consisting of the UHPM peak assemblage: pyrope-rich garnet ($Prp_{76}Alm_{22}Grs_2$) + jadeite ($Jd_{82}Di_{18}$) + glaucophane (close to the pure Mg-Al end member) + phengite (Si = 3.6) + coesite + rutile. This lithotype is quite interesting, since it supports experimental data that indicate that at least in the NMASH system, pure glaucophane is stable at ~33 kbar and ~750°C.

Jadeite-almandine orthogneiss. This lithotype, really a phengite-jadeite-almandine-quartz/(coesite) granofels, occurs as decimeter- to meter-thick layers within both the Monometamorphic and the Polymetamorphic complexes. The UHPM peak assemblage is coesite + jadeite (Jd_{95}) + almandine-rich garnet ($Alm_{70\text{-}80}Prp_{10\text{-}20}Grs_{5\text{-}20}$) + minor phengite (Si = 3.25–3.41) + accessory apatite, rutile, and zircon. This rock is distinguished from the jadeite-kyanite-quartzite by a lack of kyanite and an almandine-rich garnet composition. The jadeite-almandine orthogneiss exhibits polyphase retrogression, described in detail by Hirajima and Compagnoni (1993), which includes albite/oligoclase, a series of Na pyroxenes progressively decreasing in the jadeite component from omphacite (J_{55}) to aegirine (Jd_0), and

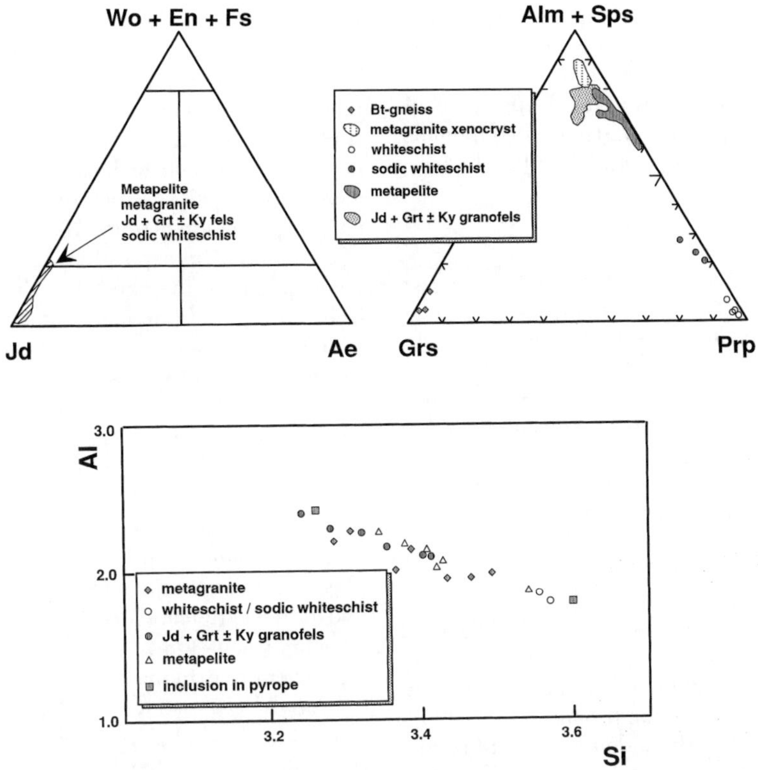

FIG. 2. Compositional range of UHP garnets, clinopyroxenes, and phengites from representative lithotypes from the Brossasco-Isasca Unit, Dora Maira Massif. Data from Chopin et al. (1991), Schertl et al. (1991), Kienast et al. (1991), Compagnoni and Hirajima (1993), Hirajima and Compagnoni (1993), and unpublished data.

zoned amphiboles ranging in composition from ferronyböite to taramite. Minor Mg-rich chlorite and zoisite also were reported by Chopin et al. (1991). On the basis of composition and field occurrence, this lithotype most likely is the UHP metamorphic product of original granitoid dikes (Chopin et al., 1991).

Metagranitoids and related rocks. Several tens of lens-like bodies of relict metagranitoids, from a few to several meters in length, that escaped polyphase Alpine deformation occur within the greenschist-facies orthogneiss of the Monometamorphic Complex. At the contact between the Monometamorphic and Polymetamorphic complexes, remnants of xenoliths within metagranite and of the Variscan thermally metamorphosed country paragneiss in the metamorphic basement are preserved locally. For perfect preservation of primary microstructures and development of only pseudomorphous and coronitic reactions, the UHP transformations of original igneous and metamorphic minerals may be studied in detail. One interesting result of such studies, carried out by Biino and Compagnoni (1992), Compagnoni et al. (1994, 1995), and Bruno (1998), is evidence that the UHP recrystallization occurred in a closed system, in agreement with a geochronologic study by Tilton et al. (1997) showing "that the minerals of the undeformed metagranites failed to reach isotopic equilibration during Alpine subduction to >100 km with temperatures >700°C."

Dabie-Sulu Region

In the Dabie-Sulu region of east-central China, most UHPM rocks described thus far are meta-igneous rocks including basaltic eclogite and garnet-bearing ultramafics (Wang et al., 1995; Liou et al., 1996). Field and petrologic evidence, however, suggests that some country rocks, such as metagranitoid, gneiss, and marble, underwent

UHPM, but lost their mineral assemblage during the subsequent exhumation history. Among the many lithologies that may have experienced UHPM, the rocks best preserving the peak assemblage are: garnet-jadeite-quartz/coesite granofels, talc-kyanite quartzites, and kyanite-bearing quartz-rich rocks.

Jadeite quartzite

The most peculiar and widespread felsic rock showing a UHPM mineral assemblage is a jadeite-quartzite, really a garnet-jadeite-quartz/coesite-granofels, since the quartz content does not exceed 75 vol%. This rock occurs in the eastern Dabie Mountains as an E-W–trending belt 1 to 2 km wide and ~40 km long. The jadeite-quartzite has been studied in a number of localities, the most important being exposures at Shuanghe/Miaozhu (Okay, 1993; Cong et al., 1995; Su et al., 1996; Liou et al., 1997) and Xinjian (Rolfo et al., 1997; Rolfo, 1998). Similar, although strongly retrogressed rocks interlayered with eclogite and minor kyanite quartzite have been described at Caihu and Hetang in the Donghai area of the Sulu region (Zhang et al., 1995).

Field occurrence. The jadeite quartzite occurs as lens-like bodies, up to hundreds of meters long, closely associated with marble and two mica–epidote gneisses and minor eclogite bands and boudins. The whole-rock association is enclosed within quartzofeldspathic granitic gneiss. Most jadeite quartzite is medium-grained and isotropic, mimicking a granitoid texture, but types with an apparent foliation also occur, defined by the preferred dimensional orientation of jadeite and monomineralic quartz stringlets. A weak compositional banding is locally evident, produced by enrichment in jadeite or garnet. Quite often, decimeter-sized quartz-bearing garnetite or quartz-rich eclogite nodules occur, as centimeter-thick variably transposed metamorphic veins rich in quartz and minor zoisite, phengite, and rutile. Even in the less retrogressed samples, the jadeite granoblasts are surrounded by a plagioclase + magnetite-rich black corona.

Petrography. The jadeite quartzite consists of quartz (45 to 60 vol%), jadeite (35 to 50 vol%), garnet (0 to 15 vol%), and accessories. The texture is generally granoblastic, but quartz is finer grained than jadeite and garnet, suggesting its inversion from former coesite. The UHP peak assemblage is coesite + jadeite + garnet ± zoisite, with accessory rutile, apatite, and zircon. Fresh coesite inclusions, first described in jadeite at Shuanghe (Su et al., 1996), are common in both jadeite and garnet at Xinjian (Rolfo et al., 1997). Inclusions of Mg-taramite, jadeite, rutile, and paragonite ± monazite occur in garnet, and inclusions of paragonite, garnet, rutile, apatite, and zircon are present in jadeite. At Xinjian (Rolfo et al., 1997), amphibole porphyroblasts zoned from glaucophane in the core through crossite to Mg-riebeckite in the rim, locally grow at the expense of jadeite. A weak S_i, defined by alignments of rutile, zircon, and apatite, is locally observed both in jadeite and in the matrix polygonal quartz aggregate. A small idioblastic epidote, zoned from clinozoisite in the core to Fe-richer epidote in the rim, locally occurs. In the jadeite quartzite of Donghai (Zhang et al., 1995), clinozoisite includes paragonite and is partially replaced by plagioclase. In the jadeite quartzite, both jadeite and garnet systematically show a polyphase retrogression (Liou et al., 1997; Rolfo, 1998), recording the entire exhumation path.

Mineral chemistry. Representative compositions of the UHPM peak minerals from the jadeite quartzite are reported in Figure 3. Garnets are rich in almandine and show no significant compositonal zoning. In the Dabie Mountains, garnets from Xinjiang ($Alm_{59\text{-}64}Prp_{27\text{-}32}Grs_{7\text{-}10}$ on average; Rolfo et al., 1997) and Shuanghe ($Alm_{59\text{-}62}Prp_{20\text{-}22}Grs_{17\text{-}19}Sps_{0.5}$; Cong et al., 1995; Liou et al., 1997) have similar compositions. Garnets from similar rocks from Hetang and Caihu in the Sulu terrane (Zhang et al., 1995) have lower almandine and higher grossular, pyrope, and spessartine components, most likely reflecting differences in both bulk chemical composition and peak metamorphic conditions, lower in the Dabie Mountains. Clinopyroxene is a jadeite, lacking any systematic compositional zoning, with a composition of $Jd_{82\text{-}85}Aeg_{1\text{-}9}Di_{3\text{-}7}$ in Xinjian (Rolfo et al., 1997) and $Jd_{82\text{-}87}Aeg_{1\text{-}4}Di_{8\text{-}9}$ in Shuanghe (Cong et al., 1995; Liou et al., 1997). The prograde blue-green amphibole, included in garnet, is a Mg-taramite with low Si (6.00 to 6.39 a.p.f.u.), X_{Mg} = 0.64 to 0.65, high Al^{IV} (1.62 to 1.99 a.p.f.u.), and Al^{VI} (1.41 to 1.52 a.p.f.u.). At Xinjian, amphibole porphyroblasts, most likely developed during the earliest decompressional phases, are zoned from glaucophane in the core through crossite to Mg-riebeckite in the rim.

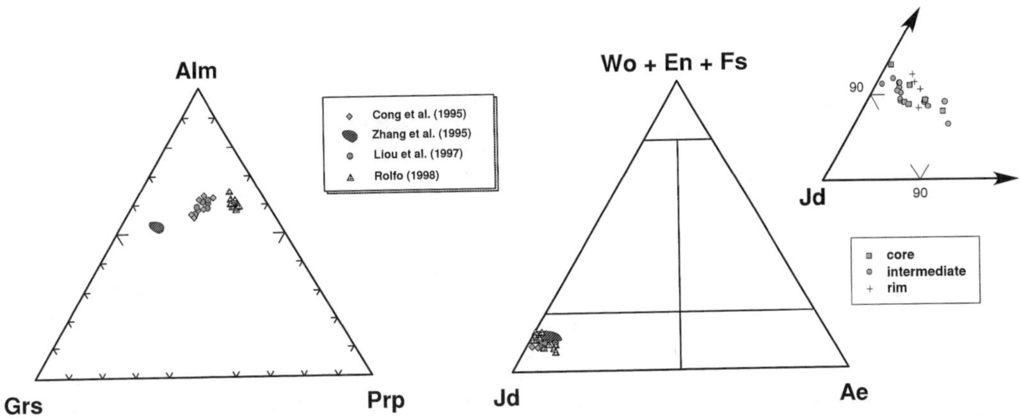

FIG. 3. Compositions of UHP peak garnet and clinopyroxene from jadeite quartzite from Shuanghe (Cong et al., 1995; Liou et al., 1997) and Xinjian (Rolfo, 1998), Dabie Shan, and Donghai Sulu, China (Zhang et al., 1995).

The protolith. Although in most discriminant chemical diagrams, the jadeite quartzites plot at the boundary between igneous and sedimentary rocks, a sedimentary protolith usually is assumed (Zhang et al., 1995)—a graywacke for Su et al. (1996), a psammitic sediment for Cong et al. (1995), and an albitized siltstone enriched in Na and depleted in K and Ca for Liou et al. (1997). The close association with paragneiss, marble, and eclogite also supports a sedimentary origin.

Talc-kyanite quartzite

Field occurrence. The talc-kyanite quartzite, really an omphacite-talc-kyanite-epidote quartzite, has been reported only from the Hualiangting Reservoir in the coesite-eclogite zone of the southern Dabie Complex, a few hundred meters north of the presumed boundary with the quartz-eclogite zone (Rolfo et al., 1997). It is a whitish, weakly deformed layer, ~30 cm thick, which cuts across a meter-sized boudin of partially amphibolitized eclogite. Along the sharp contacts, the host eclogite typically is rich in randomly oriented epidote porphyroblasts. The eclogite boudin is interlayered with meter-thick white mica + epidote schists, locally rich in magnetite.

Petrography. The talc-kyanite quartzite is fine-grained and consists of quartz (~70 vol%), minor kyanite, talc, epidote, omphacite, and accessory rutile, apatite, and zircon. Lozenge-shaped Fe-rich (Fe^{3+} = 0.9 to 1.4 a.p.f.u.) epidote idioblasts include abundant quartz (typically polycrystalline aggregates after coesite) and less abundant kyanite, omphacite, and rutile. Talc occurs as randomly oriented porphyroblasts, several millimeters long, rimmed by tremolite with an outer thin rim of cummingtonite. Talc often includes kyanite and epidote. Fresh omphacite is commonly preserved only as inclusions in kyanite and epidote, being otherwise replaced by a very fine-grained symplectite of amphibole (magnesio- to actinolitic hornblende) + quartz ± oligoclase. Kyanite may be found either as: (1) millimeter-long idioblasts, including polycrystalline quartz aggregates after coesite, epidote, and minor omphacite, or (2) prismatic aggregates, clearly pseudomorphous after a pre-existing mineral, consisting of smaller crystals associated with quartz ± epidote. Kyanite commonly is rimmed by a thin corona of oligoclase ($An_{12\text{-}25}$) ± epidote. Kyanite aggregates and idioblastic epidote define a rough foliation, parallel to the talc-kyanite quartzite selvages; this foliation is crosscut by talc flakes.

Mineral chemistry. Representative compositions of minerals are plotted in Figure 4. Clinopyroxene is a quite homogeneous omphacite with average composition $Di_{25}Jd_{52}Aeg_{10}$, the jadeite content varying in the range X_{Jd} = 0.44-0.60. Omphacite is replaced by a clinopyroxene + plagioclase symplectite, where the pyroxene is an aegirine-rich omphacite with an average composition $Di_{60}Jd_{15}Aeg_{10}$. Epidote is iron-rich (Fe^{3+} = 0.78–1.15 a.p.f.u.). Strong compositional zoning is common, with an idioblastic core richer in iron (Fe^{3+} >1.1 a.p.f.u.) than is the rim (Fe^{3+} <1.0 atoms p.f.u.). Usually, the epidote cores,

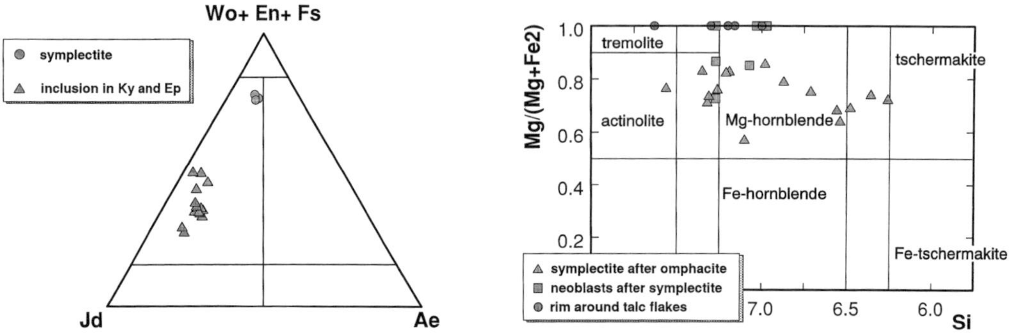

FIG. 4. Compositions of peak and retrograde clinopyroxenes from the talc-kyanite quartzite of Dabie Shan in the jadeite-aegirine-(wollastonite + enstatite + ferrosilite) diagram of Morimoto et al. (1988), and compositions of amphiboles from the talc-kyanite quartzite of Dabie Shan in the diagram of Leake (1978) for calcic amphiboles.

which include quartz pseudomorphs after coesite, contain 3 to 5 wt% REE, especially Nd and Ce. Talc is homogeneous, with low Al (0.02 to 0.05 a.p.f.u.), Na (<0.06 a.p.f.u.), and Fe (0.10-0.07 a.p.f.u.).

The protolith. On the basis of the bulk-rock chemical composition, it is impossible to discriminate between a magmatic or a sedimentary origin (Rolfo, 1998). However, some minor-element concentrations are consistent with a magmatic origin. These ambiguous geochemical characteristics indicate that the talc-kyanite quartzite does not reflect the actual protolith composition, but suggests that its chemistry has been modified by a metasomatic and/or partial melting process. This conclusion appears to be supported by the occurrence of the kyanite + quartz aggregates, clearly pseudomorphs after a former Al-bearing mineral, possibly a feldspar, which lost alkalies during UHPM. This interpretation is consistent with field observations, which suggest the talc-kyanite quartzite to have been derived from a probable felsic dike.

Other kyanite-bearing quartzites

At Xinxian, in the western Dabie Mountains, kyanite quartzites are typically associated with eclogites, and consist of the UHPM peak assemblage quartz/coesite + kyanite + phengite + rutile ± garnet. Inclusions of coesite and polycrystalline quartz aggregates after coesite have been reported from garnet and kyanite (Li et al., 1993).

From the Donghai area of the southern Sulu terrane, two types of kyanite quartzites have been described by Zhang et al. (1995): (1) kyanite quartzites, from one to tens of meters thick, interlayered with eclogite; and (2) kyanite quartzites, hosting the eclogite pods. Type 1 kyanite quartzite from Qinglongshan consists of quartz (70 vol%), epidote (12 vol%), kyanite and omphacite (13 vol%), garnet (3 vol%), phengite (2 vol%), and accessory rutile. Epidote aggregates are locally abundant and include quartz pseudomorphs after coesite. Foliated kyanite quartzites have a mineralogy similar to massive ones, but lack omphacite and show a strong preferred orientation of kyanite, phengite, and epidote. Kyanite occurs either as coarse-grained idioblasts or as fine-grained inclusions in phengite and epidote. Polycrystalline quartz aggregates after coesite generally occur in both kyanite and epidote. Type 2 kyanite quartzites from Hetang consist of quartz (60 to 65 vol%), kyanite (25 to 30 vol%), minor clinozoisite, garnet, and traces of pyrite. Quartz pseudomorphs after coesite occur in kyanite and garnet. In the matrix, quartz is always finer grained than the other minerals, suggesting its inversion from former coesite. Kyanite is altered mostly to fine-grained multi-phase aggregates composed of an outer albite corona and a core including: (1) Zo ± Ky relics; (2) Pg ± Bt; or (3) Ky relics + Zo + Pg. Biotite + amphibole + albite aggregates are interpreted as pseudomorphs after former omphacite.

In the Sulu region, two other kyanite-bearing lithotypes are found, which may have recrystallized under UHPM conditions, although direct evidence of coesite has not been found thus far. They are: (1) kyanite + garnet + zoisite + phengite + quartz schists; and (2) kyanite-bearing gneisses, consisting of quartz + garnet + zoisite +

phengite + K-feldspar + biotite + epidote + amphibole + rutile. Because most kyanite-bearing rocks have concordant contacts with the associated coesite-eclogite layers, a consistent tectonometamorphic evolution was suggested by Zhang (1992).

In many localities of southernmost Sulu, including Hanshan, Wanshan, and Houzui (Xu, 1998), kyanite-bearing schists are exposed that show mineral assemblages including rutile and the Al-rich minerals topaz, phengite, woodhouseite [$CaAl_3(PO_4)(SO_4)(OH)_6$], pyrophyllite, and lazulite. In such lithologies, kyanite typically contains polycrystalline quartz aggregates after coesite, and phengite is highly celadonitic (Si $\leq$3.5 a.p.f.u.).

Aegirine orthogneiss

In the Dabie Mountains, aegirine-bearing two-feldspar orthogneiss is widely exposed throughout the UHPM terrane. This lithotype, first described by Compagnoni et al. (1996) within the quartz-eclogite-bearing belt at Caochong, west of Taihu, is characterized by the peculiar association of an aegirine-rich pyroxene + albite. These rocks occur within the country biotite gneiss as deformed bodies and layers of various thickness, usually with sharp contacts, often recalling original dikes. In hand specimen, only magnetite and locally garnet may be recognized embedded in a granoblastic aggregate of quartz and feldspars.

Petrography and mineral chemistry. The orthogneiss consists of K-feldspar, albite, quartz, and scarce (<5 vol%) aegirine-rich pyroxene, riebeckitic amphibole, and Fe-rich biotite (lepidomelane). Accessory magnetite, rare ilmenite (locally Mn-rich), and very rare REE (Nb-Y)-oxide, titanite, and Mn-sulfide (hauerite?) also are present. Garnet, phengite, and rutile also have been described by Xu et al. (1998). The pyroxene is a moderately zoned aegirine and/or aegirine-augite with X_{Jd} ~10, included in albite as green skeletal crystals or bundles of nematoblasts up to 0.3 mm across. Locally, diopside with up to X_{Jd} ~17 (Fig. 5) also occurs as stumpy prismatic crystals ~0.4 mm long, typically with inclusions of K-feldspar. Some aegirine grains are replaced by a needle-like riebeckitic amphibole. Scattered garnet grains, 0.2 to 1.5 mm across, are zoned with a core composed of almandine (up to 46 mole%), grossular (up to 43 mole%), minor spessartine, and scarce pyrope, and a rim characterized by a significant increase in the spessartine component (Xu et al., 1998). Some garnet grains are intimately associated with epidote and muscovite and are included in albite. In the same samples, phengite flakes (up to Si = 3.33 atoms p.f.u.) are rimmed by muscovite or biotite. Plagioclase is almost pure albite (An_2) and includes, or is rimmed by, aegirine. K-feldspar occurs as either laths 0.1 to 3 mm across or as finer grained interstitial microcline. Rutile occurs as rare small grains, mostly rimmed with titanite. Since the aegirine-rich pyroxene + albite association is a common retrogression product of jadeite in most HP and UHP metamorphic belts from around the world, this assemblage may be interpreted as evidence for earlier HP or UHP metamorphism. Similar Na-pyroxenes and Na-amphiboles, recently discovered in gneisses and schists surrounding eclogites in the Zhucheng area, Shandong Province, also have been interpreted as evidence for former eclogite-facies metamorphism (Lai et al., 1995).

Other gneisses and micaschists

In the Dabie-Sulu region, eclogites commonly are associated and interlayered with orthogneisses, marbles, and micaschists. The orthogneisses have the average mineralogy K-feldspar + plagioclase + biotite + muscovite + epidote + garnet + rutile + quartz/coesite in the Dabie Mountains, and K-feldspar + plagioclase + biotite + muscovite/phengite + epidote + garnet + quartz + aegirine + hornblende + titanite in Sulu; the average mineralogy of micaschists is garnet + phengite + epidote + rutile + quartz/coesite in the Dabie Mountains, and garnet + kyanite + zoisite + quartz/coesite + phengite + rutile ± titanite in Sulu (Wang et al., 1995).

From the more than 1000 boreholes drilled in the southern Sulu belt, many thin layers of strongly retrogressed biotite gneisses and micaschists have been recovered together with eclogites. According to Zhang et al. (1998), the peak metamorphic assemblage should have been garnet + phengite + quartz for garnet micaschists, quartz/coesite + phengite for quartz micaschists, jadeite/omphacite + quartz + phengite for biotite gneiss, and jadeite/omphacite + quartz + phengite + garnet for epidote gneiss. Thus far, however, unambiguous evidence that all these rocks underwent a common metamorphic evolution together with the associated coesite-eclogites has not been

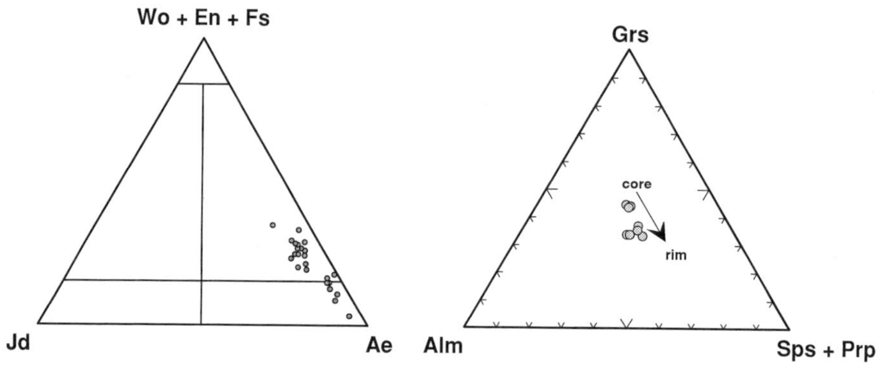

FIG. 5. Composition of pyroxenes and garnets from the aegirine-bearing orthogneiss from Dabie Shan (Compagnoni et al., 1996; Xu et al., 1998).

found. Zhang et al. (1996) described somewhat different metapelites associated with eclogites in Hetang and Yangzhuan of the Sulu region. These rocks, usually extensively retrogressed, consist of quartz (55 to 60 vol%), kyanite (20 vol%), and other minerals (20 vol%), including coarse-grained clinozoisite, garnet, and minor pseudomorphs after omphacite and rutile. Except for relict kyanite preserved as inclusions in coarse-grained clinozoisite, matrix kyanite and omphacite are totally altered to an aggregate composed of an albite corona surrounding a core including: (1) Zo; (2) Pg ± Bt; (3) Ky + Zo; and (4) Ky + Zo + Pg.

Paragneiss and micaschist associated with jadeite quartzites

The jadeite quartzites are surrounded by biotite and two mica–epidote gneisses. In Shuanghe, the jadeite quartzite + marble + eclogite association is enclosed in a biotite gneiss consisting of coarse-grained quartz strips and domains of finer-grained plagioclase, biotite, quartz, and epidote (Cong et al., 1995; Liou et al., 1997). This rock association as a whole forms an UHPM slab surrounded by granitic orthogneiss, as further suggested by the main metamorphic foliation within the UHPM slab commonly discordant with respect to the regional amphibolite-facies foliation of the country orthogneiss (Liou et al., 1997; Rolfo et al., 1997).

In Shuanghe, epidote–two mica schist and garnet-biotite gneiss have been identified by Cong et al. (1995). Their most significant mineral compositions are reported in Figure 6. The epidote–two mica schist consists of quartz, biotite, phengite, epidote and/or zoisite, and minor amounts of garnet, kyanite, titanite, rutile, amphibole, and plagioclase (An_{16-29}). Phengite generally is partially replaced by biotite-feldspar symplectites. The intergrowth of amphibole + Na-plagioclase suggests the former occurrence of omphacite. The garnet-biotite gneiss consists of biotite, quartz, plagioclase, garnet, phengite, and epidote with accessory titanite and/or rutile, and ilmenite. High-Si phengite (Si = 3.4 to 3.5) is best preserved as inclusions within garnet. In both schist and gneiss, garnet is homogeneous and consists of almandine (X_{Alm} = 46 to 52), grossular (X_{Grs} = 22 to 37), and pyrope (X_{Prp} = 13 to 21), with minor spessartine ($X_{Sps} < 4$). Epidote has a limited compositional range (Al_2O_3 = 24 to 28 wt%; CaO = 23 to 25 wt%; Fe_2O_3 = 6 to 12 wt%), with Fe increasing from core to rim.

In Xinjian, the paragneiss associated with jadeite quartzite consists of epidote, micas, plagioclase, K-feldspar, rare garnet, and accessory rutile, titanite, apatite, and zircon (Rolfo, 1998). Quartz and feldspars are locally concentrated in millimeter-thick domains, as well as micas, the white mica commonly prevailing over green-brown biotite. Epidote poikiloblasts, up to several millimeters long, are zoned from clinozoisite with relict allanite in the core, to more Fe-rich compositions in the rim. Skeletal garnet is strongly retrogressed and replaced by biotite and chlorite. Rutile has been found only as an inclusion within large titanite crystals. Close to the contacts with marble, the Xinjian paragneiss is characterized by an abundance of carbonates, suggesting a primary sedimentary transition.

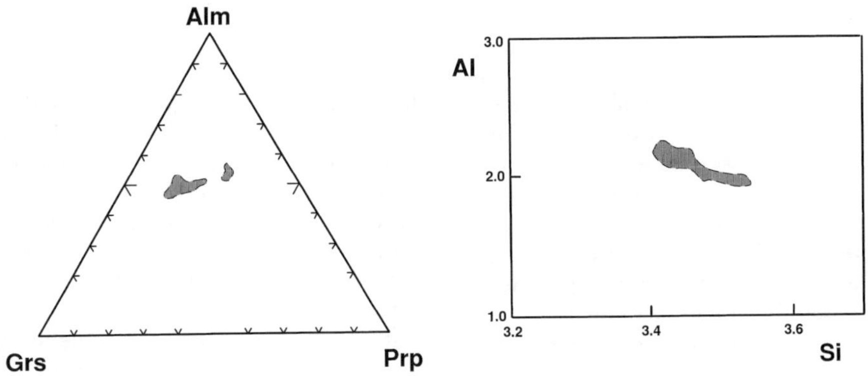

FIG. 6. Compositions of garnets in the almandine-grossular-pyrope diagram and of phengites in the Si versus total Al diagram from the epidote–two mica schist and garnet-biotite gneiss from Shuanghe (Cong et al., 1995).

Kokchetav Massif

The Kokchetav massif, in the Caledonides of northern Kazakhstan, is the first locality where microdiamonds have been found in situ within metamorphic crustal rocks (Sobolev and Shatsky, 1987), which are garnet-biotite gneisses and schists, garnet-pyroxene and pyroxene-carbonate-garnet rocks, and whiteschists (Dobretsov et al., 1995). In the most famous diamond-bearing locality at Kumdy-kol, these rocks occur as small lenticular, banded, or vein-like bodies within plagioclase gneiss of the Zerendin series (Sobolev and Shatsky, 1990). The diamondiferous gneisses and eclogites occur interlayered with granitic gneiss and plagiogneiss. In addition to microdiamond, independent indicators of UHP are K in clinopyroxene, Al in titanite, Na in garnet (Enami et al., 1995), the diopside-magnesite assemblage, coesite inclusions in zircon (Sobolev et al., 1991), and garnets (Korsakov et al., 1998) and Cr-bearing kyanite in whiteschist (Zhang et al., 1997).

Diamond-bearing gneisses and garnet-carbonate rocks

A wide range of diamondiferous rocks have been described from the Kumdy-kol site and a new locality 17 km to the northwest (Shatsky et al., 1995), which include prevailing gneisses with cataclastic to mylonitic fabric, leucocratic gneisses, a variety of fine-grained gneisses, pyroxene-carbonate and garnet-pyroxene rocks, and eclogites. The UHP peak conditions estimated by Dobretzov et al. (1995) are ~40 kbar and 900 to 1000°C. However, pressures up to 70 kbar have been inferred by Maruyama (1998) on the basis of quartz or K-feldspar lamellae exsolved from omphacite.

Petrography. Low La/Yb and Sm/Yb ratios of diamondiferous gneiss compared to diamond-free gneiss have been interpreted as evidence for partial anatexis and expulsion of the melt from the rocks by Shatsky et al. (1995). In the diamondiferous rocks, diamonds only occur as inclusions in garnet (or its mica-chlorite pseudomorphs), zircon, pyroxene, and kyanite (Shatsky and Sobolev, 1993). The diamonds are cubo-octahedral (Sobolev and Shatsky, 1990) and average 12 μm in size. They are distributed heterogeneously inside garnet and zircon, and seem to be concentrated in zones or layers otherwise undistinguishable mineralogically. Microdiamond included in garnet is often intergrown with euhedral biotite, zircon, and rutile (Sobolev and Shatsky, 1990), but almost 90% of diamond occurrence is in association with mica. A diamond/coesite intergrowth has also been reported from inside a ziron (Sobolev et al., 1994). In zoned zircons, inclusions of albite + quartz + graphite, jadeite + quartz, coesite + graphite, and diamond + coesite have been recognized by Katayama et al. (1998) from core to rim, respectively, recording the prograde metamorphic evolution. Graphite occurs in all the diamond-bearing rocks as inclusions in garnet (~50 μm long), locally intergrown with mica or coating diamond crystals, as intergranular secondary crystals (up to 300 μm long), or in fractures cutting the main minerals.

Mylonite to blastomylonite gneisses are the most widespread rock type and show a wide range in the overall mineral assemblage, which consists of quartz, garnet, K-feldspar ± plagioclase ± biotite ± muscovite ± clinopyroxene ± chlorite ± zoisite ± amphibole ± kyanite ± titanite ± carbonate ± tourmaline. Ubiquitous accessory minerals are apatite, rutile, sulfides, zircon, and graphite. Biotite replaces clinopyroxene, garnet, muscovite, and K-feldspar. Diamondiferous garnets also contain biotite/phlogopite, K-feldspar, quartz, phengite, clinopyroxene, apatite, rutile, titanite, zircon, calcite, and kyanite. Quartz and K-feldspar show, as a rule, an undulatory extinction, whereas plagioclase and biotite exhibit no signs of deformation. Cataclastic gneisses consist of garnet, clinopyroxene, quartz, K-feldspar, biotite, tourmaline ± muscovite + chlorite, with the orthoclase and plagioclase content being in inverse modal proportions to pyroxene and garnet. Leucocratic gneisses are composed of quartz, orthoclase, plagioclase, biotite, and muscovite. The fine-grained gneisses consist of a number of minerals, the most widespread being quartz, amphibole, chlorite, and locally, tourmaline. Biotite, muscovite, and porphyroblastic zoisite also are present, such as scarce garnet mostly replaced by amphibole, chlorite, and micas. Locally, large grains of orthoclase also occur.

In pyroxene-carbonate and garnet-pyroxene rocks, garnet may occur as vein-like or schlieren-like segregations, and include phlogopite, phengite, rutile, and pyroxene. Garnet-pyroxene rocks are coarse-grained with euhedral to subhedral garnet (20 to 40 vol%) and clinopyroxene (50 to 80%), interstitial calcite, accessory rutile and titanite, and secondary chlorite, quartz, and K-feldspar. Pyroxene-carbonate-garnet rocks consist mainly of dolomite, garnet, diopside, and phlogopite ± talc with a small amount of calcite. They are characterized by porphyroblasts of clinopyroxene, which are set in a slightly cataclastic matrix of mica and calcite. Garnet grains, up to 1 mm in size and 1 to 3 vol%, occur randomly throughout the rock. Microdiamonds are always found in coexistence with graphite in garnet, diopside, and phlogopite. In a diamond-free rock, Ti-clinohumite has been found (Ogasawara et al., 1998).

Mineral compositions. Garnets from diamondiferous gneisses (Fig. 7) are commonly rich in grossular with a low Fe/(Fe+Mg) ratio (grospydite-type), and have a small but significant Na content (up to 0.2 wt% Na_2O). They are similar to garnets from eclogites, but differ significantly from garnets of pyroxene-carbonate rocks, which are richer in grossular and have a higher Mg/(Mg+Fe) ratio. Zoning is present only in pink and pinkish brown garnets, with Ca and Fe increasing from core to rim, the deeper color corresponding to compositions richer in Ti and Fe^{3+} (Shatsky et al., 1995).

Pyroxenes from all lithologies are diopside-hedenbergite solid solutions with very low Al, and jadeite usually less than 10 mole%. In pyroxene-carbonate rocks, pyroxene is almost pure diopside with low jadeite and high K_2O (up to 1.15 wt%). Their zoning pattern is commonly of two types: in the matirx crystals, Al and K decrease from core to rim; in the crystals included in zircons, Al increases and K decreases from core to rim, respectively. The maximum K_2O content (1.14 wt%) was found in pyroxenes included in garnet from the pyroxene-carbonate rocks, whereas a significant K_2O content (up to 0.68%) and the maximum jadeite content (up to Jd_{50}) was found in pyroxenes included in zircon from mylonitic gneisses. Furthermore, the K-free pyroxenes have higher Fe/(Fe+Mg) ratios.

Micas considered to have crystallized at UHPM conditions (both phengite and biotite) are included in garnets and form intergrowths with diamond and graphite. In phengite, Si varies from 3.22 to 3.56 a.p.f.u., and Ti (up to 0.28 a.p.f.u.) is inversely correlated to Si. In all rock types, Fe and Al^{VI} of biotites are inversely correlated. Biotites included in garnets have lower Fe/(Fe+Mg) ratios than those in the matrix. Moreover, the Fe content of the host garnets is correlated with Fe in biotite. Some biotites contain up to ~5 wt% of F and Cl. Both rutile and titanite are regarded as stable phases under UHPM conditions (Sobolev and Shatsky, 1990). Rutile included in garnet contains up to 0.17 wt% Al_2O_3 and titanite up to 13.5 wt% Al_2O_3 and ~4 wt% F.

Whiteschists

Whiteschists with the assemblage talc + kyanite + garnet + phengite + rutile ± apatite ± amphibole, or zoisite + kyanite + garnet + talc occur in the Kulet area (Dobretsov et al., 1995). According to Zhang et al. (1997) and Massonne (1998), the whiteschists underwent UHPM with peak

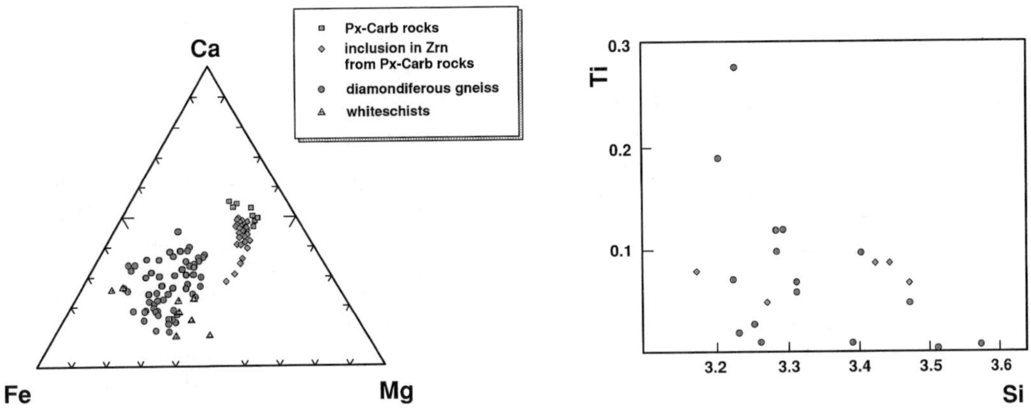

FIG. 7. Garnet and mica compositions from the UHPM gneiss and whiteschists from the Kokchetav massif (Shatsky et al., 1995; Zhang et al., 1997).

conditions of ~28 to 35 kbar and 730 to 780°C. Such estimates have been supported further by a recent find of microdiamond inclusions in garnet (Massonne, quoted in Zhang et al., 1997). Talc is usually the main sheet silicate, but phengite may occur locally instead of talc. Quartz, if present, is only a minor phase. Rutile and zircon are ubiquitous. Some garnet grains, which include kyanite and/or talc, exhibit an atoll- or half-atoll-like texture, suggesting growth from talc and kyanite. Phlogopite, amphibole, and part of phengite is considered to be part of the retrograde assemblage. Garnets contain inclusions of margarite, zoisite, kyanite, chlorite, and quartz in the core, and talc, kyanite, and phlogopite in the mantle, which are used to infer the prograde evolution (Zhang et al., 1997). The garnet composition is characterized by a detectable Cr_2O_3 content (0.18 to 0.48 wt%). In the most massive whiteschists, garnets show a well developed zonation with pyrope increasing (Prp_{25-45}) and almandine (Alm_{60-44}), and grossular (Grs_{28-10}) decreasing from core to rim. Homogeneous garnets are pyrope-rich and compositionally similar to the rims of zoned garnets. Garnet zonation is not correlated with the talc content of the whiteschists, which is in the range between 50 and 0 vol%. In the cores of euhedral Mg-rich garnets up to 1 mm across, chlorite, chloritoid, quartz, and staurolite inclusions have been recognized (Massonne, 1998). Kyanite exhibits a marked blue pleochroism, and contains 0.9 to 1.36 wt% Cr_2O_3. Pre-peak talc included in garnet has higher Al (0.17 a.p.f.u.) than does matrix talc (0.09 a.p.f.u.) formed at UHP peak conditions.

The whiteschists are considered to represent metasomatic rocks, most likely formed during a polyphase process in the presence of an active hydrous fluid phase, as a consequence of significant introduction of fluids rich in K. The extensive replacement of talc by phengite crowded with zircons suggests introductions of fluids rich in K and HFSE, such as Zr (Massonne, 1998).

Other Localities with UHPM Felsic Rocks

Among other localities, quartz pseudomorphs afer coesite also have been discovered and described by Tagiri and Bakirov (1990) in garnet-chloritoid-talc schists from the northern Tien Shan, Kyrgyzstan. Fresh coesite inclusions also have been reported in omphacite of impure marbles from the Pan-African belt of northern Mali (Caby, 1994). Recently, cuboid graphite aggregates with an edge length of up to 13 mm have been discovered by Leech and Ernst (1998) inside graphite-phengite schists from the Maksyutov Complex, southern Urals, Russia. A carbon-isotope and spectroscopic study suggests that the cuboid aggregates are diamond pseudomorphs, although coesite has not been reported thus far. In the Western Gneiss Region of the southwestern Caledonides, despite the widespread occurrence of eclogites, reports of UHPM assemblages in the host felsic rocks are scanty. Only recently, Wain (1997) described from the UHPM part of the

Nordfjord area gneisses closely associated with coesite-eclogites, which give evidence for the former existence of coesite. The gneisses consist of either quartz/coesite + garnet + clinozoisite + kyanite + phengite, or quartz/coesite + phengite + garnet and accessory kyanite and/or clinozoisite. Pressures in excess of 2.8 GPa and temperatures higher than 790°C have been estimated based on different geothermobarometers.

Conclusions

Minerals stable at UHP conditions

From this review, it appears that the UHPM minerals in felsic rocks are coesite, clinopyroxenes (including jadeite, omphacite, and diopside), garnet, epidotes, kyanite, phengite, talc, aragonite, dolomite, and accessory rutile, titanite, apatite, tourmaline, and zircon. In addition to such widespread, almost ubiquitous minerals, unique or rare phases may also occur; these are usually too sensitive to retrogression to be preserved during exhumation, and therefore to be used as key minerals (e.g., ellenbergerite). Microdiamond can also be found in rocks recrystallized at higher P, but its growth is strongly controlled by the composition of the fluid-phase (Ogasawara et al., 1998) and its presence is difficult to detect in rocks with only a few carats per ton.

Coesite is the most widespread and representative mineral of UHPM, but its occurrence is limited to the inclusions in other more competent peak minerals, such as garnet, jadeite and omphacite, kyanite, epidotes, and tourmaline. A special container for coesite and microdiamond, which inhibits retrogression, is zircon (e.g., see Sobolev et al., 1994), but its study requires techniques more sophisticated than the optical polarizing microscope. In diamond-bearing rocks or in rocks having experienced pressures in the diamond stability field, the extreme P-T conditions experienced by the rocks are further suggested by the unusual composition of common minerals, such as garnet high in Si, Y, and Yb or Na-pyroxenes rich in K.

It is not yet clear if orthoclase is a common mineral in UHPM, although its stability at extremely high P is suggested by experimental petrology. However, on the basis of careful examination of natural occurrences, it seems that, under eclogite-facies conditions and in the presence of a hydrous fluid phase, orthoclase is replaced by phengite and quartz (e.g., see Biino and Compagnoni, 1992).

Chemical compositions of UHP minerals

Garnets have a wide range of composition in the pyralspite series and their chemistry is controlled by both bulk-rock chemical composition and physical conditions, grossular and pyrope contents being proportional to P and T, respectively. This effect also is evident in Mn-rich rock compositions, where Mn-garnet is enriched in the calderitic component, an end member requiring high pressure to be stabilized (Lattard and Schreyer, 1983). Likewise, the pyroxenes are typically members of jadeite-diopside solid solutions, but their composition is strictly controlled by the rock-bulk chemistry, jadeite growing only in rocks with Na and Al, but devoid of Ca and Mg, and diopside in rocks lacking Na. All members of the epidote group are stable at UHP conditions, piemontite included, which grows in Mn-rich compositions. A common component of the Sulu paraschists is an epidote (both zoisite and epidote), characterized by the presence of significant Sr (up to SrO = 3.2 wt%) in the prograde crystal core, generally decreasing toward the crystal margin (Nagasaki and Enami, 1998). Furthermore, most ortho- and clinoepidotes from UHPM have a core distinctively rich in REE, especially La, Ce, and Nd. Micas are widespread in UHPM: they are phengites with a variable but significant celadonite (mainly Mg-Al) content. On the basis of a preliminary TEM study on phengites from the BIU, Dora-Maira massif, where talc and quartz exsolution lamellae have been identified by Ferraris (1998), it is possible that the original UHP white mica had a still higher celadonite content. It is not yet unambiguously ascertained whether the amphibole is stable at peak UHP conditions, since microstructural relationships mainly indicate its growth during the prograde and post-climactic P-T trajectory. If so, the most probable candidate is an amphibole with a composition close to pure glaucophane or nyböite.

Did a partial melt form in felsic rocks during UHPM?

The metamorphic peak conditions of the Lago di Cignana metasediments were very close to the "wet" solidus determined for "red clay" by

Nichols et al. (1994), but melting may have occurred only provided that an intergranular hydrous fluid phase was still present at this stage. This seems not to have occurred, since "neither in the field nor in thin section indications for partial melting were found" (Reinecke, 1998, p. 184).

For the Brossasca-Isasca Unit of the Dora-Maira massif, the possible presence of partial melts during UHPM has been discussed by Schreyer et al. (1987), Philippot (1993), and Sharp et al. (1993). Conditions of the UHPM are usually above the high-P "wet" solidus for granitic melts (Huang and Wyllie, 1975), but thus evidence for an unequivocal presence for melt segregation into veins is poor (Stöckhert and Renner, 1998, p. 82). The absence of veins or migmatitic structures in UHPM rocks of the Dora-Maira massif suggests that melt, wherever formed, was not easily segregated. The water-undersaturation in some lithologies is supported by a thermodynamic modeling of UHPM reactions in metagranite (Bruno, 1998), which confirms that the only fluid participating in the reactions was hydrous fluid released from the breakdown of igneous biotite to produce garnet, and immediately consumed to grow phengite. Furthermore, even in the presence of a free hydrous fluid phase during UHPM, the activity of water is strongly reduced because of the presence of highly saline brines (cf. Philippot, 1993; Philippot et al., 1995). On the contrary, the low La/Yb and Sm/Yb ratios of some diamondiferous gneisses compared to diamond-free gneisses from the Kokchetav massif was considered by Shatsky et al. (1995) as evidence that partial melting and melt expulsion did occur during UHPM.

The lack of melt in UHPM collisional orogenic belts is not striking, inasmuch as most continental crust showing UHPM derives from an already metamorphosed and dehydrated crystalline basement—i.e., consists of rocks partly dehydrated during a previous tectonometamorphic cycle. This conclusion is further supported by the consideration that the development of a partial melting during UHPM should reduce the rock mechanical strength, which is at odds with preservation of pre-metamorphic microstructures, such as igneous structures in granitoids. In any case, it should be very difficult to recognize the former presence of a partial melting in rocks that experienced a complex polyphase post-climactic tectonometamorphic evolution.

Acknowledgments

The authors are very grateful to W. G. Ernst for reviewing the paper scientifically and improving the English. T. Reinecke is thanked for the helpful revision and sharing unpublished data on the Lago di Cignana rocks. This work was carried out with financial support from the Italian Consiglio Nazionale delle Ricerche (C.N.R.), Centro Studi Geodinamica delle Catene Collisionali, Turin, and Ministero dell'Università e della Ricerca Scientifica e Tecnologica, Grant ex-40%.

REFERENCES

Amato, J., Baumgartner, L., Johnson, C., and Beard, B., 1997, Tertiary high-pressure metamorphism at Lago di Cignana, Italy, as determined by Sm-Nd dating of eclogite [abs.]: 5th Int. Eclogite Conf., Ascona, Switzerland, 1–5 Sept. 1997, Abs. suppl. no. 1, Terra Nova, v. 9, p. 1.

Biino, G., and Compagnoni, R., 1992, Very high-pressure metamorphism of the Brossasco coronite metagranite, southern Dora Maira Massif, western Alps: Schweiz. Mineral. Petrog. Mitteil., v. 72, p. 347–363.

Bruno, M., 1998, Petrographic study and thermodynamic modelling of the metamorphic reactions in a metagranodiorite from the UHPM Brossasco-Isasca Unit, Dora-Maira Massif: Unpubl. M.Sc. thesis, University of Turin, 97 p. (in Italian).

Caby, R., 1994, Precambrian coesite from northern Mali: First record and implications for plate tectonics in the trans-Saharan segment of the Pan-African belt: Eur. Jour. Mineral., v. 6, p. 235–244.

Chopin, C., 1984, Coesite and pure pyrope in high-grade blueschists of the western Alps: A first record and some consequences: Contrib. Mineral. Petrol., v. 86, p. 107–118.

———, 1987, Very-high-pressure metamorphism in the western Alps: Implications for subduction of continental crust: Philos. Trans. Roy. Soc. Lond., v. A/321, p. 183–197.

Chopin, C., Brunet, F., Gebert, W., Medembach, O., and Tillmanns, E., 1993, Bearthite, $Ca_2Al[PO_4]_2(OH)$, a new mineral from high-pressure terranes of the Western Alps: Schweiz. Mineral. Petrog. Mitteil., v. 73, p. 1–9.

Chopin, C., Henry, C., and Michard, A., 1991, Geology and Petrol. of the coesite-bearing terrane, Dora Maira massif, Western Alps: Eur. Jour. Mineral., v. 3, p. 263–291.

Chopin, C., and Schertl, H.-P., 1999, The UHP unit in the dora-Maira massif, Western Alps: Int. Geol. Rev., v. 41, in press.

Chopin, C., and Sobolev, N., 1995, Principal mineralogic indicators of UHP in crustal rocks, *in* Coleman, R. G., and Wang X., eds., Ultrahigh pressure metamorphism: New York, Cambridge University Press, p. 96–132.

Compagnoni, R., and Hirajima, T., 1992, Occurrence and significance of superzoned garnet in the coesite-bearing whiteschist of southern Dora-Maira Massif, Western Alps [abs.]: 29th Int. Geol. Cong., Kyoto, Abs. vol., v. 2, p. 599.

———, 1993, The occurrence of a superzoned garnet in the coesite-bearing unit of Dora-Maira Massif and the origin of the whiteschists [abs.]: 4th Int. Eclog. Conf., Fuscaldo (Italy), Terra Abs., v. 4, p. 5.

Compagnoni, R., Hirajima, T., and Chopin, C., 1995, Ultra-high-pressure metamorphic rocks in the Western Alps, *in* Coleman, R. G., and Wang, X., eds. Ultrahigh pressure metamorphism: New York, Cambridge University Press, p. 206–243.

Compagnoni, R., Hirajima, T., Turello, R., and Castelli, D., 1994, Ultrahigh pressure metamorphism in the continential crust: the Brossasco-Isasca Unit of Dora-Maira massif, *in* Compagnoni, R., Messiga, B., and Castelli, D., eds., High pressure metamorphism in the Western Alps: Guidebook to the B1 field excursion of the 16th General IMA Meeting, Pisa, 4–9 September 1994, p. 87–105.

Compagnoni, R., Rolfo, F., Xu, S., Jiang, L., Su, W., and Liu, Y., 1996, Aegirine-rich pyroxene in felsic gneisses of Dabie Shan, East Central China: Evidence for a former high-pressure metamorphism? [abs.]: 30th Int. Geol. Cong., Beijing, Abs. vol., v. 2/3, p. 608.

Cong, B., Zhai, M., Carswell, D. A., Wilson, R. N., Wang, Q., Zhao, Z., and Windley, B. E., 1995, Petrogenesis of ultrahigh-pressure rocks and their country rocks at Shuanghe in Dabieshan, Central China: Eur. Jour. Mineral., v. 7, p. 119–138.

Dobretsov, N. L., Sobolev, N. V., Shatsky, V. S., Coleman, R. G., and Ernst, W. G., 1995, Geotectonic evolution of diamondiferous paragneisses, Kokchetav Complex, northern Kazakhstan—the geologic enigma of ultra-high-pressure crustal rocks within a Phanerozoic fold-belt: The Island Arc, v. 4, p. 267–279.

Enami, M., Cong, B., Yoshida, T., and Kawabe, I., 1995, A mechanism for Na incorporation in garnet: An example from garnet in orthogneiss from the Su-Lu terrane, eastern China: Amer. Mineral., v. 80, p. 475–482.

Ferraris, C., 1998, HRTEM evidence for retrograde exsolution in phengite-3T from the Dora-Maira massif (Italian western Alps) [abs.]: 17th IMA Meeting, August 9–14, 1998, Toronto, Canada, Abs. vol., p. A101.

Gebauer, D., Schertl, H.-P., Brix, M., and Schreyer, W., 1997, 35 Ma old ultrahigh-pressure metamorphism and evidence for very rapid exhumation in the Dora Maira Massif, Western Alps: Lithos, v. 41, p. 5–24.

Hirajima, T., and Compagnoni, R., 1993, Petrology of a jadeite-quartz/coesite-almandine-phengite fels with retrograde ferro-nyböite from the Dora-Maira Massif, Western Alps: Eur. Jour. Mineral., v. 5, p. 943–955.

Hirajima, T., Wallis, S. R., Zhai, M., and Ye, K., 1993, Eclogitized metagranitoid from the Su-Lu Ultra-High-Pressure (UHP) Province, eastern China: Proc. Japan Acad., v. 69-B(10), p. 249–254.

Huang, W. L., and Willie, P. J., 1975, Melting reactions in the system $NaAlSi_3O_8$-$KAlSi_3O_8$-SiO_2 to 35 kilobars, dry and with excess water: Jour. Geol., v. 83, p. 737–748.

Hunziker, J. C., Desmons, J., and Hurford, A. J., 1992, Thirty-two years of geochronological work in the Central and Western Alps: A review on seven maps: Mém. de Géol., Lausanne, v. 13, 59 p.

Katayama, I., Zayachkovska, A., and Maruyama, S., 1998, A window to prograde metamorphism of the Kokchetav Massif, northern Kazakhstan, *in* Int. Workshop on UHP Metamorphism and Exhumation: Stanford, CA, Stanford University, Abs. vol., p. A91–A95.

Kienast, J. R., Lombardo, B., Biino, G., and Pinardon, J. L., 1991, Petrology of very high-pressure eclogitic rocks from the Brossasco-Isasca Complex, Dora-Maira Massif, Italian western Alps: Jour. Metamor. Geol., v. 9, p. 19–34.

Korsakov, A. V., Shatsky, V. S., and Sobolev, N. V., 1998, The first finding of coesite in the eclogites of the Kokchetav Massif: Doklady Akad. Nauk, v, 360, p. 77–81 (in Russian).

Kretz, R., 1983, Symbols for rock-forming minerals: Amer. Mineral., v. 68, p. 277–279.

Lai, X., Su, S., and Cheng, Z., 1995, High-pressure relics in gneisses surrounding the eclogite in Zhucheng area, Shandong, China: Chinese Sci. Bull., v. 40 (suppl.), p. 83–85.

Lattard, D., and Schreyer, W., 1983, Synthesis and stability of the garnet calderite in the system Fe-Mn-Si-O: Contrib. Mineral. Petrol., v. 84, p. 199–214.

Leake, B. E., 1978, Nomenclature of amphiboles: Amer. Mineral., v. 63, p. 1023–1052.

Leech, M. L., and Ernst, W. G., 1998, Graphite pseudomorphs after diamond? A carbon-isotope and spectroscopic study of graphite cuboids from the Maksyutov Complex, south Ural Mountains, Russia: Geochim. et Cosmochim. Acta, v. 62, p. 2143–2154.

Li, X., Kang, W., and Xu, P., 1993, Discovery of coesite in eclogite and kyanite quartzite from the Dabie Mountains, Henan Province: Jour. Changchun Univ. Earth Sci., v. 23, p. 256–261.

Liou, J. G., Zhang, R. Y., Eide, E. A., Ernst, W. G., and Maruyama, S., 1996, Metamorphism and tectonics of high-P and ultrahigh-P belts in the Dabie-Sulu region, eastern central China, *in* Harrison, T. M., and Yin, A., eds., The tectonic evolution of Asia: New York, Cambridge University Press, p. 300–344.

Liou, J. G., Zhang, R. Y., and Jahn, B. M., 1997, Petrology, geochemistry and isotope data on a ultrahigh-pressure jadeite quartzite from Shuanghe, Dabie Mountains, East-central China: Lithos, v. 41, p. 59–78.

Maruyama, S., 1998, The Kokchetav UHP-HP metamorphism: Current state of our research, *in* Int. Workshop on UHP Metamorphism and Exhumation: Stanford, CA, Stanford University, Abs. vol., p. A58–A59.

Massonne, H.-J., 1998, Whiteschists from the Kokchetav Massif, Kazakhstan, and the Dora Maira Massif, W Alps—a comparison of their metamorphic and metasomatic evolution, *in* International Workshop on UHP Metamorphism and Exhumation: Stanford, CA, Stanford University, Abs. vol., p. A204–A205.

Michard, A., Henry, C., and Chopin, C., 1995, Structures in ultrahigh-pressure metamorphic rocks, *in* Coleman, R. G., and Wang, X., eds., Ultrahigh pressure metamorphism: New York, Cambridge University Press, p. 132–158.

Morimoto, N., Fabries, J., Ferguson, A. K., Ginzburg, I. V., Ross, M., Seifert, F. A., Zussman, J., Aoki, K., and Gottardi, G., 1988, Nomenclature of pyroxenes: Schweiz. Mineral. Petrog. Mitteil., v. 68, p. 95–111.

Nagasaki, A., and Enami, M., 1998, Sr-bearing zoisite and epidote in ultra-high pressure (UHP) metamorphic rocks from the Su-Lu province, eastern China: An important Sr reservoir under UHP conditions: Amer. Mineral., v. 83, p. 240–247.

Nichols, G. T., Wyllie, P. J., and Stern, C. R., 1994, Subduction zone melting of pelagic sediments constrained by melting experiments: Nature, v. 371, p. 785–788.

Ogasawara, Y., Ohta, M., Katayama, I., and Maruyama, S., 1998, Stability relations of microdiamond and carbonates in the marbles from the Kokchetav Massif, northern Kazakhstan, *in* International Workshop on UHP Metamorphism and Exhumation: Stanford, CA, Stanford University, Abs. vol., p. A62–A65.

Okay, A. I., 1993, Petrology of a diamond- and coesite-bearing metamorphic terrane: Dabie Shan, China: Eur. Jour. Mineral., v. 5, p. 659–675.

Philippot, P., 1993, Fluid-melt-rock interaction in mafic eclogites and coesite-bearing metasediments: Constraints on volatile recycling during subduction: Chem. Geol., v. 108, p. 93–112.

Philippot, P., Chevallier, P., Chopin, C., and Debussy, J., 1995, Fluid composition and evolution in coesite-bearing rocks, Dora-Maira massif, Western Alps: Implications for element recycling during subduction: Contrib. Mineral. Petrol., v. 121, p. 29–44.

Reinecke, T., 1991, Very-high-pressure metamorphism and uplift of coesite-bearing metasediments from the Zermatt-Saas zone, Western Alps: Eur. Jour. Mineral., v. 3, p. 7–17.

———, 1998, Prograde high- to ultrahigh-pressure metamorphism and exhumation of oceanic sediments at Lago di Cignana, Zermatt-Saas zone, Western Alps: Lithos, v. 42, p. 147–190.

Reinecke, T., van der Klaw, S. N., and Stöckert, B., 1994, UHP metamorphic oceanic crust of the Zermatt-Saas Zone (Piemontese Zone) at Lago di Cignana, Valtournenche, Italy: Guidebook to the B1 field excursion of the 16th General IMA Meeting, Pisa, 4–9 September 1994, p. 15–16.

Rolfo, F., 1998, Tectonometamorphic evolution of collisional orogens: The example of North-West Himalaya (Northern Pakistan) and of Dabie Shan (East-Central China): Unpubl. Ph.D. thesis, Torino University, 157 p. (in Italian with English abstract).

Rolfo, F., Compagnoni, R., Xu, S., Wu, W., Jiang, L., Su, W., and Liou, Y., 1997, Garnet-jadeite-quartz/coesite granofels in the Dabie Mountains (eastern China): Further evidence for the occurrence of "in situ" ultrahigh-pressure metamorphism [abs.]: Terra Nova Abs. Suppl., v. 9/1, p. 105.

Rubatto, D., Gebauer, D., and Fanning, M., 1998, Jurassic formation and Eocene subduction of the Zermatt-Saas Fee ophiolites: Implications for the geodynamic evolution of the Central and Western Alps: Contrib. Mineral. Petrol., v. 132, p. 269–287.

Schertl, H.-P., Schreyer, W., and Chopin, C., 1991, The pyrope-coesite rocks and their country rocks at Parigi, Dora Maira massif, western Alps: Detailed petrography, mineral chemistry and P-T path: Contrib. Mineral. Petrol., v. 108, p. 1–21.

Schreyer, W., Massonne, H.-J., and Chopin, C., 1987, Continental crust subducted to mantle depths near 100 km: Implications for magma and fluid genesis in collision zones, *in* Mysen, O. O., ed., Magmatic processes: Physicochemical principles: The Geochem. Soc., Spec. Publ. no. 1, p. 155–163.

Sharp, Z. D., Essene, E. J., and Hunziker, J. C., 1993, Stable isotope geochemistry and phase equilibria of coesite-whiteschists, Dora Maira Massif, Western Alps: Contrib. Mineral. Petrol., v. 114, p. 1–12.

Shatsky, V. S., and Sobolev, N. V., 1993, Some aspects of origin of diamonds in metamorphic rocks: Doklady Akad. Nauk, v. 331, p. 1217–1219 (in Russian).

Shatsky, V. S., Sobolev, N. V., and Vavilov, M. A., 1995, Diamond-bearing metamorphic rocks of the Kokchetav Massif (northern Kazakhstan), *in* Coleman, R. G., and Wang, X., eds., Ultrahigh pressure metamorphism: New York, Cambridge University Press, p. 427–455.

Sobolev, N. V., and Shatsky, V. S., 1987, Inclusions of carbon mineral in the garnets from metamorphic rocks: Soviet Geol. Geophys., v. 27A, p. 1–8.

———, 1990, Diamond inclusions in garnets from metamorphic rocks: Nature, v. 343, p. 742–746.

Sobolev, N. V., Shatsky, V. S., Vavilov, M. A., and Goryainov, S. V., 1991, Coesite inclusion in zircon from diamondiferous gneiss of the Kokchetav massif—first find of coesite in metamorphic rocks in the USSR: Doklady Akad. Nauk, v. 321, p. 184–188 (in Russian).

———, 1994, Zircon from ultrahigh-pressure metamorphic rocks of folded regions as an unique container of inclusions of diamond, coesite, and coexisting minerals: Doklady Akad. Nauk, v. 334, p. 488–492 (in Russian with extended English abstract).

Stöckhert, B., and Renner, J., 1998, Rheology of crustal rocks at ultrahigh pressure, *in* Hacker, B., and Liou, J. G., eds., When continents collide: Geodynamics and geochemistry of ultrahigh-pressure rocks: Dordrecht, Kluwer Acad. Publ., p. 57–96.

Su, W., Xu, S., Jiang, L., and Liu, Y., 1996, Coesite from quartz jadeitite in the Dabie Mountains, Eastern China: Mineral. Mag., v. 60, p. 659–662.

Tagiri, M., and Bakirov, A., 1990, Quartz pseudomorph after coesite in garnet from a garnet from a garnet-chloritoid-talc schist, northern Tien-Shan, Kirghiz, SSR: Proc. Japan Acad., v. 66, ser. B, no. 7, p. 135–139.

Tilton, G. R., Ames, L., Schertl, H.-P., and Schreyer, W., 1997, Reconnaissance isotopic investigations on rocks of an undeformed granite contact within the coesite-bearing unit of the Dora Maira Massif: Lithos, v. 41, p. 25–36.

van der Klauw, S. N. G. C., Reinecke, T., and Stöckhert, B., 1997, Exhumation of ultrahigh-pressure metamorphic oceanic crust from Lago di Cignana, Piemontese zone, western Alps: The structural record in metabasites: Lithos, v. 41, p. 79–102.

Wain, A., 1997, New evidence for coesite in eclogite and gneisses: Defining an ultrahigh-pressure province in the Western Gneiss Region of Norway: Geology, v. 25, p. 927–930.

Wang, X., Zhang, R., and Liou, J. G., 1995, UHPM Terrane in East Central China, *in* Coleman, R. G., and Wang, X., eds., Ultrahigh pressure metamorphism: New York, Cambridge University Press, p. 356–390.

Xu, H., 1998, On the HP-UHP metamorphic belt in the Donghai Region, Jangsu province, China, *in* International Workshop on UHP Metamorphism and Exhumation: Stanford, CA, Stanford University, Abs. vol., p. A132–A133.

Xu, S., Wu, W., Su, W., Jiang, L., and Liu, Y., 1998, Metagranitoid from the high-ultrahigh pressure metamorphic belt in the Dabie Mountains and its tectonic significance: Acta Petrol. Sinica, v. 14/1, p. 42–59.

Zhang, R. Y., 1992, Petrogenesis of high-pressure metamorphic rocks in the Sulu and Dianxi regions, China: Unpubl. Ph.D. thesis, Kyoto University.

Zhang, R. Y., Hirajima, T., Banno, S., Cong, B., and Liou, J. G., 1995, Petrology of ultrahigh-pressure rocks from the southern Su-Lu region, eastern China: Jour. Metamor. Geol., v. 13, p. 659–675.

Zhang, R. Y., Liou, J. G., Coleman, R. G., Ernst, W. G., Sobolev, N. V., and Shatsky, V. S., 1997, Metamorphic evolution of diamond-bearing and associated rocks from the Kokchetav Massif, northern Kazakhstan: Jour. Metamor. Geol., v. 15, p. 479–496.

Zhang, R. Y., Liou, J. G., and Ye, K., 1996, Petrography of UHPM rocks and their country rock gneisses, *in* Cong, B., ed., Ultrahigh-pressure metamorphic rocks in the Dabieshan-Sulu Region in China: Beijing, Science Press, p. 49–68.

Zhang, Z., Xu, Z., and Xu, H., 1998, Petrology of ultra-high-pressure eclogites from ZK703 drillhole in the Donghai, eastern China, *in* Int. Workshop on UHP Metamorphism and Exhumation: Stanford, CA, Stanford University, Abs. vol., p. A128–A131.

EXPERIMENTAL ULTRAHIGH-PRESSURE RECRYSTALLIZATION

Experimental Aspects of UHP Metamorphism: Granitic Systems

WERNER SCHREYER

Research Group for High-Pressure Metamorphism, Institut für Mineralogie, Ruhr-Universität, D-44780 Bochum, Germany

Abstract

The outstanding development of recent years in the experimental petrology of granite systems with water is the in situ determination of critical curves in pressure-temperature diagrams, along which the hydrous melts and silicate-bearing vapors coexisting at low pressures merge compositionally, and form a single-phase supercritical fluid stable at higher pressures. Their locations, extending from pressures of ~15 kbar near 1000°C to ~25 kbar near 700°C, suggest that previous determinations of excess-water granite melting at higher pressures were influenced by quench problems. Under the conditions of ultrahigh-pressure metamorphism, therefore, granites in the presence of water do not melt, but are more or less gradually dissolved in the supercritical fluid with rising temperature, i.e., without any sharp "melting temperature." The anhydrous compositions of the supercritical fluids change from strongly incongruent, mainly potassium silicatic at low temperatures via ultrapotassic, leucititic compositions to congruent granitic near the temperature of complete dissolution. The presence of hydrous supercritical fluids during ultrahigh-pressure metamorphism, possible at all temperatures whenever water is available, has a decisive weakening effect on the rheological properties of crustal rocks during deep subduction.

Introduction

IN MOST ultrahigh-pressure (UHP) metamorphic terranes, rocks of granitic and related compositions are quite common, although their fabrics and states of deformation may vary drastically. In the coesite-bearing unit of the Dora-Maira massif, Western Alps, for example, these range from nearly preserved plutonic igneous textures through gneisses with or without feldspar augen to fine-grained phyllitic textures (Chopin et al., 1991). Most of these rocks do not show any obvious mineralogical and petrological features characteristic of UHP metamorphism, so that there has been a long and vivid debate as to whether they have in fact been subjected to the UHP conditions. Thus, if they had experienced such conditions, it must be clear that their earlier UHP features were largely destroyed during retrograde metamorphic events. Indeed, it is now widely accepted that granitoid systems—compared to the well-established basic rock types of the eclogite facies and to typical UHP metamorphic rocks such as pyrope quartzites—react particularly sensitively to late-stage retrogression during exhumation.

One of the difficulties encountered in properly evaluating the metamorphic history of granitoid rocks during and after ultradeep subduction is still—strangely enough—a considerable lack of knowledge on the stable phase relations in granitic systems under UHP conditions. This applies both to the so-called subsolidus conditions and to what is generally termed the melting relations. The enormous experimental efforts by P. J. Wyllie and his co-workers that started during the late 1960s and continued up to recently (Wyllie et al., 1996) represent pioneering work on granitic and related systems concentrating mainly on melting relations up to ~30 to 40 kbar, whereas the lower-temperature, melt-absent equilibria were only of interest as direct starting materials for melting. Despite the attainment of pressures in the UHPM range and the frequent presence of the component H_2O, the Wyllie school found no evidence for supercritical behavior in the various rock and synthetic systems studied. On the contrary, it was usually emphasized that—in the presence of excess water—granitic melts containing dissolved water coexist with hydrous fluids (e.g., Huang and Wyllie, 1981), thus proving a subcritical behavior. However, it must be borne in mind that all these experiments were peformed using the quenching method, which implies that the high-pressure, high-temperature assemblages synthesized during run conditions are cooled as rapidly as possible at the end of the runs, generally with only slightly decreasing pressures. Thus, it may not be certain that the phase assemblages observed optically and determined by x-ray powder diffraction afterward are identical with those present at the experimental conditions, and this may apply in particular to the fluid phases melt and vapor (gas).

Most recently, novel experimental techniques were introduced that allow direct in situ measurements and observations under run conditions. Although these techniques may not be fully and generally appreciated and accepted, their application leads to results that are considerably different from those obtained by the quenching method, especially concerning the problem of supercritical behavior in granitic systems. Because this problem is of fundamental importance for both the phase relations and the rheological behavior of granitic rocks undergoing prograde UHP metamorphism, it is discussed here in considerable detail. On the other hand, the mechanism of dehydration melting of metasedimentary and igneous rocks taking place at UHP depths only at high temperatures of 900°C and more to produce granitic magmas has been addressed recently by Patiño Douce and McCarthy (1998) (without taking supercritical behavior into consideration), and the reader is referred to that publication.

The Problem of Supercritical Behavior in Hydrous Granitic Systems

For investigating the possibility of supercritical conditions developing from the coexistence of hydrous melts and vapors, Shen and Keppler (1997) as well as Bureau and Keppler (1999) employed externally heated diamond-anvil cells, in which they could visually observe and distinguish silicate melts and vapors on account of their drastically different optical refraction, but also notice any changes finally leading to homogeneous fluids by the combination of the two. In this way they were able to determine in P-T plots for various silicate systems the critical curves that originate from the so-called "second critical end points." In physical-chemical theory, these are unique points in P-T-X space where the three-phase curves of coexisting solid + melt + vapor (also known as hydrous melting curves) end. It should be remembered here that as early as some 40 years ago the second critical end point of the system SiO_2-H_2O, where the three-phase curve quartz + melt + vapor ends, was determined experimentally by Kennedy et al. (1962) and found to lie near 9.7 kbar, 1080°C, at a composition of ~75 wt% SiO_2 and 25 wt% H_2O.

The system albite-H_2O

Figure 1 shows the critical curve for the system albite-water as reported by Shen and Keppler (1997), together with the melting curve of albite (here low-albite) in the presence of excess water vapor as determined by Goldsmith and Jenkins (1985), and other reaction curves of the system relevant for high pressures. It is along this critical curve that the coexisting phases hydrous melt (liquid) of the composition albite + minor water and hydrous vapor of the composition water + minor albite that are stable below the curve homogenize to form a single-phase supercritical fluid that is stable under all P-T conditions above the curve. Throughout this range extending to UHP conditions, there is thus complete miscibility of fluids starting from albite composition to pure water. The critical curve determined by Shen and Keppler (1997) intersects the "wet melting curve" of albite near 16 kbar, 670°C, which marks the location of the second critical end point (SCEP) in the albite-water system. It should be noted that this is virtually identical to the location of SCEP as predicted from solubility work by Paillat et al. (1992), whose approximate critical curve is shown in Figure 1 as well. At pressures above SCEP, albite actually no longer melts but is gradually dissolved in the fluid with rising temperature.

Because of the location of SCEP at relatively low pressures still within the stability field of albite, the disappearance, with rising temperature, of the high-pressure assemblage jadeite + SiO_2, so common in UHP metamorphism, can no longer be seen as being caused by a "wet melting curve" leading to a eutectic "liquid," as implied by Huang and Wyllie (1975) and shown in Figure 1. The supercritical fluid, perhaps but not necessarily of albite + water composition, is probably not only present above the curve of Huang and Wyllie but also below, where, however, its compositional range no longer covers the entire join albite-water. It is this fluid that gradually dissolves both jadeite and the stable SiO_2 phase, the more so the

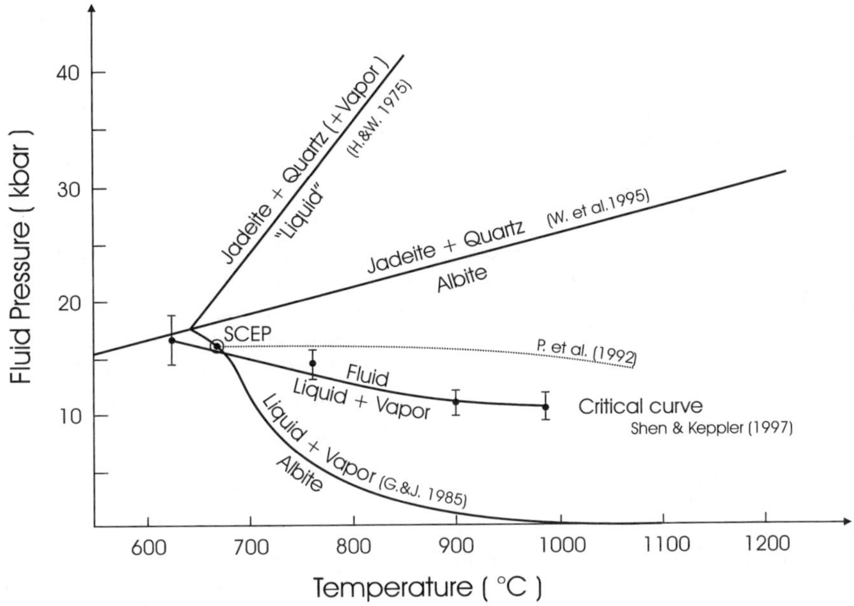

FIG. 1. Pressure-temperature plot showing (as solid line with error bars) the critical curve of the system albite-water as determined experimentally by Shen and Keppler (1997). The dotted line represents the estimated position of this curve following the earlier work of Paillat et al. (1992). The remaining curves exhibit the "wet melting" of albite, as determined by Goldsmith and Jenkins (1985), the upper pressure limit of albite after Waterwiese et al. (1995), and the "wet melting" of its high-pressure equivalent jadeite + SiO_2 taken from Huang and Wyllie (1975). SCEP is the second critical end point of the system and implies that the "wet melting curves" shown at higher pressures do not exist in this form.

higher the water content of the system. In the absence of water, the two solid phases finally produce a melt along their dry melting curve in the P-T plot. The "wet melting curve" of Figure 1 may only define the disappearance of jadeite + SiO_2 for the specific amount of water added by Huang and Wyllie (1975) to their experiments. Because both solid phases are anhydrous, any water present in the system must enter the fluid, which becomes increasingly silicate-rich with rising temperatures, whereas rising pressures will have the opposite effect (Paillat et al., 1992). Unfortunately, no solubility data are available as yet for what was formerly called the "subsolidus range" jadeite + quartz + vapor (see Fig. 1), but the dissolution trends as a function of increasing temperature and decreasing pressure should, in principle, be similar to those shown in the sketches of Figure 2. Thus, the dissolution behavior of a jadeite quartzite with interstitial water undergoing prograde UHP metamorphism must be governed by the local P-T-time path of the crustal unit. Nevertheless, it seems clear that the amounts of dissolved solids will be high enough to cause fluid properties quite different from those of pure water.

As indicated before, it seems possible, or even likely, that the low-temperature fluids coexisting with jadeite and the SiO_2 phase are not of hydrous albite composition, but are incongruent relative to albite. They may, in fact, not even lie within the plane jadeite-SiO_2-water but be shifted toward more sodic compositions. Because of the disappearance of the phase "melt" or "liquid" at pressures above SCEP, the system has—according to the phase rule—gained one degree of freedom. Therefore, in the four-component system Na_2O-Al_2O-SiO_2-H_2O jadeite may coexist stably over a divariant P-T range with up to two other solid phases plus the supercritical fluid. Such additional solid phases could be kyanite and/or corundum (in the absence of SiO_2). If the amounts of water present in the rock are low and the temperatures relatively high, as is usual in UHP metamorphism, the coexisting supercritical fluids may not be very hydrous (compare Fig. 2), so that their physical properties can approach those of "melts."

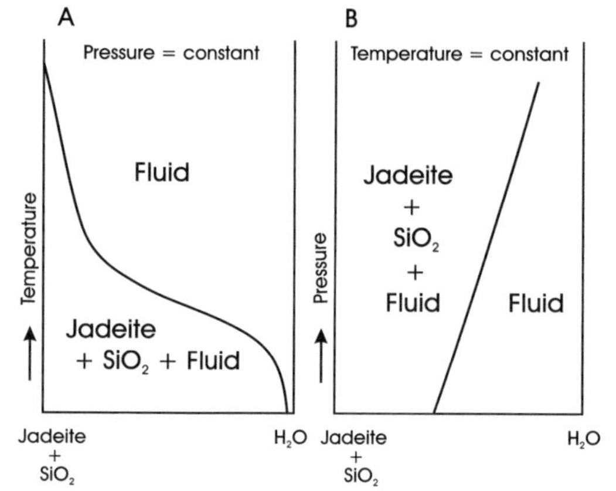

FIG. 2. Schematic phase relations between fluids and solids in the system albite-water at high pressures under supercritical conditions, shown in a TX (A) and a PX (B) section. The intersection of the curve in A with the temperature axis is the dry melting point of the assemblage jadeite + SiO_2. Note the variable ranges of fluid compositions.

Various granite-H_2O systems

In a quite recent study Bureau and Keppler (1999) applied their diamond-anvil method to several other systems, including the minerals jadeite and nepheline and two different types of granite. From their results these authors concluded that, in general, the critical curves determined for these compositions lie at consecutively higher P-T conditions the higher their SiO_2 contents are. However, they actually present exceptions to this rule as well. The nepheline critical curve lies at the lowest conditions, considerably below that of albite, as shown in Figure 1 (the curve passing through a point near 550°C at 15 kbar). On the other hand, the critical curve for jadeite lies above, and not below (as it should) that of albite, and passes through a point near 800°C at 15 kbar. The critical curves for the granites follow at still higher P-T conditions.

Figure 3 shows the critical curve for the haplogranite studied by Bureau and Keppler (1999), compared to relevant melting curves for an alkali granite in the system $NaAlSi_3O_8$-$KAlSi_3O_8$-SiO_2-H_2O published by Huang and Wyllie (1975) and two other univariant curves relevant for (U)HP metamorphism. As in the albite-H_2O system, the critical curve for haplogranite has a negative slope but with a somewhat stronger curvature so that it approximately parallels the low-pressure wet melting curve of alkali granite over the fluid-pressure range of 15 to 20 kbar. It lies between 3 and 10 kbar above the critical curve for albite-H_2O (Fig. 1). I have tentatively extrapolated the critical curve in Figure 3 to lower temperatures so that it meets the wet solidus curve of alkali granite after Huang and Wyllie (1975) and 26 kbar and 670°C. This could be taken as a kind of "second critical end point" in the simplest possible granite system; however, because of the larger number of components in the requisite system K_2O-Na_2O-Al_2O_3-SiO_2-H_2O (KNASH), this simplistic view does not hold, as will be shown later.

The relatively low pressure derived for the initiation of critical phenomena, as suggested in the plot of Figure 3, is surprising because it had been emphasized by Stern and Wyllie (1973) that a miscibility gap between silicate magmas and aqueous vapor phases persists up to at least 35 kbar. The discrepancy cannot be reconciled by a steeper extrapolation of the critical curve of Bureau and Keppler (1999) at temperatures below ~750°C, because Stern and Wyllie (1973) extended their postulation of subcritical behavior in the granite-water system up to a temperature of 850°C. Thus, the question arises as to whether the coexistence of melts and vapors in granite systems up to 35 kbar, as found by the Wyllie school (e.g., Boettcher and Wyllie 1968; Huang and Wyllie, 1975, 1981; Stern and Wyllie, 1981) is not the result of quench problems. It should be noted that quenching of a supercritical fluid from high temperatures at more or less constant pressures leads automatically into a metastable two-phase region melt + vapor on account of the negative slopes of the critical curves (see Figs. 1 and 3). Using a double-capsule technique, Adam et al. (1997) presented convincing evidence for subcritical behavior in a trondhjemite-water system at 20 kbar, 900°C, which is clearly above the critical curve for haplogranite shown in Figure 3. This may not contradict the results of Bureau and Keppler (1999), but probably indicates that the critical curve for the K-poor trondhjemite composition lies well above that of haplogranite. At any rate, for the conditions of UHP metamorphism within the coesite field (Fig. 3), there is a high likelihood that the fluid/rock interaction in granite-water systems is of a supercritical nature.

The phase relations in the five-component KNASH system are certainly more complicated

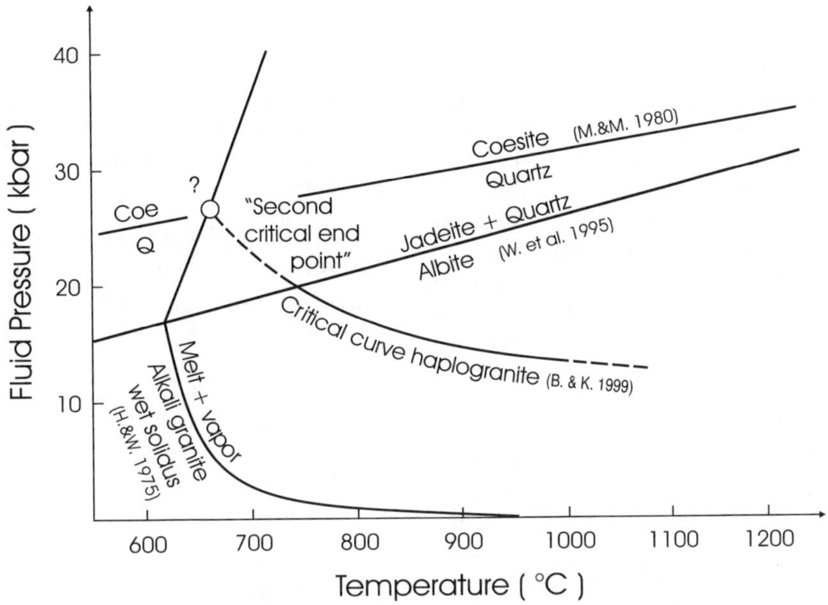

FIG. 3. Pressure-temperature plot showing the critical curve of a haplogranite as determined by Bureau and Keppler (1999), together with other relevant reaction curves. Upper pressure stability limits of albite are after Waterwiese et al. (1995) and of quartz after Mirwald and Massonne (1980); the "wet melting curve" of an alkali granite was taken from Huang and Wyllie (1975). Note that the latter curve is interesected by the extrapolated critical curve (dashed line). However, because of the larger number of components present in this system, the resulting point cannot be compared directly with the second critical end point shown in Figure 1 for albite-water. Nevertheless, supercritical behavior of a more complex nature takes place at higher pressures, and the "wet melting curve" of Huang and Wyllie (1975) should no longer be present there.

than for albite because of the additional solid phases present—e.g., K-feldspar, K-cymrite (Massonne, 1995)—but also because of incongruent compositions of fluids or melts that already arose from earlier experiments. Even if one neglects the possibility of supercritical behavior, it is clear from the results of Huang and Wyllie (1975) in the alkali-granite system and from their own considerations that at high pressures granite no longer exhibits a eutectic behavior as at crustal pressures. Therefore, the anhydrous compositions of the "hydrous melts" developing are no longer granitic but—based on Figure 5A of Huang and Wyllie (1975)—they become less siliceous (that is more syenitic) with increasing fluid pressure. In a run at 35 kbar, 720°C—that is, just above their "wet solidus" (see Fig. 3 of this paper)—Huang and Wyllie (1975, Table 2) obtained the solids jadeite and coesite, indicating that all the potassium of their starting mixture was present in the "melt." If the data are viewed in the context of supercritical behavior, this means that the supercritical fluid is strongly incongruent relative to the rock starting material. For the metamorphism of granites under UHP metamorphic conditions in the presence of water, it must be concluded that their partial dissolution leads to a strong fractionation of potassium into the fluid phase. Obviously, this may be very important for any questions of potassium metasomatism in the vicinity of metagranites or in overlying units.

Potassium enrichment of coexisting fluids already was evident from early work on the melting of a biotite granite up to 30 kbar by Boettcher and Wyllie (1968), who reported complete dissolution of biotite and K-feldspar even in subsolidus runs at 20 kbar. It should be considered, however, that a natural biotite granite represents a still more complicated system than the haplogranite discussed previously because of the many additional components present (FeO, Fe_2O, MgO, TiO_2, etc.). Therefore, the disappearance of the potassic-phase biotite also may be the result of chemical reactions occurring in the subsolidus assemblage. The foremost example to be noted here is the instability of the assemblage K-feldspar + phlogopite in the presence of quartz and excess water in the system K_2O-MgO-Al_2O_3-SiO-H_2O (Schreyer et al., 1987; Massonne, 1992) at subsolidus temperatures and pressures of 15 to 20 kbar. K-feldspar and phlogopite react to form phengite plus a highly potassic Mg/Si-bearing fluid that is distinctly different from granitoid melt compositions that appear at higher temperatures. Unfortunately, no experiments were conducted since these works were completed to test the influence of iron oxides on this reaction. In addition, the relations between the low-temperature KMgSi fluids and the "melts"(?) developing above ~700°C remain unclear (Massonne, 1992). The above examples show that the construction of a single "second critical end pont" in Figure 3 for haplogranite must be seen as a gross simplification, and this is even more pertinent to the multicomponent granite systems.

With granites being no longer of eutectic composition at high fluid pressures, their phase relations for subcritical conditions valid for temperatures above the wet solidus curves are typified by a series of consecutive curves in the P-T plots, along which solid phases disappear (mineral-out curves) as partners of the pairs melt + vapor (e.g., Huang and Wyllie, 1981). If supercritical conditions prevail, these curves should be regarded as dissolution curves with their locations being governed by the amounts of water present in the experiments (see Fig. 2A). A notable feature for UHP conditions is that potassic white micas of the muscovite-phengite series remain stable in the presence of excess fluid up to rather high temperatures (e.g., Schmidt, 1996; Schmidt and Poli, 1998). According to Huang and Wyllie (1981, Fig. 2), the "muscovite-out" curve for 35 kbar lies near 800°C—that is, about 100° above their "wet solidus" curve for a muscovite granite. With smaller amounts of water present than required for saturation of the subcritical melt, but also just with less water present in the supercritical fluid, the P-T range for these mineral-out curves expands toward higher temperatures until the liquidus, or the total dissolution surface, for a given water content is reached. In the various papers by the Wyllie school, these relations are illustrated in T-H_2O(wt%) diagrams for a given pressure (e.g., Huang and Wyllie 1981, Figs. 4–5). For subcritical conditions, this is the region where no vapors can coexist with the melts. Above the critical curve for haplogranite in Figure 3, the locations of these mineral-out curves may be rather similar, as the properties of the less hydrous fluids are probably close to those of the undersaturated melts.

Compositions of phases in granite-water systems in the supercritical range

In the high-pressure studies on granite melting up to 35 kbar by the Wyllie school, no attention had been paid to the compositions of the solid phases existing under subsolidus conditions or in the melting range. However, it was noted that new phases not present in the initial granitic starting materials formed during run conditions. These are, in particular, kyanite, garnet, jadeite, K-feldspar-hydrate (= K-cymrite; Massonne, 1995), and, of course, coesite. In a current yet unfinished study by M. Burchard at the Institut für Mineralogie at Bochum, all these phases plus, wherever possible, the quench-products of the coexisting fluids were analyzed with the aid of an electron microprobe. In the present context I am quoting preliminary results from two extended abstracts (Burchard and Schreyer, 1995a, 1995b) and also some selected information made known to the author by M. Burchard in personal communications.

The starting material used by Burchard is the country rock gneiss of the Dora-Maira pyrope quartzite (Schertl et al., 1991), which is essentially of an S-type granitic composition (Tilton et al., 1991), but exhibits a greenschist-facies mineral assemblage. The run conditions ranged from 25 to 45 kbar, 600 to 900°C with 5 to 20 wt% of water added to the rock powder (Burchard and Schreyer, 1995a, 1995b). Run durations of seven days were necessary to reach equilibrium, although locally relic inclusions of starting minerals could still be detected within the newly formed phases by the probe work. However, the minerals albite, biotite, and chlorite, initially present, had always disappeared during the runs, so that the new assemblages contained quartz (or coesite), phengite, clinopyroxene, garnet, occasionally epidote, and always a quenched fluid.

According to Burchard and Schreyer (1995a) the clinopyroxene formed at high pressures was never pure $NaAlSi_2O_6$, but always contained up to 30 mole% of the other pyroxene components Ca(Fe,Mg,Mn)[Si_2O_6] and $(Fe,Mg,Mn)_2[Si_2O_6]$. In general, the jadeite component decreased with increasing pressure and decreasing temperature. Phengites contain up to 7–8 wt% (FeO + MgO + MnO), with iron dominating (molar Fe/(Fe + Mg + Mn) about 0.70–0.77), and they exhibit Si-values between ~3.2 and 3.5 per formula unit (22 cation valences). In general, these Si values representing the phengite component in the muscovite decrease with increasing temperature, but the pressure influence may be either positive or negative as a function of the amount of water added to the capsule. In the garnets, the almandine component dominates strongly over grossular and pyrope. A typical composition formed at 30 kbar, 900°C (5% water added) was almandine 71, grossular 14, pyrope 11, and spessartine 3, neglecting small amounts of TiO_2, P_2O_5, and Na_2O.

Glasses as quench products of fluids with melt-like physical properties were only found in runs at 750°C and higher. Based on analytical totals of the probe analyses, such glasses contain 12 to 13 wt% H_2O (Burchard and Schreyer, 1995a), which, however, has to be taken as a minimum because of their instability under the electron beam, perhaps also reflecting quench problems (see above). When recalculated on a 100% oxide basis (Burchard and Schreyer, 1995b), the glass quenched from 30 kbar, 750°C (10% water added to the capsule) is very alkali rich, with 8.0 wt% K_2O and 2.9% Na_2O. In the presence of 74 wt% SiO_2 this makes an extremely alkali rich granite composition plotting at the edge of the alkali granite field in a total alkali versus SiO_2 diagram (Wilson, 1989, Fig. 1.2). The glass quenched and analyzed from a 900°C run, also at 30 kbar with 10% water added, has about the same amount of SiO_2 but less alkalies (5.3% K_2O and 2.7% Na_2O) and thus attained a more normal granitic composition.

Unusual non- or microcrystalline quench aggregates were obtained in runs at 30 kbar, 600°C with 20 wt% water added—that is, well below what previously was called "wet solidus" in the granite-water system. The microprobe analysis of this material yielded ~88 wt% SiO_2, 7.1% Al_2O_3, 2.2% Na_2O, and 1.5% K_2O, plus ~1% oxides of Fe, Mg, and Ca, when recalculated to 100% (Burchard and Schreyer, 1995b). However, it is doubtful that these values are representative for the initial composition of the fluid, because presumably not only water, but also alkalies were lost from the aggregates during quenching. This is corroborated by more recent microprobe work (Burchard, pers. commun.) on quench products from runs at 40 to 45 kbar, 700 to 750°C with 3 to 10 wt% water added. These products contain crystals of a carbonate that probably was formed from the primary quench product, an alkaline watery liquid, after opening the capsule through interaction with the CO_2 of the air. Bulk analyses on these products yield totals between only 32 and 55 wt%, but they show impressively high values of K_2O between 17 and 26 wt%. If one recalculates the analyses to 100% of metal oxides, K_2O varies from ~31 to 65 wt%, the rest being SiO_2 (22 to 67 wt%), Na_2O (0.7 to 5.5 wt%), Al_2O_3 (0.7 to 14 wt%) and minor amounts of other oxides. Thus, these analyses present the first direct, although certainly semiquantitative, evidence for incongruent, strongly potassic fluids forming from granite in the presence of water at high pressures and relatively low temperatures, in agreement with the early observations by Boettcher and Wyllie (1968).

Figure 4 is a ternary plot of the selected components K_2O, Al_2O_3, and the combined oxides (Mg,Fe)O in the presence of SiO_2 and H_2O, but ignoring other components and not projecting from additional phases such as clinopyroxenes. It shows the plotting points of some relevant mineral phases, but also of five "fluids" (F1–F5) analyzed by Burchard (pers. commun.). The phengite composition plotted is that of a phengite synthesized by Burchard and Schreyer (1995a) from the Dora-Maira "granite" starting material (Tilton et al., 1991), which is plotted as well. The important reason for this type of plot is that it can be used as an illustration for a reaction that occurred during the high-pressure run relative to the biotite-bearing starting material gneiss. Modifying the conclusions of Massonne (1992) for the Fe-free system, the following reaction is envisaged:

$$\text{biotite} + \text{K-feldspar} \pm SiO_2 = \text{phengite} + \text{potassic fluid}.$$

The crossing tie lines for the existence of phengite with fluids support its validity in four (F1–F4) out of five cases of fluid composition.

Fluid F5 was analyzed from the 45 kbar, 750°C (10% water) run product—that is, from the run at the higgest temperature of the series. It is much less potassic, but more aluminous than the fluids analyzed from the 700 to 725°C runs, thus appar-

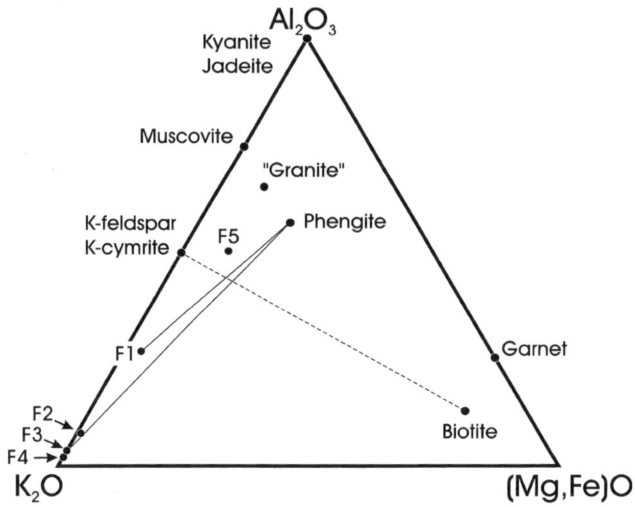

FIG. 4. Ternary plot of selected components of the Dora-Maira granite starting material (Tilton et al., 1991, here shown as "Granite") used by M. Burchard (pers. commun.) for his experiments. It shows the partial compositions of some relevant solid and fluid phases appearing in the presence of silica and water. Clinopyroxene was ignored. The phengite composition is taken from Burchard and Schreyer (1995a, GN1); biotite is an idealized phlogopite or annite. F1–F5 are the fluid compositions as analyzed from quench products of runs in the range from 40 to 45 kbar, 700 to 750°C, with 3 to 10 wt% water added to the capsule (Burchard, pers. commun.). The dotted line refers to the solid assemblage present in the starting material, whereas the solid lines represent resulting assemblages under run conditions. Note that this plot cannot strictly be taken as a compatibility triangle.

ently getting closer to the "granite" composition of Figure 4. When recalculated on a dry 100% basis, however, this fluid has a composition with ~54 wt% SiO_2, 21% Al_2O_3, 4% (FeO + MgO), 2% Na_2O, and 17% K_2O. Except for Fe and Mg, this is close to the composition of leucite! Thus, the petrogenetic possibility arises that UHP metamorphism of wet granites may lead directly to ultapotassic, silica-undersaturated fluids and thus magmas(?). The importance of potassic fluids derived from deeply subducted crustal rocks and subsequently interacting with ultramafic mantle material has been discussed by Schreyer et al. (1987) as a mechanism for producing basic ultrapotassic igneous rocks such as lamproites. Burchard's results could mean that ultrapotassic rocks poor in mafic components could actually be derived from deeply subducted crust without interaction with neighboring mantle rocks simply by percolating water through its units of granitic composition.

Summary and Rheological Consequences

The new data reviewed in this article and compared with results of earlier studies strongly suggest that granitic rocks of the continental crust when subducted to UHP depths of metamorphism show supercritical behavior during interaction with free water. This water may originally have been present as fluid inclusions or incorporated in hydrous minerals of the rocks themselves, but also may have been released from neighboring units undergoing dehydration reactions (Ernst et al., 1998) so that fluids can percolate through large volumes of rock. The supercritical behavior implies that the granites or derivative metamorphic rocks of granitic composition do not melt any more when their individual solidus temperatures, valid only for subcritical conditions, are surpassed. However, supercritical hydrous fluids are present—in varying quantities and with changing compositions—at all temperatures up to the dry melting temperature when water is absent. Thus, under UHP conditions, granitic rocks dissolve gradually in their coexisting fluids as temperature rises.

Importantly, the low-temperature fluids are strongly incongruent relative to granite chemistry, with potassium and silica dominating their compositions. With increasing temperatures, the fluids gradually dissolve other granite components as well, especially alumina, so that their compositions approach more and more those of granitic melts—interestingly, however, with intervening ultrapotassic types.

The presence of supercritical fluids with gradually changing compositions in granite-water systems under all UHP conditions is probably of paramount importance for the rheological behavior of such rocks during the tectonic processes of deep subduction. Quite independent of chemical composition, it is clear that UH pressures alone will lead to denser fluids with vastly different properties than those of vapors (gases) present under subcritical conditions. Stöckhert and Renner (1998) have emphasized that, on the basis of microstructural studies, many UHP metamorphic rocks "reveal evidence for weak or even essentially absent deformation," which seems to be caused by low strengths of these rocks under UHP conditions. The authors argue that this could be the result of the presence of interstitial fluids with high solute contents leading to small dihedral angles and efficient wetting of grain boundaries. Thus, grain-boundary sliding may take place, and the shapes of the crystals can be accommodated by "dissolution precipitation, analogous to liquid phase sintering in ceramics."

For future programs, experimental quantification is necessary in several directions. First of all, the fluid-granite dissolution equilibria will have to be defined chemically as a function of pressure and temperature (see Fig. 2). Moreover, granite-water experiments should be carried out in deformation apparatus similar to the ones mentioned by Stöckhert and Renner (1998) in order to define quantitatively the weakening influence of supercritical fluids on solid rocks, again as a function of pressure and temperature, but also of the amounts of H_2O present.

Acknowledgments

I thank Peter Wyllie (Pasadena, California) for up-to-date information on his experimental granite research, Hans Keppler (Bayreuth) for providing a preprint of the Bureau and Keppler paper, and Michael Burchard (Bochum) for permission to use some of his unpublished data.

REFERENCES

Adam, J., Green, T. H., Sie, S. H., and Ryan, C. G., 1997, Trace element partitioning between aqueous fluids, silicate melts, and minerals: Eur. Jour. Mineral., v. 9, p. 569–584.

Boettcher, A. L., and Wyllie, P. J., 1968, Melting of granite with excess water to 30 kilobars pressure: Jour. Geol., v. 76, p. 235–244.

Burchard, M., and Schreyer, W., 1995a, Melting experiments on the granitic country rock of the Dora Maira pyrope quartzites: Bochum. Geol. Geotech. Arb., v. 44, p. 23–27.

———, 1995b, Coexisting phases in the granitic country rock of the Dora Maira pyrope-quartzites reacted at 30 kbar, 600–900°C with variable water quantities [abs.]: Berichte der Deutschen Mineralogischen Gesellschaft, Beih. z. Eur. Jour. Mineral., v. 7, p. 41.

Bureau, H., and Keppler, H., 1999, Complete miscibility between silicate melts and hydrous fluids in the upper mantle: Experimental evidence and geochemical implications: Earth Planet. Sci. Lett., v. 165, p. 187–196.

Chopin, C., Henry, C., and Michard, A., 1991, Geology and petrology of the coesite-bearing terrain, Dora Maira massif, Western Alps: Eur. Jour. Mineral., v. 3, p. 263–291.

Ernst, W. G., Mosenfelder, J. L., Leech, M. L., and Liu, J., 1998, H_2O recycling during continental collision: Phase-equilibria and kinetic constraints, in Hacker, B. R., and Liou, J. G., eds., When continents collide: Geodynamics and geochemistry of ultra-high-pressure rocks: Dordrecht, Kluwer Acad., Publ., p. 275–295.

Goldsmith, J. R., and Jenkins, D. M., 1985, The hydrothermal melting of low and high albite: Amer. Mineral, v. 70, p. 924–933.

Huang, W.-L., and Wyllie, P. J., 1975, Melting reactions in the system $NaAlSi_3O_8$-$KAlSi_3O_8$-SiO_2 to 35 kilobars, dry and with excess water: Jour. Geol., v. 83, p. 737–748.

———, 1981, Phase relationships of S-type granite with H_2O to 35 kbar: Muscovite granite from Harney Peak, South Dakota: Jour. Geophys. Res., v. 86, p. 10,515–10,529.

Kennedy, G. C., Wasserburg, G. J., Heard, H. C., and Newton, R. C., 1962, The upper three-phase region in the system SiO_2-H_2O: Amer. Jour. Sci., v. 260, p. 501–521.

Massonne, H.-J., 1992, Evidence for low-temperature ultrapotassic siliceous fluids in subduction zone environments from experiments in the system K_2O-MgO-Al_2O_3-SiO_2-H_2O (KMASH): Lithos, v. 28, p. 421–434.

———, 1995, Experimental and petrogenetic study of UHPM, in Coleman, R. G., and Wang, X., eds., Ultrahigh pressure metamorphism: New York, Cambridge University Press, p. 33–95.

Mirwald, P. W., and Massonne, H.-J., 1980, The low-high quartz and quartz-coesite transition to 40 kbar between 600° and 1600°C and some reconnaissance data on the effect of $NaAlO_2$ component on the low quartz-coesite transition: Jour. Geophys. Res., v. 85, p. 6983–6990.

Paillat, O., Elphick, S. C., and Brown, W. L., 1992, The

solubility of water in $NaAlSi_3O_8$ melts: A reexamination of Ab-H_2O phase relationships and critical behavior at high pressures: Contrib. Mineral. Petrol., v. 112, p. 490–500.

Patiño Douce, A. E., and McCarthy, T. C., 1998, Melting of crustal rocks during continental collision and subduction, in Hacker, B. R., and Liou, J. G., eds., When continents collide: Geodynamics and geochemistry of ultrahigh-pressure metamorphism: Dordrecht, Kluwer Acad. Publ., p. 27–55.

Schertl, H.-P., Schreyer, W., and Chopin, C., 1991, The pyrope-coesite rocks and their country rocks at Parigi, Dora-Maira Massif, Western Alps: Detailed petrography, mineral chemistry and P-T path: Contrib. Mineral. Petrol., v. 108, p. 1–21.

Schmidt, M. W., 1996, Experimental constraints on recycling of potassium from subducted oceanic crust: Science, v. 272, p. 1927–1930.

Schmidt, M. W., and Poli, S., 1998, Experimentally based water budgets for dehydrating slabs and consequences for arc magma generation: Earth Planet. Sci. Lett., v. 163, p. 361–379.

Schreyer, W., Massonne, H.-J., and Chopin, C., 1987, Continental crust subducted to depths near 100 km: Implications for magma and fluid genesis in collision zones, in Mysen, B., ed., Magmatic processes: Physicochemical principles: The Geochem. Soc., Spec. Publ. 1, p. 155–163.

Shen, A. H., and Keppler H., 1997, Direct observation of complete miscibility in the albite-H_2O system: Nature, v. 385, p. 710–712.

Stern, C. R., and Wyllie, P. J., 1973, Water-saturated and undersaturated melting relations of a granite to 35 kilobars: Earth Planet. Sci. Lett., v. 18, p. 163–167.

———, 1981, Phase relationships of I-type granite with H_2O to 35 kilobars: The Dinkey Lakes biotite-granite from the Sierra Nevada Batholith: Jour. Geophys. Res., v. 86, p. 10,412–10,422.

Stöckhert, B., and Renner, J., 1998, Rheology of crustal rocks at ultrahigh pressure, in Hacker, B. R., and Liou, J. G., eds., When continents collide: Geodynamics and geochemistry of ultrahigh-pressure rocks: Dordrecht, Kluwer Acad. Publ., p. 57–95.

Tilton, G. R., Schreyer, W., and Schertl, H.-P., 1991, Pb-Sr-Nd isotopic behavior of deeply subducted crustal rocks from the Dora-Maira Massif, Western Alps: What is the age of the ultrahigh-pressure metamorphism?: Contrib. Mineral. Petrol., v. 108, p. 22–33.

Waterwiese, T., Chatterjee, N. D., Dierdorf, I., Göttlicher, J., and Kroll, H., 1995, Experimental and thermodynamic study of heterogeneous and homogeneous equilibria in the system $NaAlSiO_4$-SiO_2: Contrib. Mineral. Petrol., v. 121, p. 61–73.

Wilson, M., 1989, Igneous petrogenesis: London, Harper Collins Acad., 466 p.

Wyllie, P. J., Wolf, M. B., and van der Laan, S. R., 1996, Conditions for formation of tonalities and trondhjemites: Magmatic sources and products, in de Wit, M. J., and Ashwahl, L. D., eds., Tectonic evolution of greenstone belts: Oxford, U.K., Oxford University Press, p. 258–267.

Experimental Aspects of UHP Metamorphism in Pelitic Systems

HANS-JOACHIM MASSONNE

Institut für Mineralogie und Kristallchemie, Universität Stuttgart, Azenbergstr. 18, D-70174 Stuttgart, Germany

Abstract

Recent experimental studies have addressed mineral phases that can be expected to occur in metasedimentary rocks at ultrahigh-pressure conditions. These studies are based mainly on the P-T location of reaction curves using reversals and the determination of compressibility data as well as the compositional change of solid solution phases with P and T in fixed mineral assemblages.

In the system SiO_2, the reactions bounding the coesite stability field as well as the rheology of coesitites and the kinetics of the coesite-quartz transformation are discussed. In the system Al_2O_3-SiO_2-H_2O, the stability relations of kyanite, piezotite, OH-topaz, and diaspore are reviewed. In MgO-Al_2O_3-SiO_2-H_2O, investigations on the enstatite polymorphs, talc, the 10 Å phase, MgMgAl-pumpellyite, and pyrope, which can coexist with coesite, are reported. Experimental studies on Mg-chloritoid, Mg-staurolite, and chlorite also are reviewed. In FeO-Al_2O_3-SiO_2-H_2O, chloritoid, staurolite, and ferrosilite are addressed. In K_2O-Al_2O_3-SiO_2-H_2O, K-feldspar, hollandite-type $KAlSi_3O_8$, K-cymrite, kalsilite, wadeite-type $K_2Si_4O_9$, and the $KAlSi_2O_6$ component in clinopyroxenes are considered. In K_2O-FeO-MgO-Al_2O_3-SiO_2-H_2O (KFMASH), studies on various white and dark micas are mainly reported, but also information on stilpnomelane and zussmanite is given. Subsequently, various borates, borosilicates, and phosphates—most of them are rich in Al and therefore have the possibility of occurring in UHP metapelites—are addressed. Finally, aragonite, dolomite, and magnesite are considered as well.

At the end of the paper a summary is presented that points to the necessity of continuing with various kinds of studies related to the determination of thermodynamic properties of end-member phases and solid solution series. Such studies will enable us to elucidate the stable topology of phase relations in KFMASH and related systems. In addition, such thermodynamic data are needed for more reliable geothermobarometry.

Introduction

THE OBJECTIVE of this paper is to review experimental work that has appeared quite recently in the literature, chiefly since 1995, addressing mineral phases that are relevant to sedimentary rocks metamorphosed at ultrahigh pressure (UHP) conditions. Under these circumstances, almost all of the reviewed studies were undertaken using high-pressure instruments, and most are related to the stability of the P range of coesite. The minerals being considered are to a major extent those of the classical model system for metapelites, K_2O-FeO-MgO-Al_2O_3-SiO_2-H_2O (KFMASH). In addition, a few sodium-bearing phases, like jadeite, are considered as well. However, Ca-bearing silicates such as lawsonite and zoisite are neglected; although they can appear in UHP metapelites, they actually play a dominant role only in metabasites. Thus, these phases are treated extensively in other papers dedicated to experimental progress in metabasic systems. For the same reason, developments in research on amphiboles are not reported in this paper, even when an amphibole like glaucophane fits the systems considered here for metapelites.

Besides the generally known silicates, some exotic phases such as borates and phosphates are considered to draw the reader's attention to the possible appearance of such phases in UHP metapelites. In addition, carbonates are not discussed. Because the relevant work on the formation of H_2O-bearing fluids at HP and UHP conditions already have been reviewed in a paper on progress in experimental research on granitic systems (Schreyer, 1999), and because the fluids treated here appear in systems that are similar to those of metasediments, there is no reason to describe these experimental investigations here.

The experimental studies that are of interest here are arranged according to the model system including the corresponding mineral. The results of these studies are not only compiled and synthesized, but criticized as well, when necessary.

Results in Various Model Systems

System SiO_2

The well-known stability relations for coesite have recently been revisited. Zhang et al. (1996) improved our knowledge of the P-T position of the coesite-stishovite boundary by an in situ X-ray diffraction study using reversals. The resulting curve is shown in Figure 1.

Bose and Ganguly (1995) studied the coesite-quartz transition by reversal experiments in a piston-cylinder apparatus. However, their results caused confusion rather than contributed to a more precise P-T location of the corresponding transition curve, which already was bracketed within the narrow P range of 1 kbar as a result of experimental studies in low-friction cells of a piston-cylinder apparatus (see Massonne, 1995). In comparison with these studies, the transition curve of Bose and Ganguly (1995) was located—for instance, at 600°C—at 1.5 kbar higher P (Fig. 1). The reason is probably the use of hexagonal BN around the sample, which does not have the characteristics of a true low-friction cell. The dynamic method applied by Bose and Ganguly (1995) to correct for pressure is not suitable for a static high-pressure experiment. Further evidence for a correct P-T location of the coesite-quartz transition by earlier experiments—e.g., Mirwald and Massonne (1980)—is given by calorimetric measurements of SiO_2 polymorphs by Akaogi et al (1995). The use of the new thermodynamic data results in calculated curves that are virtually identical to the coesite-quartz boundary of Mirwald and Massonne (1980) and the stishovite-coesite curve of Zhang et al. (1996).

Most experimental studies in the SiO_2 system recently have been devoted to the rheology of coesite and the kinetics of the coesite-quartz transition. Renner et al. (1997) synthesized coesitites from SiO_2 glass applying various P-T trajectories before annealing the product at static conditions. The fabrics of some coesitites produced were believed by the authors to represent natural coesitites. Mosenfelder and Bohlen (1997) studied the kinetics of the transformation of coesite to quartz. They observed quartz fabrics around partially replaced coesite similar to that seen in natural examples, an observation already reported by Wirth and Stöckhert (1995). Zinn et al. (1997) investigated the kinetics for the formation of coesite from quartz in a multi-anvil press, applying an *in situ* method with synchrotron radiation. The rheology of coesitites was studied by Renner et al. (1995) in a solid-medium press with molten-salt technology. These authors presented rheological parameters applying the powder flow law, and Stöckhert and Renner (1998) improved the data.

System Al_2O_3-SiO_2-H_2O

In the water-free ASH system, kyanite is the stable phase that coexists with SiO_2 over a wide P range. At ~15 GPa it breaks down to form stishovite + corundum, according to the experimental study of Schmidt et al. (1997). For instance, these authors located the corresponding curve at 1000°C and 14 GPa.

In the presence of H_2O, the three phases diaspore, piezotite ($Al_3Si_2O_7(OH)_3$), and OH-topaz have to be considered in addition to kyanite (see Fig. 1). The stability fields of piezotite and OH-topaz have already been experimentally studied by Wunder et al. (1993a, 1993b). Recently, new experimental data for ASH were provided by Theye et al. (1997), who investigated the metastable extension of the curve of the reaction diaspore + quartz/coesite = pyrophyllite toward high T. Subsequently, thermodynamic calculations were undertaken by Theye et al. (1997) to relocate the reactions diaspore + quartz or coesite = kyanite + H_2O. However, the reaction with coesite is believed by Wunder et al. (1993a) to be metastable in regard to the appearance of piezotite (see Fig. 1). New expansivity data, such as those for diaspore (Pawley et al., 1996), but mainly new compressibility data for ASH phases (see Table 1), will support efforts to determine more precisely the P-T position of reaction curves by thermodynamic calculations in the future. However, the new data on diaspore in Table 1 require further comment, as the values of Xu et al. (1994) and Grevel et al. (1996) differ significantly. On the one hand, the high-pressure study of Xu et al. was far beyond the maximum P of 6.5 GPa applied by Grevel et al., and thus should be preferred. On the other hand, the experiments of Xu et al. (1994) were conducted using a diamond-

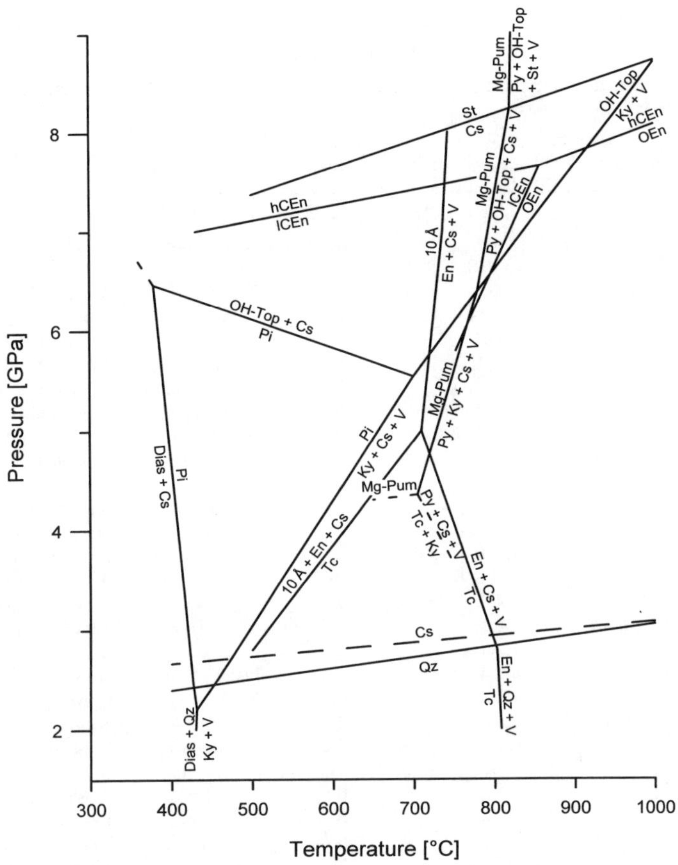

FIG. 1. Experimental results in subsystems of MASH with SiO_2 in excess shown in a P-T diagram. The quartz (Qz)-coesite (Cs) transition curve was taken from Mirwald and Massonne (1980) and from Bose and Ganguly (1995, long dashes). Additional sources (see also text): Fockenberg (1998a) for MgMgAl-pumpellyite (Mg-Pum); Pawley and Wood (1995) for talc (Tc) and 10 Å phase (10 Å); Theye et al. (1997) for diaspore (Dias); Woodland and Angel (1997) for the triple point of orthoenstatite (OEn), low-P clinoenstatite (lCEn), and high-P clinoenstatite (hCEn); Wunder et al. (1993a, 1993b) for OH-topaz (OH-Top) and piezotite (Pi); Zhang et al. (1996) for stishovite (St). The curves shown by short dashes are thermodynamically calculated curves (see e.g. Massonne, 1995) such as talc + kyanite (Ky) = pyrope (Py) + coesite + hydrous fluid (V).

anvil apparatus without pressure medium and, thus, there might be a clear difference between true pressure and that computed by these authors.

In the system without SiO_2 in excess, the reaction diaspore = corundum + H_2O was studied by Fockenberg et al. (1996) up to hydrous fluid pressures of 5 GPa. This study confirmed an earlier bracket by Vidal et al. (1994) obtained close to 3 GPa. It is a surprise that the resulting reaction curve shows a clear increase of the $\delta P/\delta T$ slope in the P range of 3 to 5 GPa (Fig. 2), which is not consistent with the results of thermodynamic calculations. This could mean that the P-T data given especially for the 5 GPa experiments were incorrect. Another explanation for the unexpected curvature of the diaspore breakdown curve could be the quality of PVT data for H_2O at high P. However, recent experimental studies by Wood et al. (e.g., 1995) at high P tend to confirm our previous knowledge about the PVT behavior of H_2O. Perhaps the dissolution of Al in H_2O increases with rising P and, thus, decreases the water activity, which could be the reason for the curvature, as well.

TABLE 1. Recent Compressibility Data on Phases Relevant for HP and UHP Metapelites and Related Rocks[1]

Phase	$V/V_0 = 1+aP+bP^2$, with P in bar[3]	Murnaghan EOS ($K_0' = 4$)[2]	P_{Max}/ GPa	Type of measurement	Authors
Phases of FMASH and related systems					
Diaspore		230	27	X-ray powd.	Xu et al. (1994)
—	−0.741, 1.48	(136.8)	6.5	X-ray powd.	Grevel et al. (1996)
Kyanite		193, 4	4.5	X-ray cryst.	Yang et al. (1997)
—		160, 4	6	X-ray cryst.	Comodi et al. (1997)
—		168.4	7	X-ray powd.	Grevel et al. (1998)
Pyrophyllite	−2.39, 10.1	(40.8)	6.5	X-ray powd.	Grevel et al. (1996)
O-enstatite		122.8, 5.6	8.5	X-ray cryst.	Hugh-Jones and Angel (1994)
L+H-C-enstatite		111.1, 6.6	8	X-ray cryst.	Angel and Hugh-Jones (1994)
O-pyroxene		115.4, 7.8	12.5	Acoustic w.	Chai et al. (1997)
Talc	−2.47, 31	(57.6)	4	Piston-displ.	Massonne et al. (1996)
Pyrope		171, 4.4	33	X-ray cryst.	Zhang et al. (1998)
Pyr50Maj50		173.1, μ = 92.3		Brillouin	Sinogeikin et al. (1997)
Majorite		165.6, μ = 88		Brillouin	Sinogeikin et al. (1997)
—		159.8, μ = 89.7		Brillouin	Pacalo and Weidner (1997)
$Mn_5Si_3O_{12}$ garnet		151.6, 6.4	12	Xy-ray cryst.	Arlt et al. (1998)
Mg-staurolite		168.4	7	X-ray powd.	Grevel et al. (1998)
Mg-carpholite	−1.23, 4.2	(84.4)	5.5	X-ray powd.	Theye (pers. commun.)
Mg-chloritoid		125.6	6.5	X-ray powd.	Grevel et al. (1997)
Mg-chlorite	−1.75, 6.5	(57.4)	4.5	Piston-displ.	Massonne et al. (1996)
—		77.6	6.5	X-ray powd.	Grevel et al. (1997)
Fe-Mg-chlorite	−1.47, 4.7	(68.3)	4.5	Piston-displ.	Massonne et al. (1996)
Cookeite		(71)	5.5	X-ray powd.	Fasshauer et al. (1998a)
K phases and Na phases for comparison					
Kalsilite		(51.2)	6	X-ray powd.	Fasshauer et al. (1998b)
Leucite		(48.0)	5	X-ray powd.	Fasshauer et al. (1998b)
Jadeite		125, 5	8	X-ray powd.	Zhao et al. (1997)
K-feldspar		58.3, 4	7	X-ray cryst.	Allan and Angel (1997)
Albite		54, 6	4	X-ray cryst.	Downs et al. (1994)
K-cymrite		(40.1)	6	X-ray powd.	Fasshauer et al. (1997)
Muscovite		62.5, 4	15	X-ray powd.	Faust and Knittle (1994)
—	−2.19, 15.5	(54.0)	4.5	Piston-displ.	Massonne et al. (1996)
Paragonite		53.0, 5.5	4	X-ray cryst.	Comodi and Zanazzi (1997)
Phlogopite	−2.22, 10.5	(45.7)	7	X-ray powd.	Schebaum and Massonne (1998)
Biotite	−2.30, 12.1	(45.6)	7	X-ray powd.	Schebaum and Massonne (1998)
—	−2.18, 9.8	(46.3)	6	X-ray powd.	Schebaum and Massonne (1998)
Ferro-stilpnomel.	−2.05, 1.1	(40.3)	6.5	X-ray powd.	Schebaum and Massonne (1998)
Ferri-stilpnomel.	−2.06, 1.8	(40.7)	6.5	X-ray powd.	Schebaum and Massonne (1997)
Zussmanite	−1.80, 5.3	(53.2)	6.5	X-ray powd.	Schebaum and Massonne (1997)
Carbonates					
Magnesite		137.5, 4	25	X-ray powd.	Fiquet et al. (1994)
—		103, 4	7	X-ray powd.	Zhang et al. (1997)
—		111, 4	7	X-ray cryst.	Ross (1997)
Dolomite		112.9, 4	11	X-ray powd.	Fiquet et al. (1994)
—		90.7, 4	7	X-ray powd.	Martinez et al. (1996)
Aragonite		64.8, 4	8	X-ray powd.	Martinez et al. (1996)

[1]Numbers in parentheses are recalculated to the Murnaghan EOS with $K_0' = 4$, considering identical values for volume according to the original and the assumed volume expression at 4 GPa, except for talc (3 GPa0). The type of measurement was related either to a specific piston displacement method undertaken in a piston-cylinder apparatus, Brillouin spectroscopy, acoustic wave velocity determination, or *in situ* studies with X-rays in various high-pressure instruments using a single crystal or powder of the phase of interest.

[2]Or other EOS (e.g., Birch-M.), K_0/GPa, K_0'.

[3]The first item is the a parameter divided by 10^6 bar; the second item is the b parameter divided by 10^{12} bar^2.

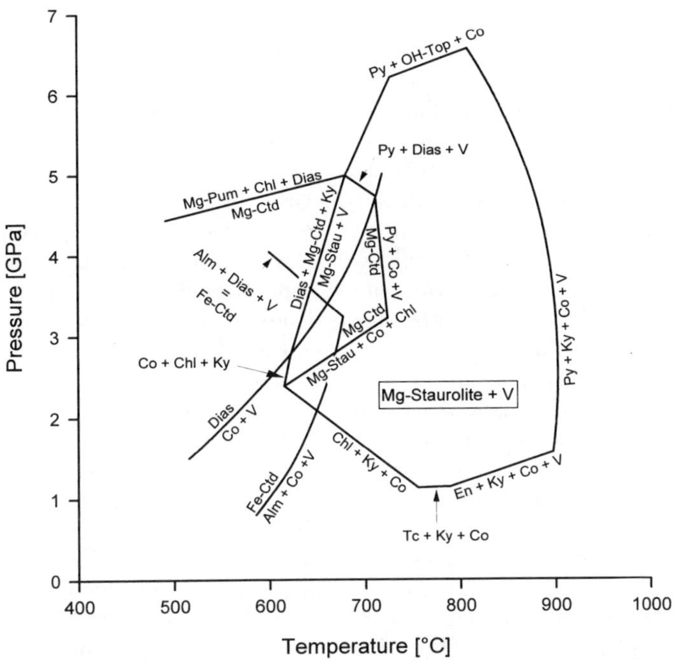

FIG. 2. P-T diagram showing experimental results in FMASH in the absence of quartz or coesite. The curve of the reaction diaspore = corundum (Co) + hydrous fluid was derived from results of Fockenberg et al. (1996). The stability fields of Mg-chloritoid (Mg-Ctd) and Mg-staurolite (Mg-Stau) were given by Fockenberg (1995) and (1998b), respectively. The stability field of Fe-chloritoid (Fe-Ctd) was determined by Vidal et al. (1994) using the NNO buffer. Further phase abbreviations not explained in Figure 1: Alm = almandine; Chl = chlorite.

System FeO-MgO-Al_2O_3-SiO_2-H_2O and related alkali-free systems

In the subsystems MgO-SiO_2 and FeO-SiO_2 orthopyroxenes coexist with SiO_2. Recent high-pressure experiments (see Woodland and Angel, 1997) have contributed to a better P-T location of the stability fields of the three polymorphs of both $MgSiO_3$ and $FeSiO_3$. The triple point of orthorhombic enstatite, low-P $P2_1/c$ monoclinic enstatite, and high-P C2/c monoclinic enstatite is shown in Figure 1 to lie at 860°C and 7.7 GPa. The corresponding triple point for ferrosilite was located by Woodland and Angel (1997) at 830°C and 4.9 GPa, with all three transition curves showing a somewhat higher $\delta P/\delta T$ slope compared to enstatites. The corresponding curves for members of the enstatite-ferrosilite solid solution series probably lie between the curves of the end members (Woodland and Angel, 1997).

New compressibility data for the above pyroxenes are summarized in Table 1. If Al is added to the system, the orthopyroxenes can introduce Al not only according to the Tschermak's substitution, but there is also a limited solid solution series toward $Mg_{0.05}Al[Si_2O_6]$, observed by Fockenberg and Schreyer (1997) through experiments at 2.5 GPa in the MAS system. The corresponding substitution should actually play a role when SiO_2 is in excess. However, Fockenberg and Schreyer (1997) were not able to demonstrate that this is a phenomenon present under stable conditions. If water is present, some can even be introduced into enstatite, a so-called "dry" phase; Rauch and Keppler (1998) determined contents of 0.1 wt% H_2O at 10 GPa and 1100°C. The presence of Al in the system increases water solubility in enstatite.

In the subsystem MgO-SiO_2-H_2O, talc and the so-called 10 Å phase, $Mg_3[Si_4O_{10}](OH)_2 \cdot xH_2O$, can coexist with SiO_2 as well. The experimental study of Pawley and Wood (1995) has proven that the 10 Å phase is stable at low temperature and high hydrous fluid pressure. Its stability field, bounded by two reaction curves that were thermodynamically calculated by Pawley and Wood

(1995), is shown in Figure 1. According to these calculations, talc is stable up to 5 GPa in the presence of H_2O. At the coesite-quartz transition, talc breakdown to enstatite + SiO_2 + hydrous fluid occurs at 800°C, or somewhat above, as already reported in earlier studies (see Massonne, 1995). In the system MgO-Al_2O_3-SiO_2-H_2O (MASH), talc can contain Al through Tschermak's substitution. The P-T dependence of the extent of this substitution in the presence of kyanite + quartz/coesite was experimentally investigated (see Massonne, 1995), demonstrating that an increase in the Al content of talc is caused by increasing T or decreasing P. These experimental data were used to derive thermodynamic data for the theoretical end member Al-talc in order to apply these data for geothermobarometry on HP and UHP metapelites. Further experimental information on Al introduction in talc was provided by Hoschek (1995).

In addition to the aforementioned phases, garnet and a pumpellyite phase, $Mg_5Al_5Si_6O_{21}(OH)_7$, can coexist with SiO_2 at UHP metamorphic conditions, and therefore these phases can be expected to occur with coesite in metapelites. New experimental data on MgMgAl-pumpellyite were provided by Fockenberg (1998a), showing breakdown temperatures of 700 to 800°C for this phase in the P interval from 4.5 to 7 GPa, as calculated previously by Massonne (1995). Because of this coincidence, a minimum pressure for MgMgAl-pumpellyite + coesite is likely to be somewhat above 4 GPa (see Fig. 1). In MASH, garnet + coesite appear above 700°C. The minimum T is reached close to 4.3 GPa, according to thermodynamic calculations of Massonne (1995). However, because of new compressibility data for talc that were experimentally determined by Massonne et al. (1996), and for pyrope (see Table 1), these calculations presently can be improved. In addition, new experimental data on the formation of pyrope from various assemblages in the absence of quartz or coesite provided by Fockenberg (1995) can contribute to this, because these data will help to refine the thermodynamic data of pyrope.

With respect to pyrope, it is worth noting that some H_2O can be introduced. Lu and Keppler (1997) observed rising H_2O contents, with P reaching 200 ppm H_2O at 1000°C and 10 GPa. In order to estimate phase relations for UHP metapelites with garnet compositions deviating from pyrope or to use natural garnet in such rocks for geothermobarometry, adequate knowledge of the thermodynamic properties of Al-garnet solid solutions is necessary. Recent experimental work has contributed to this, involving, for instance, calorimetric measurements such as those reported by Bosenick et al. (1996), the study of the P-T displacement of a particular reaction curve as a function of garnet composition as undertaken by Ganguly et al. (1996) and Koziol (1996), and the determination of cation partitioning among garnet and another mineral (e.g., Wood et al., 1994). Moreover, Geiger and Feenstra (1997) reported new volume data for synthetic garnets.

Although the absence of quartz or coesite is very rare in metapelites, it is worth reviewing corresponding experimental studies—for instance, on chloritoid or staurolite in FMASH—since these investigations contribute at least to the improvement of the thermodynamic properties of such phases, as do HP in situ measurements to determine the equation of state (EOS) for Mg-chloritoid and Mg-staurolite (Grevel et al., 1997, 1998; see Table 1). In MASH, the P-T location of several reactions bounding the stability fields of Mg-chloritoid (see Fig. 2) and chlorite toward high P and T were investigated by Fockenberg (1995). The main problems with the corresponding bracketing experiments are related to the uncertain topology in the complex MASH system and the unknown composition of the reacting phases, especially when a phase with an extensive solid solution such as chlorite, reflecting Tschermak's substitution, is participating.

The same problems exist with regard to the stability field of Mg-staurolite (Fockenberg, 1998b; see Fig. 2) that has the ideal composition $Mg_4Al_{18}Si_8O_{46}(OH)_2$ but deviates from this toward lower Si and higher OH contents. Moreover, the topological proposals of Fockenberg (1995, 1998a, 1998b) contain inconsistencies that are discernible when attempts are made to reconstruct the P-T position of metastable invariant points by intersecting metastable extensions of Fockenberg's curves. Despite this, the many experimental results of Fockenberg (1995, 1998a, 1998b) help to elucidate the correct phase relations in MASH at UHP conditions. In FASH, chloritoid stability is limited to lower fluid pressures compared to Mg-chloritoid, even when the Ni-NiO buffer is considered (Vidal et al., 1994;

see Fig. 2). Thermodynamic properties for Fe-staurolite partially derived from HP experiments were provided by Holdaway et al. (1995). Recent synthesis and reversed experiments on Fe-rich staurolite, including detailed characterization of the run products (Koch-Müller, 1997; Koch-Müller et al., 1997), resulted in new thermodynamic data for the staurolite solid solutions.

New information also exists regarding thermodynamic properties of chlorite solid solutions as a result of experimentation at HP conditions. Böving and Massonne (1994) as well as Roots (1994) presented volume data for synthetic MASH chlorites. Compressibility data for various chlorites (Table 1) were provided by Massonne et al. (1996) and Grevel et al. (1997). Baker and Holland (1996) presented experimental information on the stable extension of the chlorite solid solution in MASH at various P and T. Holland et al. (1998) used these data to derive a peculiar thermodynamic solid solution model for chlorite.

Systems Na_2O-Al_2O_3-SiO_2-H_2O and K_2O-Al_2O_3-SiO_2-H_2O

In both H_2O-free systems NAS and KAS, the upper pressure stability limits of the feldspars were recently investigated again by HP experiments and subsequent thermodynamic considerations. Waterwiese et al. (1995) presented a reaction curve for jadeite + quartz = albite that is shown in Figure 3. Urukawa et al. (1994) as well as Yagi et al. (1994) determined the P-T curve of the reaction wadeite-type $K_2Si_4O_9$ + kyanite + coesite = sanidine to lie between 6 and 7 GPa at high T. Fasshauer et al. (1998b) regarded this reaction to occur stably, but only below 530°C and 5.2 GPa. For higher T, these authors suggested another reaction that should be coesite + kalsilite = K-feldspar (see Fig. 3) as a result, for instance, of their calorimetric measurements on $K_2Si_4O_9$ and new compressibility data on kalsilite (see Table 1). $K_2Si_4O_9$ + kyanite + coesite should become stable at 7 GPa and 1100°C (Fasshauer et al., 1998b). The main problem with these suggestions is that the thermodynamic calculations were undertaken for pure phases. However, phases such as kalsilite can significantly deviate from the ideal composition as reported, for example, by Hovis and Roux (1998). At ~9 GPa, hollandite-type $KAlSi_3O_8$ forms from $K_2Si_4O_9$ + kyanite + coesite according to Urukawa et al. (1994) and Yagi et al. (1994). The corresponding reaction curve lies very close to the stishovite-coesite transition, as determined by Zhang et al. (1996).

The above summary shows that there is no potassium analogue to jadeite or any other phase of the composition $KAlSi_2O_6$ that is stable under the conditions of the coesite stability field. However, $KAlSi_2O_6$ can be a significant component in clinopyroxenes of the diopside-jadeite join. Up to 11 GPa, Luth (1997) observed an increase of potassium introduction into diopside up to 1.0 wt% K_2O. At similar T around 1400°C, even higher contents were reported by Harlow (1997) at P of 10 and 11 GPa. With potassium carbonate added to various pyroxenes showing compositions between diopside, jadeite, and kosmochlor, up to 4 wt% K_2O could be introduced, although the maximum was only 1 wt% for jadeite. At the moment, the complex relations of pyroxene composition including K content, excess phases, and P and T are not yet resolved.

In the wet system KASH, a phase with cymrite structure was reported in several papers published in the 1960s and 1970s and called K-cymrite, $KAlSi_3O_8 \cdot xH_2O$. A re-examination of the P-T location of the reaction K-cymrite = K-feldspar + H_2O by Massonne (1995), Fasshauer et al. (1997), and Thompson et al. (1998) led virtually to the same result (see Fig. 3); when we consider that the latter authors corrected their P-T data according to the coesite-quartz curve of Bose and Ganguly (1995), their usefulness is called into question. The water content of K-cymrite might be somewhat below the ideal value of x = 1.0 (Thompson et al., 1998). Expansivity and compressibility of K-cymrite (see Table 1) were measured by Fasshauer et al. (1997). For NASH, new compressibility data were reported by Comodi and Zanazzi (1997) for paragonite.

System K_2O-FeO-MgO-Al_2O_3-SiO_2-H_2O

In the more complex system KFMASH, micas play an important role because they are more or less omnipresent in low-temperature UHP metasediments. Reconnaissance studies on the maximum P of the appearance of phengites in such rocks were undertaken by Domanik and Holloway (1996, 1998), who employed natural starting materials. In the bulk composition close to an average metapelite, potassic white mica was still present in 11 GPa and 900°C, whereas at this

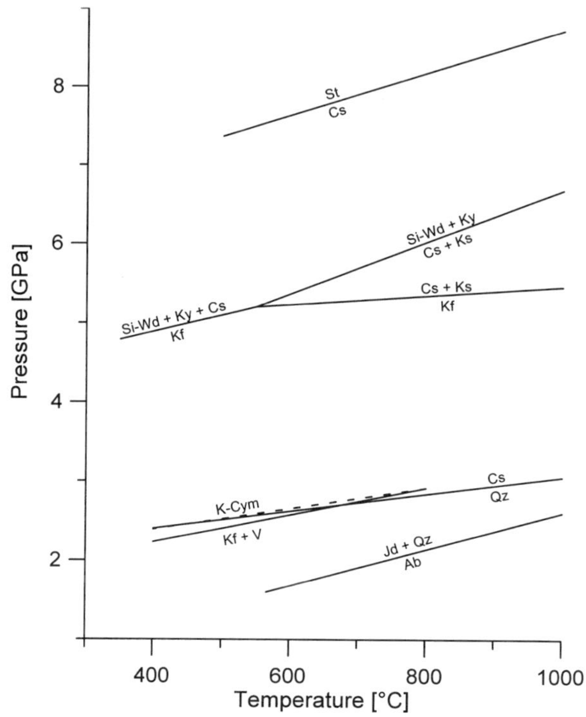

FIG. 3. Experimental and thermodynamically calculated results for NAS and KASH with and without H_2O in a P-T diagram. The curve for jadeite (Jd) + quartz = albite (Ab) was given by Waterwiese et al. (1995). The phase relations of K-feldspar (Kf), kalsilite (Ks), and wadeite-type $K_2Si_4O_9$ (Si-Wd) were taken from Fasshauer et al. (1998b). The curves bounding the coesite stability field are those of Figure 1. The curve of the reaction K-cymrite (K-Cym) = K-feldspar + hydrous fluid was taken from Massonne (1995) and Thompson et al. (1998, dashed). For phase abbreviations see Figure 1.

T and above 9 GPa a phengitic muscovite broke down to form K-hollandite, among other phases. Muscovite, however, decomposes at lower P. Faust and Knittle (1994) gave a surprisingly low value of 3 GPa at 800°C. At low T and hydrous fluid pressures above 3 GPa, phengite + coesite can coexist with talc or 10 Å phase. At a T of 710 to 800°C, depending on pressure, Mg-rich orthopyroxene replaces the latter phases (see Fig. 1). The minimum hydrous fluid pressure for phengite + SiO_2 + enstatite is ~2.7 GPa according to Massonne (1995). The P-T location of the KMASH reaction phengite + enstatite = phlogopite + pyrope + coesite is shown in Figure 4. The P-T dependence of the compositional variation of KMASH and FKASH phengites, which is mainly the result of Tschermak's substitution, was experimentally studied by Massonne and Szpurka (1997) for the assemblage with garnet, kyanite, and coesite or quartz. Some results of this study are shown in Figure 4. Massonne and Szpurka used their new and earlier experimental results to determine the thermodynamic properties of phengites and thus of the muscovite-Al-celadonite series. Then Massonne (1997) used natural garnet-phengite pairs to derive thermodynamic properties for the MgAl-celadonite–FeAl-celadonite join and to improve the data for FeAl-celadonite. Both experimental studies and research on natural rocks resulted in virtually the same data for this end member, which confirms the assumption of Massonne and Szpurka (1997), that their synthesis data represent those of stable mineral assemblages. New compressibility data for natural muscovites (see Table 1) were presented by Faust and Knittle (1994) as well as Massonne et al. (1996).

In KMASH, Rosenbleck and Massonne (1993) and Rosenbleck (1996) synthesized pure micas of the phlogopite solid solution series, and assemblages of these micas either with pyrope and en-

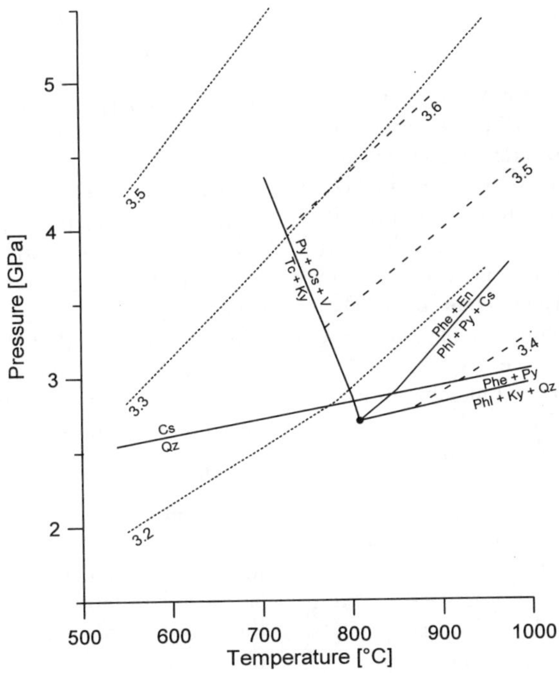

FIG. 4. Si isopleths for phengite coexisting with garnet, kyanite, and quartz or coesite in KMASH (long dashes) and KFASH (short dashes) shown as function of P and T. The reaction curves bounding the stability field of the above KMASH assemblage were taken from Massonne and Szpurka (1997). In addition, the curve of the reaction phengite (Phe) + enstatite = phlogopite (Phl) + pyrope + coesite is shown, as determined by Massonne (1995). For abbreviations and coesite-quartz curve see Figure 1.

statite or with forsterite and spinel at different P-T conditions up to 5 GPa. The compositional change of phlogopite with P and T in these assemblages and the volume data obtained from various phlogopite compositions were used to determine thermodynamic properties of the phlogopite solid solution series toward $KMg_{2.5}[Si_4O_{10}](OH)_2$ and the theoretical end member $KAl_3[Al_4O_{10}](OH)_2$ (Rosenbleck, 1996). In KFASH, a similar study was undertaken by Schebaum and Massonne (1996), resulting in thermodynamic properties for the ternary annite solid solution series in this system. The determination of thermodynamic properties of the annite-siderophyllite series by HP experimentation also was the objective of Dachs (1994) and Benisek et al. (1996). Redhammer et al. (1995) investigated the join phlogopite-annite for characterizing such micas. The EOS of various dark micas was determined by Schebaum and Massonne (1997) through *in situ* measurements up to 7 GPa (see Table 1). The same type of investigation was conducted by Schebaum and Massonne (1998) on two different stilpnomelanes and a zussmanite up to 6.5 GPa (Table 1) in order to improve thermodynamic data and, thus, the opportunity to estimate the P-T conditions of the stability field for these potential UHP phases occurring in low-temperature Fe-rich metasediments (see also Massonne and Szpurka, 1997).

Potential UHP phases for metasediments are, in principle, the newly characterized phase X, which is a K-Mg-silicate that was synthezised at 16 GPa by Konzett (1998), and $Na_2Mg_2Si_2O_7$ (Gasparik and Litvin, 1997). At least the latter phase can coexist with garnet + SiO_2 according to phase relations studied experimentally by Gasparik and Litvin up to 22 GPa. But it seems that the occurrence of both exotic phases in metasediments, if at all, is limited to pressure conditions of the stishovite stability field. At the moment, our knowledge about phase relations relevant to metapelites is extremely poor at pressures above 7 GPa.

Al- and/or Mg-bearing borates, phosphates, and related silicates

Very rare minerals that are Al-rich borates and phosphates occur in the UHP pyrope quartzite of the Dora-Maira massif, Western Alps (see Chopin and Sobolev, 1995). This indicates that in Al-rich metamorphic rocks, and thus in metapelites, those Al-rich phases can occur especially at high or even ultrahigh pressures. Such minerals may even be indicative of UHP metamorphism. Therefore, it is important to review recent experimental studies on such kinds of minerals.

The stability of the OH end member of the Al-borate jeremejevite, $Al_6(BO_3)_5(OH)_3$, was studied by Stachowiak and Schreyer (1998). This mineral is stable up to at least 5 GPa, and decomposes at a T of ~750°C (see Fig. 5). Mg-borates such as kotoite, $Mg_3(BO_3)_2$, and szaibelyite, $MgBO_2(OH)$, could be synthesized at 5 GPa and 4 GPa, respectively (see Werding and Schreyer, 1996). The reaction of pseudosinhalite, $Mg_4Al_6O_2(BO_4)_4(OH)_2$, to sinhalite, $Mg_5Al_5(BO_4)_5$, corundum, and H_2O was studied by Krosse (1995) up to 4.5 GPa, and the corresponding curve is shown in Figure 5. Various borosilicates also were recently investigated at high P. The upper thermal stability limit of dumortierite, $\sim Al_{6.67}BSi_3O_{17}(OH)$, was studied by Werding and Schreyer up to 4.5 GPa and was determined to lie close to 1000°C at this P. The tourmaline phase dravite, $NaMg_3Al_6Si_6O_{18}(BO_3)_3(OH)_4)$, which is stable up to at least 6 GPa (Werding and Schreyer, 1996) breaks down at similar T (see Fig. 5). The stability field of prismatine of the composition 3.67 MgO · 3.02 Al_2O_3 · 3.74 SiO_2 · 0.43 B_2O_3 · 1.3 H_2O was determined by Krosse (1995), who applied reversal experiments in a piston-cylinder apparatus. Preliminary experimental results of Ufer presented by Schreyer and Werding (1997) addressed the upper pressure limit of reedmergnerite, $NaBSi_3O_8$, the boron analogue of albite. It breaks down at 3.5 GPa (see Fig. 5) to form coesite and the boron analogue of nepheline. Despite the many recent studies on borates and borosilicates, the complicated phase relations, especially those that could be of relevance to metapelites, are not yet sufficiently well known.

In principle, the same is true concerning phosphates, although some recent experimental studies addressed this mineral group. In the MgO-P_2O_5-H_2O system, $Mg_3(PO_4)_2$ and Mg_2PO_4OH are stable up to at least 2.3 GPa and 3.0 GPa, respectively (Brunet et al., 1998). This applies to phosphoellenbergerite, which is close to $Mg_7P_4O_{15}(OH)_4$, as well. Surprisingly, the stability limit of this phase lies at P as low as 0.5 GPa (see Fig. 5) according to the reversal experiments of Brunet et al. (1998), whereas the silicate analogue to phosphoellenbergerite—whether Ti or Zr bearing (Burchard and Schreyer, 1996)—shows a lower P limit close to 2.5 GPa. Brunet and Chopin (1995) demonstrated that the phosphates trolleite, $Al_4(PO_4)_3(OH)_3$, and bearthite, $Ca_2Al(PO_4)_2(OH)$, are stable at pressures clearly below 1 GPa, but claim that these phases exist at very high pressures, as well.

Ca-Mg carbonates

Recently the reaction aragonite + magnesite = dolomite was studied by Martinez et al. (1996) by in situ measurements in a multi-anvil apparatus. The resulting curve that was finally calculated by these authors using their newly determined EOS of aragonite and dolomite (see Table 1) is shown in Figure 6 to lie, for instance, at 1000°C and 5.5 GPa. The same figure exhibits the newly determined curve of the reaction magnesite + quartz = enstatite + CO_2 by Koziol and Newton (1995). Both curves show a similar $\delta P/\delta T$ slope, which also holds true for early studies of the reaction diopside + magnesite = enstatite + dolomite by Kushiro et al. (1975) and Brey et al. (1983). Under these circumstances, it is impossible to examine which topology in the CaO-MgO-SiO_2-CO_2 system is the stable one. In every case, it can be concluded that coesite coexists with aragonite + magnesite at very high pressures. This assemblage reacts to form diopside + CO_2. Therefore, the possible appearance of coesite + aragonite + magnesite in carbonate-bearing metasediments at UHP conditions is a clear function of the CO_2 activity. New experimental studies might help to elucidate the true topology of the CaO-MgO-SiO_2-CO_2 system at UHP conditions, as do thermodynamic data such as the newly determined compressibility data for carbonates presented in Table 1.

The rheology of synthetic polycrystalline aragonite as a model substance for carbonate rocks was studied by Rybacki et al. (1995) in a solid-medium press with molten salt technology. The flow parameters were found to be similar to calcite rocks. Rheological parameters for polycrys-

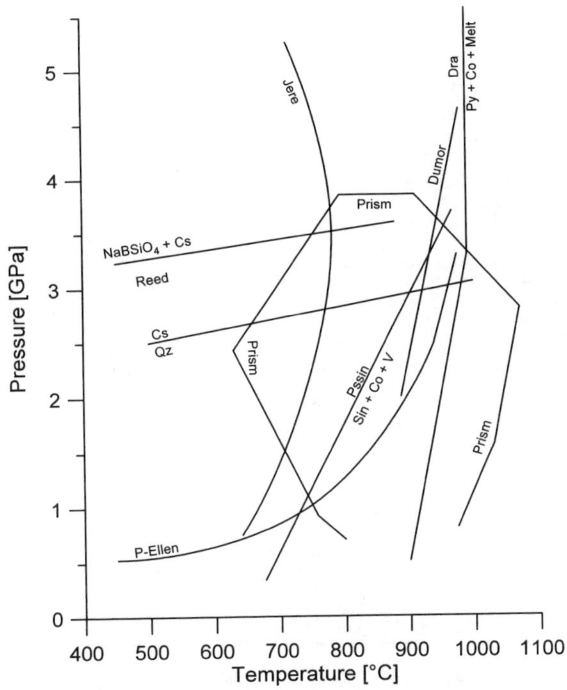

FIG. 5. The P-T stability limits of various borates, borosilicates, and phosphates according to reversal studies. For composition of minerals and sources of reaction curves see text. Abbreviations not mentioned in the previous figures are: Dra = dravite; Dumor = dumortierite; Jere = jeremejevite; P-Ellen = phosphoellenbergerite; Prism = prismatine; Pssin = pseudosinhalite; Reed = reedmergnerit; Sin = sinhalite.

talline aragonite were given by Stöckhert and Renner (1998).

Final Evaluation and Prospects

Despite the many new experimental results, reviewed above, on phases of the metapelitic model system KFMASH and related systems, there is only rudimentary information on phase relations of metasediments at UHP conditions. Even for the simple subsystems KAS and MASH, we cannot be certain about the topology of stable reactions in the range from 3 to 8 GPa. This was discussed above in the context of studies of Fasshauer et al. (1998b) or Fockenberg (1995, 1998a, 1998b). Of course, critical phase relations in model systems can be investigated currently by reversal experiments in order to arrive at final topological conclusions. However, this makes less sense when the participating minerals that are frequently solid solution phases are not compositionally well characterized. This is necessary if attempts to derive thermodynamic properties from such reversal studies are to be successful. Accurate thermodynamic properties of mineral end members and solid solution series will finally lead to a correct topological relation of a model system. Moreover, these relations then can be extrapolated to more complex systems in order to approach natural systems.

In addition, we need the thermodynamic properties of solid solutions for geothermobarometry necessary to investigate the metamorphic evolution of sedimentary rocks at HP and UHP conditions. Under these circumstances, other types of measurements in addition to reversal studies are very important. These concern the determination of compressibility for rock-forming minerals, especially when we are interested in phase relations at very high pressures. Therefore, the above review has strongly emphasized the recent high-pressure studies on the β_T-function or even the EOS of minerals. However, the study of α_P- and c_P-functions contribute to the understanding of

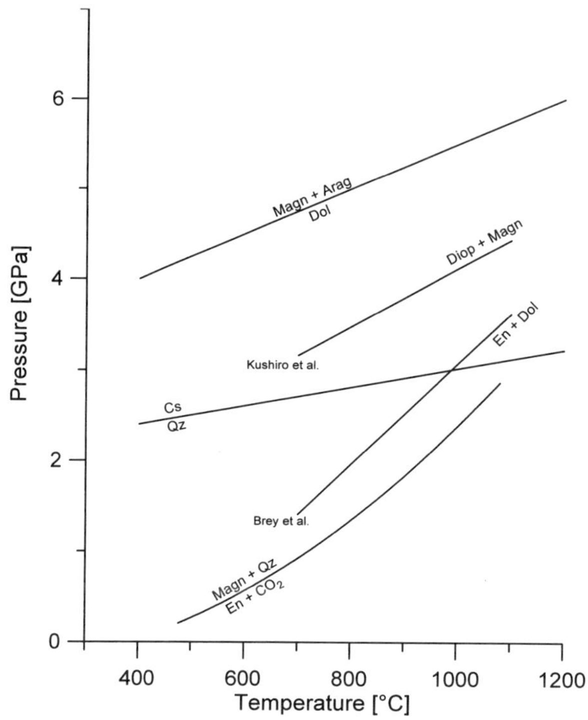

FIG. 6. P-T diagram showing reaction curves that involve aragonite (Arag), dolomite (Dol), and magnesite (Magn). For sources, see text. Abbreviations: Diop = diopside.

thermodynamic properties as well. Cation distribution experiments and experiments aimed at the study of the displacement of a reaction curve also are valuable. These concern the study of a particular equilibrium, in general, in which the least amount of phases possible should participate. At the moment, we have some clues about thermodynamic properties of the solid solution series in garnet, pyroxene, and micas. Limited information exists also on other phases. In addition to phase-equilibrium studies relevant to metapelites, the experimental study of the kinetics of reactions and the rheology of model rocks should be investigated more intensively.

REFERENCES

Akaogi, M., Yusa, H., Shiraishi, K., and Suzuki, T., 1995, Thermodynamic properties of α-quartz, coesite, and stishovite and equilibrium phase relations at high pressures and high temperatures: Jour. Geophys. Res., v. 100B, p. 22,337–22,347.

Allan, D. R., and Angel, R. J., 1997, A high-pressure structural study of microcline ($KAlSi_3O_8$) to 7 GPa: Eur. Jour. Mineral., v. 9, p. 263–275.

Angel, R. J., and Hugh-Jones, D. A., 1994, Thermodynamics of enstatite at high pressures [abs.]: Terra Abs., v. 6, no. 1, 5th Int. Symp. Exp. Mineral. Petrol. Geochem., p. 1.

Arlt, T., Armbruster, T., Miletich, R., Ulmer, P., and Peters, T., 1998, High pressure single-crystal synthesis, structure, and compressibility of the garnet $Mg^{2+}{}_3Mn^{3+}{}_2[SiO_4]_3$: Phys. Chem. Minerals, v. 26, p. 100–106.

Baker, J., and Holland, T. J. B., 1996, Experimental reversals of chlorite compositions in divariant $MgO + Al_2O_3 + SiO_2 + H_2O$ assemblages: Amer. Mineral., v. 81, p. 676–684.

Benisek, A., Dachs, E., Redhammer, G., Tippelt, G., and Amthauer, G., 1996, Activity-composition relationship in Tschermak's substituted Fe biotites at 700°C, 2 kbar: Contrib. Mineral. Petrol., v. 125, p. 85–99.

Böving, R., and Massonne, H.-J., 1994, Thermodynamische eigenschaften synthetischer chlorite des systems $MgO-Al_2O_3-SiO_2-H_2O$ (MASH) [abs.]: Ber. Deuts. Mineral. Ges., Beih./Eur. Jour. Mineral., v. 6, no. 1, p. 36.

Bose, K., and Ganguly, J., 1995, Quartz-coesite transition

revisited: Reversed experimental determination at 500–1200°C and retrieved thermochemical properties: Amer. Mineral., v. 80, p. 231–238.

Bosenick, A., Geiger, C. A., and Cemic, L., 1996, Heat capacity measurements of synthetic pyrope-grossular garnets between 320 and 1000 K by differential scanning calorimetry: Geochim. et Cosmochim. Acta, v. 60, p. 3215–3227.

Brey, G., Brice, W. R., Ellis, D. J., Green, D. H., Harris, K. L., and Ryabchikov, I. D., 1983, Pyroxene-carbonate reactions in the upper mantle: Earth Planet. Sci. Lett., v. 62, p. 63–74.

Brunet, F., and Chopin, C., 1995, Bearthite, $Ca_2Al(PO_4)_2OH$: Stability, thermodynamic properties, and phase relations: Contrib. Mineral. Petrol., v. 121, p. 258–266.

Brunet, F., Chopin, C., and Seifert, E., 1998, Phase relations in the MgO-P_2O_5-H_2O system and the stability of phosphoellenbergerite: petrological implications: Contrib. Mineral. Petrol., v. 131, p. 54–70.

Burchard, M., and Schreyer, W., 1996, The lower pressure and upper thermal stability of zirconium ellenbergerite [abs.]: Terra Abstr., v. 8, no. 1, 6th Int. Symp. Exp. Mineral. Petrol. Geochem., p. 10.

Chai, M., Brown, J. M., and Slutsky, L. J., 1997, The elastic constants of an aluminous orthopyroxene to 12.5 GPa: Jour. Geophys. Res., v. 102B, p. 14,779–14,785.

Chopin, C., and Sobolev, N. V., 1995, Principal mineralogical indicators of UHP in crustal rocks, in Coleman, R., and Wang, X., eds., Ultrahigh pressure metamorphism: New York, Cambridge University Press, p. 96–131.

Comodi, P., and Zanazzi, P. F., 1997, Pressure dependence of structural parameters of paragonite: Phys. Chem. Minerals, v. 24, p. 274–280.

Comodi, P., Zanazzi, P. F., Poli, S., and Schmidt, M. W., 1997, High-pressure behavior of kyanite: Compressibility and structural deformations: Amer. Mineral., v. 82, p. 452–459.

Dachs, E., 1994, Annite stability revised. 1. Hydrogen-sensor data for the reaction annite = sanidine + magnetite + H_2: Contrib. Mineral. Petrol., v. 117, p. 229–240.

Domanik, K. J., and Holloway, J. R., 1996, The stability and composition of phengitic muscovite and associated phases from 5.5 to 11 GPa: Implications for deeply subducted sediments: Geochim. et Cosmochim. Acta, v. 90, p. 4133–4150.

———, 1998, Experimental determination of the stability and phase relations of phengitic muscovite in a calcareous metapelite from the Dabie Mountains, China [abs.], in International Workshop on UHP Metamorphism and Exhumation: Stanford, CA, Stanford University, Abs. vol., p. A206–A207.

Downs, R. T., Hazen, R. M., and Finger, L. W., 1994, The high-pressure crystal chemistry of low albite and the origin of the pressure dependency of Al-Si ordering: Amer. Mineral., v. 79, p. 1042–1052.

Fasshauer, D. W., Chatterjee, N. D., and Marler, B., 1997, Synthesis, structure, thermodynamic properties and stability relations of K-cymrite, $K[AlSi_3O_8] \times H_2O$: Phys. Chem. Minerals, v. 24, p. 455–462.

Fasshauer, D. W., Chatterjee, N. D., and Cemic, L., 1998a, A thermodynamic analysis of the system $LiAlSiO_4$-$NaAlSiO_4$-Al_2O_3-SiO_2-H_2O based on new heat capacity, thermal expansion, and compressibility data for selected phases: Contrib. Mineral. Petrol., v. 133, p. 186–198.

Fasshauer, D. W., Wunder, B., Chatterjee, N. D., and Höhne, G. W. H., 1998b, Heat capacity of wadeite-type $K_2Si_4O_9$ and the pressure-induced stable decomposition of K-feldspar: Contrib. Mineral. Petrol., v. 131, p. 210–218.

Faust, J., and Knittle, E., 1994, The equation of state, amorphization, and high-pressure phase diagram of muscovite: Jour. Geophys. Res., v. 99B, p. 19,785–19,792.

Fiquet, G., Guyot, F., and Itie, J. P., 1994, High-pressure X-ray diffraction study of carbonates: $MgCO_3$, $CaMg(CO_3)_2$, and $CaCO_3$: Amer. Mineral., v. 79, p. 15–23.

Fockenberg, T., 1995, New experimental results up to 100 kbar in the system MgO-Al_2O_3-SiO_2-H_2O (MASH): Preliminary stability fields of chlorite, chloritoid, staurolite, MgMgAl-pumpellyite, and pyrope: Bochumer Geol. Geotech. Arb., v. 44, p. 39–44.

———, 1998a, An experimental study of the pressure-temperature stability of MgMgAl-pumpellyite in the system MgO-Al_2O_3-SiO_2-H_2O: Amer. Mineral., v. 83, p. 220–227.

———, 1998b, An experimental investigation on the P-T stability of Mg-staurolite in the system MgO-Al_2O_3-SiO_2-H_2O: Contrib. Mineral. Petrol., v. 130, p. 187–198.

Fockenberg, T., and Schreyer, W., 1997, Synthesis and chemistry of unusual excess-Si aluminous enstatite in the system MgO-Al_2O_3-SiO_2 (MAS): Eur. Jour. Mineral., v. 9, p. 509–518.

Fockenberg, T., Wunder, B., Grevel, K.-D., and Burchard, M., 1996, The equilibrium diaspore-corundum at high pressures: Eur. Jour. Mineral., v. 8, p. 1293–1299.

Ganguly, J., Cheng, W., and Tirone, M., 1996, Thermodynamics of aluminosilicate garnet solid solution: New experimental data, an optimized model, and thermometric applications: Contrib. Mineral. Petrol., v. 126, p. 137–151.

Gasparik, T., and Litvin, Y. A., 1997, Stability of $Na_2Mg_2Si_2O_7$ and melting relations on the forsterite-jadeite join at pressures up to 22 GPa: Eur. Jour. Mineral., v. 9, p. 311–326.

Geiger, C. A., and Feenstra, A., 1997, Molar volumes of mixing of almandine-pyrope and almandine-spessartine garnets and the crystal chemistry and thermodynamic-mixing properties of the aluminosilicate garnets: Amer. Mineral., v. 82, p. 571–581.

Grevel, K.-D., Burchard, M., and Fasshauer, D., 1996, In situ X-ray diffraction investigations of diaspore, lawsonite, and pyrophyllite with MAX 80 using synchrotron radiation [abs.]: Terra Abs., v. 8, no. 1, 6th Int. Symp. Exp. Mineral. Petrol. Geochem., p. 25.

Grevel, K.-D., Fasshauer, D. W., and Erzner, S., 1997, New compressibility data for clinochlore, kyanite, Mg-chloritoid, and Mg-staurolite [abs.]: Ber. Deuts. Mineral. Ges., Beih./Eur. Jour. Mineral., v. 9, no. 1, p. 138.

Grevel, K.-D., Fasshauer, D. W., Rohling, S., Gollerthan, S., and Schebaum, K., 1998, In situ X-ray diffraction studies on kyanite and Mg-staurolite at high pressure and temperatures: Hamburger Synchrotronstrahlungslabor HASYLAB am Deutschen Elektronen-Synchrotron DESY, Annual Report 1997, Part 1, p. 595–596.

Harlow, G. E., 1997, K in clinopyroxene at high pressure and temperature: An experimental study: Amer. Mineral., v. 82, p. 259–269.

Holdaway, M. J., Mukhopadhyay, B., and Dutrow, B. L., 1995, Thermodynamic properties of stoichiometric staurolite $H_2Fe_4Al_{18}Si_8O_{48}$ and $H_6Fe_2Al_{18}Si_8O_{48}$: Amer. Mineral., v. 80, p. 520–533.

Holland, T., Baker, J., and Powell, R., 1998, Mixing properties and activity-composition relationships of chlorites in the system MgO-FeO-Al_2O_3-SiO_2-H_2O: Eur. Jour. Mineral., v. 10, p. 395–406.

Hoschek, G., 1995, Stability relations and Al content of tremolite and talc in CMASH assemblages with kyanite + zoisite + quartz + H_2O: Eur. Jour. Mineral., v. 7, p. 353–362.

Hovis, G. L., and Roux, J., 1998, Effects of excess silicon on the thermodynamics of nepheline and kalsilite solid solutions [abs]: Terra Abs., v. 10, no. 1, 7th Int. Symp. Exp. Mineral. Petrol. Geochem., p. 27.

Hugh-Jones, D. A., and Angel, R. J., 1994, A compressional study of $MgSiO_3$ orthoenstatite up to 8.5 GPa: Amer. Mineral., v. 79, p. 405–410.

Koch-Müller, M., 1997, Experimentally determined Fe-Mg exchange between synthetic staurolite and garnet in the system MgO-FeO-Al_2O_3-SiO_2-H_2O: Lithos, v. 41, p. 185–212.

Koch-Müller, M., Langer, K., Behrens, H., and Schuck, G., 1997, Crystal chemistry and infrared spectroscopy in the OH-stretching region of synthetic staurolites: Eur. Jour. Mineral., v. 9, p. 67–82.

Konzett, J., 1998, Structure and composition of phase X, a hydrous alkali-rich high pressure silicate [abs.]: EOS (Trans. Amer. Geophys Union), v. 79, no. 45, p. F996 (suppl.).

Koziol, A. M., 1996, Quaternary (Ca-Fe-Mg-Mn) garnet: Displaced equilibrium experiments and implications for current garnet mixing models: Eur. Jour. Mineral., v. 8, p. 453–460.

Koziol, A. M., and Newton, R. C., 1995, Experimental determination of the reactions magnesite + quartz = enstatite + CO_2 and magnesite = periclase + CO_2, and enthalpies of formation of enstatite and magnesite: Amer. Mineral., v. 80, p. 1252–1260.

Krosse, S., 1995, Hochdrucksynthese, stabilität und eigenschaften der borsilikate dravit und kornerupin sowie darstellung und stabilitätsverhalten eines neuen Mg-Al-borates: Unpubl. Ph.D. thesis, Ruhr-Universität Bochum, 131 p.

Kushiro, I., Satake, H., and Akimoto, S., 1975, Carbonate-silicate reactions at high pressures and possible presence of dolomite and magnesite in the upper mantle: Earth Planet. Sci. Let., v. 28, p. 116–120.

Lu, R., and Keppler, H., 1997, Water solubility in pyrope to 100 kbar: Contrib. Mineral. Petrol., v. 129, p. 35–42.

Luth, R. W., 1997, Experimental study of the system phlogopite-diopside from 3.5 to 17 GPa: Amer. Mineral., v. 82, p. 1198–1209.

Martinez, I., Zhang, J., and Reeder, R. J., 1996, In situ X-ray diffraction at high pressure and high temperature of aragonite and dolomite: Evidence for dolomite breakdown to aragonite and magnesite: Amer. Mineral., v. 81, p. 611–624.

Massonne, H.-J., 1995, Experimental and petrogenetic study of UHPM, in Coleman, R., and Wang, X., eds., Ultrahigh pressure metamorphism: New York, Cambridge University Press, p. 33–95.

Massonne, H.-J., 1997, An improved thermodynamic solid solution model for natural white micas and its application to the geothermobarometry of metamorphic rocks [abs.]: Geol. Surv. Finland, Guide 46, Mineral equilibria and databases, Abs., p. 49.

Massonne, H.-J., 1995, Andersen, G., Böving, R., Burchard, M., and Theye, T., 1996, Compressibility of some rock-forming minerals up to 7 GPa [abs.]: Terra Abs., v. 8, no. 1, 6th Int. Symp. Exp. Mineral. Petrol. Geochem., p. 42.

Massonne, H.-J., and Szpurka, Z., 1997, Thermodynamic properties of white micas on the basis of high-pressure experiments in the systems K_2O-MgO-Al_2O_3-SiO_2-H_2O and K_2O-FeO-Al_2O_3-SiO_2-H_2O: Lithos, v. 41, p. 229–250.

Mirwald, P. W., and Massonne, H.-J., 1980, The low-high quartz and quartz-coesite transition to 40 kbar between 600° and 1600°C and some reconnaissance data on the effect of $NaAlO_2$ component on the low quartz-coesite transition: Jour. Geophys. Res., v. 85, p. 6983–6990.

Mosenfelder, J. L., and Bohlen, S. R., 1997, Kinetics of the coesite to quartz transformation: Earth Planet. Sci. Lett., v. 153, p. 133–147.

Pacalo, R. E. G., and Weidner, D. J., 1997, Elasticity of majorite, $MgSiO_3$ tetragonal garnet: Phys. Earth Planet. Int., v. 99, p. 145–154.

Pawley, A. R., and Wood, B. J., 1995, The high-pressure stability of talc and 10 Å phase: Potential storage sites for H_2O in subduction zones: Amer. Mineral., v. 80, p. 998–1003.

Pawley, A. R., Redfern, S. A. T., and Holland, T. J. B.,

1996, Volume behavior of hydrous minerals at high pressure and temperature: I. Thermal expansion of lawsonite, zoisite, clinozoisite and diaspore: Amer. Mineral., v. 81, p. 335–340.

Rauch, M., and Keppler, H., 1998, Water solubility in enstatite to 100 kbar [abs.]: Terra Abs., v. 10, no. 1, 7th Int. Symp. Exp. Mineral. Petrol. Geochem., p. 51.

Redhammer, G. J., Dachs, E., and Amthauer, G., 1995, Mössbauer spectroscopic and X-ray powder diffraction studies of synthetic micas on the join annite $KFe_3AlSi_3O_{10}(OH)_2$–phlogopite $KMg_3AlSi_3O_{10}(OH)_2$): Phys. Chem. Minerals, v. 22, p. 282–294.

Renner, J., Rummel, F., and Stöckhert, B., 1995, Experimental studies on the rheology of coesite under ultrahigh pressure metamorphic conditions: Bochumer Geol. Geotech. Arb., v. 44, p. 160–165.

Renner, J., Zerbian, A., and Stöckhert, B., 1997, Microstructures of synthetic polycrystalline coesite aggregates: The effect of pressure, temperature, and time: Lithos, v. 41, p. 169–184.

Roots, M., 1994, Molar volumes on the clinochlore-amesite binary: Some new data: Eur. Jour. Mineral., v. 6, p. 279–284.

Rosenbleck, C., 1996, Phlogopitmischkristalle im System K_2O-MgO-Al_2O_3-SiO_2-H_2O: Hochdrucksynthesen, physikochemische Eigenschaften und geothermobarometrische Anwendung: Unpubl. Ph.D. thesis, Ruhr-Universität Bochum, 235 p.

Rosenbleck, C., and Massonne, H.-J., 1993, Thermodynamische eigenschaften von phlogopit-mischkristallen im system K_2O-MgO-Al_2O_3-SiO_2-H_2O (KMASH) und ihre geothermobarometrische anwendung [abs.]: Ber. Deuts. Mineral. Ges., Beih./Eur. Jour. Mineral., v. 5, no. 1, p. 71.

Ross, N., 1971, The equation of state and high-pressure behavior of magnesite: Amer. Mineral., v. 82, p. 682–688.

Rybacki, E., Rummel, F., and Stöckhert, B., 1995, Synthesis and experimental deformation of polycrystalline aragonite: Bochumer Geol. Geotech. Arb., v. 44, p. 189–194.

Schebaum, K., and Massonne, H.-J., 1996, Hochdrucksynthesen und eigenschaften von biotiten im system K_2O-FeO-Al_2O_3-SiO_2-H_2O [abs.]: Ber. Deuts. Mineral. Ges., Beih./Eur. Jour. Mineral., v. 8, no. 1, p. 237.

———, 1997, Zustandsgleichungen für die gesteinsbildenden minerale zussmanit und stilpnomelan: Hamburger Synchrotronstralungslabor HASYLAB am Deutschen Elektronen-Synchrotron DESY, Annual Report 1996, p. 583–584.

———, 1998, Zustandsgleichungen für die gesteinsbildenden minerale biotit und stilpnomelan: Hamburger Synchrotronstralungslabor HASYLAB am Deutschen Elektronen-Synchrotron DESY, Annual Report 1997, Part I, p. 598–599.

Schmidt, M. W., Poli, S., Comodi, P., and Zanazzi, P. F., 1997, High-pressure behavior of kyanite: Decomposition of kyanite into stishovite and corundum: Amer. Mineral., v. 82, p. 460–466.

Schreyer, W., 1999, Experimental aspects of UHP metamorphism: Granitic systems: INT. GEOL. REV., v. 41, in press.

Schreyer, W., and Werding, G., 1997, High-pressure behaviour of selected boron minerals and the question of boron distribution between fluids and rocks: Lithos, v. 41, p. 251–266.

Sinogeikin, S. V., Bass, J. D., O'Neill, B., and Gasparik, T., 1997, Elasticity of tetragonal end-member majorite and solid solutions in the system $Mg_4Si_4O_{12}$-$Mg_3Al_2Si_3O_{12}$: Phys. Chem. Minerals, v. 24, p. 115–121.

Stachowiak, A., and Schreyer, W., 1998, Synthesis, stability and breakdown products of the hydroxyl end member of jeremejevite in the system Al_2O_3-B_2O_3-H_2O: Eur. Jour. Mineral., v. 10, p. 875–888.

Stöckhert, B., and Renner, J., 1998, Rheology of crustal rocks at ultrahigh pressure, in Hacker, B. R., and Liou, J. G., When continents collide: Geodynamics and geochemistry of ultrahigh-pressure rocks: Dordrecht, Kluwer Acad. Publ., p. 57–95.

Theye, T., Chopin, C., Grevel, K.-D., and Ockenga, E., 1997, The assemblage diaspore + quartz in metamorphic rocks: A petrological, experimental and thermodynamic study: Jour. Metamor. Geol., v. 15, p. 17–28.

Thompson, P., Parsons, I., Graham, C. M., and Jackson, B., 1998, The breakdown of potassium feldspar at high water pressure: Contrib. Mineral. Petrol., v. 130, p. 176–186.

Urakawa, S., Kondo, T., Igawa, N., Shimomura, O., and Ohno, H., 1994, Synchrotron radiation study on the high-pressure and high-temperature phase relations of $KAlSi_3O_8$: Phys. Chem. Minerals, v. 21, p. 387–391.

Vidal, O., Theye, T., and Chopin, C., 1994, Experimental study of chloritoid stability at high pressure and various fO_2 conditions: Contrib. Mineral. Petrol., v. 118, p. 256–270.

Waterwiese, T., Chatterjee, N. D., Dierdorf, I., Göttlicher, J., and Kroll, H., 1995, Experimental and thermodynamic study of heterogeneous and homogeneous equilibria in the system $NaAlSiO_4$-SiO_2: Contrib. Mineral. Petrol., v. 121, p. 61–73.

Werding, G., and Schreyer, W., 1996, Experimental studies on borosilicates and selected borates, in Grew, E. S., and Anovitz, L. M., eds., Boron: Rev. Mineral., v. 33, p. 117–163.

Wirth, R., and Stöckhert, B., 1995, Transformation of coesite to quartz: Microstructures in nature and experiment: Bochumer Geol. Geotech. Arb., v. 44, p. 260–262.

Wood, B. J., Hackler, R. T., and Dobson, D. P., 1994, Experimental determination of Mn-Mg mixing properties in garnet, olivine, and oxide: Contrib. Mineral. Petrol., v. 115, p. 438–448.

Wood, B. J., Frost, D., and Pawley, A. R., 1995, Properties

of H_2O and CO_2 at high P and T: Bochumer Geol. Geotech. Arb., v. 44, p. 264–266.

Woodland, A. B., and Angel, R. J., 1997, Reversal of the orthoferrosilite–high P clinoferrosilite transition: A phase diagram for $FeSiO_3$ and implications for the mineralogy of the Earth's upper mantle: Eur. Jour. Mineral., v. 9, p. 245–254.

Wunder, B., Medenbach, O., Krause, W., and Schreyer, W., 1993a, Synthesis, properties and stability of $Al_3Si_2O_7(OH)_3$ (phase Pi), a hydrous high-pressure phase in the system Al_2O_3-SiO_2-H_2O (ASH): Eur. Jour. Mineral., v. 5, p. 637–649.

Wunder, B., Rubie, D. C., Ross, C. R., II, Medenbach, O., Seifert, F., and Schreyer, W., 1993b, Synthesis, stability, and properties of $Al_2SiO_4(OH)_2$: A fully hydrated analogue of topaz: Amer. Mineral., v. 78, p. 285–297.

Xu, J., Hu, J., Ming, L., Huang, E., and Xie, H., 1994, The compression of diaspore, AlO(OH) at room temperature up to 27 GPa: Geophys. Res. Lett., v. 21, p. 161–164.

Yagi, A., Suzuki, T., and Akaogi, M., 1994, High pressure transitions in the system $KAlSi_3O_8$-$NaAlSi_3O_8$: Phys. Chem. Minerals, v. 21, p. 12–17.

Yang, H., Downs, R. T., Finger, L. W., Hazen, R. M., and Prewitt, C. T., 1997, Compressibility and crystal structure of kyanite, Al_2SiO_5, at high pressure: Amer. Mineral., v. 82, p. 467–474.

Zhang, J., Li, B., Utsumi, W., and Liebermann, R. C., 1996, In situ X-ray observations of the coesite-stishovite transition: Reversed phase boundary and kinetics: Phys. Chem. Minerals, v. 23, p. 1–10.

Zhang, J., Martinez, I., Guyot, F., Gillet, P., and Saxena, S. K., 1997, X-ray diffraction study of magnesite at high pressure and high temperature: Phys. Chem. Minerals, v. 24, p. 122–130.

Zhang, L., Ahsbahs, H., and Kutoglu, A., 1998, Hydrostatic compression and crystal structure of pyrope to 33 GPa: Phys. Chem. Minerals, v. 25, p. 301–307.

Zhao, Y., Von Dreele, R. B., Shankland, T. J., Weidner, D. J., Zhang, J., Wang, Y., and Gasparik, T., 1997, Thermoeleastic equation of state of jadeite $NaAlSi_2O_6$: An energy-dispersive Rietveld refinement study of low symmetry and multiple phases diffraction: Geophys. Res. Lett., v. 24, p. 5–8.

Zinn, P., Hinze E., Lauterjung, J., and Wirth, R., 1997, Kinetic and microstructural studies of the quartz-coesite phase transition: Phys. Chem. Earth, v. 22, p. 105–111.

H_2O and Ultrahigh-Pressure Subsolidus Phase Relations for Mafic and Ultramafic Systems

W. G. ERNST

Department of Geological and Environmental Sciences, Stanford University, Stanford, California 94305-2115

Abstract

Experimental studies show that aqueous fluid–mediated mineralogic solution/redeposition mechanisms are orders faster than solid-solid transformations; hence the presence of a separate aqueous fluid markedly enhances reaction rates, whereas its total absence impedes mineralogic transformations. Where does this volatile component come from? For typical subduction-zone P-T trajectories, amphibole constitutes the major OH-bearing phase in most deep-seated metamorphic rocks of basaltic composition; other hydrous minerals are of minor abundance. Clinoamphiboles dehydrate at pressures of ~2.0 to 2.4 GPa, but devolatilization may be delayed slightly by pressure overstepping; thus mafic blueschists and barroisitic amphibolites expel H_2O at arc melt-generation depths of ~100 km, and commonly achieve the stable eclogitic phase configuration. Serpentinized mantle beneath the oceanic crust devolatilizes at comparable conditions. Only where metagabbroic rocks are completely dry and coarse grained are low-pressure assemblages metastably preserved. For realistic subduction-zone geothermal gradients, white micas ± biotites remain stable in sialic crust to pressures exceeding 3.5 GPa. Accordingly, under conditions attending descent to great depths, mica-rich quartzofeldspathic schists and gneisses that constitute the continental crust fail to evolve substantial amounts of H_2O, and transform incompletely to stable eclogite-facies assemblages. The deep underflow of partly hydrated oceanic lithosphere thus generates most of the deep-seated volatile flux—and consequent partial melting to produce the calc-alkaline suite along and above a subduction zone; where large volumes of micaceous sialic materials are carried down to extreme depths, volatile flux severely diminishes.

Role of H_2O in Metamorphism

General statement

CONSUMPTION OF an oceanic crust–capped plate allows the subduction of entrained sedimentary material and underlying basaltic crust, either off-loaded into the accretionary prism, as is most of the low-density sedimentary debris, or returned to the deep upper mantle, as is most of the relatively high density oceanic crust. Where microcontinental entities constitute portions of the converging lithospheric plates, conduction to the site of subjuction leads ultimately to collisonal suturing; in such cases, continental crust may be carried beneath the stable, nonsubducted plate to depths exceeding 100 km. Regardless of whether the descending materials are unconsolidated sediments or old sialic crust, pressure within the slab rises with subduction depth, while temperature only gradually increases. Volatiles, principally H_2O, may be added to relatively dry portions of the downgoing plate, and driven off from the more volatile-rich parts (Schreyer et al., 1987; Peacock, 1990, 1995; Bebout, 1996). Shallow-level, pre-eclogitic stages of these reactions have been treated quantitatively by Frey (1987) and Ernst (1990). Attending the deep subduction of crustal lithologies, retained H_2O is sequestered exclusively in relatively refractory silicate phases (Schmidt and Poli, 1998). Figure 1 (Liou et al., 1998; Okamoto and Maruyama, 1999) illustrates typical P-T conditions appropriate for the common metamorphic facies assemblages of the continental and oceanic crust, as well as the realms of high-pressure (HP) and ultrahigh-pressure (UHP) recrystallization; selected individual reactions also are shown for completeness.

Eventually, some of the low-density subducted quartzofeldspathic material, recrystallized under high-pressure or ultrahigh-pressure conditions, may separate from the descending slab and rise to normal continental crustal levels. Generally, weakly consolidated trench mélange characteristic of many Pacific-type junctions decouples at shallow depths (15 to 30 km), whereas old continental crust–capped

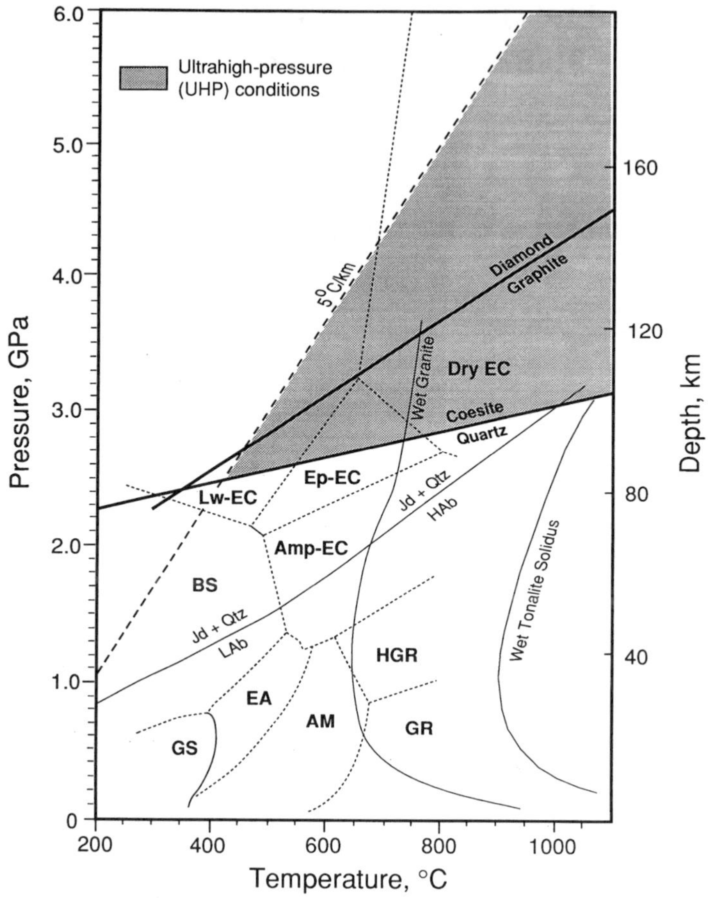

FIG. 1. Petrogenetic grid for metabasaltic bulk-rock compositions, after Liou et al. (1998, Fig. 1) and Okamoto and Maruyama (1999). An extremely low subduction-zone geothermal gradient of 5°C/km also is shown. For literature references to experimental data, both specific subsolidus phase transitions and the onset of melting curves, see Liou et al. (1998). Metamorphic facies abbreviations: AM = amphibolite; Amp-EC = amphibolite-eclogite; BS = blueschist; EA = epidote amphibolite; EC = eclogite; EP-EC = epidote-eclogite; GS = greenschist; HGR = granulite; P = prehnite; PA = pumpellyite-actinolite; PP = prehnite-pumpellyite; ZE = zeolite.

slabs continue sinking to greater depths. During exhumation, volatiles gain access to the rising, buoyant mass, or diffuse away from it, depending on the composition, average grain size, permeability, and extent of shearing of the various constitutent units. Some evolved fluids rise into and through the overlying mantle wedge, whereas the rest migrates back up the conduit provided by the subduction channel. In any case, decompression is generally accompanied by backreaction.

This section briefly summarizes experimental phase-equilibrium constraints on H_2O-bearing minerals stable in subducting crustal sections (e.g., Massone, 1995) and the overlying mantle wedge (Mysen et al., 1998). Laboratory studies demonstrate that reaction rates are strongly influenced by the presence or absence of an aqueous fluid phase (Rubie, 1986, 1990). Beyond relatively shallow depths where compaction and hydrous mineral devolatilization have been completed (25 ± 15 km), aqueous fluid is probably not present as a separate phase, and free H_2O becomes available only to the extent that OH-bearing minerals undergo dehydration. Three principal lithologic assemblages constitute most of the spatially associated, noncarbonate, volatile-bearing subducted, recrystallized crust:

metabasalts + metagabbros; serpentinized mantle materials; and metagraywackes ± pelitic schists, high-grade gneisses, and metagranitoids.

Hydrated mafic rocks

Phase-equilibrium studies have been performed on metabasaltic bulk compositions over the P-T range 300–1050°C and 0–3.5 GPa (Essene et al., 1970; Helz, 1973, 1979; Liou et al., 1974; Spear, 1981; Gilbert et al., 1982; Apted and Liou, 1983; Moody et al., 1983; Poli, 1993; Liu et al., 1996; Poli and Schmidt, 1997). As shown by long-term experiments and reaction reversals for unaltered, chemically pristine MORB, barroisitic to pargasitic amphibole is the major hydrous mineral at and above ~650°C. The Ca-amphibole stability field contracts drastically at high pressures, and disappears at or below 1.6 to 2.4 GPa, depending on the temperature. Phase relations for H_2O-saturated MORB are illustrated in Figure 2 (Schmidt and Poli, 1998); this diagram is applicable to relatively unmetasomatized basalts and gabbros. Substantial amounts of lawsonite and sodic amphibole characterize such rocks at moderate pressures and at temperatures below 500 to 600°C (Maresch, 1977), but the glaucophane stability field is limited by elevated pressure, analogous to the P-T behavior of Ca-amphiboles (Koons, 1982; Carmen and Gilbert, 1983; Welch and Graham, 1992).

Lawsonite evidently remains stable in rocks of MORB composition as an important mineral under UHP conditions (Ono, 1998; Okamoto and Maruyama, 1999), but only at unrealistically low temperatures characterized by a geothermal gradient of ~2 to 5°C, as shown in Figure 2. In synthesis experiments involving more aluminous mafic igneous rocks, epidote and chlorite possess broad P-T ranges at temperatures below ~600 to 650°C (Pawley and Holloway, 1993; Schmidt and Poli, 1994, 1998; Poli and Schmidt, 1997), where they apparently coexist with actinolite at low pressures and with blue amphibole at intermediate pressures; these phases markedly decrease in amount with elevated pressures. However, attainment of chemical equilibrium was not demonstrated in these studies. Where metabasalts are more potassic than MORB, small amounts of white mica persist to relatively elevated pressures, and may be a minor associate of garnet, omphacite, and rutile in the HP/UHP paragenesis of altered mafic rocks (Fig. 2). Chloritoid has been reported in some synthesis experiments, but this mineral probably does not represent an important phase in subducted MORBs because of its unusual Al_2O_3 + FeO–rich composition and restricted thermal stability.

Ca- and Na-amphiboles all decompose at pressures approaching UHP values, so metabasaltic rocks generate modest amounts of H_2O by dehydration, conducive to the production of eclogitic assemblages; accordingly, the prograde reaction is thought to be favored except in dry, impermeable protoliths such as coarse-grained granulite-facies metagabbros (Austrheim, 1990; Hirajima et al., 1993; Zhang and Liou, 1997).

Hydrated ultramafic rocks

At temperatures above the lowest metamorphic grades, antigoritic serpentinites dehydrate to orthopyroxene or talc + olivine, liberating large amounts of H_2O at ~550 to 600°C and 2.0 to 3.0 GPa, depending on the geothermal gradient (Johannes, 1975; Evans et al., 1976; Ulmer and Trommsdorff, 1995; Wunder and Schreyer, 1997; Pawley, 1998). Similar to the behavior of clinoamphiboles, antigorite is pressure limited in its stability range. Equilibrium phase relations for antigorite + H_2O are presented in Figure 3, after Wunder and Schreyer (1997); see also the summary by Mysen et al. (1998).

But to what extent is the dunite-harzburgite-lherzolite mantle lithosphere underlying the oceanic crust hydrated? Most recovered specimens of suboceanic mantle are partly serpentinized. However, sampling thus far has chiefly involved oceanic fracture zones, transform faults, and on-land ophiolites, sections characterized by unusually intense deformation and fluid circulation (e.g., Bonatti, 1976). Fully hydrated mantle would have a bulk density of less than 2.5, and should be virtually unsubductable. Thus, only 10 to 20% serpentinized peridotite seemingly could descend along a convergent plate junction. This conclusion is supported by seismic transmission velocities of 8.1 km/sec measured for the oceanic mantle, in contrast to Vp = 5.5 km/sec for pure serpentinite (Coleman, 1971; Christensen and Salisbury, 1975). Nevertheless, modest amounts of hydrated ultramafics seem to be present in the uppermost part of the oceanic lithosphere, so subduction and devolatilization of hydrated peridoite probably contributes to the aqueous fluid flux of a descending plate (Ulmer and Trommsdorff, 1995).

Hydrated quartzofeldspathic rocks

Laboratory studies of metapelites, metagraywackes, schists, gneisses, and metagranitoids

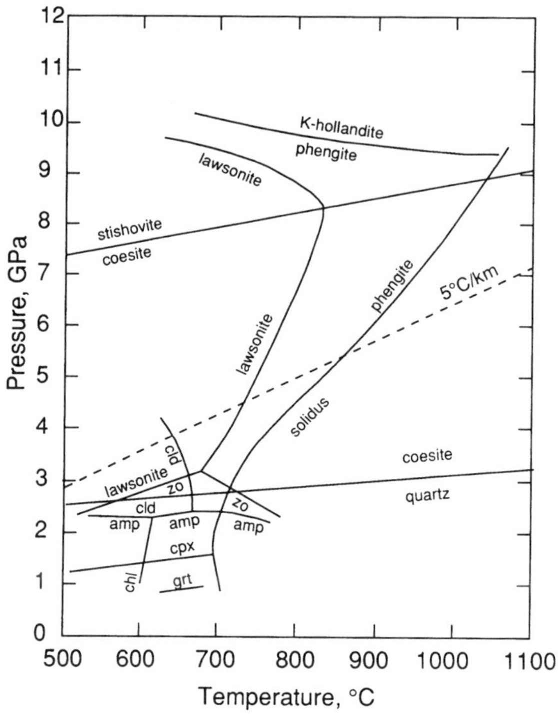

FIG. 2. Synthesis fields and presumed stability relations in the basalt-H_2O system, after Schmidt and Poli (1998, Fig. 1). An extremely low subduction-zone geothermal gradient of 5°C/km also is shown. Abbreviations for phase fields: amph = amphibole; chl = chlorite; cld = chloritoid; cpx = clinopyroxene; grt = garnet; zo = zoisite.

over a wide range of experimental conditions demonstrate that chlorites dehydrate above ~550 to 600°C, whereas potassic white micas ± biotites remain stable to considerably higher temperatures (Le Breton and Thompson, 1988; Vielzeuf and Holloway, 1988; Vielzeuf and Montel, 1994; Gardien et al., 1995; Patiño Douce and Beard, 1996; Patiño Douce and McCarthy, 1998; Nichols et al., 1996; Skjerlie and Johnston, 1996; Luth, 1997; Massone and Szpurka, 1997). In metapelitic rocks, the assemblage biotite + kyanite is replaced by muscovite + garnet at pressures above ~1.7 GPa, but in more quartzofeldspathic compositions, biotite persists to much higher pressures. For subduction-zone geothermal gradients, muscovite-phengite solid solutions remain stable in peraluminous lithologies at least to 3.5–4.0 GPa (e.g., Massone and Szpurka, 1997), and for metaluminous rocks, biotite is also present to similar pressures. Because phengitic micas ± biotites constitute the volumetrically most important hydroxyl-bearing phases in meta-shaly and quartzofeldspathic equilibrium assemblages, units that dominate the continental crust may fail to evolve significant amounts of H_2O during UHP metamorphism at depths of 100 km or more; thus, in the absence of a catalytic aqueous fluid, such lithologies might retain metastable feldspar + quartz in micaceous rocks, and fail to develop the stable prograde UHP mineralogic assemblage.

Convergent plate junctions and the evolution of H_2O

Subducting, low-grade, hydrous metamorphic rocks lose most of their volatiles at shallow depths, and by the stage in which they reach HP/UHP conditions, micaceous quartzofeldspathic and metapelitic lithologies should be relatively deficient in H_2O-producing phases. Thus, the clinoamphibole-bearing oceanic crust ± serpentinized peridotite—not mica-bearing sialic crust—probably represents the most significant dehydrating rock type at depths approaching or exceeding 100 km. Devolatilization of clinoamphiboles ± serpentine minerals should begin at ~70 to 80 km, but if reactions are sluggish on the time scale of lithospheric plate descent, aqueous fluid evolution probably reaches a maximum at somewhat greater

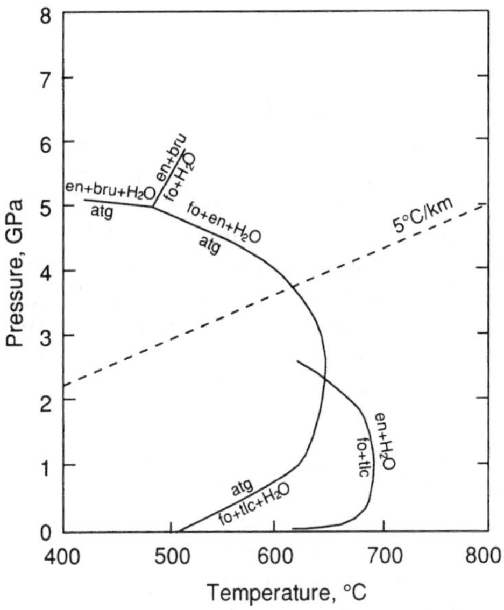

FIG. 3. Synthesis fields and presumed stability relations in the $MgO-SiO_2-H_2O$ system, after Wunder and Schreyer (1997); see also compilation by Mysen et al. (1998, Fig. 12). An extremely low subduction-zone geothermal gradient of 5°C/km also is shown. Abbreviations for phases: atg = antigorite; bru = brucite; en = enstatite; fo = forsterite; tlc = talc.

subduction depths. Such pressure overstepping appears to be characteristic of some UHP terranes (Austrheim, 1998). The longcontinued descent of thousands of kilometers of Pacific-type oceanic lithosphere may well provide the greatest proportion of deep-seated aqueous fluid rising through the subduction channel as well as diffusing into the mantle wedge. Partial fusion of the eclogitic oceanic crust and/or the refertilized mantle hanging wall evidently results in generation of the metaluminous calc-alkaline suite (Drummond and Defant, 1990; Kushiro, 1990; Hawkesworth et al., 1993; Morris, 1995).

The fluid flux under UHP conditions is probably curtailed or terminated by the arrival, profound underflow, and suturing of sialic crust. Continental collision, therefore, would not be expected to generate a substantial calc-alkaline arc, but might involve the partial melting of overthickened, accretionary quartzofeldspathic crust, producing peraluminous granitoids. True continental growth, however, appears to be the result of amphibole dehydration within the descending oceanic crust, followed by volatile-induced partial melting of oceanic crust and/or the overlying, metasomatized mantle wedge.

Hydration-dehydration rates

Evidence of widespread disequilibriurn during eclogite-facies metamorphism (e.g., Rubie, 1990; Hacker, 1996; Austrheim, 1990) highlights the need for more quantitative experimental data on reaction rates. From the study of HP and UHP complexes, it appears that the availability of fluid, rather than P-T stability relations, controls the rate at which reactions occur and the extent to which they run to completion. Although the rate-enhancing effect of H_2O is well known and specific mechanisms for catalysis have been proposed (Rubie, 1986; Mosenfelder and Bohlen, 1997), little is known quantitatively about the dependence of reaction rates and mechanisms on fluid concentrations and chemistries in systems with the low to very low fluid-rock ratios attending the profound subduction of accretionary prism material or coherent continental crust.

Quantification of the rates for reactions that consume or evolve H_2O seems to be the key to understanding deep-seated mineralogic transformations in Alpine- and Pacific-type plate junctions. It is generally assumed that dehydration either occurs rapidly or is controlled by the rate of enthalpy production (Ridley, 1985). Most laboratory experiments, employing powdered starting materials, yield rates that appear to be too rapid when extrapolated to geologic conditions (e.g., Wegner and Ernst, 1983; Schramke et al., 1987; Jové and Hacker, 1997). In contrast, the investigation of dehydration reactions employing rocks as starting materials yields slower rates and different mechanisms (Brearley, 1987; Rubie and Brearley, 1987; Brearley and Rubie, 1990; Hacker, 1990, 1991). Extrapolation of solid-rock results to longer time scales and to temperatures close to actual values is difficult at present. However, because dehydration reactions generate a rate-enhancing constituent, prograde transformations proceed far more rapidly by solution/redeposition mechanisms than does the equivalent backreaction involving solid-solid transformations. This phenomenon—coupled with the fact that mafic lithologies carry substantial amounts of a pressure-limited hydrous phase, clinoamphibole, whereas micaceous felsic rocks are not so pressure constrained—yields an important conclusion: the subduction of altered oceanic lithosphere is characterized by the deep-seated evolution of aqueous fluids, whereas the deep-seated underflow of continental lithosphere in general remains dry. It is quite probable that the H_2O given off from a rapidly sinking, relatively cool oceanic slab is only released over a range of depths greater than predicted by phase equilibria—say, 100, rather than 70 to 80 km.

The ease with which aqueous fluids gain access to, or egress from, subducting and ascending low-density crustal masses is a complex function of rock compositions, relative permeabilities, extent of shearing, aggregate grain sizes, and availability of volatiles. These quantities reflect the past geologic histories and dynamics of the lithotectonic units moving along a convergent plate junction. What is not known with any certainty is the rates at which mineralogic transformations take place, and therefore the conditions and tectonic realms under which volatile constituents are evolved and consumed. Given locally anhydrous circumstances, low-pressure assemblages can be retained under prograde HP/UHP conditions (Austrheim, 1990), and HP/UHP relics can be preserved under retrograde low-pressure conditions (Hirajima et al., 1993; Liou and Zhang, 19%). Quantification of the kinetics for the controlling reactions is the key, and these data are not available except for a few, chemically simple mineral systems. However, solution/redeposition mechanisms are orders faster than solid-solid transformations, and thus it is clear that the presence of a separate aqueous fluid markedly enhances reaction rates, whereas its total absence impedes mineralogic transformations.

Discussion

As described above, how do the experimentally determined phase equilibria and rate studies, involving the presence or absence of H_2O, constrain the course of UHP metamorphism? Laboratory studies show that, for reasonable subduction-zone geothermal gradients of approximately 5 to 8°C/km, glucophane (at low temperatures) and barroisite (at higher temperatures) constitute the major hydroxyl-bearing phase in deep-seated metamorphic rocks of MORB chemistry. Other hydrous minerals such as lawsonite, clinozoisite, and chloritoid are of minor abundance and/or are confined to unreasonably low P-T trajectories. Na- and Ca-amphiboles dehydrate at pressures of ~2.0 to 2.4 GPa, but devolatilization probably is delayed by modest pressure overstepping at relatively low temperatures. Low-pressure assemblages may persist metastably in dry, coarse-grained metagabbroic rocks, but mafic blueschists and barroisitic amphibolites expel H_2O, and commonly achieve stable eclogitic phase assemblages. Partly serpentinized mantle beneath the oceanic crust devolatilizes under similar P-T conditions. In both cases, the release of H_2O promotes the partial fusion of the downgoing, transformed oceanic crust and/or the overlying, refertilized mantle wedge, generating arc magmas in the process. For plausible subduction-zone P-T gradients, white micas ± biotites remain stable in continental crust to pressures >3.5 GPa. Therefore, where mica-rich quartzofeldspathic schists and gneisses that constitute the continental crust are carried to depths of 100 km or more, they generally fail to evolve substantial amounts of H_2O, and transform only incipiently to the stable eclogite-facies phase configurations. The deep underflow of partly hydrated oceanic lithosphere (altered mafic + ultramafic materials) thus generates most of the deep-seated volatile flux—and resultant partial melting to produce arc magmas—along and above a subduction zone; where large volumes of micaceous continental crust are carried down to great depths, volatile flux diminishes severely.

Acknowledgments

This study was supported by Stanford University, and by the U.S. National Science Foundation through grant no. EAR97-25347 to J. G. Liou. Ruth Y. Zhang helped to finalize the illustrations. An early-draft manuscript was reviewed and improved by J. G. Liou.

REFERENCES

Apted, M. J., and Liou, J. G., 1983, Phase relations among greenschist, epidote-amphibolite, and amphibolite in a basaltic system: Amer. Jour. Sci., v. 283A, p. 328–353.

Austrheim, H., 1990, The granulite-eclogite facies transition: A comparison of experimental work and a natural occurrence in the Bergen Arcs, western Norway: Lithos, v. 25, p. 163–169.

———, 1998, The influence of fluid and deformation on metamorphism of the deep crust and consequences for geodynamics of collision zones, *in* Hacker, B. R., and Liou, J. G., eds., When continents collide: Geodynamics and geochemistry of ultrahigh-pressure rocks: Dordrecht, Kluwer Acad. Publ., p. 297–323.

Bebout, G. E., 1996, Volatile transfer and recycling at convergent margins: Mass-balance and insights from high-P/T metamorphic rocks: Amer. Geophys. Union, Geophys. Monog. 96, p. 179–193.

Bonattie, E., 1976, Serpentinite protrusions in the oceanic crust: Earth Planet. Sci. Lett., v. 32, p. 107–113.

Brearley, A. J., 1987, An experimental and kinetic study of the breakdown of aluminous biotite at 800°C: Reaction microstructures and mineral chemistry: Bull. Mineral., v. 110, p. 513–532.

Brearley, A. J., and Rubie, D. C., 1990, Effects of H_2O on the disequilibrium breakdown of muscovite + quartz: Jour. Petrol., v. 31, p. 925–956.

Carmen, J. H., and Gilbert, M. C., 1983, Experimental studies of glaucophane stability: Amer. Jour. Sci., v. 283A, p. 414–437.

Christensen, N. I., and Salisbury, M. H., 1975, Structure and constitution of the lower oceanic crust: Rev. Geophys. Space Phys., v. 13, p. 57–86.

Coleman, R. G., 1971, Petrologic and geophysical nature of serpentinites: Geol. Soc. Amer. Bull., v. 82, p. 897–918.

Drummond, M. S., and Defant, M. J., 1990, A model for trondhjemite-tonalite-dacite genesis and crustal growth via slab melting: Archean to modern comparisons: Jour. Geophys. Res., v. 95, p. 21,503–21,521.

Ernst, W. G., 1990, Thermobarometric and fluid-expulsion history of subduction zones: Jour. Geophys. Res., v. 95, p. 9047–9053.

Essene, E., Henson, B. J., and Green, D. H., 1970, Experimental study of amphibolite and eclogite stability: Phys. Earth Planet. Inter., v. 3, p. 378–384.

Evans, B. W., Johannes, W., Oterdoom, H., and Trommsdorff, V., 1976, Stability of chrysotile and antigorite in the serpentinite multisystem: Schweiz. Mineral. Petrog. Mitteil., v. 56, p. 79–93.

Frey, M., ed., 1987, Low temperature metamorphism: Glasgow, Blackie Publ., 351 p.

Gardien, V., Thompson, A. B., Grujic, D., and Ulmer, P., 1995, Experimental melting of biotite + plagioclase + quartz ± muscovite assemblages and implications for crustal melting: Jour. Geophys. Res., v. 100, p. 15,581–15,591.

Gilbert, M. C., Helz, R. T., Popp, R. K., and Spear, F. S., 1982, Experimental studies of amphibole stability, *in* Veblen, D. R., and Ribbe, P. H., eds., Amphiboles: Petrology and experimental phase relations: Mineral. Soc. Amer., Rev. Mineral., v. 9B, p. 229–353.

Hacker, B. R., 1990, Amphibolite-facies to granulite-facies reactions in experimentally deformed unpowdered amphibolite: Amer. Mineral., v. 75, p. 1349–1361.

______, 1991, The role of deformation in the formation of metamorphic field gradients: Ridge subduction beneath the Oman ophiolite: Tectonics, v. 10, p. 455–473.

______, 1996, Eclogite formation and the rheology, buoyancy, seismicity, and H_2O content of oceanic crust: Amer. Geophys. Union, Geophys. Monog. 96, p. 171–178.

Hawkesworth, C. J., Gallagher, K., Hergt, J. M., and McDermott, F., 1993, Mantle and slab contributions in an arc magma: Ann. Rev. Earth Planet. Sci., v. 21, p. 175–204.

Helz, R. T., 1973, Phase relations of basalts in their melting range at P_{H2O} = 5 kb as a function of oxygen fugacity: Jour. Petrol., v. 14, p. 249–302.

______, 1979, Alkali exchange between hornblende and melt: A temperature-sensitive reaction: Amer. Mineral., v. 64, p. 953–965.

Hirajima, T., Wallis, S., Zhai, R., and Ye, K., 1993, Eclogitized metagranitoid from the Su-Lu ultrahigh-pressure (UHP) province, eastern China: Proc. Japan Acad., v. 69, section B, p. 249–254.

Johannes, W., 1975, Zur synthese und thermischen stabilität von antigorit: Fortschrit für Mineral., v. 53, p. 36.

Jové, C., and Hacker, B. R., 1997, Experimental investigation of laumontite → wairakite + H_2O: A model diagenetic reaction: Amer. Mineral., v. 82, p. 781–789.

Koons, P. O., 1982, An experimental investigation of the behaviour of amphibole in the system Na_2O-MgO-$A1_2O_3$-SiO_2-H_2O at high pressures: Contrib. Mineral. Petrol., v. 79, p. 258–267.

Kushiro, I., 1990, Partial melting of mantle wedge and evolution of island arc crust: Jour. Geophys. Res., v. 95, p. 15,929–15,939.

Le Breton, N., and Thompson, A. B., 1988, Fluid-absent (dehydration) melting of biotite in metapelites in the early stages of crustal anatexis: Contrib. Mineral. Petrol., v. 99, p. 226–237.

Liu, J., Bohlen, S. R., and Ernst, W. G., 1996, Stability of hydrous phases in subducting oceanic crust: Earth Planet. Sci. Lett., v. 143, p. 161–171.

Liou, J. G., Kuniyoshi, S., and Ito, K., 1974, Experimental studies of the phase relations between greenschist and amphibolite in a basaltic system: Amer. Jour. Sci., v. 274, p. 613–632.

Liou, J. G., and Zhang, R. Y., 1996, Occurrences of intergranular coesite in ultrahigh-P rocks from the Sulu region, eastern China: Implications for lack of fluid during exhumation: Amer. Mineral., v. 81, p. 1217–1221.

Liou, J. G., Zhang, R. Y., Ernst, W. G., Rumble, D., III, and Maruyama, S., 1998, High pressure minerals from deeply subducted metamorphic rocks: Rev. Mineral., v. 37, p. 33–96.

Luth, R. W., 1997, Experimental study of the system phlogopite-diopside from 3.5 to 17 GPa: Amer. Mineral., v. 82, p. 1198–1209.

Maresch, W. V., 1977, Experimental studies of glaucophane: An analysis of present knowledge: Tectonophys., v. 43, p. 109–125.

Massone, H. J., 1995, Experimental and petrogenetic study of UHPM, *in* Coleman, R. G., and Wang, X., eds., Ultrahigh pressure metamorphism: New York, Cambridge University Press, p. 33–95.

Massone, H. J., and Szpurka, Z., 1997, Thermodynamic properties of white micas on the basis of high-pressure experiments in the systems K_2O-MgO-Al_2O_3-SiO_2-H_2O and K_2O-FeO-Al_2O_3-SiO_2-H_2O: Lithos, v. 41, p. 229–250.

Moody, J. B., Meyer, D., and Jenkins, J. E., 1983, Experimental characterization of the greenschist/amphibolite boundary in mafic systems: Amer. Jour. Sci., v. 283, p. 48–92.

Morris, P. A., 1995, Slab melting as an example of Quaternary volcanism and aseismicity in southwest Japan: Geology, v. 23, p. 395–398.

Mosenfelder, J. L., and Bohlen, S. R., 1997, Kinetics of the coesite to quartz transformation: Earth Planet. Sci. Lett., v. 153, p. 133–147.

Mysen, B. O., Ulmer, P., Konzett, J., and Schmidt, M. W., 1998, The upper mantle near convergent plate boundaries: Rev. Mineral., v. 37, p. 97–138.

Nichols, G. T., Wyllie, P. J., and Stern, C. R., 1996, Experimental melting of pelagic sediment, constraints relevant to subduction: Amer. Geophys. Union, Geophys. Monog. 96, p. 293–298.

Okamoto, K., and Maruyama, S., 1999, The-high pressure stability limits of lawsonite in the MORB + H_2O system: Amer. Mineral., v. 84, p. 362–373.

Ono, S., 1998, Stability limits of hydrous minerals in sediment and mid-ocean ridge basalt compositions: Implications for water transport in subduction zones: Jour. Geophys. Res., v. 103, p. 18,253–18,267.

Patiño Douce, A. E., and Beard, J. S., 1996, Effects of P, $f(O_2)$, and Mg/Fe ratio on dehydration-melting of model metagraywackes: Jour. Petrol., v. 37, p. 999–1024.

Patiño Douce, A. E., and McCarthy, T. C., 1998, Melting of crustal rocks during continental collision and subduction, *in* Hacker, B. R., and Liou, J. G., eds., When continents collide: Geodynamics and geochemistry of ultrahigh-pressure rocks: Dordrecht, Kluwer Acad. Publ., p. 27–55.

Pawley, A. R., 1998, The reaction talc + forsterite = enstatite + H_2O: New experimental results and petrological implications: Amer. Mineral., v. 83, p. 51–57.

Pawley, A. R., and Holloway, J. R., 1993, Water sources for subduction zone volcanism: New experimental constraints: Science, v. 260, p. 664–667.

Peacock, S. M., 1990, Fluid processes in subduction zones: Science, v. 248, p. 329–337.

______, 1995, Ultrahigh-pressure metamorphic rocks and the thermal evolution of continent collision belts: The Island Arc, v. 4, p. 376–383.

Poli, S., 1993, The amphibolite-eclogite transformation: An experimental study on basalt: Amer. Jour. Sci., v. 293, p. 1061–1107.

Poli, S., and Schmidt, M. W., 1997, The high-pressure stability of hydrous phases in orogenic belts: An experimental approach on eclogite-forming processes: Techtonophys., v. 273, p. 169–184.

Ridley, J., 1985, The effect of reaction enthalpy on the progress of a metamorphic reaction, *in* Thompson, A. B., and Rubie, D. C., eds., Metamorphic reactions: Kinetics, textures, and deformation: New York, Springer-Verlag, p. 80–97.

Rubie, D. C., 1986, The catalysis of mineral reactions by water and restrictions on the presence of aqueous fluid during metamorphism: Mineral. Mag., v. 50, p. 399–415.

______, 1990, Role of kinetics in the formation and preservation of eclogites, *in* Carswell, D. A., ed., Eclogite facies rocks: New York, Chapman and Hall, p. 111–140.

Rubie, D. C., and Brearley, A. J., 1987, Metastable melting during the breakdown of muscovite + quartz at 1 kbar: Bull. Mineral., v. 110, p. 533–549.

Schmidt, M. W., and Poli, S., 1994, The stability of lawsonite and zoisite at high pressures: Experiments in CASH to 92 kbar and implications for the presence of hydrous phases in subducted lithosphere: Earth Planet. Sci. Lett., v. 124, p. 105–118.

______, 1998, Experimentally based water budgets for dehydrating slabs and consequences for arc magma generation: Earth Planet. Sci. Lett., v. 163, p. 361–379.

Schramke, J. A., Kerrick, D. M., and Lasaga, A. C., 1987, The reaction muscovite + quartz ↔ andalusite + K-feldspar + water: Part 1, growth kinetics and mechanism: Amer. Jour. Sci., v. 287, p. 517–559.

Schreyer, W., Massone, H. J., and Chopin, C., 1987, Continental crust subducted to depths near 100 km: Implications for magma and fluid genesis in collision zones, *in* Mysen, B. O., ed., Magmatic processes: Physicochemical principles: The Geochem. Soc. Spec. Publ. no. 1, p. 155–163.

Skjerlie, K. P., and Johnston, A. D., 1996, Vapor-absent melting from 10 to 20 kbar of crustal rocks that contain multiple hydrous phases: Implications for anatexis in the deep to very deep continental crust and active continental margins: Jour. Petrol., v. 37, p. 661–691.

Spear, F. S., 1981, An experimental study of hornblende stability and compositional variability in amphibolite: Amer. Jour. Sci., v. 281, p. 697–734.

Ulmer, P., and Trommsdorff, V., 1995, Serpentine stability to mantle depths and subduction-related magmatism: Science, v. 268, p. 858–861.

Vielzeuf, D., and Holloway, J. R., 1988, Experimental determination of the fluid-absent melting relations in the pelitic system. Consequences for crustal differentiation: Contrib. Mineral. Petrol., v. 98, p. 257–276.

Vielzeuf, D., and Montel, J. M., 1994, Partial melting of metagraywackes. 1. Fluid-absent experiments and phase relationships: Contrib. Mineral. Petrol., v. 117, p. 375–393.

Wegner, W. M., and Ernst, W. G., 1983, Experimentally determined hydration and dehydration reaction rates in the system $MgO-SiO_2-H_2O$: Amer. Jour. Sci., v. 283A, p. 151–180.

Welch, M. D., and Graham, C. M., 1992, An experimental study of glaucophanic amphiboles in the system $Na_2O-MgO-Al_2O_3-SiO_2-SiF_4$ (NMASF): Some implications for glaucophane stability in natural and synthetic systems at high temperatures and pressures: Contrib. Mineral. Petrol., v. 111, p. 248–259.

Wunder, B., and Schreyer, W., 1997, Antigorite: High-pressure stability in the system $MgO-SiO_2-H_2O$ (MSH): Lithos, v. 41, p. 213–227.

Zhang, R. Y., and Liou, J. G., 1997, Partial transformation of gabbro to coesite-bearing eclogite from Yangkou, the Sulu terrane, eastern China: Jour. Metamor. Geol., v. 15, p. 183–202.

CLASSIC AREAS OF ULTRAHIGH-PRESSURE METAMORPHISM

The UHP Unit in the Dora-Maira Massif, Western Alps

CHRISTIAN CHOPIN

Laboratoire de Géologie, URA 1316 du CNRS, Ecole normale supérieure, 24 rue Lhomond, 75005 Paris, France

AND HANS-PETER SCHERTL

Institut für Mineralogie, Ruhr-Universität Bochum, 44780 Bochum, Germany

Abstract

New results acquired in the coesite-bearing terrane of the Dora-Maira massif, Italian Western Alps, indicate that the UHP unit extends over less than 45 km^2 and is bounded by two lower-grade continental units. The contact to the footwall unit is a late, low-angle, top-to-the-SW extensional one, which probably cuts across the original thrust contact. The protolith of the pyrope "whiteschist" of the UHP unit is shown to be the metasomatic transformation product of granitic rock. Near–end member magnesiochloritoid and magnesiostaurolite occur in pyrope megablasts. Ordering in magnesiostaurolite leads to cross-hatched twinning and superstructures with either a doubled *a* or doubled *c* parameter. The $Si_{3.5}$ phengite-3*T* from the same whiteschist contains nano- to microscale quartz platelets and talc layers parallel to (001) of the host mica, interpreted as in situ decompression products of the mica. Some of the pyrope megablasts contain late, crack-related granulite-facies mineral assemblages, which reflect very low H_2O activities as a result of local closed-system evolution at temperatures as low as 550 to 600°C. Thermobarometry on successive mineral generations in the dark- and light-colored UHP eclogites (probably former dikes and/or volcanic tuffs with sedimentary components) better constrains the prograde path and confirms that decompression took place at temperatures just slightly higher than during burial. The geochemistry of the main rock types in this piece of subducted continental crust shows no evidence of interaction with mantle material. Local-scale fluid evolution and control by deformation seem to be a general feature. New geochronological and fission-track data imply very high uplift rates, on the order of 2.2 cm/y.

Introduction

WHEN THE ILP task-group III-6 was formed in 1993, the UHP terrane of the Dora-Maira massif was probably the best-studied one known at that time, the relatively poor exposure conditions being counterbalanced by the small size of the terrane (~5 km x 10 km), the long tradition of Alpine geological studies, and the early finding of coesite. The important results already achieved at that time were: the coherency and main boundaries of a UHP metamorphic unit (Chopin et al., 1991), a P-T path for the metamorphic evolution (Schertl et al., 1991), the young age of the UHP event (~38 Ma) (Tilton et al., 1991), and an idea of fluid behavior through stable-isotope study (Sharp et al., 1993). These results were summarized in several chapters of the book *Ultrahigh Pressure Metamorphism* (Coleman and Wang, 1995) and addressed in the review papers by Schreyer (1995) and Harley and Carswell (1995).

Meanwhile, this terrane has been used as a natural UHP laboratory, so that a number of recent studies now have refined and augmented the earlier-existing picture. The results of these studies are briefly reviewed in this paper: they concern field relations as well as UHP mineralogy; the petrology of prograde, climax, and retrograde mineral assemblages; the geochemistry of the main rock types; fluid composition and behavior; cooling ages; and uplift rates.

Geological Setting: How Is the UHP Unit Bounded?

A general problem in defining the boundaries of any coherent UHP unit is that clear evidence for pressure climax conditions is preserved in only a few volumetrically minor rock types, which are scattered throughout a pile of country rock (metagranitic gneiss) in which a regional low-pressure overprint is ubiquitous. Detailed mapping (1:25,000) was pursued in Dora-Maira to better constrain the outline provided by Michard et al. (1993) and Henry et al. (1993), in particular the extension of the UHP unit to the north and its relation to the underlying San-

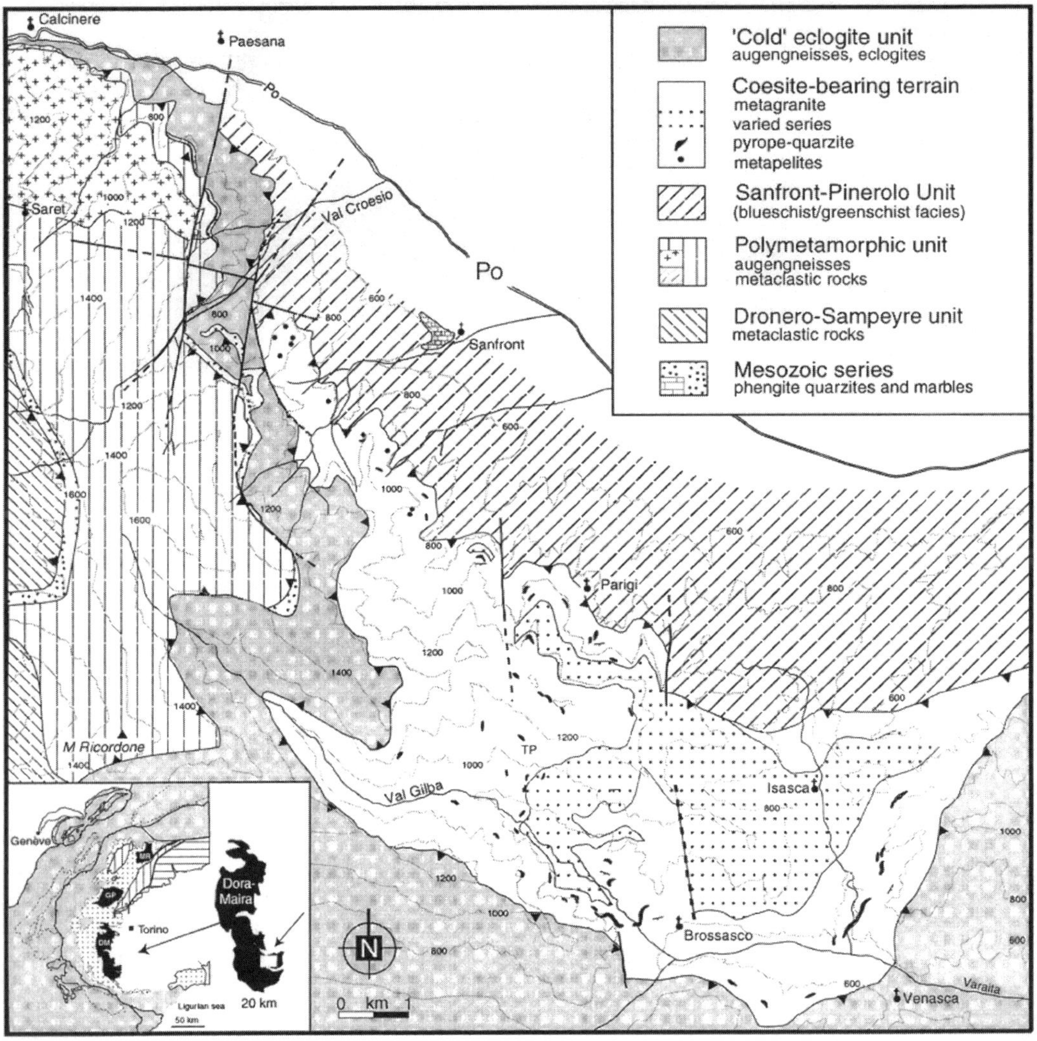

FIG. 1. Geological outline of the coesite-bearing terrane in the southern Dora-Maira massif, after Henry (1990), Michard et al. (1993), Simon et al. (1997a), and unpublished data from I. Schnieders, H. Klock, and the authors.

front-Pinerolo unit, the lowest structural crustal unit of the Western Alps. The UHP unit pinches out to the north and its thickness decreases from ~800 m to less than 200 m west of Sanfront, where the northernmost UHP rock sample was found (almandine–kyanite metapelite). It disappears there through the complex interplay of late N-S and NE-SW steeply dipping faults, which bring the underlying Sanfront-Pinerolo Unit into direct contact with the overlying "cold eclogite" HP unit (Fig. 1). Nowhere is the UHP unit in contact with oceanic or mantle-derived units.

Within the pile of more or less mylonitic orthogneiss that forms the country rock in both the HP and UHP units, the contact of the UHP unit with its hanging wall is not well defined, but simply bracketed within 50 to 200 m by the occurrences of either HP or UHP rocks. The contact with the footwall is better defined because of a lithological contrast (orthogneiss overlying a metasedimentary series) and a more localized deformation evolving from ductile throughout the pile to semi-brittle near the contact. The gently SW-dipping fault contact is a relatively late normal fault, formed in relation to the top-to-the-SW extensional phase that affected the entire massif; this late normal fault probably cuts across the original thrust contact concordantly with the shallow-dipping foliation of both units (Avigad et al., in progress).

As to the formations constituting the UHP terrane, two significant discoveries must be mentioned. One is that of relict coesite included in omphacite within a xenolith in scarcely deformed metagranite (near Bastoneri, Val Gilba) (D. R. Snoeyenbos, pers. commun., 1995). The other is that of a jadeitic clinopyroxene inclusion in zircon from the country-rock orthogneiss near Parigi (M. Burchard, pers. commun., 1999). Assuming equilibrium between garnet, pyroxene, and phengite, all present as inclusions within the same zircon crystal, yields a range of 730 to 830°C and 35 to 42 kbar for the equilibration conditions (using Green and Hellmann, 1982 and Waters and Martin, 1993 thermobarometers). Both findings provide indisputable evidence that the country-rock metagranite also underwent UHP conditions—confirming an expectation held by earlier workers on a more tenuous basis (Chopin et al., 1991; Biino and Compagnoni, 1992) and thus the coherency of the UHP unit.

UHP Mineralogy of the "Pyrope Whiteschist": MASH in Nature

Recent petrological work has focused on eclogites (see below) and on the characteristic coesite-bearing "pyrope quartzite" or "pyrope whiteschist"—a quartz-rich pyrope–kyanite–phengite–schist that occurs as meter-sized layers or trails of boudins within the country-rock orthogneiss. This highly magnesian pyrope-bearing rock is one of the best natural approximations of the MgO-Al_2O_3-SiO_2-H_2O (MASH) experimental system at high pressure (see the section on "Geochemistry" below). The phase relations and petrological evolution of the quartz-rich matrix were well understood previously (Chopin, 1984, 1986; Schertl et al., 1991; Sharp et al., 1993), but the pyrope megablasts (5 to 20 cm in size) locally present in this matrix had not revealed all their mineralogical characteristics, except for the new high-pressure Mg-Al-(Ti, Zr)-silicate ellenbergerite and Mg-Al-borosilicate magnesiodumortierite (see Enami and Banno, 1999). The new findings concern the Mg end members of chloritoid and staurolite, and the occurrence of granulite-facies assemblages in these megablasts, which has an important bearing on fluid behavior and derivation of the exhumation path. Before addressing the findings made with respect to megapyrope, we first mention a significant discovery resulting from the TEM study of phengite crystals in the matrix.

Exsolution in UHP phengite?

High-resolution transmission electron microscopy (HRTEM) of the $Si_{3.55}$ phengite-3*T* in the pyrope whiteschist revealed the existence of some defect areas (Ferraris and Chopin, 1998) in a perfect 3*T* matrix (Amisano Canesi, 1994; Pavese et al., 1997, 1999). These areas contain platelets of single-crystal quartz, 100 to 700 Å thick, bounded by the (001) planes of mica, and extending laterally over 5,000 to at least 10,000 Å with a rounded termination against the mica matrix (Ferraris and Chopin, 1998). In their vicinity and only there, the 29.7 Å periodicity of the 3*T* mica matrix is faulted by a few, up to six-unit-thick sequences of 18.9 Å periodicity, interpreted as talc-2*M* on the basis of analytical microscopy. These observations and modal estimates suggest that the high-Si UHP phengite became unstable upon decompression and reacted to form talc and quartz, "exsolved" in a less celadonitic mica, according to a reaction of the type

$$3 \text{ Mg-Al-celadonite} = 1 \text{ muscovite} + 5 \text{ quartz} + 1 \text{ talc} + 2 \text{ KOH},$$

which is actually not an exsolution mechanism proper since potassium was removed from the system. Optical examination of 50 to 60 μm thick petrographic sections of phengite flakes parallel to (001) revealed the presence of thin platelets interlayered within (001) of mica, showing a characteristic lobate, amoebic contour. On the basis of their low birefringence, the presence of fluid inclusions, and an enhanced Si signal under the electron probe, these platelets were identified as quartz and are thought to represent the same object as revealed by transmission electron microscopy (TEM), despite the different scale (Ferraris et al., 1999).

Most interestingly, optical reexamination of phengite from several HP or UHP localities (Monte Mucrone, Sesia Zone, Dabie Shan) revealed the presence of the same lobate lamellae. Obviously, micas thus far considered to be homogeneous may not be. The implications of this intracrystalline reorganization upon decompression are important with respect to thermobarometry (regarding the resolution of the analytical method employed) and thermochronometry as well, especially with respect to the closure of the system and degassing behavior (cf. Harrison and Fitz Gerald, 1986).

Magnesiochloritoid

The finding of near–end member magnesiochloritoid (Simon et al., 1997a) was unexpected on theoretical grounds because the MASH assemblage clinochlore + kyanite (± talc), common in the pyrope megablasts, precluded the stability of the higher-pressure alternative pair Mg-chloritoid + talc (Chopin and Schreyer, 1983). The discussion below is based on Simon et al. (1997a).

Magnesiochloritoid (up to 97 mol% end member) occurs along with ellenbergerite, talc, chlorite, or kyanite as prograde inclusions in a few pyrope megablasts from Val Gilba. Whereas most pyrope megablasts are relatively homogeneous (Prp90–Prp98), the magnesiochloritoid-bearing ones commonly preserve a clear prograde Fe-Mg zonation pattern (from Prp70 to Prp98). The nearly homogeneous chloritoid inclusions also record this zonation pattern; their X_{Mg} (= Mg/(Fe+Mg)) increase from 0.70 to 0.96 from garnet core to garnet rim. These zonations in garnet and chloritoid were shown to be prograde and to record increasing P and T conditions up to ~700°C, 30 kbar. The occurrence of magnesiochloritoid in some garnets simply reflects a locally slightly more Fe-rich system, which stabilizes the chloritoid + talc pair with respect to chlorite + kyanite. Interestingly, the relevant reaction is water conserving and thus is independent of fluid composition; it provides an upper pressure limit of ~32 kbar to the entire metamorphic evolution through the instability of chloritoid + talc in the most Mg-rich systems.

Fe-Mg partitioning between chloritoid and zoned host-garnet changes from normal (i.e., $X_{Mg}^{Cld} > X_{Mg}^{Grt}$) to reverse ($X_{Mg}^{Cld} < X_{Mg}^{Grt}$) from garnet core to garnet rim, or from rim to core in other megablasts of similar composition. In the event of reverse partitioning between inclusion and garnet, a diffusion zone of a few tens of micrometers exists in garnet: X_{Mg}^{Grt} decreases toward the inclusion and reaches a normal partitioning value at the contact. These features were interpreted in terms of disequilibrium and tentatively ascribed to slow Fe-Mg diffusion in chloritoid with respect to the garnet growth rate, or to two competing garnet-forming reactions, with the metastable one, which produces a more Mg-rich garnet, being kinetically favored.

Magnesiochloritoid locally breaks down into symplectites of chlorite and an extremely low-Si magnesiostaurolite (down to 7.0 Si per formula unit). The latter is probably metastable with respect to Si-rich magnesiostaurolite + corundum (cf. Fockenberg, 1995).

Magnesiostaurolite

This mineral, approximately $(Mg,Fe,Li,\square)_4Al_{18}Si_8O_{44}(OH)_4$, occurs exclusively as inclusions in pyrope megablasts. After Simon et al. (1997b) and Simon (1998), the prograde magnesiostaurolite inclusions have X_{Mg} in the range from 0.78 to 0.95 and are structurally characterized by the near absence of tetrahedral Al, whereas Mg preferentially occupies the tetrahedral T2 site (Hawthorne et al., 1993a). In contrast to the sample of Hawthorne et al. (S41), all the staurolites studied by Simon (1998) have low Li contents (250 to 390 ppm by ion microprobe). They exhibit a consistent Fe-Mg reversal with garnet— i.e., $X_{Mg}^{St} < X_{Mg}^{Grt}$, which is interpreted as an equilibrium feature at high pressure (and high X_{Mg}) (Koch-Müller, 1997).

Some crystals exhibit a "cross-hatched" twinning in thin section, due to two submicroscopic lamellar systems, a prominent one parallel to (100), the other parallel to (001) (Simon et al., 1997b). A TEM study revealed two major findings. First, lamellae parallel to (100) are from ~10 nm to 0.3 μm thick, with straight boundaries to the host staurolite with which they are a twin. The difference in the β angle (Δβ = 0.4°) indicates departure from ideal orthorhombic symmetry. Moreover, satellite diffraction spots parallel to **a*** indicate that the lamellae represent a new structural variant of staurolite with a superstructure having $\mathbf{a} = 2\ \mathbf{a}_{st}$ (Czank et al., 1997).

Second, lamellae parallel to (001) have a thickness of a few unit cells and are irregularly distributed, with the clear presence of $00l$ ($l = 2n+1$) diffraction spots that are pseudoextinct in the host matrix. Interestingly, the diffraction contrast between lamellae and matrix disappears under exposure to the electron beam, suggesting that different proton contents or proton orderings could be responsible for the occurrence of these lamellae. However, single-crystal X-ray diffraction patterns show three sets of diffraction spots, one pseudo-orthorhombic and two monoclinic, among which one has a doubled *c* axis, thus pointing to a further new variant of staurolite (Simon, 1998). Since these effects also are revealed by X-ray diffraction, not only protons but also heavier atoms must be involved in the process—through variable short- or long-range ordering (since no compositional difference could be detected by analytical microscopy), as can be expected after the study of Hawthorne et al. (1993b).

Phosphates

The occurrence of several uncommon or new phosphates in the pyrope quartzite, like bearthite, OH-rich wagnerite, and phosphoellenbergerite (Chopin and Sobolev, 1995), was the incentive for an experimental study of the $CaO-Al_2O_3-P_2O_5-H_2O$ and $MgO-P_2O_5-H_2O$ systems under a geological pressure range (Brunet and Chopin, 1995; Brunet et al., 1998, respectively). The latter article also provides petrographic and analytical data on Mg-phosphates in UHP rocks, to which the reader is referred for further details.

Metamorphic Evolution of the Pyrope Whiteschist

Fluid behavior

In their stable-isotope study, Sharp et al. (1993) demonstrated that the pyrope quartzite failed to achieve oxygen-isotope equilibrium with the enclosing orthogneiss, thus providing evidence against fluid flow across lithologic layers. They also explored the possibility (or necessity!) of low H_2O activities in the pyrope quartzite. They demonstrated that CO_2 cannot have been the main dilutant of an aqueous fluid at peak metamorphism because extensive carbonatation would have resulted; they therefore settled for the idea of partial melting controlling and lowering H_2O activity in this rock—a hypothesis that depends on the interpretation (as a partial melt or not) of the thin jadeite-kyanite-quartzite layers that occur within pyrope quartzite (Schreyer et al., 1987; see Geochemistry section below). Actually, an alternative, still unexplored, possibility was that of a highly saline fluid.

In their fluid-inclusion study of the pyrope quartzite and megablasts in Parigi, Philippot et al. (1995) reported several fluid generations, depending on host mineral and rock type—a low-salinity H_2O-CO_2 fluid, CO_2-rich fluids coexisting with brines (up to 84 wt% equivalent NaCl), and a low-salinity aqueous phase. This again is evidence for a limited fluid mobility and short-range equilibration. They also noted the precipitation of a number of daughter minerals like magnesium phosphates and carbonate, in addition to more common chlorides, in some inclusions. The observations are evidence that at some stage—during decompression-related overprinting, according to Philippot et al. (1995)—highly saline fluids were present in these rocks (or part of them). This is corroborated by the occurrence of Na-phlogopite filling some cracks within millimeter-sized garnet in jadeite-kyanite-quartzite at Parigi; this Na-phlogopite is rimmed by potassium phlogopite, but no Na-phlogopite hydrate was found, which under high H_2O activity would have formed inevitably (Carman, 1974; Schreyer et al., 1980). In summary, no percolating fluids coursed through the UHP rocks at the peak of metamorphism, although fluids existed at some stage and were *locally* highly saline. The data and results reported in the next paragraph confirm this conclusion at an even more local scale, where buffering by the solids leads to extreme fluid compositions.

Crack-related "granulitic" assemblages

Throughout the UHP unit, the pyrope megacrysts contain the typical assemblage clinochlore–kyanite–talc–rutile–zircon ± phlogopite ± ellenbergerite as prograde inclusions. However, in the upper part of the UHP unit in Val Gilba, some garnet megacrysts (X_{Mg} = 0.89 to 0.98) in addition contain polymineralic inclusions consisting of various combinations of enstatite, gedrite, sapphirine, clinochlore, talc, magnesiostaurolite, and rare corundum or spinel, which are reminiscent of the granulite facies. Simon and Chopin (1999) demonstrated that these assemblages are developed along cracks running across the megablasts, and therefore are of late origin, post-dating the highest-pressure stage (unlike the inclusions described in UHP garnet from Dabie-Shan by Okay, 1994).

Enstatite (X_{Mg} = 0.94 to 0.99) contains 0.7 to, typically, 3 wt% Al_2O_3, but up to 8.4 wt% in the paragenesis with sapphirine. Sapphirine (X_{Mg} = 0.96 to 0.998, Be-free) shows the largest compositional variations, with Si contents ranging from 1.7 to at least 2.1 atoms per formula unit (a.p.f.u.), thereby clearly exceeding the 2:2:1 stoichiometry. The late-stage talc contains up to 4 wt% Al_2O_3, 0.35 wt% Na_2O, and 0.6 wt% F; gedrite is 1.1 to 2.9 wt% Na_2O and up to 0.36 wt% F.

The succession within pyrope of alternative hydrous assemblages involving enstatite + an Al-rich phase (kyanite, sapphirine, magnesiostaurolite) ± clinochlore, a gedrite compositionally close to pyrope, and then talc + an Al-rich phase (sapphirine, corundum) is a clear record of decompression.

However, the temperature conditions implied under the assumption of high H_2O activity are 100 to 150°C higher than, and also inconsistent with, existing constraints on the decompression path, in particular the stability of talc + phengite (Massone and Schreyer, 1989) in the pyrope quartzite matrix assemblage—which shows no evidence of a granulite-facies event. This inconsistency can be resolved only if H_2O activity inside the garnet megablast was (or became) considerably reduced with respect to that in the matrix. Fluid influx into an opening fracture in garnet, sealing of the fracture by breakdown (hydration) products of pyrope, and continued reaction under nearly closed-system conditions may have led to increasing solute concentration and thus low H_2O activity within the garnet megablast. Micrometer-size inclusions of Ca-sulfate and crandallite-type compounds in minerals of these reactive areas could be taken as evidence for such residual brines and suggest they were phosphate- and sulfate- rather than halide-dominated.

It is important to note that the selective occurrence and development of granulite-facies assemblages within pyrope megablasts was triggered by garnet fracturing upon decompression (see also Austrheim, 1998, p. 308–309, in the Western Gneiss Region) and reflects a local decrease of H_2O activity ("desiccation"; see Kullerud, 1996; Markl and Bucher, 1998) rather than high temperatures. The resulting low-temperature origin proposed by Simon and Chopin (1999) for minerals commonly found in granulite-facies rocks conveys the same implication as the occurrence of the enstatite + andalusite assemblage (replacing cordierite), stabilized by low H_2O activity (0.1!) as reported by Ballèvre et al. (1997) from Australia: the occurrence of phases such as sapphirine or enstatite + Al-silicate thus are dependent on H_2O activity and by themselves are not an indication that typical granulite-facies temperature conditions (700 to 800°C) were reached. This may be of particular relevance for the formation of sapphirine during decompression of many kyanite-eclogites (e.g., Liati and Seidel, 1994; Möller, 1998), especially when evidence for granulite-facies metamorphism is lacking in the country rock.

New Petrological Results on UHP Eclogites

Kienast et al. (1991) investigated eclogites from five different localities of the coesite-bearing Brossasco-Isasca Unit and derived pressures of 27 to 29 kbar at temperatures close to 700°C for the UHP metamorphic event. The overlying "cold eclogite unit" only experienced conditions near 15 kbar and 550°C (Chopin et al., 1991). A recent study of eclogites from 13 different localities throughout the entire coesite-bearing unit was initiated in order to derive detailed P-T data not only for the peak metamorphic conditions, but particularly for deriving information on the prograde and retrograde paths. The data presented here are based mainly on the Ph.D. thesis of Nowlan (1998).

The eclogites that occur as boudins or lenses of decimeters to several meters in size can be subdivided into dark- and light-colored types. The mineral constituents of the dark type are garnet, omphacite, and amphibole, with minor amounts of quartz, phengite, paragonite, zoisite, biotite, rutile, and titanite. The light samples typically contain additional kyanite and are rich in mica, whereas the dark ones contain only minor amounts of mica. In one light-colored sample, coesite was observed as an inclusion within omphacite.

Garnet

Analyzed garnets show a wide compositional range; they can be either Fe-rich (group I), more Ca-rich (group II), or more Mg-rich (group III). Dark eclogites consistently contain either Ca-rich (group II) or Fe-rich (group I) garnets, plotting in field C (Alpidic type) according to Coleman et al. (1965). On the other hand, light-colored eclogites only contain Mg-rich garnets of group III, thus falling into field B, which is typical for gneissic terrains (Coleman et al., 1965). Prograde zonation patterns of garnet show increasing Mg from core to rim. Concomitantly, Ca decreases from core to rim, whereas Fe can be either more or less constant or positively correlated to Mg. Retrograde garnet growth is expressed by a decreasing pyrope and increasing grossular component. The almandine component either stays more or less constant or it decreases slightly with Mg.

Omphacite

In general, the jadeite component of omphacites decreases from core to rim. Omphacite inclusions in garnet vary in composition: early formed ones have rather low jadeite contents and occur in the cores of garnet, whereas those with the highest jadeite contents are enclosed in garnet rims characterized by high pyrope contents. These latter high-jadeite compositions essentially correspond to the core compositions of matrix omphacites.

Phengite

Si contents of phengites vary from 3.2 to 3.6 pfu. Large phengite crystals contain the highest Si concentration in their cores, whereas rim compositions as well as small phengite flakes can be distinguished by generally lower Si concentrations. Phengite inclusions in early-formed garnet zones contain low amounts of Si, whereas phengite inclusions with the highest Si contents can be found in garnet zones with highest Mg concentration, formed during peak-metamorphic conditions.

Other minerals

Locally, kataphoritic to taramitic and barroisitic amphiboles occur, forming inclusions within garnet and omphacite. On the other hand, the matrix amphiboles of a retrogressively formed second generation are of edenitic to pargasitic and actinolitic to tschermakitic compositions. Although zoisite, kyanite, rutile, and coesite are of prograde origin or formed during peak metamorphic conditions, titanite, paragonite, and, locally, plagioclase, chlorite, and biotite are formed retrogressively.

Metamorphic Conditions

For P-T calculations different geothermobarometers were applied; the most consistent results were attained using the Gt-Cpx-Phe geobarometer of Waters and Martin (1993) in combination with the Gt-Cpx geothermometer (Carswell et al., 1997). On the basis of garnet zonation, three main stages of metamorphism are distinguished—a prograde stage (I), a peak metamorphic stage (II), and a retrograde stage (III). Early, low-Si phengite and low-jadeite omphacite inclusions within garnet cores are correlated with the prograde stage I. In the pyrope-rich portions of garnet (stage II), phengite inclusions are characterized by high amounts of Si, whereas omphacite exhibits high amounts of jadeite component. P-T data of the retrograde path are derived using rim compositions of matrix phengites and omphacites.

Considering all P-T data calculated, the prograde path for the eclogites investigated could be described by the conditions of 15 kbar/500°C, 25 kbar/570°C, 32 kbar/650°C, and, at the peak of metamorphism, 37 kbar/790°C (Fig. 2). The exhumation of the rocks took place under slightly higher

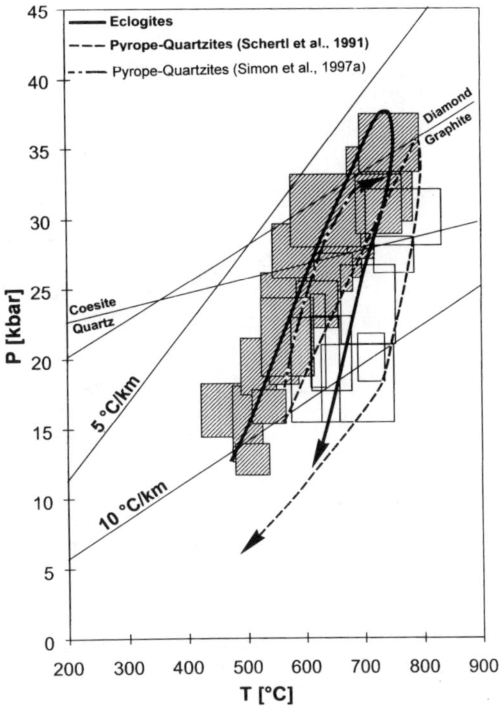

FIG. 2. P-T path for eclogites of the coesite-bearing unit compared to a P-T path for pyrope quartzites according to Schertl et al. (1991) and a prograde section for pyrope quartzites according to Simon et al. (1997a). The boxes indicate the P-T data calculated for each sample using mineral analyses representative of stages I to II (hatched boxes) and II to III (open boxes).

temperatures compared with subduction; representative P-T data are 24 kbar/650°C and 14 kbar/620°C. These data are in good agreement with data derived from pyrope quartzites forming lenses within the same metamorphic unit. The P-T path presented by Schertl et al. (1991) lies at slightly higher temperatures when compared to the one derived from the eclogites. This might be caused by overestimation of temperature by Schertl et al. (1991), who used a water activity of 1 for calculating the reaction talc + kyanite = pyrope + coesite + H_2O. The prograde section of the P-T path determined for the pyrope quartzites of the Val Gilba by Simon et al. (1997a) is in excellent agreement with the prograde section derived from eclogites. The data obtained strengthen the conclusion that the eclogites, similar to the pyrope quartzites, were subducted to mantle depths of more than 100 km and exhumed afterward as one coherent structural unit together with their country-rock gneisses.

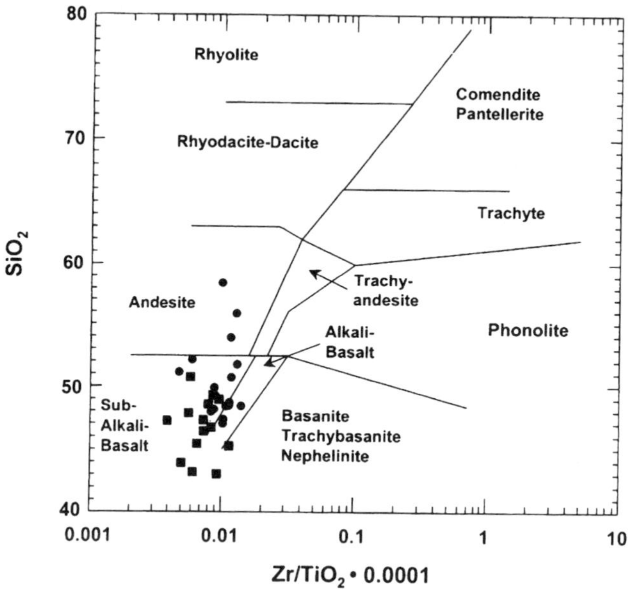

FIG. 3. Nomenclature diagram of volcanic rocks (Winchester and Floyd, 1977) for dark- (filled boxes) and light-colored eclogites (filled circles) of the UHP unit in the Dora-Maira massif.

Geochemical Investigations

The dark- and light-colored eclogites, the pyrope coesite rocks, their phengite-schist and jadeite-rich rock inclusions, as well as biotite-phengite gneiss country rocks and undeformed metagranites were investigated for their whole-rock chemistry. The data presented in this paper mainly represent a compilation of results gathered from Nowlan (1998), Nowlan et al. (1999), and Schertl and Schreyer (1999).

Eclogites

In general, the light-colored eclogites (kyanite-bearing, rich in mica) contain higher amounts of SiO_2, MgO, Al_2O_3, K_2O, Ba, and Rb than do the dark-colored samples. On the other hand, the dark eclogites typically are characterized by higher amounts of TiO_2, FeO, P_2O_5, Zr, Nb, Ni, Cu, and Zn. Transitions between dark- and light-colored types are gradual, both forming portions of the mono- as well as the polymetamorphic complex (Compagnoni et al., 1994, 1995) of the UHP unit (or Brossasco-Isasca unit). Admittedly, the process of metamorphism could have generated a movement of elements, thus producing different types of eclogites. However, most interestingly, the compositional patterns of the elements for both types mirror not only a considerable variation of mobile elements (e.g., K, Ba) but also of immobile ones like Nb, P, Zr, and Ti.

In order to obtain information concerning possible protoliths of the eclogites, the analyses were first plotted in a classification diagram for magmatic rocks (concerning the amounts of SiO_2 and $Zr/TiO_2 \times 0.0001$) (Winchester and Floyd, 1977; see Fig. 3): The compositions lie within the field of sub-alkali basalts and andesites; some of the light-colored eclogites are somewhat higher in SiO_2, thus plotting in the andesitic field. According to Irvine and Baragar (1971), the light-colored eclogites exhibit a calc-alkaline, the dark eclogites a tholeiitic character. Therefore, in the course of a potential magmatic evolution, the precursor of the dark eclogites could represent a first generation, while the light-colored ones could be related to a younger calc-alkaline generation. Nevertheless, there is no convincing proof for that genetic model because the eclogites are not very widespread and occur as isolated lenses within their adjacent country rocks. The majority of the light-colored eclogites exhibit a "nonoceanic character" according to the P_2O_5-K_2O-TiO_2-discrimination diagram of Pearce and Cann (1973), whereas the dark samples almost plot within the field typical for an oceanic origin. This feature would point toward different magmatic environments of the protolith, which is doubtful because of the nature of occurrence of the eclogites as small lenses. Taking the elements Zr, Ti, and Y into account (Pearce and Cann, 1973), most of the eclogites show compositions that are typical for intraplate basalts. Since the eclogites contain high amounts of TiO_2 and low amounts of K_2O, Rb, and Sr, a sedimentary protolith like a marl is unlikely (compare Matthes et al., 1975).

In the "Münchberger Gneismasse" (Bavaria, Germany), dark- and light-colored eclogites also are developed. In all probability, the protolith of the light-colored type was an Al-rich basalt, whereas the dark type could represent former continental tholeiitic material, poor in potassium; Matthes et al. (1975) favored the idea of a genetic relationship between the two protoliths as a result of differentiation of a basaltic magma. Differentiation processes leading to TiO_2- and P_2O_5-rich and TiO_2- and P_2O_5-poor magmas, respectively, are well known from basalts from the Parana province in Brazil (Bellieni et al., 1986; Fodor, 1987) and from the Karoo province in the southern part of Africa (Cox et al., 1967; Duncan and Green, 1987; Marsh, 1987). However, in contrast to the aforementioned basalts, where the amount of potassium is positively correlated with TiO_2 and P_2O_5, the Dora-Maira eclogites that are rich in TiO_2 and P_2O_5 contain only low amounts of potassium. The different concentrations of iron relative to magnesium in the dark- and light-colored eclogites also could point to a differentiation trend (Cox, 1980), provided that the chemical composition of the protolith was not affected by metasomatism. Similarly to the dark eclogites investigated, several eclogites of the Monte-Rosa and Gran-Paradiso massifs also contain high concentrations of iron and titanium (e.g., Dal Piaz and Lombardo, 1986).

Although extensive geochemical data are available for the UHP eclogites of the Dora-Maira massif, various questions remain open. In particular, considering the possibility of metasomatic processes during metamorphism, it is difficult to argue for or against one distinct model of origin. However, it is quite certain that the protolith is of magmatic origin. Because of the very limited spatial occurrence of the eclogite lenses, a plutonic origin is out of the question. Furthermore, the possibility of differentiational magmatic processes leading to the protolith of the eclogites also is disputable. According to the very restricted mode of occurrence, models like basaltic dikes or pyroclastic processes leading to a deposition of volcanic tuffs that may also contain sedimentary components (especially for the light-colored eclogites surrounded by parametamorphic country rocks) appear to be a more likely key for explanation. Nevertheless, detailed isotope- and age-determination work should be carried out in order to resolve these problems in the future.

Pyrope quartzites and related rocks

A full range of geochemical data for pyrope quartzites and various rock inclusions is fundamental to an understanding of the nature of their protoliths and initial relationship to the surrounding gneisses. The first whole-rock analyses on the bulk chemistry of pyrope quartzites (or "whiteschists") and their country-rock gneisses were presented by Tilton et al. (1991). They demonstrated that the compositions of pyrope quartzites largely fall in the system K_2O-MgO-Al_2O_3-SiO_2-H_2O and that the biotite-phengite gneiss country rocks have essentially granitic compositions. Whereas the derivation of the gneisses from a former Hercynian granite is undisputed, two contrasting genetic models concerning the protolith of pyrope quartzite—sedimentary evaporitic (Chopin, 1984) versus metasomatic (Compagnoni and Hirajima, 1992, 1993)—exist. Recently Gebauer et al. (1997) presented—with respect to cathodoluminescence and SHRIMP studies on zircons—a combination of both models: a metasomatic version where metasomatizing fluids were derived from evaporitic sediments. Stable-isotope data were already presented by Sharp et al. (1993), indicating that pyrope quartzites and surrounding country rocks are in equilibrium with respect to hydrogen isotopes but not to oxygen isotopes. In order to test the different hypotheses regarding the origin of the pyrope quartzites, several samples of leucophyllite (partly named "Kornstein") and their country-rock gneisses from the Eastern Alps (for details see Schertl and Schreyer, 1999) were analyzed. Finally, truly metasedimentary whiteschist analyses (Schreyer, 1977) were used for comparative studies.

Fine-grained pyrope quartzites. All samples investigated are typically high in MgO and contain low amounts of iron oxides, CaO, and Na_2O. Retrograde samples show analogous compositions except for higher H_2O contents due to secondary Mg chlorite. Concerning minor and trace elements, the pyrope quartzites and their retrograde products are consistently depleted in Cu, Zn, Rb, Ba, Sr, and P relative to the various granitic rocks, whereas Zr seems to be somewhat enriched in the white-schist-type rocks.

Orthogneisses. The country-rock gneisses of the pyrope quartzite lenses are essentially of granitic composition. Most interestingly, the orthogneisses of the coesite-bearing Brossasco-Isaca unit contain a comprehensive suite of different textural appearances like metagranite (having largely preserved their original granitic texture), coarse-grained augengneiss, and fine-grained biotite-phengite gneiss.

Phengite schist inclusions within pyrope quartzite. At the locality near Parigi, a series of phengite-schist intercalations, 3 to ~80 cm thick, is developed, in which a compositional banding is aligned parallel to the mica orientation of the surrounding pyrope-quartzite lens (Schertl et al., 1991), thus suggesting a common deformational and metamorphic history—which is documented by coesite inclusions within kyanite in the schist. Three groups of compositional layers in the phengite-schist intercalations must be differentiated: the first group is chemically rather similar to the biotite-phengite gneiss country rocks; the second is enriched in Fe, Mn, Ca, Na, Al, and Si relative to the third, which contains higher amounts of K, P, and H_2O than the second. From a mineralogical point of view, this is the result of higher amounts of phengite in group three relative to group two, and to the occurrence of kyanite and pseudomorphs after jadeite, forming major constituents only in the second group.

Jadeite-rich rocks. Garnet-jadeite-quartzite layers as well as a jadeite quartzite boudin differ from the surrounding pyrope quartzite essentially in their lower K and Mg but higher Na, Fe, and Ca contents. Most interestingly, jadeite-rich rocks from the classical locality near Parigi are distinctly lower in K_2O than are samples from a second locality in Val Gilba. This is because phengite occurs only in accessory amounts at Parigi. Concerning trace elements, the jadeite-rich rocks from Parigi exhibit a particularly strong depletion in Rb, Ba, and Sr; nevertheless, their trace-element spectra are essentially similar to those of the pyrope quartzites. However, the origin of the jadeite-rich rocks is still a matter of debate (former melts?) (Schreyer et al., 1987).

Discussion. The granitic country rocks exhibit an S-type-character (see Chappell and White, 1974) and peraluminous chemistry (molar $Al_2O_3/(CaO + Na_2O + K_2O) > 1.1$) (Shand, 1950). In a diagram of SiO_2 versus $Na_2O + K_2O$, according to Cox et al.

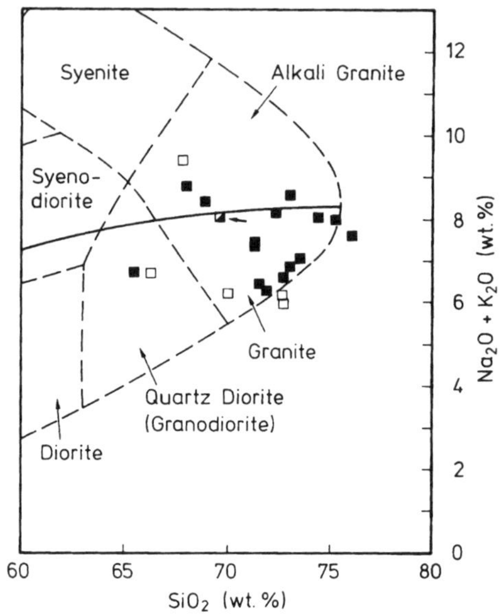

FIG. 4. Nomenclature diagram of plutonic rocks (Cox et al., 1979) for phengite-biotite gneiss (filled boxes), augen-gneiss (half-filled box with arrow), and undeformed metagranite (open boxes) from the UHP unit.

(1979; see Fig. 4), the variety of analyses ranges from quartz diorite to granites and alkali granites. Following the trace-element plot of Rb versus Y + Nb of Pearce et al. (1984), the compositions lie near the border of syncollisional, within-plate, and volcanic-arc granites, which is typical for postcollisional granites. Without any doubt, this mirrors the tectonic position of the Dora-Maira granitoids within the pre-Alpine Variscan basement (Compagnoni et al., 1994, 1995).

The discrimination of pyrope quarzites from the adjacent country rocks can best be described using the elements Mg and K on the one hand, and Fe, Ca, and Na on the other. The simple MgO versus FeO_{tot} (total iron as FeO) plot already shows clear distinctions and also yields the first possible genetic indications (Fig. 5). The orthogneisses, apart from their textural variations, always plot at low amounts of MgO and high FeO_{tot}, whereas the pyrope quartzites and their retrograde derivatives are strictly discriminated by consistently high MgO and generally low FeO_{tot}. Most interestingly, the leucophyllites and their country rocks from the Eastern Alps (Fig. 6; see also Modjtahedi and Wieseneder, 1974; Demény et al., 1997) fall with their respective clusters of compositions well into the ranges defined by the

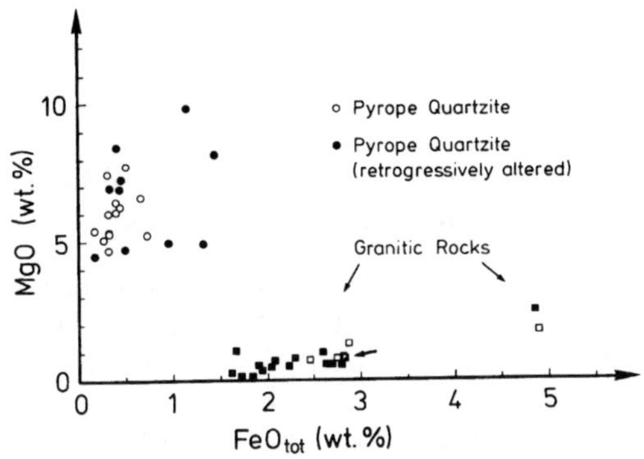

FIG. 5. MgO versus FeO_{tot} diagram for fresh and overprinted pyrope quartzite as well as phengite-biotite gneiss (filled boxes), augen-gneiss (half-filled box with arrow), and undeformed metagranite (open boxes) from the UHP Unit.

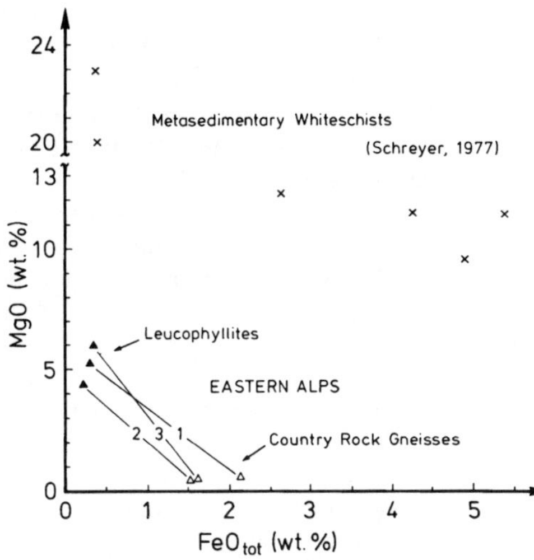

FIG. 6. MgO versus FeO_{tot} diagram for leucophyllites and their country-rock gneisses from the Eastern Alps (locality 1 = Miesenbachtal; 2 and 3 = Rabenwald) as well as for whiteschists of evaporitic origin according to Schreyer et al. (1977).

Dora-Maira rocks. Therefore, strong genetic links are advocated. However, whiteschist analyses taken from Schreyer (1977) plot clearly outside the range of pyrope quartzites and leucophyllites because of their still higher MgO and FeO_{tot} (Fig. 6). It seems that a simple criterion like the Mg/Fe ratio can be used for distinguishing whiteschists of evaporitic origin from whiteschist compositions that possibly formed through metasomatic processes.

In an MgO versus FeO_{tot} diagram of the type of Figure 5, the intercalated phengite schists partly overlap the field of granitic rocks; however, layers of group three show an obvious tendency of Mg enrichment at the expense of Fe, thus pointing in the direction of the pyrope quartzites. Most interestingly, the combined bulk chemistry of groups two and three is again close to granitic compositions. For that reason, the compositional layering with its element fractionation may be a result of some sort of metamorphic differentiation of the original granite during UHP conditions. Also, the trace-element spectra of the phengite schists support a close relationship to the granitic country rocks of the pyrope quartzites; however, internal variations of Na + Ca versus K + Mg are matched by Rb, Ba, and to a lesser extent Sr.

As a resulting picture, the pyrope quartzites can be clearly distinguished chemically from metasedimentary evaporitic rocks from Afghanistan, Tanzania, and Zambia as described by Schreyer (1977), thus making a sedimentary origin unlikely. Also the field relations of the pyrope quartzites, forming trails of lenses within a former granite, point to a different type of protolith. On the other hand, leucophyllites from the Eastern Alps, as described by Modjtahedi and Wieseneder (1974), clearly formed metasomatically and, occurring along shear zones within granitic gneisses, are chemically very similar to the pyrope quartzite rocks (for more details see Schertl and Schreyer, 1999).

New Geochronological Data

Concerning the age of UHP metamorphism of the Dora-Maira rocks, the previously accepted Cretaceous U/Pb ages on zircon (121 Ma) (Paquette et al., 1989) were first questioned by Tilton et al. (1991). The latter presented U/Pb–zircon multigrain data with a chord intersecting the concordia at 38 ± 1.4 Ma and 304 ± 10 Ma, thus bringing an Eocene age of UHP metamorphism into the discussion. Moreover, their U/Pb and Th/Pb investigations on monazite and ellenbergerite also led to young ages in the range of 30 to 34 Ma. The discovery of excess Ar in micas of the pyrope quartzite (Hammerschmidt et al., 1992) even questioned the Ar/Ar ages of 105 Ma on phengites presented by Monié and Chopin (1991). Further evidence for excess Ar leading to geologically meaningless plateau ages was found in several minerals of the nearly undeformed Brossasco UHP metagranite (Arnaud and Kelley, 1995) and in phengites of eclogites and Mg-rich micaschist from one of the overlying HP units (Scaillet et al., 1992; Scaillet, 1996). For a more general treatment of the UHP behavior of K-Ar sytems, the reader is referred to Scaillet (1998).

SHRIMP analysis of zircons from different rock types of the whiteschist lens near Parigi led to a weighted mean age of 35.4 ± 1.0 Ma for the zircon overgrowths, interpreted to have formed under UHP conditions and thus to date the UHP metamorphism (Gebauer et al., 1997), thus confirming the Eocene zircon multigrain results of Tilton et al. (1991). Interestingly, no 35 m.y.-old zircon domains were found in the country-rock gneisses, which, however, should not be taken as evidence that the gneisses were not subjected to UHP metamorphism. In contrast to the whiteschist-type rocks, the less hydrous nature of the granitic rocks could explain the apparent absence of 35 m.y.-old metamorphic zircon domains (Gebauer et al., 1997). Even Tilton et al. (1997) demonstrated that minerals of the Brossasco metagranite, although having experienced subduction to more than 100 km with temperatures exceeding 700°C, failed to reach isotopic equilibrium. Likewise Paquette et al. (1999) showed that the U-Th-Pb system of monazite in this undeformed metagranite was hardly affected by the UHP event, whereas it was completely reset in the pyrope quartzite (Tilton et al., 1991). Fission-track data derived from zircons of fine-grained pyrope quartzites and their biotite-phengite gneiss country rocks yield a weighted mean age of 29.9 ± 1.4 Ma (Gebauer et al., 1997), which indicates that the pyrope quartzite and the surrounding gneiss cooled simultaneously to below the closure temperature for fission tracks in zircon (~290°C). Thus, the exhumation of the UHP metamorphic rocks from depths of ~120 km must have occurred within ~5 to 6 Ma, leading to uplift rates of ~20 to 24 km/Ma or 2 to 2.4 cm/a. Recently, Duchêne et al. (1997) derived a whole-rock/garnet age of 32.8 ± 1.2 Ma on pyrope quartzite rocks, using Lu-Hf dating. Since the closure temperature for the Lu-Hf system in garnet-bearing assemblages most likely exceeds 600°C, this age might represent an early stage of exhumation (Duchêne et al., 1997).

An important feature concerning SHRIMP analyses of zircons was brought into discussion by Schertl and Schreyer (1996). They demonstrated the importance of the precise location of mineral inclusions relative to distinct domains of zircon grains, made visible by backscattered electrons or cathodoluminescence. If it is possible to ascribe the mineral inclusions to discrete stages of metamorphism, SHRIMP data obtained on the surrounding domains thus represent a valuable constraint for deriving a P-T-time path.

Continental Crust into the Mantle and Back: Problems and Perspectives

In light of these studies, the following picture arises for this terrane at the beginning of Alpine orogeny: a piece of European continental crust consisted of a Paleozoic metasedimentary series (interlayered with some metavolcanic products) that was intruded by granite bodies (producing contact aureoles) at ~300 Ma. This margin of the European plate, most likely after deposition of a Mesozoic cover series and rifting, was dragged during the Eocene to ~100 km depth into the subduction zone that consumed the Ligurian ocean (see Michard et al., 1996). It now seems clear that much greater depths were not reached by the terrane presently exposed, because of the absence of diamond and the instability of chloritoid + talc in the most Mg-rich systems (which sets a maximum pressure limit at definitely less than 35 kbar). However, it is not yet clear when the protolith of the pyrope quartzite was formed—i.e., whether the metasomatic transformation of the granitic rocks (along shear zones?) occurred at a post-magmatic stage during the Late Paleozoic, or during Early Mesozoic rifting, with possible involvement of brines from the (now missing) cover series (Gebauer et al., 1997), or at peak UHP metamorphic conditions (Compagnoni and Hirajima, 1992, 1993).

In our opinion, one of the most important conclusions, on which the field, petrological, and geochemical studies concur, involves the scale and extent of mass transfer under UHP conditions. Despite the formidable P-T conditions experienced by this crustal terrane that was brought to mantle depth, no evidence was found for any type of interaction with any mantle-derived material—even for such a small crustal sliver, as compared to the large tracts of UHP terrains in China or Norway. On the contrary, there is ample evidence for local-scale evolution of distinct systems in isolation from one another, as if each rock "stewed in its own juice" (if any) during most of the metamorphic evolution—a conclusion reached long ago for the HP Monte-Rosa massif (Frey et al., 1976). This is illustrated by the local control and/or preservation of different fluid compositions at both a regional and a thin-section scale: the same rock type (whiteschist) that contained phosphate-rich brines in Val Gilba, leading to local development of very low H_2O-activity conditions with crack-related "granulite-facies" assemblages and phosphates (Simon and Chopin, 1999), was characterized by other fluids in Val Po (a few kilometers or hundreds of meters away) because of scapolite formation in pseudomorphs after pyrope and the absence of late "granulite-facies" minerals or abundant phosphates (Schertl et al., 1991; Philippot et al., 1995). This closed-system impression is obviously enhanced by the absence or localization of deformation, as beautifully shown by the comparison of isotopic data on an undeformed (Brossasco metagranite) and a completely reworked rock during Alpine times (the pyrope quartztite). This triggering effect of deformation, allowing mass transfer in otherwise closed or unreactive systems regardless of the P-T conditions reached, seems to be a key feature in other UHP terranes as well (Austrheim, 1998; Wain, 1997). One may wonder whether this impression is not biased by our focus on the UHP relics, which are of course best preserved in rocks escaping later deformation or fluid ingression. Would the picture be different if the focus was on country-rock gneiss regionally deformed and equilibrated under low P-T conditions? Or if looking along lithologic layers rather than across them? This might be an incentive for further studies (see Rumble, 1998).

Another major feature is the rate of exhumation derived for these UHP rocks (2 to 2.4 cm/y) (Gebauer et al., 1997), which compares with plate velocities rather than uplift or erosion rates. In addition, such high values are not easily reconciled with the petrologic constraints of continuous cooling during decompression, because very rapid exhumation favors nearly isothermal (cf. adiabatic) decompression. Refinements in thermal modeling, better constraints on the retrograde P-T path, and a better understanding of exhumation mechanisms still appear to be needed.

Acknowledgments

Our thanks go to the many persons who contributed to the results reviewed here and shared unpublished data: Dov Avigad, Michael Burchard, Cristiano Ferraris, Heicke Klock, Uschi Nowlan, Volker Schenk, Immo Schnieders, and Gilla Simon. We also are grateful for inspiring discussions with Werner Schreyer and the help of Thomas Fockenberg, of the analytical chemistry team at Ruhr Universität Bochum, and of A. Fischer and R. Lehman for figure preparation.

REFERENCES

Amisano Canesi, A. 1994, Studio cristallografico di minerali di altissima pressione del complesso Brossasco-Isasca (Dora-Maira meridionale): Unpubl. doctoral thesis, University of Torino, 185 p.

Arnaud, N. O., and Kelley, S. P., 1995, Evidence for excess argon during high pressure metamorphism in the Dora Maira Massif (western Alps, Italy), using an ultra-violet laser ablation microprobe ^{40}Ar-^{39}Ar technique: Contrib. Mineral. Petrol., v. 121, p. 1–11.

Austrheim, H., 1998, Influence of fluid and deformation on metamorphism of the deep crust and consequences for the geodynamics of collision zones, *in* Hacker, B. R., and Liou, J. G., eds., When continents collide: Geodynamics and geochemistry of ultrahigh-pressure rocks: Dordrecht, Kluwer Acad. Publ., p. 297–323.

Avigad, D., Chopin, C., and Le Bayon, in progress, Exhumation of the Dora-Maria UHP field unit: Field relations to footwall and hanging wall [abs.]: Terra Abs.

Ballèvre, M., Hensen, B. J., and Reynard, B., 1997, Orthopyroxene-andalusite symplectites replacing cordierite in granulites from the Strangways Range (Arunta Block, central Australia): A new twist to the P-T history: Geology, v. 25, p. 215–218.

Bellieni, G., Comin-Chiaramonti, P., Marques, S., Melfi, A. J., Nardy, A. J. R., Papatrechas, C., Piccirillo, M., Roisenberg, A., and Stolfa, D., 1986, Petrogenetic aspects of acid and basaltic lavas from the Parana Plateau (Brazil): Geological, mineralogical, and petrological relationships: Jour. Petrol., v. 27, p. 915–944.

Biino, G., and Compagnoni, R., 1992, Very high pressure metamorphism of the Brossasco coronite metagranite, southern Dora-Maira massif, western Alps: Schweiz. Mineral. Petrog. Mitteil., v. 72, p. 347–362.

Brunet, F., and Chopin, C., 1995, Bearthite, $Ca_2Al[PO_4]_2(OH)$: Synthesis, stability, and thermodynamic properties: Contrib. Mineral. Petrol., v. 121, p. 258–266.

Brunet, F., Chopin, C., and Seifert, F., 1998, Phase relations in the system MgO-P_2O_5-H_2O and the stability of phosphoellenbergerite: Petrological implications: Contrib. Mineral. Petrol., v. 131, p. 54–70.

Carman, J. H., 1974, Synthetic sodium and its two hydrates: Stabilities, properties, and mineralogic implications: Amer. Mineral., v. 59, p. 261–273.

Carswell, D. A,, O'Brien, P. J., Wilson, R. N., and Zhai, M., 1997, Thermobarometry of phengite-bearing eclogites in the Dabie Mountains of central China: Jour. Metamor. Geol., v. 15, p. 239–252.

Chappell, B., W., and White, A. J. R., 1974, Two contrasting granite types: Pacific Geol., v. 8, p. 173–174.

Chopin, C., 1984, Coesite and pure pyrope in high-grade blueschists of the Western Alps: A first record and some consequences: Contrib. Mineral. Petrol., v. 86, p. 107–118.

_______, 1986, Phase relationships of ellenbergerite, a new high-pressure Mg-Al-Ti silicate in pyrope-coesite-quartzite from the Western Alps: Geol. Soc. Amer. Mem. 164, p. 31–42.

Chopin, C., Henry, C., and Michard, A., 1991, Geology and petrology of the coesite-bearing terrain, Dora Maira massif, Western Alps: Eur. Jour. Mineral., v. 3, p. 263–291.

Chopin, C., and Schreyer, W., 1983, Magnesiocarpholite and magnesiochloritoid: Two new index minerals in pelitic blueschists and their preliminary phase relations in the model system MgO-Al_2O_3-SiO_2-H_2O: Amer Jour. Sci., v. 283-A, p. 72–96.

Chopin, C., and Sobolev, N. V., 1995, Principal mineralogic indicators of UHP in crustal rocks, *in* Coleman, R. G., and Wang, X., eds., Ultrahigh-pressure metamorphism: New York, Cambridge University Press, p. 96–131.

Coleman, R. G., Lee, D. E., Beatty, L. B., and Brannock, W. W., 1965, Eclogites and eclogites: Their differences and similarities: Geol. Soc. Amer. Bull., v. 76, p. 483–508.

Coleman, R. G., and Wang, X., eds., 1995, Ultrahigh pressure metamorphism: New York, Cambridge University Press, 528 p.

Compagnoni, R., and Hirajima, T., 1992, Occurrence and significance of superzoned garnet in coesite-bearing whiteschist of southern Dora Maira, Western Alps [abs.]: 29th Int. Geol, Cong., Kyoto, Abs. v. 2, p. 599.

______, 1993, The occurrence of a superzoned garnet in the coesite bearing unit of Dora Maira Massif and the origin of the whiteschists [abs.]: 4th Int. Eclogite Conf., Fuscaldo, Terra Abs., v. 4, p. 5.

Compagnoni, R., Hirajima, T., and Chopin, C., 1995, Ultra-high-pressure metamorphic rocks in the Western Alps, *in* Coleman, R. G., and Wang, X., eds., Ultra-high-pressure metamorphism: New York, Cambridge University Press, p. 206–243.

Compagnoni, R., Hirajima, T., Turello, R., and Castelli, D., 1994, The Brossasco-Isasca unit of the Dora-Maira Massif, *in* Compagnoni, R., and Messiga, B., eds., High pressure metamorphism in the Western Alps. Guidebook to the field excursion B1: 16th Gen. Meeting Int. Mineral. Assoc., Pisa, p. 87–105.

Cox, K. G., 1980, A model for flood basalt volcanism: Jour. Petrol., v. 21, p. 629–650.

Cox, K. G., Bell, J. D., and Pankhurst, R. J., 1979, The interpretation of igneous rocks: London, Allen and Unwin, 450 p.

Cox, K. G., MacDonald, R., and Hornung, G., 1967, Geochemical and petrographic provinces in the Karoo basalts of southern Africa: Amer. Mineral., v. 52, p. 1451–1474.

Czank, M., Simon, G., and Chopin, C., 1997, Lamellae in magnesiostaurolite: Budapest, First EMU School "Modular Mineralogy", Abs. vol.

Dal Piaz, G. V., and Lombardo, B., 1986, Early Alpine eclogite metamorphism in the Penninic Monte Rosa-Gran Paradiso basement nappes of the northwestern Alps: Geol. Soc. Amer. Mem. 164, p. 249–265.

Demény, A., Sharp, Z. D., and Pfeifer, H.-R., 1997, Mg-metasomatism and formation conditions of Mg-chlorite-muscovite-quartzphyllites (leucophyllites), of the Eastern Alps (W. Hungary) and their relations to Alpine whiteschists: Contrib. Mineral. Petrol., v. 128, p. 247–260.

Duchêne, S., Blichert-Toft, J., Luais, B., Télouk, P., Lardeaux, J.-M., and Albarède, F., 1997, The Lu-Hf dating of garnets and the ages of the Alpine high-pressure metamorphism: Nature, v. 387, p. 586–589.

Duncan, R. A., and Green, D. H., 1987, The genesis of refractory melts in the formation of oceanic crust: Contrib. Mineral. Petrol., v. 96, p. 326–342.

Enami, M., and Banno, S., 1999, Major rock-forming minerals in UHP metamorphic rocks: INT. GEOL. REV., v. 41, in press.

Ferraris, C., and Chopin, C., 1998, HRTEM evidence for retrograde exsolution in phengite-3T from Dora-Maira massif (Italian Western Alps) [abs.]: 17th Int. Mineral. Assoc., Abs. vol., p. A101.

Ferraris, C., Chopin, C., and Wessicken, R., 1999, Nano- to microscale decompression products in ultra-high-pressure phengite: Amer. Mineral., submitted.

Fockenberg. T., 1995, Synthesis and chemical variability of Mg-staurolite in the system MgO-Al_2O_3-SiO_2-H_2O as a function of water pressure: Eur. Jour. Mineral., v. 7, p. 1373–1380.

Fodor, R. V., 1987, Low- and high-TiO_2 flood basalts of southern Brazil: Origin from picritic parentage and a common mantle source: Earth Planet. Sci. Lett., v. 84, p. 423-430.

Frey, M., Hunziker, J. C., O'Neil, J. R., and Schwander, H. W., 1976, Equilibrium-disequilibrium relations in the Monte Rosa granite, Western Alps: Petrological, Rb-Sr, and stable isotope data: Contrib. Mineral. Petrol., v. 55, p. 147-179.

Gebauer, D., Schertl, H.-P., Brix, M., and Schreyer, W., 1997, 35 Ma old ultrahigh-pressure metamorphism and evidence for very rapid exhumation in the Dora-Maira-Massif, Western Alps: Lithos, v. 41, p. 5-24.

Green, D. H., and Hellmann, P. L., 1982, Fe-Mg partitioning between coexisting garnet and phengite at high pressures, and comments on a garnet-phengite geothermometer: Lithos, v. 15, p. 253-266.

Hammerschmidt, K., Schertl, H.-P., Friedrichsen, H., and Schreyer, W., 1992, Überschuss-argon während der retrograden metamorphose der ultrahochdruck-gesteine des Dora-Maira Massivs, italienische Westalpen:

Ber. der DMG, Beih Z./ Eur. Jour. Mineral., v. 4, p. 112.

Harley, S. L., and Carswell, D. A., 1995, Ultradeep crustal metamorphism: A prospective view: Geophys. Res., v. 100, p. 8367-8380.

Harrison, T. M., and Fitz Gerald, J. D., 1986, Exsolution in hornblende and its consequences for $^{40}Ar/^{39}Ar$ age spectra and closure temperature: Geochim. et Cosmochim. Acta, v. 50, p. 247–253.

Hawthorne, F. C., Ungaretti, L., Oberti, R., Caucia, F., and Callegari, A., 1993a, The crystal chemistry of staurolite. I. Crystal structure and site populations: Can. Mineral., v. 31, p. 551–582.

_______, 1993b, The crystal chemistry of staurolite. II. Order-disorder and the monoclinic to orthorhombic phase transition: Can. Mineral., v. 31, p. 583–595.

Henry, C., 1990, L'unité à coesite du massif Dora-Maira dans son cadre métamorphique et structural (Alpes occidentales, Italie): Mém. Sci. de la Terre, v. 90, no. 22, p. 1–453 (unpubl. doctoral thesis, University of Paris 6).

Henry, C., Michard, A., and Chopin, C., 1993, Geometry and structural evolution of ultra-high-pressure and high-pressure rocks from the Dora-Maira massif, Western Alps, Italy: Jour. Struct. Geol., v. 15, p. 965–982.

Irvine, T. N., and Baragar, W. R. A., 1971, A guide to the chemical classification of the common rocks: Can. Jour. Earth Sci., v. 8, p. 523–548.

Kienast, H. R., Lombardo, B., Biino, G., and Pinardon, J. L., 1991, Petrology of very-high-pressure eclogitic rocks from the Brossasco-Isasca Complex, Dora-Maira Massif, Italian Western Alps: Jour. Metamor. Geol., v. 9, p. 19–34.

Koch-Müller, M., 1997, Experimentally determined Fe-Mg exchange between synthetic staurolite and garnet in the system MgO-FeO-Al_2O_3-SiO_2-H_2O: Lithos, v. 41, p. 185–212.

Kullerud, K., 1996, Chlorine-rich amphiboles: Interplay between amphibole composition and an evolving fluid: Eur. Jour. Mineral., v. 8, p. 355–370.

Liati, A., and Seidel, E., 1994, Sapphirine and högbomite in overprinted kyanite-eclogites of Central Rhodope, N. Greece: First evidence of granulite-facies metamorphism: Eur. Jour. Mineral., v. 6, p. 733–738.

Markl, G., and Bucher, K., 1998, Composition of fluids in the lower crust inferred from metamorphic salt in lower crustal rocks: Nature, v. 391, p. 781–783.

Marsh, J. S., 1987, Basalt geochemistry and tectonic discrimination within continental flood basalt provinces: Jour. Volcanol. Geotherm. Res., v. 32, p. 35–49.

Massonne, H.-J., and Schreyer, W., 1989, Stability field of the high-pressure assemblage talc + phengite and two new phengite barometers: Eur. Jour. Mineral., v. 1, p. 391–410.

Matthes, S., Richter, P., and Schmidt, K., 1975, Die eklogitvorkommen des kristallinen grundgebirges in NE-Bayern. IX. Petrographie, geochemie und petrogenese der eklogite des Münchberger gneisgebietes: Ns. Jahrb. Mineral., v. 126, p. 45–86.

Michard, A., Chopin, C., and Henry, C., 1993, Compression versus extension in the exhumation of the Dora-Maira coesite-bearing unit, Western Alps, Italy: Tectonophys., v. 221, p. 173–193.

Michard, A., Goffé, B., Chopin, C., and Henry, C., 1996, Did the Western Alps develop through an Oman-type stage? The geotectonic setting of high-pressure metamorphism in two constrasting Tethyan transects: Eclogae Geol. Helv., v. 89, p. 43–80.

Modjtahedi, M,, and Wieseneder, H., 1974, Entstehung und Zusammensetzung der leukophyllite (weisssschiefer) in den Ostalpen: Archiv Lagerstätten. Ostalp, sdb. 85, p. 6983–6990.

Möller, C., 1998, Decompressed eclogites in the Sveconorwegian (–Grenvillian) orogen of SW Sweden: Petrology and tectonic implications: Jour. Metamor. Geol., v. 16, p. 641–656.

Monié, P., and Chopin, C., 1991, $^{40}Ar/^{39}Ar$ dating in coesite bearing and associated units of the Dora-Maira-Massif, Western Alps: Eur. Jour. Mineral., v. 3, p. 239–262.

Nowlan, E. U., 1998, Druck-temperatur-entwicklung und geochemie von eklogiten des Dora-Maira-Massivs, Westalpen: Unpubl. Ph.D. thesis, Ruhr-Universität Bochum, parts I/II, 167 p.

Nowlan, E. U., Schertl, H.-P., and Schreyer, W., 1999, P-T evolution of eclogites from the coesite-bearing unit of the southern Dora Maira Massif, Western Alps: Lithos, submitted.

Okay, A. I., 1994, Sapphirine and Ti-clinohumite in ultra-high-pressure garnet-pyroxenite and eclogite from Dabie Shan, China: Contrib. Mineral. Petrol., v. 116, p. 145–155.

Paquette, J.-L., Chopin, C., and Peucat, J. J., 1989, U-Pb zircon, Rb-Sr and Sm-Nd geochronology of high- to very-high-pressure meta-acidic rocks from the western Alps: Contrib. Mineral. Petrol., v. 101, p. 280–289.

Paquette, J.-L., Montel, J.-M., and Chopin C., 1999, U-Th-Pb dating of the Brossasco ultrahigh-pressure metagranite, Dora-Maira massif, western Alps: Eur. Jour. Mineral., v. 11, p. 67–74.

Pavese, A., Feraris, G., Pischedda, V., and Mezouar, M., 1999, Synchrotron powder diffraction study of phengite-3T from Dora-Maira massif: P-V-T equation of state and petrological consequences: Phys. Chem. Minerals, in press.

Pavese, A., Ferraris, G., Prencipe, M., and Ibberson, R., 1997, Cation site ordering in phengite-3T from the Dora-Maira massif (western Alps): Eur. Jour. Mineral., v. 9, p. 1183–1190.

Pearce, J. A., and Cann, J. R., 1973, Tectonic setting of basic volcanic rocks determined using trace element analyses: Earth Planet Sci. Lett., v. 19, p. 290–300.

Pearce, J. A., Harris, N. B. W., and Tindle, A. G., 1984, Trace element discrimination diagrams for the tectonic interpretation of granitic rocks: Jour. Petrol., v. 25, p. 956–983.

Philippot, P., Chevallier, P., Chopin, C., and Dubessy, J., 1995, Fluid composition and evolution in coesite-bearing rocks (Dora-Maira massif, Western Alps): Implications for element recycling during subduction: Contrib. Mineral. Petrol., v. 121, p. 29–44.

Rumble, D., 1998, Stable isotope geochemistry of ultrahigh-pressure rocks, *in* Hacker, B. R., and Liou, J. G., eds., When continents collide: Geodynamics and geochemistry of ultrahigh-pressure rocks: Dordrecht, Kluwer Acad. Publ., p. 241–259.

Scaillet, S., 1996, Excess ^{40}Ar transport scale and mechanism in high-pressure phengites: A case study from an eclogitized metabasite of the Dora-Maira nappe, western Alps: Geochim. et Cosmochim. Acta, v. 60, p. 1075–1090.

______, 1998, K-Ar (^{40}Ar/^{39}Ar) geochronology of of ultrahigh-pressure rocks, *in* Hacker, B. R., and Liou, J. G., eds., When continents collide: Geodynamics and geochemistry of ultrahigh-pressure rocks: Dordrecht, Kluwer Acad. Publ., p. 161–202.

Scaillet, S., Féraud, G., Ballèvre, M., and Amouric, M., 1992, Mg/Fe and [(Mg,Fe)Si-Al_2] compositional control on argon behavior in high-pressure white micas: A ^{40}Ar/^{39}Ar continuous laser-probe study from the Dora-Maira nappe of the internal western Alps, Italy: Geochim. et Cosmochim. Acta, v. 56, p. 2851–2872.

Schertl, H. P., and Schreyer, W., 1996, Mineral inclusions in heavy minerals of the ultrahigh-pressure metamorphic rocks of the Dora-Maira massif and their bearing on the relative timing of petrological events, *in* Basu, A., and Hart, S., eds., Earth processes: Reading the isotopic code: Washington, DC, Geophys. Monog., p. 331–342.

______, 1999, Geochemistry of coesite-bearing pyrope quartzites and related rocks of the southern Dora Maira Massif, Western Alps: I—major, minor, and trace elements, in prep.

Schertl, H. P., Schreyer, W., and Chopin, C., 1991, The pyrope-coesite rocks and their country rocks at Parigi, Dora Maira Massif, Western Alps: Detailed petrography, mineral chemistry and P-T path: Contrib. Mineral. Petrol., v. 108, p. 1–21.

Schreyer, W., 1977, Whiteschists: Their compositions and pressure-temperature regimes based on experimental, field, and petrographic evidence: Tectonophys., v. 43, p. 127–144.

______, 1995, Ultradeep metamorphic rocks: The retrospective viewpoint: Geophys. Res., v. 100, B5, p. 8353–8366.

Schreyer, W,, Abraham, K., and Kulke, H., 1980, Natural sodium phlogopite coexisting with potassium phlogopite and sodian aluminium talc in a metamorphic evaporite sequence from Derrag, Tell Atlas, Algeria: Contrib. Mineral. Petrol., v. 74, p. 223–233.

Schreyer, W., Massonne, H.-J., and Chopin, C., 1987, Continental crust subducted to mantle depths near 100 km: Implications for magma and fluid genesis in collision zones: Geochem Soc. Spec Publ. 1, p. 155–163.

Shand, S.J., 1950, Eruptive rocks: London, T. Murby, 488 p.

Sharp, Z. D., Essene, E. J., and Hunziker, J. C., 1993, Stable isotope geochemistry and phase equilibria of coesite-bearing whiteschists, Dora Maira massif, western Alps: Contrib. Mineral. Petrol., v. 114, p. 1–12.

Simon, G., 1998, The pyrope megablasts of the coesite-bearing terrain, Dora-Maira massif, western Alps: Inclusion mineralogy and petrology: Unpubl. Ph.D. thesis, University of Kiel, 101 p.

Simon, G., and Chopin, C., 1999, Granulite-facies (?) crack-related mineral assemblages in ultrahigh-pressure pyrope megablasts, Dora-Maira massif, Italian Western Alps: Contrib. Mineral. Petrol., submitted.

Simon, G., Chopin, C., and Schenk, V., 1997a, Near-end-member magnesiochloritoid in prograde-zoned pyrope, Dora-Maira massif, western Alps: Lithos. v. 41, p. 37–57.

______, 1997b, Magnesiostaurolite in pyrope megacrysts, Dora-Maira massif, Western Alps [abs.]: Terra Nova, v. 9, Abs. suppl., no. 1, p. 34.

Tilton, G. R., Ames, L., Schertl, H.-P., and Schreyer, W., 1997, Reconnaissance isotopic investigations on rocks of an undeformed granite contact within the coesite-bearing unit of the Dora-Maira Massif: Lithos, v. 41, p. 25–36.

Tilton, G. R., Schreyer, W., and Schertl, H. P., 1991, Pb-Sr-Nd isotopic behavior of deeply subducted crustal rocks from the Dora-Maira Massif, Western Alps, Italy-II: What is the age of ultrahigh-pressure metamorphism? Contrib. Mineral. Petrol., v. 108, p. 22–33.

Wain, A., 1997, Plagioclase metastability and phase transformations in the coesite-eclogite province of western Norway: Tectonic or kinetic implications for UHPM [abs.]: Terra Nova 9, Abs. suppl., no, 1, p. 39.

Waters, D., and Martin, H. N., 1993, The garnet-clinopyroxene-phengite barometer [abs.]: Terra Abs., v. 5, p. 410–411.

Winchester, J. A., and Floyd, P. A., 1977, Geochemical discrimination of different magma series and their differentiation products using immobile elements: Chem. Geol., v. 20, p. 325–343.

Ultrahigh-Pressure Metamorphism in the Western Gneiss Region of the Norwegian Caledonides

D. A. CARSWELL,

Department of Earth Sciences, University of Sheffield, Dainton Building, Brookhill, Sheffield S3 7HF, United Kingdom

S. J. CUTHBERT,

Department of Civil Engineering–Geology Section, University of Paisley, High Street, Paisley PA1 2BE, United Kingdom

AND E. J. KROGH RAVNA

Institute of Geology, University of Tromsø, 9037-Tromsø, Norway

Abstract

The distribution and characterization of UHP rocks within the Western Gneiss Region (WGR) of the Norwegian Caledonides is reviewed. While recent studies have documented a significantly increased number of eclogite localities preserving mineralogical evidence for Scandian-aged UHP metamorphism, much uncertainty remains over the regional extent of any UHP province because of the widespread overprinting by retrograde amphibolite-facies assemblages (especially in the dominant gneisses) during exhumation of the terrain. Based on current observations, the UHP metamorphic province may be limited to a northwest region of only ~4000 km^2, although an enigmatic mixed zone of HP (quartz-stable) and UHP (coesite-stable) eclogites extends a minimum of 5 km farther south and east in the Outer Nordfjord area.

Quantitative P-T evaluation of key mineral reaction equilibria for eclogites sampled across the WGR indicates an overall regional trend of increased T and P to the northwest. This is consistent with Baltic plate rocks in the northwestern part of the WGR having been subducted to greatest depths during the Scandian plate collision. The distribution of garnet peridotites within the WGR and their significance to understanding the nature, location, and timing of crust-mantle interaction within a major continental-plate subduction zone also is briefly considered.

Definition of UHPM

ULTRAHIGH PRESSURE METAMORPHISM (UHPM) refers to metamorphism that has occurred at sufficiently high pressures for the stability of coesite rather than alpha quartz (Coleman and Wang, 1995). The most recent evaluation of P-T conditions for coesite stability (Hemingway et. al., 1998) indicates the minimum pressure required for UHPM to be ~2.63 GPa at 600°C and 2.73 GPa at 700°C.

It is generally believed that such pressures are achieved essentially by lithostatic loading (Coleman and Wang, 1995) and that, at deep crustal levels, rock and mineral strengths are insufficient to sustain significant deviatoric stresses (tectonic overpressures) or intracrystalline overpressures as considered by Smith (1995). Hence the UHPM of crustal rocks is widely accepted as indicative of deep-level subduction, a minimum of ~90 km at 700°C, in major plate-collision belts.

Geological Background

Since the early documentation by Eskola (1921) of the spectacular occurrences of eclogites in the coastal region of western Norway between Bergen and Trondheim, the Western Gneiss Region (WGR) has been recognized as an archetypal high-grade, eclogite-bearing metamorphic terrain (Fig. 1). More recent reviews of the occurrences of eclogites and other cofacial high-pressure (HP) metamorphic rocks in this region, within the context of the tectono-metamorphic development of the Scandinavian Caledonides, include those by Cuthbert et al. (1983, 1999), Cuthbert and Carswell (1990), Griffin et al. (1985), Krogh and Carswell (1995), and Smith (1988, 1995).

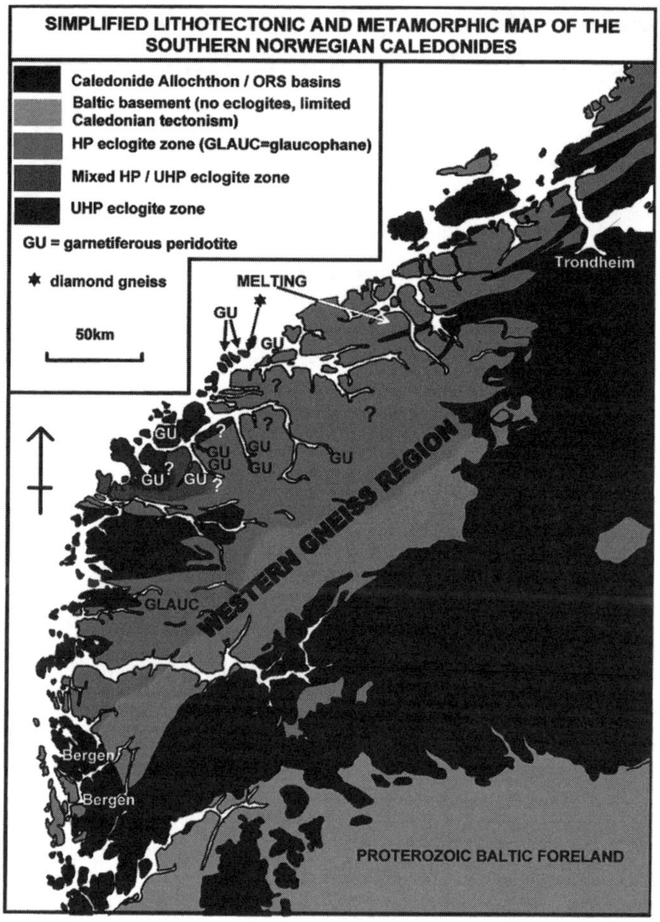

FIG. 1. Simplified geological and metamorphic map for western Norway, roughly between the towns of Bergen (B) and Trondheim (T), indicating the generalized locations of occurrences of UHP and HP lithologies within the large Western Gneiss Region basement window structurally below the Caledonide allochthon.

The main issue of controversy, not yet adequately resolved, has been whether or not the predominant "country-rock" gneisses enclosing the eclogites, garnet peridotites, and other recognized HP rocks within this region also experienced Scandian-aged (Ordovician–Silurian) HP metamorphism during the major Caledonian collision between the Baltic and Laurentian continental plates. There has therefore been an ongoing debate over whether the HP rocks were stabilized "in situ" within the encompassing gneisses in a more-or-less structurally coherent metamorphic terrain or alternatively represent "exotic" blocks or lenses within some sort of highly disrupted tectonic mélange.

Evidence and arguments in favor of an essentially "in situ" development of most crustal protolith eclogites in the WGR have been documented, for example, by Griffin et al. (1985) and Cuthbert and Carswell (1990). As the lowest exposed structural level in the Scandinavian Caledonides, the WGR is thought to comprise mostly Proterozoic autochthonous to para-autochthonous basement rocks belonging to the leading edge of the Baltic plate, which underwent transient deep-level subduction beneath the Laurentian plate during the Scandian phase of the Caledonian orogeny. Relatively passive exhumation leading to the subsequent exposure of this HP/UHP metamorphic terrain without major tectonic

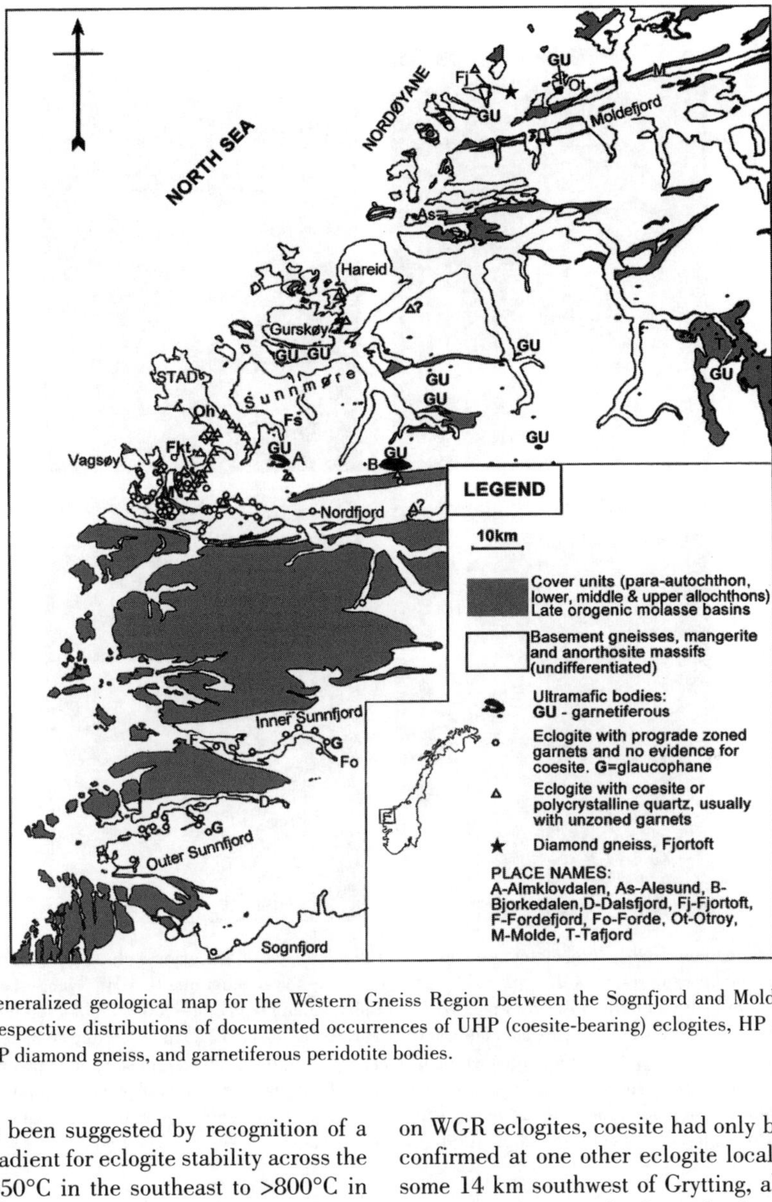

FIG. 2. Generalized geological map for the Western Gneiss Region between the Sognfjord and Moldefjord areas, showing the respective distributions of documented occurrences of UHP (coesite-bearing) eclogites, HP (coesite-free) eclogites, UHP diamond gneiss, and garnetiferous peridotite bodies.

disruption has been suggested by recognition of a temperature gradient for eclogite stability across the WGR from ~550°C in the southeast to >800°C in the coastal areas to the northwest (Krogh, 1977; Griffin et al., 1985).

First Recognition of UHPM in Norway

Smith (1984) was the first to recognize, as well as confirm by Raman spectroscopy, the occurrence of relict coesite within a WGR eclogite sample from the locality of Grytting on the southwestern corner of the Stadlandet Peninsula (STAD on Figure 2). By the time of Smith's (1988) expansive review article on WGR eclogites, coesite had only been positively confirmed at one other eclogite locality, Straumen, some 14 km southwest of Grytting, although petrographic evidence in the form of polycrystalline or at least multicrystalline quartz inclusions within garnet or omphacite (interpreted as likely to have pseudomorphed earlier coesite) was recorded from five additional, more widely scattered, localities.

On this basis, Smith (1988) recognized the existence within the Western Gneiss Region of the Norwegian Caledonides of a specific coesite-eclogite province containing rocks that had experienced UHP metamorphic conditions but was unable to establish any boundaries for such a UHP province.

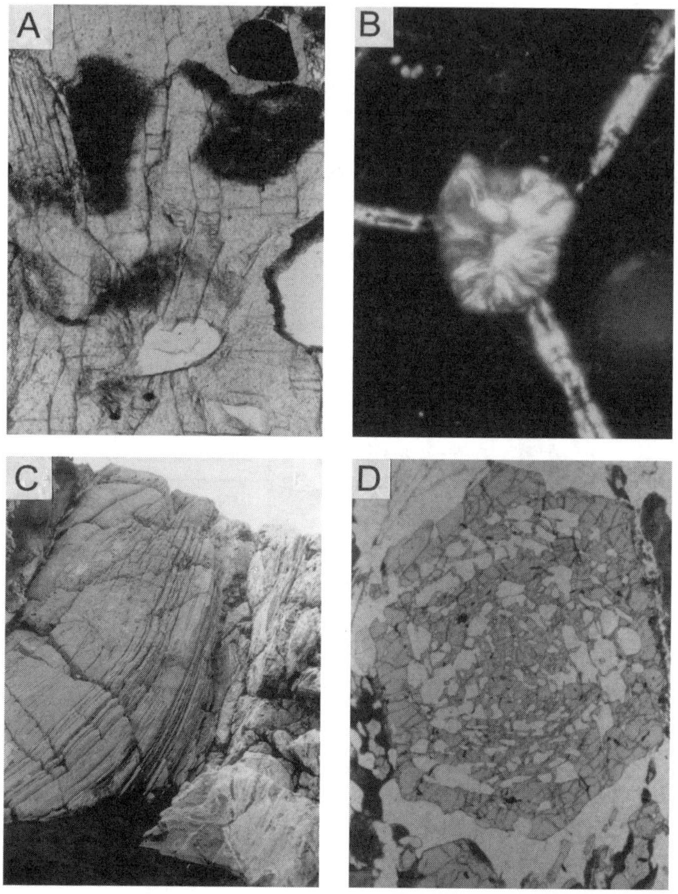

FIG. 3. A. Plane-polarized-light photomicrograph of a ~100 micron–sized inclusion within omphacite, comprising a central higher-relief coesite relict partly replaced by an aggregate of lower-relief quartz. UHP kyanite-bearing eclogite sample A181 from Flatraket harbor. B. Cross-polarized-light photomicrograph of a ~100 micron–sized inclusion within garnet comprising a polycrystalline aggregate of quartz after earlier coesite. UHP kyanite-bearing eclogite sample A226 from Liset, Stadlandet Peninsula. C. Field photograph of sheared and retrograded margin of eclogite pod in quartzofeldspathic gneisses in coastal exposures on Flemsøy, Nordøyane. D. Plane-polarized light photomicrograph of a ~4.5 mm–sized garnet showing prograde compositional zoning with a darker, more Fe-rich core, and numerous single quartz-grain inclusions. HP kyanite-bearing eclogite sample A296 from Nordpollen.

In addition, Smith (1988) emphasized a lack of mineralogical evidence for the intervening gneisses between the various WGR eclogite localities having experienced HP (quartz-eclogite stable), let alone UHP (coesite-eclogite stable), conditions. Accordingly, he proposed that the WGR comprises a highly imbricated tectonic mélange of UHP and HP eclogites encompassed within lower-pressure lithologies, rather than a regionally extensive, structurally coherent UHP province/terrain.

Recent Discoveries of UHP Rocks in the WGR

More recent studies by Wain (1997) and Cuthbert et al. (1999) have greatly increased the number of WGR eclogite localities now classified as having experienced UHPM on the basis of the recognition of relict coesite inclusions preserved within garnets, kyanites, or omphacites (Fig. 3A) or of petrographically distinctive polycrystalline quartz pseudomorphs after coesite (Fig. 3B). Wain (1997) documented discoveries of relict coesites from 5 new eclogite localities and of quartz pseudomorphs after coesites at a further 12 localities. All of these newly reported localities are from the Outer Nordfjord and Stadlandet areas of the WGR (Fig. 2) and in the general vicinity of the two coesite-bearing eclogite localities first recognized by Smith (1988).

Cuthbert et al. (1999, and unpubl.) have discovered an additional relict coesite-bearing eclogite locality at Flister in Outer Nordfjord, and also convincing polycrystalline quartz pseudomorphs after coesite at seven other eclogite localities. Importantly, one of the latter localities is in Stigedalen, to the south of the garnetiferous peridotite outcrops around Bjorkedalsvatnet (Fig. 2) and is ~35 km east of any previously recognized UHP eclogite within the WGR. Another is on the island of Fjørtoft, ~85 km northeast of the UHP eclogite localities in the Nordfjord and Stadlandet areas previously recognized by Smith (1988) and Wain (1997). Smith (1988, 1995) has, however, also reported finding poly- or multicrystalline quartz thought to have replaced earlier coesite in the large Ulstein-Dimnøy eclogite body on Hareidlandet and in the Hessdalen eclogite body on the opposite side of Vartdalsfjorden, these localities being roughly half way between Stadlandet and Fjørtoft (Fig. 2). It is also significant that the UHP eclogite locality discovered on the eastern side of Fjørtoftvika on Fjørtoft is only ~2 km along strike from a garnet peridotite locality (Jamtveit et al., 1991) and also from outcrops of the garnet-kyanite-biotite gneiss with common matrix graphite, from which scarce microdiamonds were recovered by dissolution by Dobrzhinetskaya et al. (1995).

Extent of UHPM in the WGR

Coleman and Wang (1995) speculated that the UHPM terrain in the Western Gneiss Region covers an area of roughly 350 × 150 km, on the basis of the WGR comprising a structurally coherent terrain that displays a regional P-T gradient related to the depth of subduction. This scenario contrasts, at the other extreme, with the view of this UHPM province expressed by Smith (1980, 1988 and in other publications) as comprising only sparsely scattered outcrops of UHP eclogites within a complex tectonic mélange of rocks of highly variable metamorphic grade. Also it does not match well with evidence from the meticulous lithostratigraphic mapping of the WGR in the Moldefjord and Nordøyane areas (Fig. 2) by Robinson (1995), which has demonstrated the existence of structurally highly attenuated synclinal downfolds into the Proterozoic basement rocks of various Caledonian cover nappe units displaying markedly contrasting metamorphic grades. Also there is an apparent lack of eclogites in the southeastern part of the WGR (see Fig. 1), where the Precambrian basement rocks beneath the Caledonide allochthon have undergone less extensive reworking during the Caledonian orogeny (Dietler et al., 1985; Milnes et al., 1997).

Throughout the WGR, the main obstacle to establishing the regional extent of any HPM or UHPM terrain is that for the most part HP or UHP mineralogies are observed only in basic or ultrabasic lithologies (eclogites and garnet peridotites, respectively), whereas the dominant quartzofeldspathic gneisses display amphibolite-facies mineral assemblages, and thus lack obvious petrographic evidence that they experienced HP, let alone UHP, metamorphic conditions.

The margins of bodies of eclogite or garnet peridotite typically show retrogression to amphibolite, typically linked to development of a deformation fabric (Fig. 3C). Although it might be tempting to link this deformation-linked retrogression with tectonic emplacement of the HP/UHP rocks into higher-structural-level, lower-pressure gneisses, not all eclogite body margins are the focus of such shear deformation, and invariably the scale of deformation is inconsistent with the notion that the HP or UHP bodies have been thrust upward some 20–50 km to account for the confining pressure contrasts with the amphibolite-facies mineralogies of the enclosing gneisses.

That being the case, it is now more widely argued, especially for crustal protolith eclogite bodies that may show unequivocal evidence of a prograde metamorphic evolution from subducted metagabbros (e.g., Mørk, 1985), that the HP or UHP mineralogies were developed essentially "in situ" in a subducted slab of continental crust that included the protoliths of the surrounding gneisses. The absence of obvious petrographic evidence that the gneisses also experienced the HP or UHP conditions is then explained away by arguing that the late-stage deformation associated with the exhumation of the HP/UHP terrain has been partitioned largely into the more ductile gneisses, resulting in the essentially pervasive metamorphic reworking and elimination of earlier HP/UHP mineralogies.

To prove that the latter interpretation is the generally valid one to explain the metamorphic grade between the eclogites and their enclosing "country-rock" gneisses, it is important to document instances—even if only rare cases— where the enclosing gneisses preserve actual mineralogical or petrographic evidence that they did in fact experience the HP or UHP conditions. In this connection, the clear documentation by Wain (1977) and Krabbendam and Wain (1997) that local sequences of peltic paragneisses adjacent to eclogites in the Outer Nordfjord and Stadlandet areas (Bryhni, 1966) contain the HP assemblage garnet + phengite + quartz + rutile ± kyanite ± zoisite, and locally even petrographic evidence in the form of polycrystalline quartz pseudomorphs after coesite that they witnessed UHP conditions, is highly significant. However, obtaining unequivocal proof that the much more voluminous Proterozoic orthogneisses (Lappin et al., 1979; Harvey, 1983) also experienced the HP or even UHP conditions remains a tantalizing challenge. These rocks need to be more widely sampled and searched for elusive petrographic evidence, but frustratingly appear to only rarely contain (or retain) garnet.

The development and preservation of coronitic garnet in relatively undeformed augen orthogneisses alongside augitic clinopyroxene has been reported by Carswell and Harvey (1985) in samples from the island of Otrøy in Moldefjord (Fig. 2). However, this garnet has only X_{Ca} of ~0.26 and both phengite and rutile are conspicuously absent, unlike the suspected UHP granitic orthogneisses in the Dabie Shan UHPM terrain recently documented by Carswell et al. (1999). Nonetheless, the retrieval of microdiamonds from dissolution of the graphite-bearing garnet-kyanite-biotite-rutile-quartz paragneisses on the nearby island of Fjørtoft (Dobrzhinetskaya et al., 1995) was both startling and challenging. At this stage, it would seem most likely, however, that the best chance of proving that the voluminous WGR granitic and granodioritic orthogneisses experienced Scandian HP or even UHP conditions will lie in searching for preserved inclusions of jadeite or coesite in zircon mineral separates—a project currently under way.

P-T Evaluation of UHPM in the WGR

A regional trend of decreasing K_D values for Fe^{2+}/Mg^{2+} partitioning between garnet and omphacite in eclogites sampled from southeast to northwest across the WGR was first noted by Krogh (1977) and translated into a temperature gradient from ~550°C to >800°C in the northwestern coastal regions by Griffin et al. (1985). For most common, essentially bimineralic, eclogites it is only possible to deduce minimum formation pressures of < 2.0 GPa for the stability of the omphacite (usually ~Jd_{40}) in the absence of albite, using the Holland (1990) model for the activity of jadeite in omphacite. Higher minimum pressures of ~2.8 GPa are, however, signaled in the few samples with petrographic evidence for coesite stability, but these data are barely sufficient to confirm the assumption of Griffin et al. (1995) and others of a regional gradient of increasing pressure that matches the deduced temperature gradient. Increased P and T conditions for eclogite formation from southeast to northwest across the WGR have generally been related to increased depth of subduction of the leading edge of the Baltic plate beneath the Laurentian plate down a NW-dipping subduction zone, as for example in the tectonometamorphic model of Cuthbert and Carswell (1990).

Importantly, for phengite-bearing eclogites Waters and Martin (1993) have proposed a barometric evaluation methodology that enables reliable quantitative evaluation of absolute equilibration pressures. This methodology combines key pressure-dependent composition features of coexisting garnet, omphacite, and phengite as expressed in the KMASH-system exchange reaction equilibrium:

pyrope + grossular + 3 ideal celadonite =
6 diopside + 3 ideal muscovite.

This methodology has been successfully applied by Carswell et al. (1997) to confirm peak pressures consistent with coesite stability in UHP eclogites from the Dabie Shan in central China, and also by Wain (1997) for UHP eclogites in the Outer Nordfjord/Stadlandet area of the WGR. Moreover, in this latter area, Wain (1997) employed this geobarometer to demonstrate the existence of an important geological boundary (see Fig. 4) between a UHPM province in the Stadlandet area containing UHP eclogites stabilized at P > 2.79 GPa and T > 790°C, and a HPM zone in Outer Nordfjord containing only HP eclogites stabilized at P = 2.0 to 2.43 GPa and T = 700 to 800°C. It also was emphasized that the HP eclogites lack petrographic evidence that coesite was ever stable and characteristically contain garnets showing conspicuous prograde compositional zoning and only monocrystalline quartz inclusions (e.g., Fig. 3D). The prograde zoned garnets also may show amphibole inclusions in cores, in

FIG. 4. Map for the Outer Nordfjord and Stadlandet area of the Western Gneiss Region, showing the distribution of documented occurrences of UHP (coesite-bearing) eclogites relative to those of HP (coesite-free) eclogites. The location of the indicated UHP-HP transition zone is after Wain (1997).

contrast to omphacite inclusions in distinct eclogite-facies rims.

Wain (1997) and Krabbendam and Wain (1997) were not, however, able to recognize a sharp boundary between the UHPM and HPM zones in this area. Instead they have indicated the existence of a ~10 km wide boundary zone (Fig. 4) containing a mix of UHP and HP eclogites. Moreover, it was emphasized that both the pressure gap between the UHP and HP eclogites and the existence of what was taken to be a tectonically disrupted boundary zone challenged the view that the WGR behaved as a coherent tectonic unit during subduction and exhumation, as implied by a smooth regional gradient in both P and T for the eclogite-forming conditions (Griffin et al., 1995).

Cuthbert et al. (1998, 1999) have extended the UHP and HP eclogite database in the Nordfjord/Stadlandet area, as well as including an updated comparative evaluation of P-T conditions for formation of phengite-bearing eclogites in the Sunnfjord area farther south (Fig. 2). Based on Fe^{2+}/Mg^{2+} partitioning between garnet and omphacite (using the updated calibration of Krogh Ravna, 1999) and the phengite barometer of Waters and Martin (1993), mean calculated P-T conditions for phengite-bearing eclogites in the Sunnfjord area and for Nordfjord-area HP (quartz-stable) and Nordfjord/

Stadlandet–area UHP (coesite-stable) samples are 543°C/1.75 GPa (*n*=7), 640°C/2.31 GPa (*n*=22), and 705°C/2.80 GPa (*n*=10), respectively. Although there is danger that such P-T averaging for sample sets may disguise local P-T variations or breaks, this northward trend of increasing T and P could be consistent with steadily increasing depth of subduction down an appropriately low geothermal gradient of ~5°C/km. In that case, the diffuse boundary between the UHP and HP eclogite zones in the Nordfjord/Stadlandet area (Fig. 4) may just be the result of kinetic controls on the efficiency of metamorphic transformations, and the stabilization of thoroughly equilibrated mineral assemblages under P-T conditions that were close to the quartz-coesite reaction boundary, rather than the result of tectonic telescoping of a significant P-T break in a mixed zone of UHP and HP eclogites (Cuthbert et al., 1999). Resolution of this issue will, however, require further close examination of the microstructures as well as petrography of eclogites from this mixed boundary zone.

The possible direct link and extension of the recognized UHP province in the Stadlandet area to include the sparse evidence for UHP eclogites farther north on Hareidlandet and the Nordøyane, including the diamondiferous gneiss locality on Fjørtoft, still needs to be properly confirmed. However, preliminary data for kyanite-bearing eclogites (see Cuthbert et al., 1999) tentatively support the extension of a consistent regional P-T gradient northward to the Nordøyane. On the basis of currently recognized UHP localities, it would appear that the UHP province within the WGR is largely confined to the northwestern coastal region between the Outer Nordfjord/Stadlandet area and the Nordøyene, as indicated on Figure 1. It remains to be established just how far southeastward the UHP province may extend across the WGR, but our preliminary investigation of eclogites in Stigedalen, ~40 km east of the UHP eclogites in the Stadlandet area, indicates that at least the mixed HP/UHP zone extends that far east, as shown on Figure 1.

In view of the uncertainty over the extent of the UHP province in the WGR, we considered that it might be instructive to reevaluate P-T estimates for the stabilization of orthopyroxene-bearing eclogite samples for which absolute pressures may be calculated from the garnet-orthopyroxene Al barometer (see, for example, Carswell and Harley, 1990 for an account of this methodology). Within the WGR, orthopyroxene eclogites occur both directly enclosed in the gneisses ("external" Opx eclogites) and enclosed within the peridotite bodies discussed in the next section ("internal" Opx eclogites, or alternatively, garnet pyroxenites).

Lappin and Smith (1978) indicated P-T conditions of 3.0 to 4.5 GPa and 700 to 850°C for formation of "external" Opx eclogite samples on the assumption of equilibrium between the garnet and the low-Al cores of orthopyroxene grains. Such conditions are certainly consistent with UHP conditions as well as with the fact that their sample suite was collected entirely within the UHP and boundary UHP/HP zones as recognized by Wain (1997). By contrast Carswell et al. (1985) calculated much lower P-T conditions, averaging ~700 to 740°C and 1.7 to 1.8 GPa, for a larger and geographically more widespread suite of "external" Opx eclogites on the basis of assumed equilibration of "peak" P-T garnet compositions with the higher Al rim compositions of orthopyroxene grains. However, it is now recognized that the Al increase toward the rims of orthopyroxene grains is most likely a retrograde feature related to the growth of secondary amphibole at the expense of garnet.

A preliminary reevaluation of P-T values for the "external" Opx eclogites (Cuthbert et al., 1999), although signaling UHP conditions appropriate for coesite stability in many samples, has failed, however, to yield evidence for any consistent regional pattern of P-T variation. This may only indicate that difficulties over recognition of the "peak" equilibrium garnet and orthopyroxene compositions in what are mostly highly retrograded samples remain a serious drawback with this P-T evaluation methodology.

Significance of Garnet Peridotites in the WGR

Peridotite bodies are quite widely distributed throughout the WGR (see, for example, Fig. 1 in Brueckner, 1998), although only relatively few retain the pyropic garnet that is diagnostic of the stabilization of early HP (or possibly UHP) mineral assemblages prior to widespread overprinting by retrograde amphibolite-facies assemblages. A complex multistage prograde and retrograde metamorphic evolution of these peridotite bodies has been documented by Carswell (1986) and Medaris and Carswell (1990). Such Mg-Cr–type garnet peridotites display upper-mantle-style mineralogies and chemistries (e.g., Carswell, 1968; Brueckner, 1977) consistent with derivation from mostly highly depleted subcontinental mantle.

By contrast, the WGR also contains some Fe-Ti–type garnet peridotites (not shown in Figs. 1 and 2 of this paper) that have had a prograde origin from the ultramafic portions of layered, low-pressure crustal intrusive bodies (e.g., Carswell et al., 1983; Jamtveit, 1987) that were subducted alongside other crustal-slab lithologies during the Scandian plate collision. The Eiksunddal eclogite complex on Hareidlandet with subordinate garnet peridotite layers is a well-documented occurrence of the latter type. Jamtveit (1987) did not consider that this large, layered eclogite body, which for the most part lacks a free-SiO_2 phase, had equilibrated within the coesite stability field, However, an updated P-T evaluation of ~675°C and 3.2 GPa for an associated garnet peridotite, sample E4—for which mineral compositions were given in Carswell et al. (1983)—is in fact indicative of UHPM conditions. Regarding the other two Fe-Ti–type garnet peridotite occurrences documented by Carswell et al. (1983), the best indications based on more recent thermobarometric calibrations are that the Kolmannskog sample H185 equilibrated at ~765°C and 2.6 GPa (just about compatible with coesite stability conditions when attaching realistic error brackets), and the Raknestangen sample U125 from Otrøy seemingly at rather lower P-T of ~675°C and 2.1 GPa. However, as in the "external" Opx eclogite samples, there may well be problems in identifying the equilibrium garnet and orthopyroxene compositions required for successful application of the garnet-orthopyroxene Al barometer to these heavily retrograded Fe-Ti–type garnet peridotites.

Numerous recent papers on Mg-Cr–type garnet peridotites in the WGR have evaluated the P-T conditions for equilibration of the coarse-grained, porphyroclastic, four-phase garnet lherzolite assemblages to be consistent with UHPM conditions for coesite stability in suitably siliceous rocks (see, for example, Medaris and Carswell, 1990; Krogh and Carswell, 1995; Brueckner and Medaris, 1998). However, Jamtveit et al. (1991) have importantly shown that these earliest UHP garnet peridotite assemblages are of mid-Proterozoic age, and thus were stabilized in the uppermost subcontinental mantle long before the Scandian plate collision. Even more startling are the very recent discoveries by van Roermund and co-workers (van Roermund, 1998; van Roermund et al., 1999) and Terry et al. (1999) of relict majoritic-composition garnets showing exsolution of pyroxene lamellae in peridotite bodies on the islands of Otrøy and Flemsøy (Nordøyane). These discoveries are taken to indicate that these peridotite bodies originally had an even higher-pressure, deeper-mantle origin, probably within a rising mantle diapir at ~1700 Ma.

These mantle-derived garnet peridotite bodies are widely interpreted (e.g., Medaris and Carswell, 1990; Krogh and Carswell, 1995) to have been tectonically emplaced into the subducted crustal slab rocks of the WGR during the Scandian plate collision. Brueckner (1998) and Brueckner and Medaris (1999) presented new dynamic models for crust-mantle interaction in major continental plate–collision zones. Different locations and timings are possible for tectonic emplacement of mantle rocks into the crustal slab either during subduction or subsequent exhumation.

Given the current limited state of knowledge on the ages of the mineral assemblages, it is unclear as to whether these particular WGR peridotite bodies (with their extant older Proterozoic garnetiferous assemblages) were entrained into the crustal slab during subduction and so were themselves taken back down to deeper levels, where they may well have experienced the Scandian UHPM conditions, or were incorporated later during the exhumation of the crustal slab. In the latter case, the Mg-Cr–type peridotite bodies would have missed the Scandian UHPM conditions and just experienced direct, deformation-induced, reequilibration into a neoblastic HP five-phase garnet + spinel peridotite assemblage, as surmised by Krogh and Carswell (1995).

It is also feasible that different individual peridotite bodies, or tectonically controlled belts of peridotite bodies, may have been entrained at different times and from different locations within the evolving subduction zone. In particular, the garnet peridotite bodies in the northernmost belt from the Nordøyane to Otrøy have somewhat distinctive mineral and whole-rock chemistries in addition to the new evidence for the stability of an early UHP majoritic garnet. Certainly these bodies, and possibly also the others, are in our opinion more likely to have been entrained from the cold, old, Baltic plate subcontinental lithospheric mantle in the footwall to the subduction zone rather than from the Laurentian plate mantle wedge in the hanging wall to the subduction zone, as suggested in the sinking-intrusion emplacement model of Brueckner (1998).

On the other hand, Griffin and Qvale (1985) documented the fact that Fe-rich eclogite lenses within the Mg-Cr–type peridotite body at Raudkleivane in Almklovdalen contain garnets that display strong prograde compositional zoning and also ferropargasite inclusions in grain cores. This strongly suggests that at least some of these mantle-derived peridotite bodies were entrained into the subducting crustal slab and experienced a Scandian prograde metamorphic evolution from amphibolite to eclogite facies in common with the subducted crustal lithologies.

Finally, the distribution of the mantle-derived garnet peridotite bodies shown in Figure 2 suggests to us that they are located in a number of discrete E-W–trending belts that appear to be close to contacts between basement gneisses and possible younger-cover paragneiss sequences. This feature, as well as the basic facts that these mantle-derived peridotite bodies must have been tectonically intercalated into the crustal gneisses and that mylonitic fabrics of both eclogite and amphibolite facies are commonplace in a variety of lithologies—including the peridotites—must indicate that the WGR has not retained complete structural coherence throughout the full subduction, then exhumation cycle associated with the Caledonian continental-plate collision, despite the evidence for a broad, subduction-related P-T gradient across the terrain.

Acknowledgments

The authors wish to acknowledge the support for their recent studies in the WGR of Norway provided by the Norwegian Research Council, the Norwegian Geological Survey, the British Council, the Carnegie Trust for Scottish Universities, and the Universities of Sheffield, Paisley, and Tromsø.

REFERENCES

Brueckner, H. K., 1977, A crustal origin for eclogites and a mantle origin for garnet peridotites: Strontium isotope evidence from clinopyroxenes: Contrib. Mineral. Petrol., v. 60, p. 1–15.

______, 1998, Sinking intrusion model for the emplacement of garnet-bearing peridotites into continent collision orogens: Geology, v. 26, p. 631–634.

Brueckner, H. K., and Medaris, G. L., 1998, A tale of two orogens: The contrasting T-P-t history and geochemical evolution of mantle in high- and ultrahigh-pressure metamorphic terranes of the Norwegian Caledonides and the Czech Variscides. Schweiz. Mineral. Petrog. Mitteil., v. 78, p. 293–307.

______, 1999, A general model for the intrusion and evolution of "mantle" peridotites in high-pressure and ultrahigh-pressure metamorphic terranes: Jour. Metamor. Geol., in press.

Bryhni, I., 1966, Reconnaisance studies of gneisses, ultrabasites, eclogites, and anorthosites in outer Nordfjord, western Norway: Norges Geol. Undersok., v. 241, p. 1–68.

Carswell, D. A., 1968, Possible primary upper mantle peridotite in Norwegian basal gneiss: Lithos, v. 1, p. 322–355.

______, 1986, The metamorphic evolution of Mg-Cr type Norwegian garnet peridotites: Lithos, v. 19, p. 279–297.

Carswell, D. A., and Harley, S. L., 1990, Mineral barometry and thermometry, *in* Carswell, D. A., ed., Eclogite facies rocks: Glasgow, Blackie, p. 83–110.

Carswell, D. A., and Harvey, M. A.,1985, The intrusive history and tectonometamorphic evolution of the Basal Gneiss Complex in the Moldefjord area, west Norway, *in* Sturt, B., and Gee. D., eds, The Caledonide orogen: New York, Wiley, p. 843–857.

Carswell, D. A., Harvey, M. A., and Al-Samman, A., 1983, The petrogenesis of contrasting Fe-Ti and Mg-Cr garnet peridotite types in the high grade gneiss complex of Western Norway: Bull. Mineral., v. 106, p. 727–750.

Carswell, D. A., Krogh, E. J., and Griffin, W. G., 1985, Norwegian orthopyroxene eclogites: Calculated equilibration conditions and petrogenetic implications, *in* Sturt, B., and Gee, D., eds, The Caledonide orogen: New York, Wiley, p. 823–841.

Carswell, D. A., O'Brien, P. J., Wilson, R. N., and Zhai, M., 1997, Thermobarometry of phengite-bearing eclogites in the Dabie Mountains of central China: Jour. Metamor. Geol., v. 15, p. 239–252.

Carswell, D. A., Wilson, R. N., and Zhai, M., 1999, Metamorphic evolution, mineral chemistry and thermobarometry of schists and orthogneisses hosting ultra-high pressure eclogites in the Dabieshan of central China: Lithos, in press.

Coleman, R. G., and Wang, X., 1995, Overview of the geology and tectonics of UHPM, *in* Coleman, R. G., and Wang, X., eds., Ultrahigh-pressure metamorphism: New York, Cambridge University Press, p. 1–32.

Cuthbert, S. J., and Carswell, D. A., 1990, Formation and exhumation of medium-temperature eclogites in the Scandinavian Caledonides, *in* Carswell, D. A., ed., Eclogite facies rocks: Glasgow, Blackie, p. 181–203.

Cuthbert, S. J., Carswell, D. A., and Krogh Ravna, E. J., 1998, Eclogites and eclogites in the Western Gneiss Region (WGR), Norwegian Caledonides [abs.]: EOS (Trans. Amer. Geophys. Union), v. 79, no. 45, V22D-12.

______, 1999, Eclogites and eclogites in the Western Gneiss Region (WGR), Norwegian Caledonides: Lithos, in press.

Cuthbert, S. J., Harvey, M. A., and Carswell, D. A., 1983, A tectonic model for the metamorphic evolution of the Basal Gneiss Complex, western south Norway: Jour. Metamor. Geol., v. 1, p. 63–90.

Dietler, T. N., Koestler, A. G., and Milnes, A. G., 1985, A preliminary structural profile through the Western Gneiss Complex, Sognefjord, southwestern Norway: Norsk Geol. Tidsskrift, v. 65, p. 233–235.

Dobrzhinetskaya, L., Eide, E. A., Larsen, R. B., Sturt, B. A., Trønnes, R. G., Smith, D. C., Taylor, W. R., and Posukhova, T. V., 1995, Microdiamond in high-grade metamorphic rocks of the Western Gneiss region, Norway: Geology, v. 23, p. 597–600.

Eskola, P., 1921, On the eclogites of Norway: Skr. Vidensk. Selsk. Christiania, Mat. -Naturv. Kl., I, v. 8, p. 1–118.

Griffin, W. L., Austrheim, H., Brastad, K., Bryhni, I., Krill, A. G., Krogh, E. J., Mørk, M.-B. E., Qvale, H., and Torudbakken, B., 1985, High-pressure metamorphism in the Scandinavian Caledonides, *in* Sturt, B., and Gee, D., eds., The Caledonide orogen: New York, Wiley, p. 783–801.

Griffin, W. L., and Qvale, H., 1985, Superferrian eclogites and the crustal origin of garnet peridotites, Almklovdalen, Norway, *in* Sturt, B., and Gee, D., eds., The Caledonide orogen: New York, Wiley, p. 804–811.

Harvey, M. A., 1983, A geochemical and Rb-Sr study of the Proterozoic augen orthogneisses on the Molde Peninsula, west Norway: Lithos, v. 16, p. 325–338.

Hemingway, B. S., Bohlen, S. R., Hankins, W. B., Westrum, E. J., and Kuskov, O. L., 1998, Heat capacity and thermodynamic properties for coesite and jadeite, reexamination of the quartz-coesite equilibrium boundary: Amer. Mineral., v. 83, p. 419–433.

Holland, T. J. B., 1980, The reaction albite – jadeite + quartz determined experimentally in the range 600–1200°C: Amer. Mineral., v. 65, p. 129–134.

Jamtveit, B., 1987, Metamorphic evolution of the Eiksunddal eclogite complex, Western Norway, and some tectonic implications: Contrib. Mineral. Petrol., v. 95, p. 82–99.

Jamtveit, B., Carswell, D. A., and Mearns, E. W., 1991, Chronology of the high-pressure metamorphism of Norwegian garnet peridotites/pyroxenites: Jour. Metamor. Geol., v. 9, p. 125–139.

Krabbendam, M., and Wain, A., 1997, Late-Caledonian structures, differential retrogression and structural position of (ultra)high-pressure rocks in the Nordfjord-Stadlandet area, Western Gneiss Region: Norges Geol. Undersok., v. 432, p. 127–139.

Krogh, E. J., 1977, Evidence for a continent-continent collision in western Norway: Nature, v. 267, p. 17–19.

Krogh, E. J., and Carswell, D. A., 1995, HP and UHP eclogites and garnet peridotites in the Scandinavian Caledonides, *in* Coleman, R. G., and Wang, X., eds., Ultrahigh pressure metamorphism, Cambridge University Press, p. 244–298.

Krogh Ravna, E. J., 1999, The garnet-clinopyroxene geothermometer—an updated calibration: Jour. Metamor. Geol., in press.

Lappin, M. A., Pidgeon, R. T., and van Breemen, O., 1979, Geochronology of basal gneisses and mangerite syenites of Stadlandet, west Norway: Norsk Geol. Tidssk., v. 59, p. 161–181.

Lappin, M. A., and Smith, D. C., 1978, Mantle-equilibrated orthopyroxene eclogite pods from the basal gneisses in the Selje district, western Norway: Jour. Petrol., v. 19, p. 530–584.

Medaris, L. G., and Carswell, D. A., 1990, Petrogenesis of Mg-Cr garnet-peridotites in European metamorphic belts, *in* Carswell, D. A., ed., Eclogite facies rocks: Glasgow, Blackie and Sons, p. 260–290.

Milnes, G., Wennberg, O. P., Skar, O., and Koestler, A. G., 1997, Contraction, extension and timing in the South Norwegian Caledonides: The Sognfjord transect, *in* Burg, J. P., and Ford, M., eds., Orogeny though time: Geol. Soc. Lond. Spec. Publ. no. 121, p. 123–148.

Mørk, M. B. E., 1985, A gabbro to eclogite transition on Flemsøy, Sunnmore, western Norway: Chem. Geol., v. 50, p. 283–310.

Robinson, P., 1995, Extension of Trollheimen tectonostratigraphic sequence in deep synclines near Molde and Brattvag, Western Gneiss Region, southern Norway: Norsk Geol. Tidsskrift, v. 75, p. 181–198.

Smith, D. C., 1980, A tectonic mélange of foreign eclogites and ultramafics in west Norway: Nature, v. 287, p. 366–367.

______, 1984, Coesite in clinopyroxene in the Caledonides and its implications for geodynamics: Nature, v. 310, p. 641–644.

______, 1988, A review of the peculiar mineralogy of the "Norwegian coesite-eclogite province" with crystal-chemical, petrological, geochemical and geodynamical notes and an extensive bibliography, *in* Smith, D. C., ed., Eclogites and eclogite-facies rocks: Amsterdam, Elsevier, p. 1–206.

______, 1995, Microcoesites and microdiamonds in Norway: An overview, *in* Coleman R. G., and Wang, X., eds., Ultrahigh pressure metamorphism: New York, Cambridge University Press, p. 299–355.

Terry, M. P., Robinson, P., Carswell, D. A., and Gasparik, T., 1999, Evidence for a Proterozoic mantle plume and a thermotectonic model for exhumation of garnet peridotites, Western Gneiss Region, Norway [abs.]: EOS (Trans. Amer. Geophys. Union), Abs. vol., in press.

van Roermund, H. L. M., 1998, An ultra-deep (d > 200 km) orogenic peridotite body in western Norway: EOS (Trans. Amer. Geophys. Union), v. 79, no. 45, V21E-03, p. F971.

van Roermund, H. L. M., and Drury, M. R., 1999, Ultrahigh pressure (P > 6 GPa) garnet peridotites in Western Norway: Exhumation of mantle rocks from >185 km depth: Terra Nova, in press.

Wain A., 1997, New evidence for coesite in eclogite and gneisses: Defining an ultrahigh-pressure province in the Western Gneiss Region of Norway: Geology, v. 25, p. 927–930.

Waters, D. J., and Martin, H. N., 1993, Geobarometry of phengite-bearing eclogites [abs.]: Terra Abs., v. 5, p. 410–411.

Regional-Scale Extensional Tectonic Pattern of Ultrahigh-Pressure and High-Pressure Metamorphic Belts from the Dabie Massif, China

ZENGQIU ZHONG, SHUTIAN SUO, AND ZHENDONG YOU

Faculty of Earth Sciences, China University of Geosciences, Wuhan 430074, China

Abstract

Recently, regional structural studies and detailed mapping at a scale of 1:10,000 in several key areas in the Dabie massif have revealed that the present regional tectonic pattern of ultrahigh-pressure and high-pressure metamorphic belts was formed mainly by extensional processes that post-date the Triassic collision between the Sino-Korean and Yangtze cratons. The tectonic pattern is characterized by an ancient metamorphic core complex, in which the most prominent feature is the development of multilayered extensional detachment zones on a crustal scale. This pattern controls the present spatial distribution of the ultrahigh-pressure metamorphic (UHPM) and high-pressure metamorphic (HPM) rocks, and obviously records large-scale subhorizontal extensional flow under amphibolite-facies conditions when the UHPM and HPM rocks were returned to lower-middle crustal levels. Based mainly on the crustal-scale tectonic scenario and combined with previous petrological and geochronological studies, the Dabie Mountains can be subdivided into six main tectonometamorphic units—the Northern Huaiyang (NHY) tectonic belt, Dabie complex, UHPM unit, HPM unit, epidote-blueschist (EBM) unit, and weakly metamorphosed sedimentary cover (SC). All units are tectonically separated by detachment ductile shear or fault zones. The Balifan-Mozitan-Xiaotian fault (BMXF), separating the NHY tectonic belt to the north from the Dabie massif (DM) to the south, is considered as the suture zone between the Sino-Korean and Yangtze cratons during the Triassic intracontinental collision. The geodynamic implications of this extensional tectonic pattern for the exhumation process of the UHPM and HPM rocks also are briefly discussed. It appears that recognition of the extensional tectonic pattern is of paramount importance for an understanding of the kinematics and dynamics of the Dabie-Sulu collisional orogen.

Introduction

DURING THE PAST decade, the widespread discovery (Xu et al., 1992; Cong et al., 1994; Zhang and Liou, 1994; Zhang et al.,1995) of rock assemblages containing characteristic UHPM and HPM index minerals from ancient complexes in the Dabie-Sulu region has attracted the extensive interest of international geoscientists, and has generated increasing study of this region. A great number of new contributions regarding the geochronology, petrology, P-T paths, and kinematic models for the creation and exhumation of UHPM and HPM rocks in the Dabie-Sulu region have been published (e.g.,Wang and Liou, 1991; Ames et al., 1993; Liou et al., 1994, 1997; Maruyama et al., 1994; Zhang and Liou, 1994; Li et al., 1997; Rowley et al., 1997). In contrast, except for a few works (Okay and Sengör, 1993; Hacker et al., 1995; Wang et al., 1995; Xue et al., 1996; Suo and Zhong, 1998), very few detailed structural analyses of the UHPM and HPM rocks have been undertaken to examine the features of crustal fabrics formed under different tectonic regimes and P-T conditions, as well as deformational processes related to the formation and exhumation of the UHPM and HPM rocks. To date, therefore, structures of the UHPM and HPM rocks are less well constrained in the Dabie-Sulu region. In fact, the least ambiguous information about geodynamic processes responsible for the formation and exhumation of the UHPM and HPM rocks is provided mainly by the structural record acquired by the crustal rocks (van der Klauw et al., 1997). Herein we describe a regional-scale tectonic pattern in many ways similar to Cordilleran metamorphic core complexes. This model is characterized by the development of multilayered detachment zones, and resulted from post-Triassic collisional extension (200 to 170 Ma) of UHPM and HPM belts in the Dabie massif, based on our regional structural observations and detailed mapping at a scale of

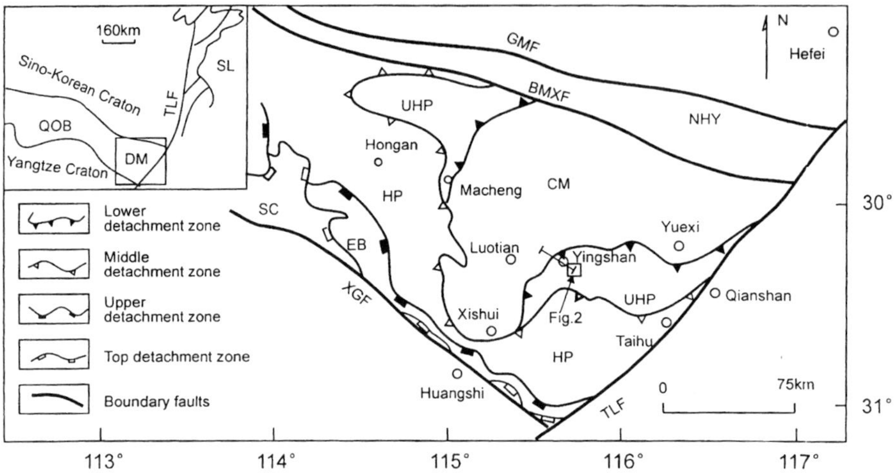

FIG. 1. Simplified tectonometamorphic map of the Dabie massif, modified slightly from Suo et al. (1999), showing the extensional tectonic pattern formed after Triassic continental collision between the Sino-Korean and Yangtze cratons. Abbreviations: GMF = Guishan-Meishan fault; BMXF = Balifan-Mozitan-Xiaotian fault; XGF = Xiangfan-Guangji fault; TLF = Tancheng-Lujiang fault; QOB = Qinling orogenic belt; DM = Dabie massif; SL = Sulu region; NHY = Northern Huaiyang tectonic belt; CC = core complex zone; UHP = ultrahigh-pressure unit; HP = high-pressure unit; EB = epidote-blueschist unit; SC = sedimentary cover. Square in inset shows the study area. The locations of Figure 2 and the cross section of Figure 3 also are shown.

1:10,000 in several key areas. It is emphasized that the determination of the regional-scale tectonic pattern related to the post-collisional extension under amphibolite-facies conditions is critical to an understanding of the exhumation process of the UHPM and HPM rocks. Structural data combined with previous petrological and geochronological studies (for reviews, see Cong et al., 1994; Hacker et al., 1996; Liou et al., 1996) have demonstrated that exhumation of the UHPM and HPM rocks from the lower-middle crust into the middle-upper crust was the result of extension, at least in the Dabie massif.

Geological Setting

The Dabie-Sulu region is the world's largest belt containing UHPM microdiamond- and coesite-bearing eclogites, and HPM coesite-free eclogites. The UHPM and HPM rocks in the belt, in general, are interpreted to be products of Triassic (240 to 210 Ma) oblique collision between the Sino-Korean and Yangtze cratons (Cong et al., 1994; Zhang et al., 1995; Hacker et al., 1996; Liou et al., 1996, 1997; Li et al., 1997). The Dabie Shan is part of the UHPM and HPM belts, in which the Balifan-Maozitan-Xiaotian fault (BMXF) separates the Northern Huaiyang (NHY) tectonic belt to the north from the Dabie massif (DM) to the south (Fig. 1). The Dabie massif comprises different metamorphic complexes with starkly contrasting metamorphic grades, which constitute the basement of the Yangtze craton. The UHPM and HPM eclogites and retrograded eclogites are enclosed in amphibolite-facies gneisses of the Yangtze basement. Upper Sinian and Paleozoic sedimentary rocks, which are locally preserved only in the southern margin of the Dabie massif, constitute the sedimentary cover of the Yangtze craton. Detailed petrological descriptions of the various rock assemblages of the Dabie massif, including possible P-T-t paths, have been provided by Suo et al. (1993), Hacker et al. (1996), Liou et al. (1996), Zhou et al. (1996), and You et al. (1997).

Extensional Tectonic Pattern

The extensional and contractional fabrics and structures in the UHPM and HPM rocks were distinguished using a variety of markers for the geometric, kinematic, and strain analysis of deformed rocks (Ramsay and Huber, 1983; Simpson, 1986), combined with petrographic and rheological observations at all scales. The rocks have undergone a complicated formation and exhumation process. The detailed structural mapping and field data for the

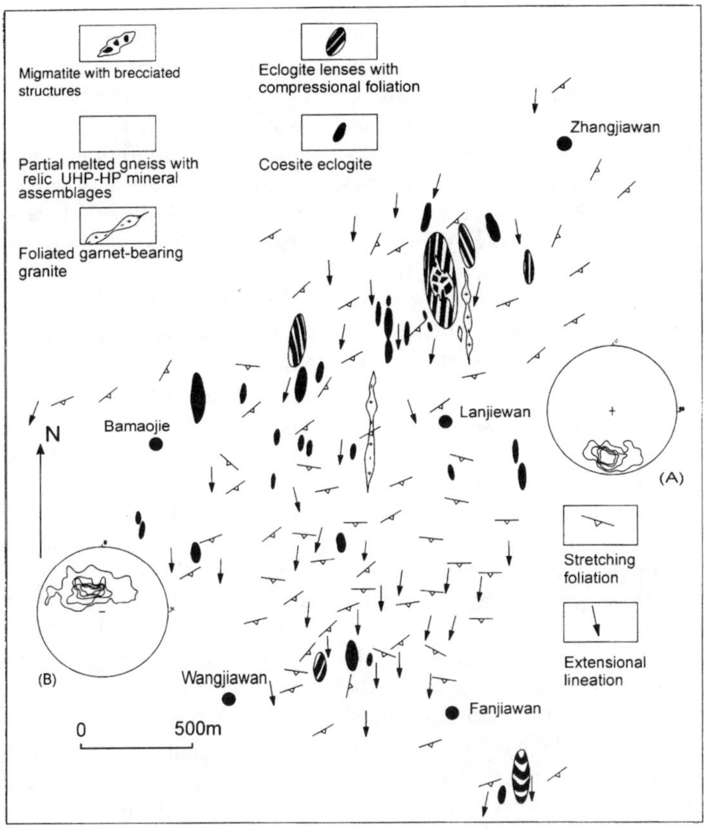

FIG. 2. Structural sketch map of the Dongchonghe area, Yingshan. See Figure 1 for location. Equal-area lower hemisphere stereoplots show stretching lineations (A) and extensional foliations (B). Contours: 1.5%, 5.8%, 9.6%, 13.5% and 2.4%, 7.1%, 11.9%, and 16.7% per 1.0% area, respectively.

Dongchonghe area ~6 km southeast of Yingshan are shown in Figure 2. It is a typical occurrence of the UHPM rocks in the Dabie massif. Here, the main rock types exposed are biotite-plagioclase gneisses with relict UHPM and HPM mineral assemblages and foliated garnet-bearing granite. Abundant eclogites and retrograded eclogites containing coesite or quartz pseudomorphs after coesite minerals (You et al., 1996) occur as lenses in the gneiss. A continuous transitional relationship exists between the eclogite or retrograded eclogite and the enclosing gneiss. Both the partially melted gneiss that contains retrograded relics of UHPM and the garnet-bearing granite exhibit gently SE-dipping, penetrative regional foliation, or compositional layering defined by a preferred alignment of mica, amphibole, plagioclase, mineral aggregates, flattened eclogite or retrograded eclogite lenses, and axial planes of A-type rootless intrafolial folds. A stretching lineation, plunging 10 to 30° to the south (Fig. 2, stereograms) in the foliation, is defined by an alignment of elongate minerals, ellipsoidal mineral aggregates, long axes of eclogite or retrograded eclogite lenses, and hinges of A-type folds (Fig. 2). Kinematic analysis, using a variety of strain and shear-sense indicators (Suo and Zhong, 1998), indicated that the development of regional foliation and lineation was dominated by a tectonic regime with subvertical shortening, associated with top-to-the-south shearing in the Dongchonghe area.

Eclogite or retrograded eclogite lenses range from tens of centimeters to tens of meters in thickness, and their shapes are characterized by a fractal system with x:y:z (x>y>z) ratios of 9:3:1 or 8:2:1, and K parameters of 0.7 to 1.47, reflecting plane strain. The pre-existing deformation fabrics formed in a contractional regime are well preserved in the large eclogite or retrograded eclogite lenses (Fig. 2). The very pronounced planar fabrics strike NNE or NNW, steeply dipping to the northwest or southwest. Detailed descriptions of the contractional fabrics including UHPM eclogite and retrograded eclogite-facies shear zones have been described elsewhere (Suo and Zhong, 1998). However, in some places the contractional fabrics have been completely obliterated by extensional deformation in association with retrometamorphism at the outcrop scale. The relics may be clearly discriminated with the aid of an optical microscope (Suo and Zhong, 1998, 1999).

A regional, unifying tectonic pattern for extensional deformation that post-dates the Triassic collision and UHP and HP metamorphism in the Dabie massif is shown approximately in Figure 1. The tectonic pattern appears to define a gneiss dome, the Luotian dome, and the pattern is similar in geometrical morphology and structure to large-scale metamorphic core complexes described in the North American Cordillera (Davies, 1983; Ames et al., 1993; Suo and Zhong, 1997, 1998), in which the most outstanding feature is the development of multilayered extensional detachment zones surrounding the core. The tectonic pattern has a significant effect on the overall geometry of the UHPM and HPM belts. The main elements of the metamorphic core complex include the core complex (CC) zone; the UHP, HP, and EB tectonometamorphic units bounded by extensional detachment zones; and the sedimentary cover (SC).

Core complex zone

The core complex zone consists mainly of the Dabie complex, in which the Neoarcheozoic–Paleoproterozoic supracrustal rocks containing layered amphibolite, biotite-plagioclase gneiss, and granulite—and a small amount of banded iron formation, garnet-bearing sillimanite gneiss, and marble of high-amphibolite to granulte-facies metamorphism—experienced polyphase deformation and migmatization (Suo et al., 1993). Large volumes of tonalitic-granodioritic-adamellitic plutons occur widely throughout the core complex zone. More than 100 mafic-ultramafic blocks, ranging from a few meters up to several hundred meters in width, are recognized; recently, relics of minor retrograded eclogite blocks also were found in the core complex zone. Xu et al. (1992, 1998) claimed that the belt containing the mafic-ultramafic blocks probably constitutes the remains of a subducted oceanic plate, and represents the suture zone between the Sino-Korean and Yangtze cratons. Thus far, however, none of the typical rock with UHPM mineral assemblages has been found in the core complex zone. The hypersthene-bearing granulite from Huangtuling, Luotian County, was metamorphosed under P = 0.89 to 1.04 GPa and T = 750 to 900°C (You et al., 1995).

UHP, HP, and EB units

On a regional scale, three tectonometamorphic units of different metamorphic P-T conditions are defined and tectonically sandwiched in between the underlying core complex zone and overlying sedimentary cover. They are, from base to top, the UHP unit, the HP unit, and the epidote-blueschist (EB) unit, bounded by four low-angle extensional detachment zones—lower, middle, upper, and top, respectively (Fig. 1). This paper focuses chiefly on the analysis of the UHP unit and the lower detachment zone. Emphasis is placed on the extension-related deformation record of the UHPM rocks and the detachment zone.

The UHP unit, with the rocks containing microdiamond, coesite, and quartz-pseudomorphs after coesite, consists dominantly of partially melted gneiss of amphibolite-facies, foliated garnet-bearing granite, or granitic gneiss, with minor eclogite, retrograded eclogite, garnet-bearing peridotite, jadeite quartzite, and marble (Wang et al., 1995; Liou et al., 1996; You et al., 1997). The UHP unit invariably overlies the core complex zone, and the lower detachment zone defines the boundary between these two units (Zhong and Suo, 1998). Rocks in the UHP unit, as mentioned earlier, underwent mostly penetrative deformation under amphibolite-facies conditions, and exhibit a gently or flat-lying regional foliation/compositional layering that delineates a domed configuration. Although cut by the Balifan-Mozitan-Xiaotian fault (BMXF) to the north, a typical domal structure is clearly displayed by the foliation orientation in the Dabie complex (Zhong and Suo, 1998). Analysis using strain markers, including deformed garnet-bearing granite dikes, pegmatite veins, symmetrical boudins, and conjugate extensional bands, indicates that development of the regional compositional layering or foliation in the UHPM unit was dominated by a tectonic regime with subvertical shortening and subhorizontal extensional flow. Generally, the vertical shortening of gneiss and foliated granite across foliation gives

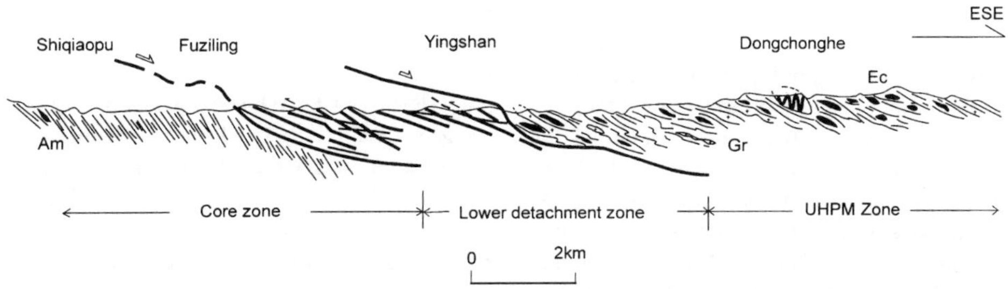

FIG. 3. Geological cross section across the lower detachment zone, showing the geometrical relationships among the core-complex zone, lower detachment zone, and UHP unit. The location of the cross section is shown in Figure 1. Abbreviations: Am = amphibolite; Gr = garnet-bearing granite; Ec = eclogite or retrograded eclogite. Note that the discrete minor shear bands with a top-to-the-west shearing formed after the extensional deformation.

values of up to 70 to 80%, and horizontal extension of about 100 to 150%, whereas the eclogite, retrograded eclogite, and amphibolite lenses exhibit vertical shortening of 50%, and horizontal extension of nearly 100%. It is clear that viscosity contrasts among different rocks have an important effect on the shape of eclogite and retrograded eclogite lenses produced during the extensional deformation, leading to the formation of a boudin-matrix or clast-matrix rheological structure (Handy, 1990). Consequently, the observed structures in the UHP unit reflect strong strain partitioning on variable scales in the lower-middle crustal levels.

The lower detachment zone designated by Wang et al. (1995) as the upper boundary of the UHP metamorphic terrain, several hundred meters (e.g., Huilonsi of Macheng) to thousand meters (e.g., Fuziling of Yingshan) in thickness, defines the contact between the core-complex zone and the UHP unit. In places, the stacking relationship among the core complex, lower detachment zone, and UHP unit can be directly observed (Fig. 3). The detachment zone is characterized by a low-angle ductile shear zone dipping outward from the core complex zone, and marked by variety of mylonites, ultramylonites, and blastomylonites. The prominent foliation in the detachment zone is subparallel to the compositional layering or foliation in the UHP unit, reflecting the fact that the extensional deformation was progressively localized in the detachment zone, which accommodated very high shear strains. Shear- and rotational-sense indicators in the detachment zone show a consistent down-dip normal sense of movement—i.e., the hanging wall (UHP unit) of the lower detachment zone is displaced downward relative to the footwall. The interpretation differs significantly from what was suggested by Wang et al. (1995). To summarize, observed structures in both the UHP unit and the lower detachment zone, combined with petrological evidence (You et al., 1997) reflect a strong subvertical shortening and subhorizontal extensional flow, associated with shearing that created a metamorphic core-complex pattern under amphibolite-facies conditions.

The HP unit and the EB unit are tectonically sandwiched in between the middle and top detachment zone. The HP unit, in general, overlies the UHP unit and locally directly contacts the core-complex zone (Fig. 1). The middle and upper detachment zones are the boundaries between the UHP and HP units, and the HP and EB units, respectively. They are similar to the UHP unit and the lower detachment zone in terms of structural geometry and kinematics. However, the pressure gap among different units indicates dramatically thinned or missed sections across the detachment zones (Ames et al., 1993; Okay and Sengör, 1993; Carswell et al., 1997).

Sedimentary cover

The sedimentary cover above the top detachment zone is composed mostly of weakly metamorphosed Upper Sinian and Paleozoic sedimentary sequences. Because of progressive, fault-bounded block uplifts and denudation of the Dabie Mountains during the Late Mesozoic and Cenozoic, the cover is partially preserved only in the southern margin of the Dabie massif. However, recently, a series of minor cover-rock outcrops was found in the internal portion of the Dabie massif; the beds probably represent remnants of the sedimentary cover. The internal parts of the sedimentary cover must have undergone contractional deformation, but intense mylonitization along its base is clearly related to extension. Rocks within the top detachment zone are characterized by the development of a fine-grained, mica-rich mylonite, or type II S-C mylonite (Lister and Snoke, 1984), crenulation cleavage, quartz veins, and pressure shadows around pyrite grains, indicating that dissolution-precipitation creep was involved. An abundance of shear-sense and kinematic indicators implies that the sedimentary cover as a whole was detached from the underlying units along the gently dipping top detachment zone, and the deformation is thin skinned, with a top-to-the-south sense of shear during the extension.

Discussion and Conclusions

The regional extension–related tectonic pattern controls the present spatial distribution of the UHPM and HPM rocks. The new data document an areal extent of the UHPM and HPM rocks that is far more widespread than previously thought. However, the northern limit of the UHP and HP units is clearly located along the Balifan-Mozitan-Xiaotian fault (BMXF) that separates the Northern Huaiyang tectonic belt to the north, comprising low-grade metavolcanic rocks and phyllites mostly of Paleozoic age, from the Dabie massif to the south, which contains a variety of UHPM and HPM rocks. The Balifan-Mozitan-Xiantian fault thus is roughly considered as a deformed suture zone formed during the Triassic intracontinental collision between the Sino-Korean and Yangtze cratons, whereas the Guishan-Meishan fault (GMF) that apparently connects to the Shangnan-Danfeng fault in the Qinling orogen on the west is interpreted as a Caledonian suture zone, and a common boundary between the Sino-Korean and Yangtze cratons (Zhou et al., 1993). In this case, the Dabie Mountains, bounded by the Guishan-Meishan fault to the north, can be broadly subdivided into six tectonometamorphic units—the Northern Huaiyang tectonic belt, the Dabie complex, the UHP unit, the HP unit, the EB unit, and very low grade metamorphic sedimentary cover (Fig. 1). All of these units are separated by faults or extensional detachment zones, as described above, and constitute a tectonic pile or stack in the crustal-scale section.

Large-scale extensional deformation has recently been reported from several collisional orogens containing UHPM and HPM rocks (Avigad, 1992; Fossen and Rykkelid, 1992; Dewey et al., 1993; Henry et al., 1993; Andersen et al., 1994; Andersen, 1998); these studies have focused on the geodynamic significance of extension during exhumation of the UHPM and HPM rocks. In the Dabie Mountains, many models for the exhumation of the UHP and HPM rocks have been suggested (e.g., Okay and Sengor, 1993; Cong et al., 1994; Maruyama et al., 1994; Wang et al., 1995; Hacker et al., 1996), but none of the models are widely accepted. Any model has to be consistent with geological and geochronological data acquired from the area studied; structural evidence is most important.

On the basis of the new structural study, and combined with previous petrological and geochronological information (Ames et al., 1993; Eide et al., 1994; Jahn et al., 1994; Hacker and Wang, 1995; Liou et al., 1996; Li et al., 1997), a three-stage evolutionary model for exhumation of the UHPM and HPM rocks in the Dabie massif can be constructed. During the first stage, UHPM and HPM rocks were quickly extruded or returned from mantle depths to lower middle crustal levels at ~210 to 200 Ma, within an overall contraction or collisional setting. In the second stage, discussed mainly in the text, the UHPM and HPM rocks and their enclosing gneisses were gradually exhumed to middle upper crustal levels in an extensional regime under amphibolite-facies conditions. The process occurred between ~200 and 170 Ma, and continued probably to 130 Ma. This conjecture is supported by Late Mesozoic (134 to 118 Ma) undeformed granite plutons cutting the regional extensional foliation. In the third stage, the UHPM and HPM rocks together with the country-gneiss rocks finally have been exhumed to the surface by differential fault-bounded block uplift, extensional collapse (Dewey, 1988), and denudation. Detailed discussion of both the first and the third stages of the exhumation process for the UHPM and HPM rocks is beyond the scope of this paper.

Extensive crustal partial melting in the sialic UHP and HP units, marked by development of large volumes of migmatites and foliated garnet-bearing granites or granitic gneisses, with whole rock–biotite U-Pb ages of 174 to 181 Ma (Jahn et al., 1994), was recognized through recent field work (Xu et al., 1998; Suo et al., 1999). The migmatites and granites exhibit the same gentle regional foliation and stretching lineations as the retrograded UHPM and HPM rocks, suggesting that decompression partial melting related to a change of thermal state in the lower-middle crustal levels, under amphioblite-facies conditions, was an important physical and chemical process. Consequently, the tectonic inversion (Ranalli, 1997)—i.e., the transition from a contractional regime to an extensional one in the tectonic evolution of the UHPM and HPM rocks—may have been driven by a combination of thermal and mechanical instabilities as a result of lithospheric thickening during Triassic collision between the Sino-Korean and Yangtze cratons. Furthermore, the eclogites and retrograded eclogites enclosed as lenses in gneiss and marble, and characterized by a boudin-matrix or clast-matrix rheological structure (Handy, 1990), are readily investigated in terms of progressive strain partitioning, decompression partial melting, and retrometamorphism during the crustal extensional process under amphibolite-facies conditions. For the Dabie massif, field occurrences and structural evidence support an in-situ model for the formation of the UHPM and HPM rocks (Zhang et al., 1995; Liou et al., 1997).

Structural study of the regional-scale tectonic pattern leads to the following conclusions. The present-day observed regional tectonic configuration of the UHPM and HPM rocks in the Dabie massif was formed mainly by extension that post-dates Triassic collision between the Sino-Korean and Yangtze cratons. Regionally, fabrics and structures related to the contraction or collision were partly destroyed by extensional deformation under amphibolite-facies conditions and produced a typical metamorphic core complex in which the development of a gently dipping penetrative foliation and multilayered extensional detachment zones are the most prominent structural features. This tectonic pattern governs the overall geometry of the UHPM and HPM rocks. Eclogite and retrograded eclogite lenses occur sporadically throughout the UHP and HP units tectonically sandwiched between the lower and upper detachment zones. The large eclogite and retrograded eclogite lenses were only moderately affected by extensional deformation, and the contractional fabrics in these lenses generally are well preserved.

The timing of the regional-scale extension in the Dabie massif may be dated at 200 to 170 Ma, continuing probably to 130 Ma in light of available geochronological data and relative age relations among tectonothermal events at variable scales. The exhumation of the UHPM and HPM rocks in the Dabie massif was a complicated process, which can be best explained by a three-stage model. In this model, the regional-scale tectonic pattern described in the paper occurred at an extensional stage as the UHPM and HPM rocks returned to the lower middle crust, and reflected vertical thinning of the previously overthickened crust and horizontal extensional flow by which a domal structure was created. Tectonic inversion may have been driven by thermal and mechanical instability, induced by the initial thickening of the lithosphere related to the Triassic collision between the Sino-Korean and Yangtze cratons. Regional extension thus is recognized as one of the most important controls on the exhumation of the UHPM and HPM rocks in the Dabie massif. This pattern seems to apply to large portions of the Sulu region as well, and may help to reconstruct the tectonic histories of the entire Dabie-Sulu collision belt.

Acknowledgements

This research was financed by the National Science Foundation of China, NSFC projects Nos. 49794041 and 49772146, and by the Ministry of Geology and Mineral Resources, MGMR project No. 9501102. The authors thank Professors Jin Z. and Gao S. from China University of Geosciences for stimulating discussions and corrections. Constructive reviews and editorial help from W. G. Ernst and J. G. Liou are gratefully acknowledged.

REFERENCES

Avigad, D., 1992, Exhumation of coesite-bearing rocks in the Dora Maira massif (western Alps, Italy): Geology, v. 20, p. 947–950.

Andersen, T. B., 1998, Extensional tectonics in the Caledonides of southern Norway, an overview: Tectonophys., v. 285, p. 333–351.

Andersen, T. B., Osmundsen, P. T., and Jolivet, L., 1994, Deep crustal fabrics and a model for the extensional collapse of the southwest Norwegian Caledonides: Jour. Struct. Geol, v. 16, p. 1191–1203.

Ames, L., Tilton, G. H., and Zhou, G., 1993, Timing of collision of the Sino-Korean and Yangtze cratons: U-Pb zircon dating of coesite-bearing eclogites: Geology, v. 21, p. 339–342.

Carswell, D. A., O'Brien, P. J., Wilson R. N., and Zhai, M., 1997, Thermobarometry of phengite-bearing eclogites in the Dabie Mountains of central China: Jour. Metamor. Geol., v. 15, p. 239–252.

Cong, B., Wang, Q., and Zhai, M., 1994, Ultra-high pressure metamorphic rocks in the Dabie-Su-Lu region, China: Their formation and exhumation: The Island Arc, v. 3, p. 135–150.

Cong, B., Wang, Q., Zhai, M., Zhang, R., Zhao, Z., and Ye, K., 1994, Review of research on ultrahigh-pressure metamorphic rocks in China: Chinese Sci. Bull.: v. 39, p. 2069–2075 (in Chinese).

Davies, G. H., 1983, Shear zone model for the origin of metamorphic core complexes: Geology, v. 11, p. 342–347.

Dewey, J. F., 1988, Extensional collapse of orogens: Tectonics, v. 7, p. 1123–1139.

Dewey, J. F., Ryan, P. D., and Andersen, T. B., 1993, Orogenic uplift and collapse, crustal thickness, fabrics and metamorphic phase changes: The role of eclogites, *in* Prichard, H. M., Alabster, T., Harris, N. B. W., and Neary, C. R., eds., Magmatic processes and plate tectonics: Geol. Soc. Lond. Spec. Publ. no. 76, p. 325–343.

Eide, E. A., McWiliams, M. O., and Liou, J. G., 1994, $^{40}Ar/^{39}Ar$ geochronology and exhumation of high-pressure to ultrahigh-pressure metamorphic rocks in east-central China: Geology, v. 22, p. 601–604.

Fossen, H., and Rykkelid, E., 1992, Postcollisional extension of the Caledonide orogen in Scandinavia: Structural expressions and tectonic significance: Geology, v. 20, p. 737–740.

Handy, M. R., 1990, The solid state flow of polymineralic rocks: Jour. Geophys. Res, v. 95, p. 8647–8661.

Henry, C., Michard, A., and Chopin, C., 1993, Geometry and structural evolution of ultra-high-pressure and high-pressure rocks from the Dora-Maira massif, Western Alps, Italy: Jour. Struct. Geol, v. 15, p. 965–981.

Hacker, B. R., Ratschbacher, L., Webb, L., and Kong, S., 1995, What brought them up? Exhumation of the Dabie Shan ultrahigh-pressure rocks: Geology, v. 23, p. 743–746.

Hacker, B. R., and Wang, Q., 1995, Ar/Ar geochronology of ultrahigh-pressure metamorphism in central China: Tectonics, v. 14, p. 994–1005.

Hacker, B. R., Wang, X., Eide, E. A., and Ratschbacher, L., 1996, The Qinling-Dabie ultrahigh-pressure collisional orogen, *in* Harrison, M. T., and Yin, A., eds., Tectonic development of Asia: Englewood Cliffs, NJ, Prentice-Hall, Rubey Vol. III, p. 345–370.

Jahn, B. M., Cornichet, J., Henin, O., Le Coz-Bouhnik, M., and Cong, B. L., 1994, Geochemical and isotopic investigation of ultrahigh pressure (UHP) metamorphic terranes in China: Su-Lu and Dabie complex, *in* First Workshop on UHP Metamorphism: Stanford, CA, Stanford University, p. A71–A74.

Li, S., Li, H., Chen, Y., Xiao, Y., and Liu, D., 1997, Ultra-high pressure metamorphic chronology of the Dabieshan Sulu terrane: II. U-Pb isotopic system of zircon: Science in China, ser. D, v. 27, p. 200–206 (in Chinese).

Liou, J. G., Zhang, R. Y., and Ernst, W. G., 1994, An introduction to ultrahigh-pressure metamorphism: The Island Arc, v. 3, p. 1–24.

Liou, J. G., Zhang, R. Y., and Jahn, B. M., 1997, Petrology, geochemistry, and isotope data on a ultrahigh-pressure jadeite quartzite from Shuanghe, Dabie Mountains, East-central China: Lithos, v. 41, p. 59–78.

Liou, J. G., Zhang, R. Y., Wang, X. M., Eide, E. A., Ernst, W. G., and Maruyama, S., 1996, Metamorphism and tectonics of high-P and ultrahigh-P belts in the Dabie-Sulu region, eastern central China, *in* Harrison, M. T., and Yin, A., eds., Tectonic development of Asia: Englewood Cliffs, NJ, Prentice-Hall, p. 300–344.

Lister, G. S., and Snoke, A. M., 1984, S-C mylonites: Jour. Struct. Geol., v. 6, p. 617–638.

Maruyama, S., Liou, J. G., and Zhang, R. Y., 1994, Tectonic evolution of the ultrahigh- pressure and high-pressure metamorphic belts from central China: The Island Arc, v. 3, p. 112–121.

Okay, A. I., and Sengör, A. M. C., 1993, Tectonics of an ultrahigh-pressure metamorphic terrane: The Dabie Shan/Tongbai Shan Orogen, China: Tectonics, v. 12, p. 1320–1334.

Ramsay, J. G., and Huber, M. I., 1983, The techniques of modern structural geology: v. 1, Strain analysis: London, Academic Press, p. 1–308.

Ranalli, G.,1997, Rheology of the lithosphere in space and time, *in* Burg, J. P., and Ford, M., eds., Orogeny through time: Geol. Soc. Lond. Spec. Publ., No. 1, v. 21, p. 19–37.

Rowley, D. B., Xue, F., Tucker, R. D., Peng, Z. X., Baker, J., and Davis, A., 1997, Ages of ultrahigh pressure metamorphism and protolith orthogneisses from the eastern Dabie Shan: U/Pb zircon geochronology: Earth Planet. Sci. Lett., v. 151, p. 21–36.

Simpson, C., 1986, Determination of movement sense in mylonites: Jour. Geol. Ed., v. 34, p. 246–261.

Suo, S., Sang, L., Han, Y., You, Z., and Zhong, Z., 1993, Petrology and tectonics of Dabie Precambrian metamorphic terranes, central China: Wuhan, China Univ. Geosci. Press, p. 259 (in Chinese).

Suo, S., and Zhong, Z., 1998, Sequence of ductile shear zones in a UHP metamorphic province within Dabie massif, China: Jour. China Univ. Geosci., v. 9, p. 189–195.

Suo, S., Zhong, Z., and You, Z., 1999, Extensional deformation of post-ultrahigh pressure metamorphism and exhumation process of ultrahigh-pressure metamorphic rocks in the Dabie massif, China: Science in China, ser. D, in press.

van der Klauw, S. N. G. C., Reineck, T., and Stöckhert., B., 1997, Exhumation of ultrahigh-pressure metamorphic oceanic crust from Lago di Cignana, Piemontese zone, western Alps: The structural record in metabasites: Lithos, v. 41, p. 79–102.

Wang, Q., Liu, X., Maruyama, S., and Cong, B., 1995, Top boundary of the Dabie UHPM rocks, central China: Jour. Southeast Asian Earth Sci., v. 11, p. 295–300.

Wang, X., and Liou, J. G., 1991, Regional ultrahigh pressure metamorphic terrane in central China: Evidence from coesite-bearing eclogite, marble, and metapelite: Geology, v. 19, p. 933–936.

Xu, S., Okay, A. I., Ji, S., Sengör, A. M. C., Su, W., Liu, Y., and Jiang, L., 1992, Diamonds from Dabie Shan metamorphic rocks and its implication for tectonic setting: Science, v. 256, p. 80–82.

Xu, S., Wu, W., Su, W., Jiang, L., and Liu, Y., 1998, Meta-granitoid from the high-ultrahigh pressure metamorphic belt in the Dabie Mountains and its tectonic significance: Acta Petrol. Sinica, v. 14, p. 42–59 (in Chinese).

Xue, F., Rowley, D. B., and Baker, J., 1996, Refolded syn-ultrahigh-pressure thrust in the South Dabie complex, China: Field evidence and tectonic implications: Geology, v. 24, p. 455–458.

You, Z., Chen, N., and Chalokwu, C. I., 1995, The metamorphism of deeper crust in the Dabie Mountains as evidenced by the study of granulites near Hulianshan, Luotian: Acta Petrol. Sinica, v. 11, p. 137–147 (in Chinese).

You, Z., Han, Y., Yang, W., Zhang, Z., and Wei, B., 1996, The high-pressure and ultra-high-pressure metamorphic belt in the eastern Qinling and Dabie Mountains, China: Wuhan, China Univ. Geosci. Press, 150 p.

You, Z., Suo, S., Zhong, Z., Zhang, Z., and Wei, B., 1997, On retrometamorphism of Dabieshan ultrahigh-pressure metamorphic rocks and their relation to tectonic evolution: Jour. China Univ. Geosci., v. 8, p. 95–99.

Zhang, R. Y., and Liou, J. G., 1994, Coesite-bearing eclogite in Henan Province, central China: Detailed petrography, glaucophane stability, and PT path: Eur. Jour. Mineral., v. 6, p. 217–233.

Zhang, R. Y., Liou, J. G., and Ernst, W. G., 1995, Ultrahigh-pressure metamorphism and decompressional P-T path of eclogites and country rocks from Weihai, eastern China: The Island Arc, v. 4, p. 293–309.

Zhong, Z., and Suo, S., 1998, Extensional tectonic framework of post high and ultrahigh pressure metamorphism in Dabieshan, China: Jour. China Univ. Geosci., v. 9, p. 1–5.

Zhou, G., Liou, J. G., and Liu, Y., 1996, High-pressure and ultrahigh-pressure metamorphic belts in northern Hebei Province, China: Wuhan, China Univ. Geosci. Press, 242 p. (in Chinese).

Zhou, G., Liu, Y., Eide, E., Liou, J. G., and Ernst, W. G., 1993, Progressive, high pressure low-temperature metamorphism in northern Hebei Province, central China: Jour. Metamor. Geol., v. 11, p. 561–574.

Exhumation of UHP Terranes: A Case Study from the Dabie Mountains, Eastern China

WANG QINCHEN AND CONG BOLIN

Institute of Geology, Chinese Academy of Sciences, Beijing 100029, China

Abstract

Two processes are suggested to explain how UHP rocks are exhumed from mantle depths. One is removal of the overburden either by erosion or by extension, whereas the other involves the uplifting of the UHP rocks through the overburden. Application of either of these mechanisms to the Dabie Mountains, however, is fraught with difficulty. When combined with previously published data, new studies on metamorphic P-T paths, regional structures, and deep upper-mantle architecture revealed by seismic tomography lend support to a multi-stage exhumation process that operated in the Dabie Mountains.

The first stage (230 to 200 Ma) is characterized by ductile deformation, produced during eclogite-facies recrystallization under a geothermal gradient as low as 10°C/km, implying a syn-subduction exhumation. Some of the UHP rocks evidently were exhumed to a depth of ~60 km, as indicated by petrological study of the Shuanghe eclogite. The second stage (200 to 170 Ma) attended ductile deformation and amphibolite-facies retrograde metamorphism. Subduction of the Yangtze block was halted by slab breakoff at a depth of ~200 km. The resultant geothermal gradient recovered to ~20°C/km. Slab breakoff permitted buoyancy-driven ascent of the UHP low-density mélange to shallow crustal levels in a diapir structure. When the UHP portion of the mountain root rose, the shallow portion was heated to a temperature higher than that of the peak metamorphic pressure. The third stage (170 to 120 Ma) is characterized by extension and thermal uplift, as well as erosion. Sedimentary basins and volcanic rocks developed on both sides of the Dabie Mountains. Gabbro-pyroxenite intruded the hanging wall of the UHP terrane, and granite, as well as migmatite, developed in that stage.

Exhumation mechanisms might include corner flow for the first stage, buoyancy-driven squeezing-up for the second stage, and crustal extension, as well as erosion, for the third. Rupture and loss of the subducted lithospheric plate generated the gravity instability that resulted in exhumation of the subducted UHP section.

Introduction

THE DISCOVERY OF coesite in metasediments from the Western Alps (Chopin, 1984) proved that buoyant supracrustal rocks could be subducted to depths of more than 100 km and experienced ultra-high-pressure (UHP) metamorphism. The processes whereby metasedimentary rocks were subducted to mantle depths and then exhumed to shallow crustal levels has challenged previous geodynamic concepts. An increasing number of UHP terranes has been discovered in the world since 1984. Among these, the Chinese Dabie-Sulu region is the largest and best exposed. However, as in the case of other UHP terranes, although the studies of petrology, mineralogy, geochronology, geochemistry, and structural geology have been fruitful, the processes and mechanisms responsible for exhumation remain unclear. In the present paper, we first review the previously published hypotheses, and then present new data and suggest a multi-stage exhumation process.

Review of Previously Published Models

The exhumation of UHP rocks from mantle depths to shallow crustal levels is a geodynamic process of grand dimensions. It also is a key topic, still hotly debated, in the study of UHP rocks. Many hypotheses have been suggested to explain the exhumation process and mechanism. It should be emphasized that "process" here is describes the procedure through which UHP rocks are exhumed, and "mechanism" refers to the dynamics working behind the process.

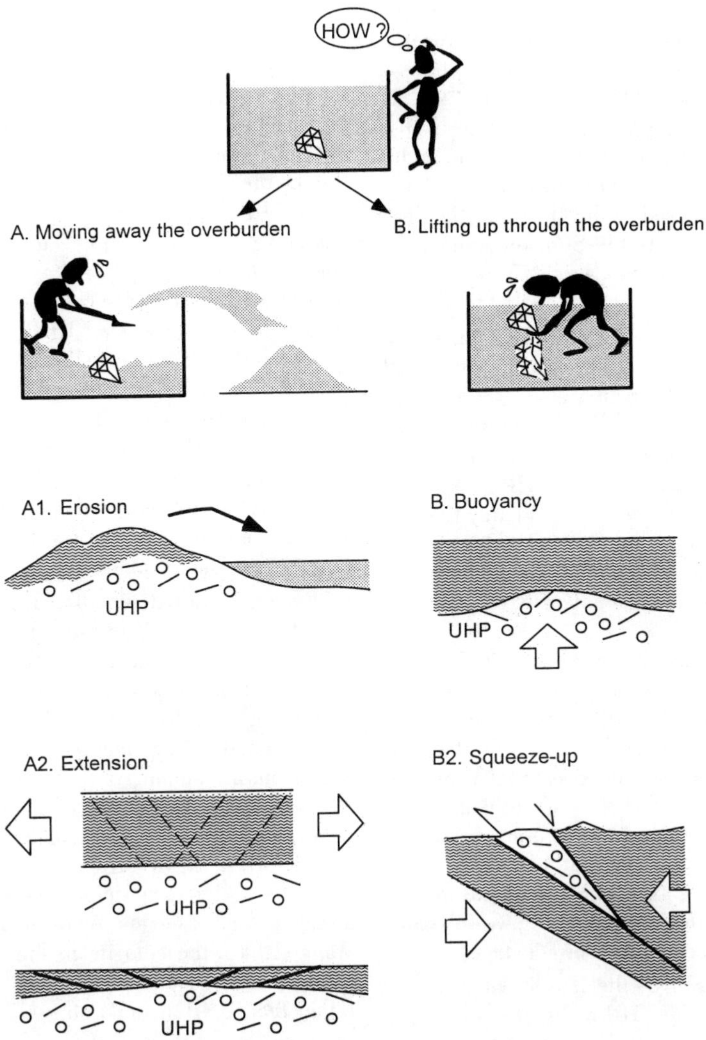

FIG. 1. Two extreme ways of exhuming UHP rocks from mantle depths.

In visualizing exhumation, a brief example is helpful. If we suppose that a diamond grain is buried in a box of sand, how can it be retrieved without breaking the box? Obviously, there are two extreme ways. One is to remove all sand overlying the diamond, whereas the other is to lift the diamond up through the overlying sand (Fig. 1). These, in fact, are the simplified cases for the exhumation of UHP rocks from mantle depths.

Yin and Nie (1993) suggested that overlying materials as thick as 100 km above the UHP rocks in the Dabie-Sulu region were eroded away and deposited in the Songpan-Ganzi basin west of the Qinling-Dabie Mountains and to the east in a basin neighboring Korea. This is an example of the first case. Extension that thinned the overburden crust (Platt, 1993) also belongs to the "remove" model. However, neither of these models is realistic for the Dabie-Sulu UHP rocks. The erosion model entails two difficulties: (1) no paleocurrent data are consistent with the suggested seaway; and (2) the thick strata composed of terrestrial debris in the Songpan-Ganzi basin accumulated in Triassic time, too early for exposure of the UHP rocks in the

Dabie-Sulu region. The extension model is also challenged by the scale of the Dabie-Sulu orogenic belt. If no material is lost when an overburden as thick as 100 km is thinned to a thickness of 1 km, its original width should be enlarged 100 times. That means an original orogenic belt of 10 km in width should reach a width of 1000 km when the UHP rocks are finally exposed at the surface. This does not correspond to the Dabie-Sulu orogenic belt, whose width is ~200 km at most.

Ernst et al. (1991) suggested that buoyancy could exhume UHP rocks from mantle depths. The corner-flow hypothesis was suggested by Cloos and Shreve (1988), and Maruyama et al. (1994) suggested that a UHP terrane could be squeezed up as a rigid wedge. All of these hypotheses belong to the "uplift" model mentioned above. UHP rocks from the Dabie Mountains are composed mainly of eclogite and metamorphosed supracrustal rocks, including jadeitic quartzite, schist, and marble (Wang et al., 1998). Their mixture has a density of ~3.0, calculated by considering the average density of metasedimentary rocks (60 vol%) as 2.75, eclogite as 3.38 (30 vol%), and peridotite (10 vol%) as 3.34. When these UHP rocks were subducted to mantle depth, they had an adequate density less than the peridotite wall rocks. Indeed, a potential buoyancy exists. However, before UHP metamorphism, those supracrustal rocks only had an overall density of ~2.55. If buoyancy was important, how could these supracrustal materials be dragged down to mantle depths? Therefore, the key point is how to explain buoyancy-driven ascent at the correct stage.

Another point involves the rate at which buoyancy could push up the UHP rocks. If too slow, all UHP mineral phases will be destroyed by thermal relaxation. Corner flow is an attractive hypothesis, allowing for the "round-trip travel" of materials in the subduction channel. However, the question for the corner-flow model is how large the subduction channel could be. The wedge-up model is not supported in the Dabie Mountains by the fact that no paired thrust-normal faults have been found. One common difficulty inherent in all of the "uplift" models is that all UHP rocks are surrounded by amphibolite-facies metamorphic rocks. This indicates that the UHP rocks in the Dabie Mountains were exhumed merely to a depth no higher than middle-crustal levels. Therefore, it seems that none of the aforementioned mechanisms could independently exhume the UHP rocks directly from mantle depths to the surface.

Various P-T Paths

P-T paths provide a record of the travel of metamorphic rocks at depth. Therefore, precise P-T estimations become increasingly important in the study of UHP terranes. The P-T paths of UHP rocks from the Dabie-Sulu region have been summarized recently by various authors (e.g., Coleman and Wang, 1995; Cong, 1996; Wang and Cong, 1996; Hacker and Liou, 1998). All exhibited a clockwise pattern of isothermal decompression, just like those in Alpine and Norwegian UHP terranes (Chopin, 1984; Smith, 1984; Carswell et al., 1990). However, these P-T paths differ in their pattern in other ways. For example, the P-T path from European Dora-Maira UHP rocks show a narrow hairpin pattern (Chopin et al., 1991), whereas the Chinese Weihai UHP rocks have an open loop (Wang et al., 1993). The difference implies that UHP rocks may have been exhumed along a different path.

Recently, we calculated P-T paths of eclogites from the Dabie UHP terrane (Liu, 1998; Liu and Wang, 1998; Wang et al., 1998). For comparison, P-T paths from coesite- and diamond-bearing eclogite (Wang et al., 1989; Xu et al., 1992) cropping out in Shima and Changpu, as well as amphibolite from Mashigou, are summarized in Figure 2. The set of P-T paths take the shape of a cock's tail, which clearly indicates that the metamorphic rocks in the Dabie terrane have varied exhumation histories. First, the peak metamorphic pressures vary even at a scale of tens of meters. As exemplified by Liu and Wang (1998), the eclogite in the southern part of Shuanghe experienced only high-pressure (HP), rather than UHP metamorphism (Fig. 2). However, there is a general trend toward higher pressure in the north (Changpu > Shuanghe/Shima > Zhujiachong), indicating that rocks cropping out in various parts of the terrane might have been subducted to, and then exhumed from, various depths. Second, these P-T–path patterns are different in the decompression section. The P-T paths in the northern areas, such as Changpu, Shuanghe, and Shima, show an isothermal decompression loop, whereas those in the south, such as cold eclogite in Zhujiachong and amphibolite in Mashigou, have a temperature-increasing decompression loop.

This implies that the northern portion of the Dabie UHP terrane has an exhumation path quite different from that of the southern portion. The initial exhumation of the hot (coesite) eclogite in the northern portion was attended by a geothermal gra-

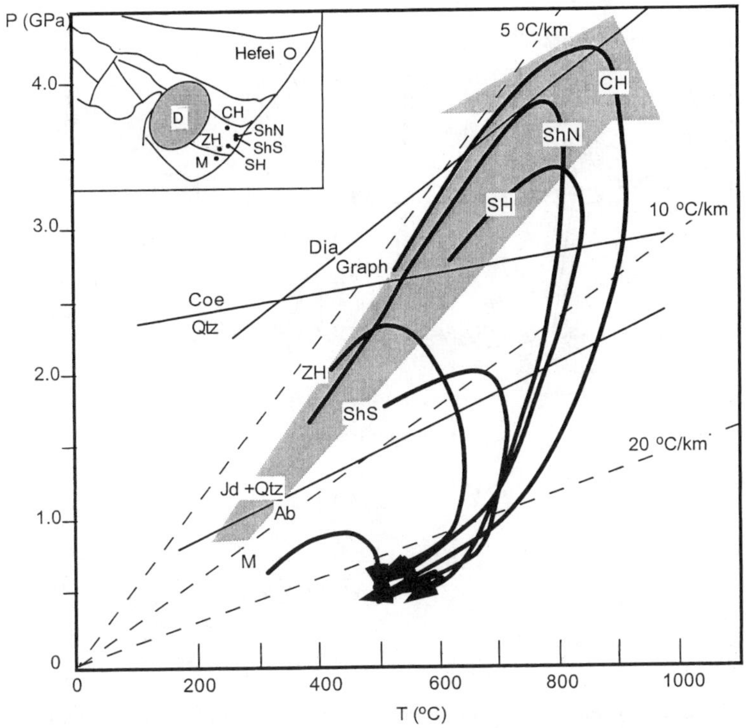

FIG. 2. P-T paths of eclogites from the Dabie Mountains. Grey arrow shows the subduction path. Abbreviations: CH = Changpu; ShN = northern Shuanghe; ShS = southern Shuanghe; SH = Shima; ZH = Zhujiachong; M = Mashigou.

dient as low as 10°C/km, which is similar to the geothermal gradient of initial subduction of continental crust, and implies a syncollisional exhumation. However, the cold eclogite and amphibolite in the southern portion started their upward travel from a shallow level when they were heated (Wang et al., 1998). Since both hot and cold eclogites experienced the same amphibolite-facies metamorphism at ~180 Ma, it is quite possible that the heat resulted from the ascent of the hot eclogite terrane from a deep level.

UHP Mélange and "Foreign" Contact

The aforementioned investigation of P-T paths indicates that not all eclogite blocks in a UHP terrane were exhumed from the same depth. In this sense, the UHP terrane is a mélange composed of rocks subducted to different depths. Such a mélange nature of the UHP terrane is manifested also by variations in lithologies and block-in-matrix structures.

As reported previously, UHP index minerals have been found not only in eclogite, but also in jadeitic quartzite, metapelite, marble, and metagranite (Okay et al., 1989; Wang et al., 1989; Hirajima et al., 1990; Wang and Liou, 1991; Xu et al., 1992; Cong et al., 1995). The protoliths have been considered as psammite, pelite, limestone, and calc-silicate rock, as well as granitoid. Such an UHP mixture of slab has been recognized in Shuanghe (Cong et al., 1995) and Yangkou (Wallis et al., 1997), where intergranular coesite was preserved in eclogite (Liou and Zhang, 1996; Ye et al., 1996). On outcrop, most of these UHP rocks show a block-in-matrix structure (Fig. 3). Foliations developed in mechanically incompetent schist, gneiss, and marble wrap around the competent eclogite blocks. Three deformation stages have been recognized in these UHP rocks, with the first stage as eclogite-facies ductile deformation (Wang and Cong, 1995). The coexistence of various rock types and high to very high pressure deformation structures imply that they were severely deformed when they

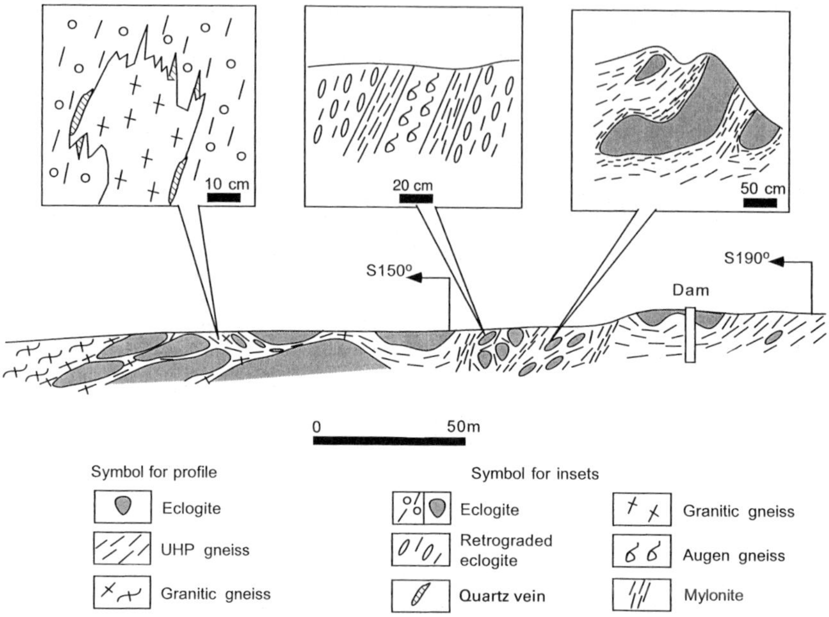

FIG. 3. Block-in-matrix structures of the UHP mélange.

traveled to great depths, perhaps in a subduction channel with corner flow (Cloos and Shreve, 1988).

On the other hand, however, not all metamorphic rocks in the Dabie terrane are UHP rocks. Whether the eclogite and its neighbor rocks are "in situ" or "foreign" has been argued for quite a long time. Detailed field mapping, as well as careful petrologic studies, have shown that the eclogites occur in two different kinds of country rocks (Cong et al., 1995). One is eclogite-facies metasedimentary rocks, including mica schist, jadeitic quartzite, and marble, all of which experienced UHP metamorphism. The other is amphibolite-facies granitic gneiss that never experienced UHP metamorphism. The two kinds of host rocks differ not only in their peak metamorphic P-T conditions (Cong et al., 1995), but also in their stable-isotope compositions (Zheng et al., 1997), as well as in their cooling history (Li et al., 1998).

Recent studies indicated an intrusive contact between the non-UHP granitic gneiss and UHP rocks (Wang, Liu et al., 1997; Xu et al., 1998; Liu, 1998). Geochemically, the non-UHP granitic gneiss is similar to the Cretaceous post-orogenic granite developing in the Dabie Mountains. Both the non-UHP granitic gneiss and post-orogenic granite show LREE-enriched REE patterns, as well as being rich in large-ion lithophile (LIL) elements and depleted in high-field-strength (HFS) elements (Liu, 1998). The high Ce/Nb, Rb/Nb, and Y/Nb values plot the non-UHP granitic gneiss in the field of Eby's (1992) A-2 type granite. The amphibolite-facies metamorphism of the non-UHP granitic age is dated as 171 to 173 Ma by the Rb-Sr isochron (biotite–whole rock) method, and 193 to 170 Ma by the Ar-Ar (amphibole) method (Chen, 1995; Li, 1998). Retrograde metamorphism of the intruded UHP mica schist has been dated as 185 to 170 Ma by the Rb-Sr isochron (phengite–whole rock) method and 181 to 169 Ma by the Rb-Sr isochron (biotite–whole rock) method (Li, 1998). According to these constraints, we deduce the following scenario: On exhumation, UHP rocks were intruded by A-2 type granite that resulted from the melting of subducted continental materials. When the UHP rocks together with the intrusive granite ascended to middle crustal levels, they experienced amphibolite-facies metamorphism. This process helps explain the "foreign" relation between the UHP rocks and their amphibolite-facies granitic gneiss host.

Dome and Extension: Late Structures

The architecture of the Dabie orogenic belt has been built up since Jurassic time (Wang et al.,

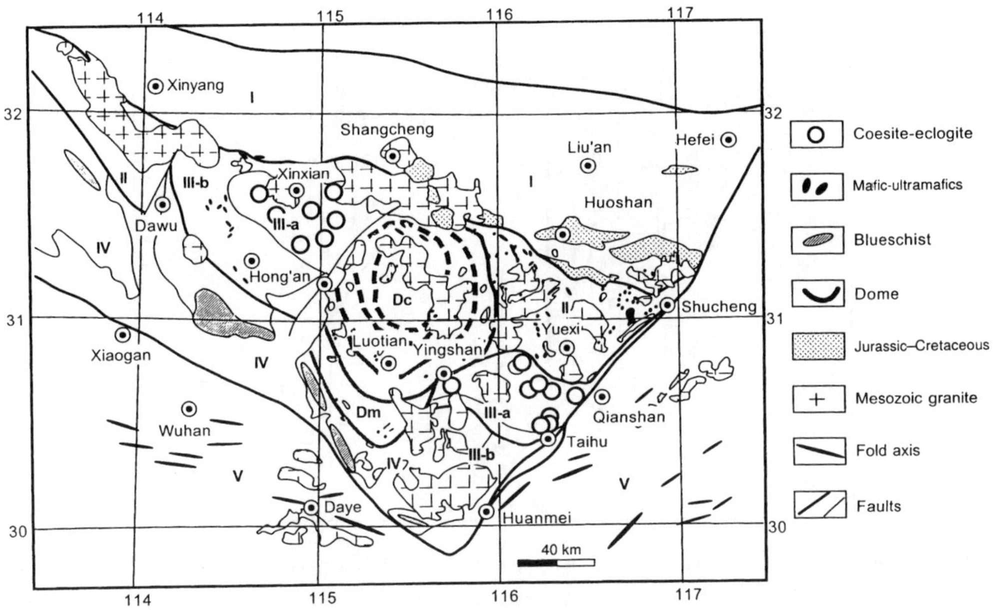

FIG. 4. Petrotectonic units in the Dabie Mountains: I = North Huaiyang backarc flysch; II = North Arc Complex; III = South Dabie Collision Complex (III-a = UHP terrane, III-b = HP terrane); IV = Sosong Metamorphic Complex; V = foreland belt. Abbreviations: Dc = dome core; Dm = dome mantle.

1998). Therefore, it is very important to differentiate late structures that developed in late stages from those produced during early stages.

Two kinds of domes have developed in the Dabie Mountains. A large dome occurs near Luotian (Fig. 4). Migmatites with exotic granulite blocks occur in the core. The north mantle is composed of grey gneiss, whereas the south mantle consists of metamorphosed banded iron formation. The difference in mantle lithology indicates an asymmetry of the Luotian dome. Several normal faults surround the dome, with fault planes dipping away from the mantle (You et al., 1998). The Luotian dome also obscured the western extension of the Shuihou-Wuhe shear zone, which marked the tectonic boundary between the South Dabie Collision Complex and the North Dabie Arc Complex. Therefore, the Luotian dome is a structure developed later than the emplacement of the UHP unit. This is further confirmed by the age of the migmatites, dated as 130 ± 1 Ma by concordant U-Pb single-zircon analyses (Wang et al., 1998). Obviously, the Luotian Dome developed not only later than the peak UHP metamorphism at ~235 Ma (Li et al., 1993), but also later than the amphibolite-facies metamorphism at ~180 Ma.

Another kind of dome structure is manifested by the UHP units themselves. The South Dabie Collision Complex in the eastern section of the Dabie Mountains is surrounded by shear zones on both the north and south. The northern boundary of the shear zone is called as the Shuihou-Wuhe fault, whereas the southern boundary is termed the Taihu-Mamiao fault. The former dips north with a shear sense of top-to-the-north (Wang et al., 1995), and the latter dips south with top-to-the-south shear sense (Wang et al., 1998). The UHP-HP rocks occur in the core of the dome. The UHP-HP dome extends to the westernmost corner of the Dabie Mountains (Liu et al., 1998). The core of the dome is composed of UHP eclogite and gneiss, whereas the mantle consists of HP eclogite and gneiss (Fig. 5). Kd (Fe^{2+}-Mg distribution coefficient between garnet and omphacite) values increase from north to south within the mantle (Liu et al., 1998), indicating a northward increase of peak metamorphic temperature. Northward subduction polarity of the HP unit is evidenced by unfolding the dome. Since both UHP eclogite and HP eclogite were overprinted by the same amphibolite-facies metamorphism, as mentioned previously, the dome structure must have

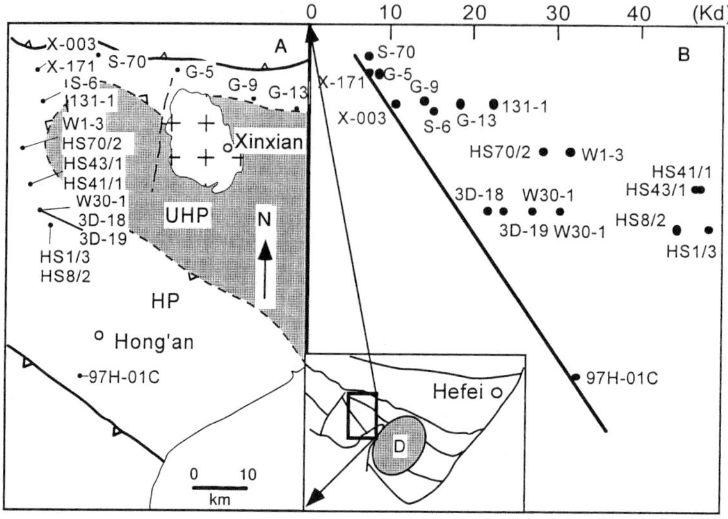

FIG. 5. Asymmetrical Xinxian dome in the western end of the Dabie Mountains (after Liu et al., 1998).

developed before 180 Ma, when the amphibolite-facies metamorphism occurred (Li et al., 1993).

In addition to the extensional faults recognized by You et al. (1998), a sedimentary basin with a basal shear zone is another kind of extensional structure. For example, the Jurassic–Cretaceous Hefei basin to the north of the Dabie Mountains is bounded by the strike-slip Xiaotian-Mozitan fault, which has a component of top-to-the-north normal shear sense. The Hefei Basin was rooted in amphibolite-facies rocks, implying that the extension structure had cut into middle-crust levels by Cretaceous time. The filling sequence of the basins recorded the unroofing history of the Dabie UHP terrane (Fig. 6). Many layers of conglomerate contain pebbles of metamorphic rocks. Taking the highest metamorphic grade recognized in the pebbles as an index of erosion depth, our estimation yields an erosion depth of ~5 km in the Early Jurassic and 15 to 20 km in the Cretaceous. Further estimation yields an average unroofing rate of ~0.20 mm/a from Early Jurassic to mid-Cretaceous. On the other hand, employing Ar/Ar radiometric data, an exhumation rate of the Dabie UHP rocks between 190 and 110 Ma was estimated as 0.15 mm/a (Chen, 1995). Consistency between the two independent estimations implies that erosion has been responsible for the exhumation since Jurassic time. All the aforementioned structures developed in Jurassic–Cretaceous time, and represent late structures.

Broken Slab: Early Structure

The structures formed in early stages are poorly preserved in the Dabie orogenic belt, except for microscopic structures indicating eclogite-facies deformation, such as elongated garnet grains in some deformed eclogites (Wang and Cong, 1995). A seismic refraction survey across the Dabie Mountains displayed a flat Moho discontinuity at a depth of ~40 km (Dong et al., 1998), and records no thickened crust.

Recent seismic tomography has revealed the deep structure of the Dabie orogenic belt (Xu, 1998). A subducted slab was shown clearly by a high-velocity anomaly (Fig. 7). The slab extends northward down to a depth of 150 km, where it is interrupted by a low-velocity anomaly. The surface extension of the slab is the metamorphosed basement and sedimentary cover of the Yangtze craton. Therefore, the seismic tomography images indicate that the Yangtze craton was once subducted beneath the Sino-Korean craton and subsequently was broken off.

Magmatic events in the Dabie orogenic belt may help in estimating the time when the subducted slab was broken off. Davies and von Blanckenburg (1995) pointed out that slab breakoff will cause mantle upwelling through the narrow rift of the broken slab and then magmatism. In the Dabie orogenic belt, the earliest magmatic record in the

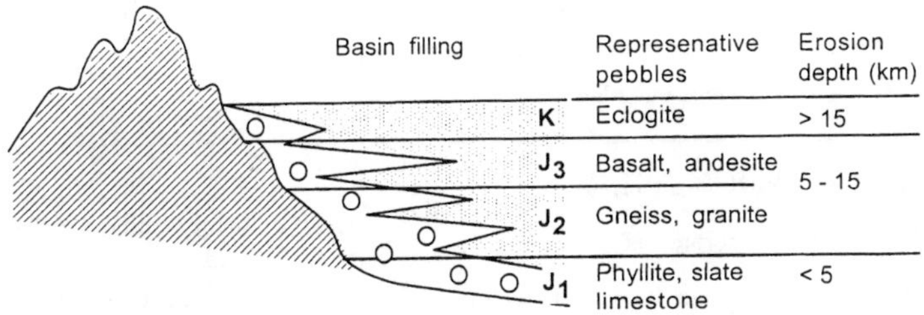

FIG. 6. Filling sequence of the Jurassic–Cretaceous basin north of the Dabie Mountains.

hanging wall of the subducted slab is found in the Jurassic volcanic rocks, followed by Cretaceous post-collision intrusion of gabbro-pyroxenite and an alkaline dike swarm. Therefore, the time of break-off should be around 190 to 200 Ma, earlier than the earliest magmatism. The depth of slab break-off was ~200 km below the Jurassic surface, if it is taken into account that crust more than 20 km thick has been eroded away since the Jurassic.

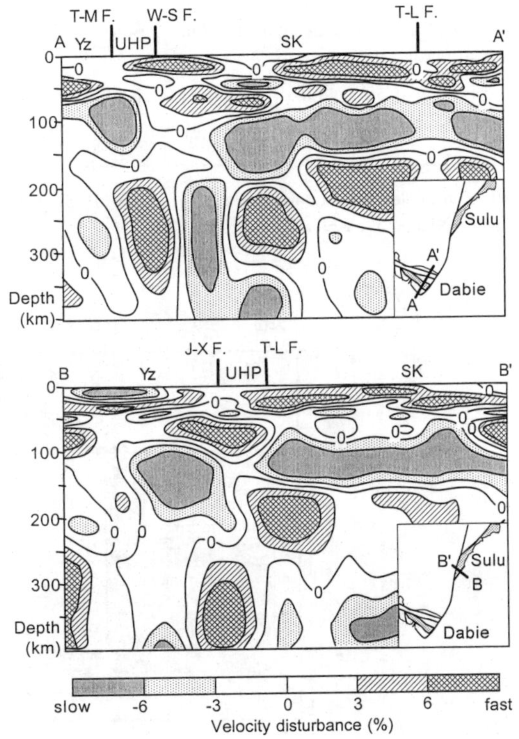

FIG. 7. Seismic tomography showing slab breakoff (after Xu, 1998).

Discussion and Conclusions: Toward a Multi-Stage Exhumation Model

The foregoing discussion has listed many tectonic events in the Dabie orogenic belt. Based on recently published geochronological data (Li et al., 1989, 1993, 1998; Ames et al., 1993; Eide et al., 1994; Ge et al., 1995; Jahn et al., 1995; Xiao et al., 1995), the following time sequence of the aforementioned tectonic events (Table 1), pointing toward a multi-stage exhumation process, can be summarized (Fig. 8).

The first stage (230 to 200 Ma) was characterized by ductile deformation, produced during eclogite-facies recrystallization under a geothermal gradient as low as 10°C/km, implying a syn-subduction exhumation. The exhumation mechanism in this stage remains to be studied, although corner flow is a viable candidate. The coexistence of eclogites with different peak metamorphic pressures, such as those studied in Shuanghe, implies a two-way travel channel. Some UHP rocks may have been exhumed to a depth of ~60 km, whereas others were still being subducted.

The second stage (200 to 170 Ma) included a series of tectonothermal events, including initial volcanic eruption, the squeezing-up of UHP terrane, and amphibolite-facies metamorphism. Subduction of the Yangtze block was halted by slab breakoff at a depth of ~200 km. Slab breakoff allowed buoyancy-driven ascent of the low-density UHP mélange to middle crustal levels as diapiric domal structures in the UHP unit. The ascending hot UHP portion heated the shallow portion, as recorded by temperature-increase sections in their P-T paths. Mantle upwelling enhanced geothermal gradients to ~25°C/km, and caused extensive amphibolite-facies metamorphism in the Dabie orogenic belt.

The third stage (170 to 120 Ma) is characterized by thermal uplift and extension, as well as by ero-

TABLE 1. Sequence of Tectonothermal Events Recognized in the Dabie Orogenic Belt[1]

Time	Events recorded in rocks	Tectonic implications
246–200 Ma	Sluggish UHP-HP metamorphism	Continent-continent collision
246–226 MA	Peak UHP metamorphism	
226–200 Ma	Ductile deformation under eclogite facies	Initial exhumation of UHP rocks
190–165 Ma	Retrograde metamorphism and deformation under amphibolite facies	Recovery of geothermal gradient, resulting from the break-off of the subducting slab, followed by mantle upwelling
Middle Jurassic	Volcanic eruptions	
Since Early Jurassic	Deposition of sedimentary basins	Extension
135–122 Ma	Gabbro-pyroxenite intrusion into the hanging wall of the UHP units; granite intrusion and migmatization	Continuous mantle upwelling and extension

[1]Geochronological data are from Li et al., 1989, 1993; Ames et al., 1993: Eide et al., 1994; Ge et al., 1995; Jahn et al., 1995; Xiao et al., 1995; Wang et al., 1998.

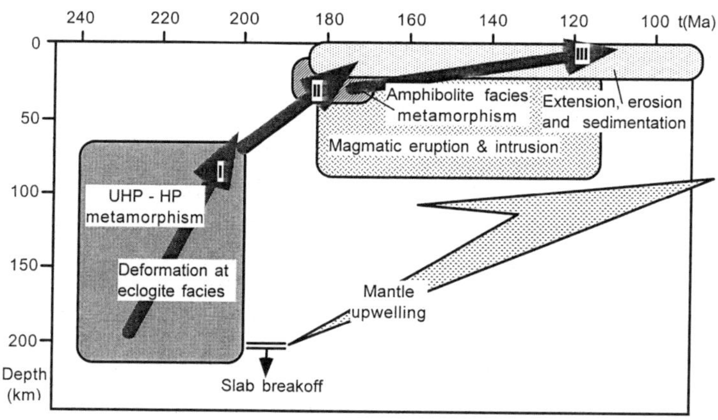

FIG. 8. Exhumation process of the Dabie UHP terrane. Grey arrows I, II, and III denote exhumation stages.

sion. Sedimentary basins and volcanic rocks developed on both sides of the Dabie Mountains. Gabbro-pyroxenite and granite intruded the hanging wall of the UHP terrane. The Luotian thermal dome and a series of extension faults developed during this stage. The mechanisms that characterized this stage were mantle upwelling, extension, and erosion. In conclusion, just as Rome was not built in a day, a UHP terrane was exhumed neither in one stage nor by a single mechanism.

Acknowledgments

This study was financially supported by the National Natural Science Foundation of China (No. 49772150 and No. 49794042) and the Chinese Academy of Sciences (No. KZ951-A1-401). The authors are grateful to Profs. Bor-ming Jahn, Li Shuguang, and H.-J. Massonne for their constructive discussions. The authors are grateful to W. G. Ernst for help in improving the English exposition of the paper.

REFERENCES

Ames, L., Tilton, G. R., and Zhou, G., 1993, Timing of collision of the Sino-Korean and Yangtze cratons: U-Pb zircon dating of coesite-bearing eclogites: Geology, v. 21, p. 239–342.

Carswell, D. A., ed., 1990, Eclogite facies rocks: Glasgow, Blackie, 396 p.

Chen, J., 1995, Cooling age of Dabie orogen, China, determined by ^{40}Ar-^{39}Ar and fission track techniques: Sci. in China, ser. B, v. 38, p. 749–757.

Chopin, C., 1984, Coesite and pure pyrope in high-grade blueschists of the western Alps: A first record and some consequences: Contrib. Mineral. Petrol., v. 86, p. 107–118.

Chopin, C., Henry, C., and Michard, A., 1991, Geology and petrology of the coesite-bearing terrane, Dora Maira massif, Western Alps: Eur. Jour. Mineral., v. 3, p. 263–291.

Cloos, M., and Shreve, R. L., 1988, Subduction-channel model of prism accretion, melange formation, sediment subduction, and subduction erosion at convergent plate margins: Pure Appl. Geophys., v. 128, p. 455–545.

Coleman, R. G., and Wang, X., eds., 1995, Ultrahigh-pressure metamorphism: New York, Cambridge University Press, 528 p.

Cong, B., ed., 1996, Ultrahigh-pressure metamorphic rocks in the Dabieshan-Sulu Region of China: Beijing and Dordrecht, Science Press and Kluwer Acad. Publ., 224 p.

Cong, B. L., Zhai, M., Carswell, D. A., Wilson, R. N., Wang, Q., Zhao, Z., and Windley, B. F., 1995, Petrogenesis of the ultrahigh-pressure rocks and their country rocks at Shuanghe in Dabieshan, central China: Eur. Jour. Mineral., v. 7, p. 119–138.

Davies, J. H., and von Blanckenburg, F., 1995, Slab breakoff; a model of lithosphere detachment and its testing in the magmatism and deformation of collisional orogens: Earth Planet. Sci. Lett., v. 129, p. 85–102.

Dong, S., Wu, X., and Gao, R., 1998, On the crust velocity levels and dynamics of the Dabieshan orogenic belt: Acta Geophys. Sinica, v. 41, p. 349–360 (in Chinese).

Eby, G. N., 1992, Chemical subdivision of the A-type granitoids: Petrogenetic and tectonic implications: Geology, v. 20, p. 641–644.

Eide, E., McWilliams, M. O., and Liou, J. G., 1994, ^{40}Ar/^{39}Ar geochronology and exhumation of high-pressure to ultrahigh-pressure metamorphic rocks in east-central China: Geology, v. 22, p. 601–604.

Ernst, W. G., Zhou, G., Liou, J. G., Eide, E., and Wang, X., 1991, High-pressure and superhigh-pressure metamorphic terranes in the Qinling-Dabie mountain belt, central China: Early- to mid-Phanerozoic accretion of the western paleo-Pacific Rim: Pacific Sci. Assoc. Inform. Bull., v. 43, p. 6–15.

Ge, N., Li, S., and Wang, Y., 1995, Sm-Nd and Rb-Sr isotopic ages of coesite-bearing eclogite in Shuanghe village, eastern Dabie Mountains: Chinese Sci. Bull., v. 40 (suppl.), p. 169.

Hacker, B., and Liou, J. G., eds., 1998, When continents collide: Geodynamics and geochemistry of ultrahigh-pressure rocks: London, Chapman and Hall.

Hirajima, T., Ishiwatari, A., Cong, B. L., Zhang, R. Y., Banno, S., and Nozaka, T., 1990, Identification of coesite in Mengzhong eclogite from Donghai county, northeastern Jiangsu Province, China: Mineral. Mag., v. 54, p. 579–584.

Jahn, B.-M., Cornichet, J., and Cong, B., 1995, Crustal evolution of the Qinling-Dabie Orogen: Isotopic and geochemical constraints from the coesite-bearing eclogites of the Su-Lu and Dabie Terranes, China: Chinese Sci. Bull., v. 40, p. 116–119.

Li, S., 1998, Chemical geodynamics of continental subduction: Earth Sci. Frontiers, v. 5, p. 211–234 (in Chinese with English abstract).

Li, S., Chen, Y., Cong, B., Zhang, Z., Zhang, R., Liou, D., Hart, S. R., and Ge, N., 1993, Collision of the North China and Yangtze blocks and formation of coesite-bearing eclogites: Timing and processes: Chem. Geol., v. 109, p. 70–89.

Li, S., Hart, S. R., Zhang, X., Guo, A., Liu, D., and Zhang, G., 1989, Sm-Nd age for collision between the Sino-Korean and Yangtze cratons, China: Acad. Sinica, v. 8, p. 312–319.

Liou, J. G., and Zhang, R. Y., 1996, Occurrence of intergranular coesite in Sulu ultrahigh-P rocks from China: Implications for fluid activity during exhumation: Amer. Mineral., v. 81, p. 1217–1221.

Liu, J., Ye, K., and Guo, L., 1998, High-pressure and ultra-high-pressure eclogite in the Hong'an region, central China: Scientia Geol. Sinica, v. 7, p. 277–283.

Liu, X., 1998, Eclogites and their related rocks in the Shuanghe area, Dabie Mountains: Their petrography, metamorphism, and tectonic implications: Unpubl. master's thesis, Institute of Geology, Chinese Academy of Sciences, 93 p. (in Chinese).

Liu, X., and Wang, Q., 1998, Quartz-ecolgite in Shuanghe area of the Dabie Mountains: Hot or cold?: Scientia Geol. Sinica, v. 7, p. 569–575.

Maruyama, S., Liou, J. G., and Zhang, R., 1994, Tectonic evolution of the ultrahigh-pressure and high-pressure metamorphic belts from central China: The Island Arc, v. 3, p. 112–121.

Okay, A. I., Xu, S., and Sengor, A. M. C., 1989, Coesite from the Dabie Shan eclogites, central China: Eur. Jour. Mineral., v. 1, p. 595–598.

Platt, J. P., 1993, Exhumation of high-pressure rocks: A review of concepts and processes: Terra Nova, v. 5, p. 119–133.

Smith, D. C., 1984, Coesite in clinopyroxene in the Caledonides and its implications for geodynamics: Nature, v. 310, p. 641–644.

Wallis, S. R., Ishiwatari, A., Hirajima, T., Ye, K., Guo, J., Nakamura, D., Kata, T., Zhai, M., Enami, M., Cong, B., and Banno, S., 1997, Occurrence and field relationships of ultrahigh-pressure metagranitoid and coesite eclogite in the Su-Lu terrane, eastern China: Jour. Geol. Soc. Lond., v. 154, p. 45–54.

Wang, J., Sun, S., and Deng, S., 1998, Geochronological constraints on the timing of migmatization in the Dabie Shan, east-central China, *in* International Workshop on UHP Metamorphism and Exhumation: Stanford, CA, Stanford University, Abs. vol., p. A-164.

Wang, Q., and Cong, B., 1995, Deformation sequence of UHP rocks from the Dabie Mountains, China: Chinese Sci. Bull., v. 40, suppl., p. 154–155 (in Chinese).

______, 1996, Tectonic implication of UHP rocks from the Dabie Mountains: Science in China (ser. D), v. 39, p. 311–318 (in Chinese).

Wang, Q., Cong, B., and Massonne, H.-J., 1998, Metamorphism and geodynamics at the south boundary of the Dabie eclogite terrane, China: Scientia Geol. Sinica, v. 7, p. 471–480.

Wang, Q., Cong, B., and Zhu, R., 1998, Geodynamics of UHP-rock-bearing continental collision zone in Central China, *in* Flower, M. F. J., Chung, S.-L., Lo, C.-H., and Lee, T.-Y., eds., Mantle dynamics and plate interactions in East Asia: Washington, DC, Amer. Geophys. Union, p. 259–268.

Wang, Q., Ishiwatari, A., Zhao, Z., Hirajima, T., Enami, M., Zhai, M., Li, J., and Cong, B. L., 1993, Coesite-bearing granulite retrograded from eclogite in Weihai, eastern China: A preliminary study: Eur. Jour. Mineral., v. 5, p. 141–152.

Wang, Q., Liu, X., and Wu, W., 1997, Intrusive contact between granitic gneiss and UHP rocks: Evidence from the Chinese Dabie Mountains and tectonic implications [abs.]: Fifth Int. Eclogite Conf., Terra Nova, v. 9, abs. suppl., p. 40.

Wang, X., and Liou, J. G., 1991, Regional ultrahigh-pressure coesite-bearing eclogitic terrane in central China: Evidence from country rocks, gneiss, marble, and metapelite: Geology, v. 19, p. 933–936.

Wang, X., Liou, J., and Mao, H. K., 1989, Coesite-bearing eclogite from the Dabie mountains in central China: Geology, v. 17, p. 1085–1088.

Xiao, Y., Li, S., Jagoutz, E., and Cheng, W., 1995, P-T-t path for coesite-bearing peridotite-eclogite association in Bixiling, Dabie Mountains: Chinese Sci. Bull., v. 40, suppl., p. 156–158.

Xu, P., 1998, Seismic tomography of the Dabie-Sulu collisional orogenic belt and its implications for the formation and exhumation of UHP rocks: Unpubl. Ph.D. thesis, Inst. Geol., Chinese Acad. Sci., Beijing, 71 p. (in Chinese).

Xu, S., Okay, A.I., Ji, S., Sengor, A. M. C., Su, W., Liu, Y., and Jiang, L., 1992, Diamond from the Dabie Shan metamorphic rocks and its implication for tectonic setting: Science, v. 256, p. 80–82.

Xu, S., Wu, W., Su, W., Jiang, L., and Liu, Y., 1998, Meta-granitoid from the high-ultrahigh pressure metamorphic belt in the Dabie Mountains and its tectonic significance: Acta Petrol. Sinica, v. 14, p. 42–59 (in Chinese).

Ye, K., Hirajima, T., Ishiwatari, A., Guo, J., and Zhai, M., 1996, Significance of interstitial coesite in eclogite at Yangkou, Qingdao City, eastern China: Chinese Sci. Bull., v. 41, p. 1407–1408.

Yin, A., and Nie, S., 1993, An indentation model for the North and South China collision and the development of the Tan-Lu and Honam fault system, eastern Asia: Tectonics, v. 12, p. 801–803.

You, Z., Suo, S., Zhong, Z., Zhang, Z., and Wei, B., 1998, The extensional tectonics and retrometamorphic processes subsequent to the high-pressure and ultrahigh-pressure metamorphic events in Dabie Shan, China: Continental Dynamics, v. 3, p. 63–71.

Zheng, Y-F., Fu, B., Xiao, Y., Gong, B., Ge, N., and Li, S., 1997, Oxygen- and hydrogen-isotope composition of eclogites from the Dabie Mountains and their geodynamic implications: Sci. in China, ser. D, v. 27, p. 121–126.

New Data Regarding Hotly Debated Topics Concerning UHP Metamorphism of the Dabie-Sulu Belt, East-Central China

CONG BOLIN, WANG QINGCHEN, AND ZHAI MINGGUO

Institute of Geology, Chinese Academy of Sciences, Beijing 100029, China

Abstract

Critical but controversial problems in the study of UHP metamorphic rocks from the Dabie-Sulu region include: (1) the possible existence of ophiolitic mélange; (2) the "in situ" versus "foreign" origin of UHP eclogites and their enclosing gneisses; (3) the possible presence and role of fluids during ultrahigh-pressure (UHP) recrystallization; (4) the timing of collision between the Yangtze and Sino-Korean continental blocks; (5) the polarity of syncollisional subduction; and (6) a single- versus multistage exhumation scenario for the UHP rocks. These questions are discussed in light of new geological, geochemical, and isotopic constraints.

Our conclusions for the Dabie-Sulu belt are as follows: (1) Mafic-ultramafic blocks are of two distinct origins: one group samples lithosphere of the suprasubduction-zone mantle wedge, whereas the second group represents post-collisional magmatic intrusions. Neither lithologic group represents true oceanic crust. (2) Quartzofeldspathic gneisses enveloping the eclogites are of two types—metasedimentary "in situ" and igneous "foreign." The paragneisses contain UHP garnets + white micas, and are uniformly older (235 ± 5 Ma) than the orthogneisses (210 ± 5 Ma), which are devoid of UHP mineralogic indicators. (3) Fluids were active under UHP conditions and allowed the formation of UHP hydrous phases such as phengite and zoisite. However, the aqueous fluids may have been restricted to certain channels/pathways during exhumation. External fluids were absent until ascent of the UHP rocks to middle-crustal levels. (4) The Yangtze and Sino-Korean continental blocks collided during 230 to 240 Ma, when supracontinental rocks experienced UHP metamorphism. The HP metamorphic event dated as >400 Ma might record a subduction of oceanic crust during the Paleozoic. (5) An ancient mantle wedge is revealed by geochemical characteristics of Mesozoic magmatic rocks developed on the southern margin of the Sino-Korean craton, the hanging wall of the UHP-rock-bearing unit. Seismic tomography images reveal that the Yangtze block extends beneath the Dabie-Sulu orogenic belt. This indicates that both oceanic and continental crust had a northward subduction polarity. (6) Taking petrologic and geochronological data into account, we prefer a multistage exhumation model. The UHP rocks were exhumed rapidly during the first stage (230 to 200 Ma), perhaps reflecting a corner-flow mechanism. Then, buoyancy and mantle upwelling brought the UHP rocks up to middle-crustal levels during the second stage (200 to 170 Ma). Extension and thermal uplift, as well as erosion, eventually exposed the UHP rocks to the surface in the third stage (170 to 120 Ma).

Introduction

GEOLOGISTS FROM many countries have studied the Dabie-Sulu ultrahigh-pressure (UHP) metamorphic belt (Fig.1), where coesite-bearing eclogite was discovered at the end of 1980s (Xu, 1987; Okay et al., 1989; Wang, 1989; Enami and Zang, 1990). Cooperative work in petrology, mineralogy, geochronology, geochemistry, and structural geology of the Dabie-Sulu UHP belt has resulted in considerable progress over the past decade. Hundreds of papers and several monographs and compendia (Coleman and Wang, 1995; Liou et al., 1995, Cong, 1996; 1998; Hacker and Liou, 1998) have been published. On the Chinese side, key projects such as "The Dabie-Sulu UHP Belt and Its Tectonic Implications" and "UHP Metamorphism and Geodynamics in a Collisional Orogenic Belt" are being financed by the National Natural Science Foundation of China (NSFC). Recently, the Science Advisory Group (SAG) of the International Continental Drilling Program (ICDP) concluded that the Dabie-Sulu UHP belt provides outstanding conditions, probably the best in the world, for study of the dynamics of continent-continent collision zones. Drilling of a borehole of 5000 m depth is now under way in order to explore the Dabie-Sulu UHP belt. Although consid-

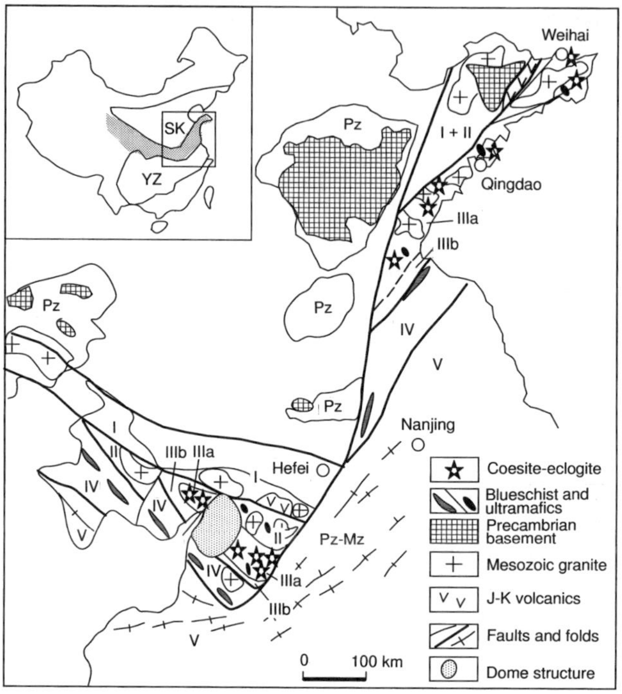

FIG. 1. Geological sketch map of the Dabie-Sulu collisional orogen, showing tectonic divisions. Symbols: I = North Huaiyang meta-flysch; II = North Dabie arc complex; III = South Dabie collision complex (IIIa = "hot" eclogite, IIIb = "cold" eclogite); IV = Susong metamorphic complex; V = foreland belt.

erable scientific progress has been made, many problems remain to be studied. Some important topics are still hotly debated. These critical but controversial problems in the study of UHP metamorphic rocks from the Dabie-Sulu region include: (1) the possible existence of ophiolitic mélange; (2) the "in situ" versus "foreign" origin of UHP eclogites and their enclosing gneisses; (3) the possible presence and role of fluids during UHP recrystallization; (4) the timing of collision between the Yangtze and Sino-Korean continental blocks; (5) the polarity of syncollisional subduction; and (6) a single- versus multi-stage exhumation scenario for the UHP rocks. These questions will be discussed in the present paper in light of new geological, geochemical, and isotopic constraints.

Relict Ophiolite or Relict Continental Mantle Slice?

More than 130 ultramafic-mafic bodies occur in the northern Dabie Mountains (Fig. 2). They have been considered as relict ophiolite or ophiolitic mélange (Xu et al., 1992, 1994). However, new geochemical data indicate that these ultramafic-mafic blocks are of two distinct origins. One group is represented by the Raobazhai peridotite. Unlike oceanic mantle, the Raobazhai peridotite was LREE enriched (Zhang et al., 1995). Its $^{87}Sr/^{86}Sr$ ratio (0.7068 to 0.7104) and ε_{Nd} value (+1.2 to –7.1) indicate an affinity for the suprasubduction-zone mantle wedge (see Fig. 5). Therefore, Zhang et al. (1995) considered the Raobozhai peridotite to be a slice of the continental lithospheric mantle above a subduction zone. Peridotite blocks with the similar origin have been reported from the Sulu region (Yang et al., 1993; Zhang et al., 1994; Chavagnac and Jahn, 1996). These ultramafic rocks were squeezed into the crust as xenoliths during the collisional orogeny.

The other group includes gabbro-pyroxenite blocks occurring at Zhujiapu, Renjiawan, and Zhongguan of the Dabie Mountains. They are neither metamorphosed nor deformed. The Zhujiapu gabbro-pyroxenite, for example, is LREE-enriched, showing no characteristics of oceanic tholeiite (Li et al., 1997). Taking into account its high ratio of Ti/Zr

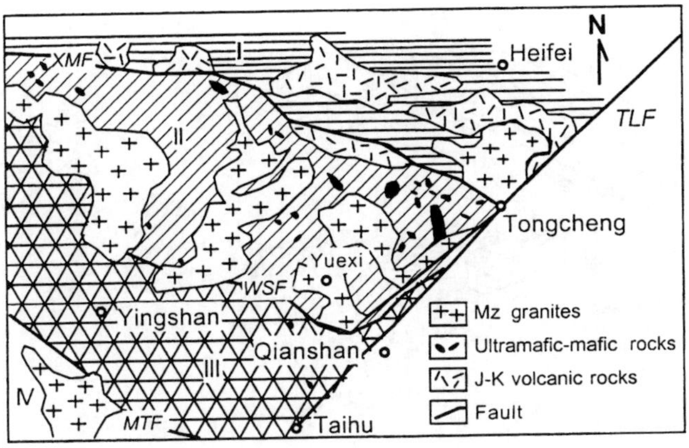

FIG. 2. Distribution of igneous rocks in the Dabie Mountains. Symbols: I = North Huaiyang meta-flysch; II = North Dabie arc complex; III = South Dabie collision complex; IV = Susong metamorphic complex. Faults: TLF = Tanlu fault; XMF = Xiaotian-Mozitan fault; WSF = Wumiao-Shuihou fault; MTF = Mamiao-Taihu fault.

and Ti/Y and negative abnormal values of high-field-strength (HFS) elements (Nb, Zr, Ti), Li et al. (1997) considered it to be an intrusive body with a magma resource from a mantle wedge affected by fluid excreted from the subducted continental crust. These intrusive gabbro-pyroxenite bodies have ages of 129 ± 2 Ma, as indicated by SHRIMP dating of zircon (Hacker et al., 1998).

Therefore, none of the above groups represents true oceanic crust. However, this does not mean that no record of oceanic crust has been preserved in the Dabie orogenic belt. An ocean might have existed between the Yangtze and Sino-Korean continental blocks during Paleozoic time. The Sujiahe cold (quartz-bearing) eclogite, with ε_{Nd} (t) > + 5.1, from the northern Dabie Mountains was dated as 422 ± 67 Ma (Sm-Nd isochron method) by Li et al. (1996). The destroyed ocean may have left its footprint in such an eclogite.

"In situ" or "Foreign" Origin?

Field mapping and petrological studies from the Shuanghe area of the Dabie Mountains have shown that quartzofeldspathic gneisses enveloping the eclogite blocks are of two types, with one being metasedimentary "in situ" and the other igneous "foreign" (Cong et al., 1995). Another example was provided from Yangkou in Qingdao (Fig. 3). However, the debate over whether the granitic gneiss has undergone UHP metamorphism together with eclogite (in situ) or not (foreign) continues.

Petrological data indicate that the orthogneisses (meta-granitoid) differ from the paragneisses (metasedimentary rocks) in many aspects (Cong et al., 1995). In contrast to the paragneisses, the orthogneisses contain Mn-enriched garnet (MnO up to 15–17%) and muscovite with a low Si content (< 3.3 pfu). The highest pressure estimate yields < 1.5 GPa (Cong et al., 1995; Liu, 1998). The UHP rocks preserved oxygen composition of meteoric water, whereas the orthogneisses as their country rocks did not (Zheng, 1997). All of these differences cannot be explained by claiming that the UHP records in the orthogneisses have been erased during later amphibolite-facies metamorphism, because the UHP rocks enclosed in the orthogneisses experienced the same overprinting of amphibolite-facies metamorphism.

Comparison of the cooling histories of both UHP rocks and the country-rock orthogneisses provides an effective way to solve the problem. Recently, precise isotopic age data were measured from the Shuanghe area. Zircon U-Pb dating yields an age of 238.4 ± 1.3 Ma in the UHP schist (Li et al., 1997), while eclogite interlayered with the UHP schist yields an Sm-Nd isochron age of 226.2 ± 2.9 Ma (Li et al., 1996). Retrograded eclogite was dated as 197.2 ± 44 Ma by the Sm-Nd isochron (Amp-Tit) method, 174.3 ± 0.5 Ma by the Rb-Sr isochron (Phn-WR) method, and 167 ± 5 Ma by the Rb-Sr isochron (Amp-WR) method (Ge et al., 1995; Li et al., 1996). The UHP schist was dated as 185 to 170 Ma by the Rb-Sr isochron (Phn-WR) method and 181 to 169 Ma by the Rb-Sr isochron (Bio-WR) method (Li et

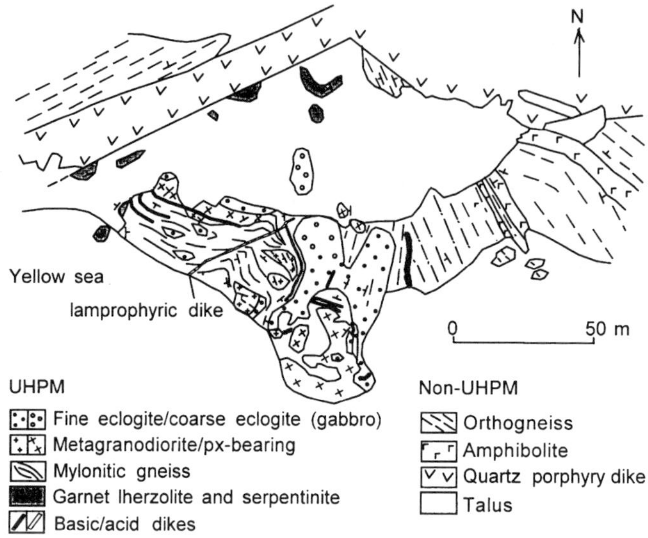

FIG. 3. Geological map of Yangkou UHP complex in the Qingdao area (modified from Wallis et al., 1997).

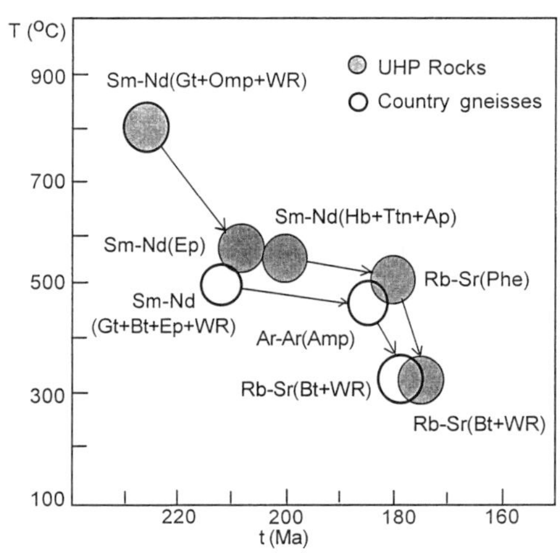

FIG. 4. Cooling history of UHP rocks and country granitic gneiss in the Shuanghe area. Mineral abbreviations: Gr = garnet, Omp = omphacite, Hb = hornblende, Ttn = titanite, Ap = apatite, Phe = phengite, Bt = biotite, Ep = epidote, WR = whole rock.

al., 1996). These data indicate that the UHP rocks from the Shuanghe area experienced two periods of rapid cooling, one before 200 Ma, and the other after 180 Ma (Fig. 4).

On the other hand, the orthogneiss that occurs as country rocks of the UHP slice in the Dabie Mountains yielded 213.3 ± 4.6 Ma using the Sm-Nd isochron method, 171 to 173 Ma by the Rb-Sr isochron (Bio-WR) method, and 193 to 170 Ma by the amphibole Ar-Ar method (Chen, Xie et al., 1995; Li et al., 1996). Obviously, the orthogneiss experienced only one period of rapid cooling during 190–170 Ma (Fig. 4). The difference in their cooling histories implies that the UHP rocks and their country orthogneiss have a "foreign" relation.

Role of Fluid in the Evolution of UHP Rocks?

One of the concepts of classical metamorphic petrology is that eclogite-facies metamorphism is a "dry" process. The fluid content drops from metamorphism of greenschist and amphibolite facies to granulite facies, with the fluid composition changing gradually from H_2O enriched to CO_2 enriched. However, this concept is challenged by the study of UHP metamorphism. In fact, some hydrous minerals such as phlogopite and pargasite have been found in mantle rocks that occur as xenoliths in basalt and kimberlite. Experimental work has shown that phengite (Schmidt, 1996), phlogopite, and pargasite remain stable at pressures of up to more than 3 GPa (Mysen, 1987).

Evidence has been reported from the Dabie-Sulu region to support the presence of water in UHP metamorphism. For example, coesite was found as inclusions not only in garnet, omphacite, and kyanite, but also in epidote and zoisite (Zhang et al., 1994). Clinohumite was reported from UHP metamorphosed ultramafic rocks (Zhang et al., 1995; Fan et al., 1996; Liou and Zhang, 1998), and magnesian staurolite from eclogite (Enami and Zang, 1988). A whiteschist assemblage was recognized in eclogite (Rolfo et al., 1997).

Furthermore, experimental results show that in the presence of excess water, many new hydrous crystalline phases can be synthesized under UHP conditions (Schreyer, 1995). For example, the stability range of Mg-chloritoid is 1.5 to 5.8 GPa at 400 to 800°C, that of Mg-staurolite >1.3 GPa at 700 to 950°C, and that of Mg-carpholite 0.7 to 5.0 GPa at <650°C (Schreyer, 1988). Glaucophane is stable over the P-T range of 1 to 3 GPa and >700°C, and "clinohumite-OH" and "chondrodite-OH" are stable at >2.9 GPa and <730°C (Yamamoto and Akimoto, 1977). Ti-ellebergerite, which is stable at 2.7 to 4.2 GPa and 650 to 725°C, has been reported in whiteschists from the Western Alps (Chopin et al., 1991).

Although water activity certainly is quite high under the UHP environment, fewer details are known about the water-rock reaction during UHP metamorphism. Studies of stable isotopes like O, C, and H seem to indicate that no fluid reacted with supracrustal rocks when they were subducted to mantle depths. Therefore, their isotopes still preserve the character of meteoric water (Yui et al., 1995; Zheng et al., 1996; Baker et al., 1997; Zhang et al., 1998). Preservation of intergranular coesite in Yangkou indicates that fluid was not present during rapid exhumation (Liou and Zhang, 1996). However, there also is evidence for the presence of water during exhumation. Most eclogites have retrograded to amphibolite because of water. Studies of O and H isotopes showed also that UHP rocks did exchange their isotope compositions with country gneiss, although limited, when they were exhumed to middle-crustal levels (Zheng, 1997).

The above evidence seems to indicate that some water was carried from the surface to mantle depths, and that fluid was limited to certain locations or channels. No doubt more work, especially systematic mapping and sampling, will help us to understand more fully the role that fluid has played during UHP metamorphism.

Age and Polarity of Continental Collision in the Dabie-Sulu Region

UHP metamorphism in the Dabie-Sulu region resulted from collision between the Sino-Korean and Yangtze cratons. However, it has been debated for a long time whether the cratons collided in Mesozoic time or much earlier, and whether the syncollisional subduction was northward or southward. If the UHP metamorphic rocks mark the beginning of the collision, geochronological data indicate that the event had started by the very beginning of the Mesozoic (Ames et al., 1993; Li et al., 1993, 1996; Eide et al., 1994; Maruyama et al., 1994; Ge et al., 1995; Jahn et al., 1995; Xiao et al., 1995). An Sm-Nd isochron dating of eclogite from Sujiahe of the Dabie Mountains (Li et al., 1996) yielded an Early Paleozoic age (420 Ma). Such Paleozoic ages (420 to 480 Ma) have been reported from the Dabie Mountains by other scientists as well (Jian et al., 1994; Zhang et al., 1995). As mentioned above, the Sujiahe eclogite geochemically exhibits an island-arc character (Li et al., 1998), and experienced only high-pressure (HP) metamorphism (Zhang et al., 1995). Therefore, it merely recorded a Paleozoic HP metamorphism earlier than the continental collision.

The following geochemical, as well as geophysical, data help to identify the subduction polarity.

1. Cretaceous intrusive bodies and dikes reflect northward subduction (Chen, Lo et al., 1995). The island-arc affinity of these magmatic rocks suggests that they were derived from a Triassic mantle wedge, which was located beneath the North Dabie Arc Complex, the former active margin of the Sino-Korean craton (Cong, 1996).

2. Paleocene tholeiite in the Hefei Basin shows also the geochemical signature of a mantle wedge above a subduction zone (Cong et al., 1996). The tholeiite is LREE-enriched, with depleted Th, Nb, Ta, and enriched Ba. The high ratio of $^{87}Sr/^{86}Sr$ (0.70583 to 0.70678) and negative ε_{Nd} values (–5.5 to –9.0) indicate an enriched mantle II (EMII) mantle resource (Fig. 5). These data imply that there was an ancient mantle wedge beneath the Hefei basin north of the Dabie Mountains. It could be inferred, therefore, that oceanic crust was subducted beneath the Sino-Korean craton.

3. Mafic rocks, such as the Zhujiapu, Renjiawan, and Zhongguan gabbro-pyroxenites, occur only in the northern Dabie Mountains (Li et al., 1997, 1998) and are considered to have been derived from a mantle wedge (see Fig. 5). This implies that the north Dabie Mountains behaved as the hanging wall, rather than the subducted plate, during continental collision.

4. The last, but not least, evidence comes from the deep upper-mantle structure. A subducted slab

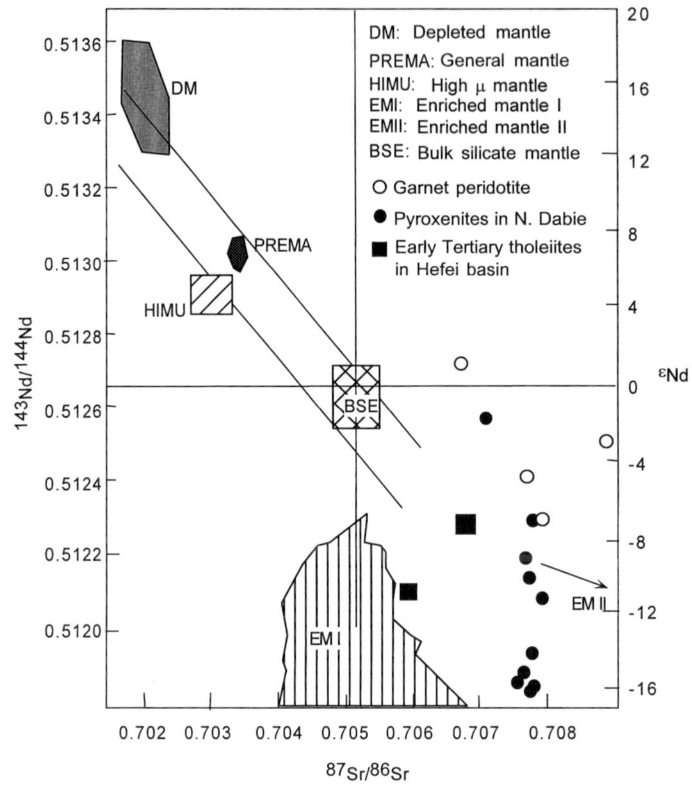

FIG. 5. $^{143}Nd/^{144}Nd$ versus $^{87}Sr/^{86}Sr$ plot of the ultramafic-mafic rocks (data cited from Zhang et al., 1995, Chavagnac and Jahn, 1996, Cong et al., 1996, and Li et al., 1998) in the northern Dabie Mountains and the Early Tertiary tholeiites in the Hefei Basin.

with high velocity (V_p) is displayed in a recent seismic tomographic study (Xu et al., 1999). The slab extends from the surface northward down to depth of 400 km in both the Dabie and Sulu regions. These data indicate clearly the northward subduction of the Yangtze craton.

Single- or Multi-stage Exhumation of the UHP Rocks?

Most workers now prefer the concept of rapid exhumation, because coesite is readily converted to quartz during slow exhumation (Hacker and Peacock, 1995; Wallis et al., 1998). It is true that a hairpin P-T path like that of the Dora-Maira UHP rocks (Chopin et al., 1991) indicates rapid exhumation. However, only the retrograde P-T section is related to exhumation. The authors' study indicates that P-T paths might differ from location to location in an UHP terrain, implying a complicated exhumation history (Wang and Cong, 1999). Therefore, more P-T paths and better resolution of paths from the one and same UHP terrain should be studied in the future.

A detailed review on the exhumation process for the Dabie-Sulu region has been proposed by Wang and Cong (1999). In this case, a multi-stage exhumation process is documented by petrologic and geochronologic studies. The first stage (230 to 200 Ma) is characterized by ductile deformation of eclogite facies developed in a geothermal gradient as low as 10°C/km, implying a synsubduction exhumation. Some UHP rocks were exhumed while others were being subducted. This explains why some coesite-bearing eclogites coexist with metagabbros that did not experience UHP metamorphism. The second stage (200 to 170 Ma) is characterized by ductile deformation and retrograde amphibolite-facies metamorphism. The UHP rocks were squeezed up into the shallow crust that occurs now as granitic rocks with or without amphibolite lenses.

From that point, both types of rock assemblage were exhumed together. Subduction of the Yangtze block ceased and the geothermal gradient recovered to ~20°C/km. The margin of the UHP terrain was heated to temperatures higher than that of peak-pressure metamorphism. The third stage (170 to 120 Ma) is characterized by extension and thermal uplift, as well as by erosion. Sedimentary basins and volcanic rocks developed on both sides of the Dabie Mountains. Gabbroic rocks intruded in the hanging wall of the UHP terrain, and granite, as well as migmatite, formed at the end of this stage (140 to 120 Ma). This might indicate the locality of slab break-off, where the hot mantle plume was upwelling. Exhumation mechanisms might include corner flow for the first stage, buoyancy-driven squeezing-up for the second stage, and extension of crust, as well as erosion, for the third stage.

Conclusions

The authors' conclusions concerning the foregoing hotly debated topics are summarized as follows.

1. The ultramafic-mafic blocks in the Dabie-Sulu region belong to two groups, one being relict continental lithosphere above a subduction zone, the other reflecting post-collision intrusion. None of these blocks exhibits ophiolitic characteristics.

2. Quartzofeldspathic gneisses enveloping the eclogites are of two types—metasedimentary "in situ" and igneous "foreign." The paragneisses contain UHP garnets + white micas, and are uniformly older (235 ± 5 Ma) than are the orthogneisses (210 ± 5 Ma), which are devoid of UHP mineralogic indicators.

3. The presence of hydrous phases such as phengite and zoisite as main-stage UHP phases indicates the attendance of H_2O during the highest pressures of metamorphism. However, aqueous fluids may have been restricted to certain channels during exhumation. External fluids were absent until ascent of the UHP rocks to middle-crustal levels.

4. The Yangtze and Sino-Korean continental blocks collided during the period from 230 to 240 Ma, when supracontinental rocks experienced UHP metamorphism. The HP metamorphic event, dated as >400 Ma, might record the subduction of oceanic crust during Paleozoic time.

5. An ancient mantle wedge is revealed by geochemical characteristics of Mesozoic magmatic rocks developed on the southern margin of the Sino-Korean craton, the hanging wall of the UHP-rock-bearing unit. Seismic tomography images depict the Yangtze block as extending beneath the Dabie-Sulu orogenic belt. Various lines of evidence demonstrate that both oceanic and continental crust reflect northward subduction.

6. Taking petrologic and geochronological data into account, we prefer a multistage exhumation model. The UHP rocks were exhumed rapidly during the first stage (230 to 200 Ma). Thereafter, buoyancy and mantle upwelling brought the UHP rocks up to middle-crustal levels during the second stage (200 to 170 Ma). Extension and thermal uplift, as well as erosion, eventually exposed the UHP rocks at the surface in the third stage (170 to 120 Ma).

Acknowledgments

Results presented here come mainly from a project that was financially supported by the National Natural Science Foundation of China (No. 49794042 and No. 49772150) and the Chinese Academy of Sciences (No. KZ951-A1-401). Research was carried out by the authors and colleagues from the Institute of Geology, Chinese Academy of Sciences, and University of Science and Technology of China. Foreign scientists from Stanford University, Kyoto University, the Tokyo Institute of Technology, Nagoya University, Waseda University, Sheffield University, the Geophysical Laboratory of the Carnegie Institution of Washington, and Stuttgart University also aided our study. The authors thank all of them for their efforts, and are grateful to W. G. Ernst for constructive suggestions that improved the paper in both scientific and English exposition.

REFERENCES

Ames, L., Tilton, G. R., and Zhou, G., 1993, Timing of collision of the Sino-Korean and Yangtze cratons: U-Pb zircon dating of coesite-bearing eclogites: Geology, v. 21, p. 239–342.

Baker, J., Matthews, A., Mattey, D., Rowley, D., and Xue, F., 1997, Fluid-rock interactions during ultra-high pressure metamorphism, Dabie Shan, China: Geochim. et Cosmochim. Acta, v. 67, p. 1685–1696.

Chavagnac, V., and Jahn, B. M., 1996, Coesite-bearing eclogites from the Bixiling Complex, Dabie Mountains, China: Sm-Nd ages, geochemical characteristics, and tectonic implications: Chem. Geol., v. 133, p. 29–51.

Chen, C. H., Lo, C. H., Teng, L. S., and Cong, B., 1995, Preliminary geochemical results of the basic pluton and basic-intermediate dike from Northern Dabieshan area, China [abs.]: Ext. Abs. Conf. on Geology across Taiwan Strait, p. 4–7.

Chen, J., Xie, Z., Liu, S., Li, X., and Foland, K. A., 1995, Cooling age of Dabie orogen, China, determined by ^{40}Ar-^{39}Ar and fission-track techniques: Sci. in China (ser. B), v. 38, p. 749–757 (in Chinese).

Chopin, C., Henry, C., and Michard, A., 1991, Geology and petrology of the coesite-bearing terrain, Dora Maira Massif, Western Alps: Eur. Jour. Mineral., v. 3, p. 263–291.

Coleman, R. G., and Wang, X., eds., 1995, Ultrahigh-pressure metamorphism: New York, Cambridge University Press, 528 p.

Cong, B., ed., 1996, Ultrahigh-pressure metamorphic rocks in the Dabieshan-Sulu Region of China: Beijing and Dordrecht, Science Press and Kluwer Acad. Publ., 224 p.

Cong, B., Wang, Q., Zhang, H., Yan, X., and Jiang, L., 1996, Petrogenesis of Cenozoic volcanic rocks in Hefei basin, China: Acta Petrol. Sinica, v. 12, p. 370–381.

Cong, B. L., Zhai, M., Carswell, D. A., Wilson, R. N., Wang, Q., Zhao, Z., and Windley, B. F., 1995, Petrogenesis of the ultrahigh-pressure rocks and their country rocks at Shuanghe in Dabieshan, central China: Eur. Jour. Mineral., v. 7, p. 119–138.

Eide, E. A., McWilliams, M. O., and Liou, J. G., 1994, $^{40}Ar/^{39}Ar$ geochronology and exhumation of high-pressure to ultrahigh-pressure metamorphic rocks in east-central China: Geology, v. 22, p. 601–604.

Enami, M., and Zang, Q., 1988, Magnesian staurolite in garnet-corundum rocks and eclogite from the Donghai District, Jiangsu Province, East China: Amer. Mineral., v. 73, p. 48–56.

______, 1990, Quartz pseudomorph after coesite in eclogite from Shandong Province, east China: Amer. Mineral., v. 75, p. 381–386.

Fan, Q., Liu, R., Ma, B., Zhao, D., and Zhang Q., 1996, The protolith and ultrahigh-pressure metamorphism of Maowu mafic-ultramafic rock block in Dabieshan Mountains: Acta Petrol. Sinica, v. 12, p. 29–46 (in Chinese).

Ge, N., Li, S., and Wang, Y., 1995, Sm-Nd and Rb-Sr isotopic ages of coesite bearing eclogite in Shuanghe village, eastern Dabie Mountains [abs.]: Chinese Sci. Bull., v. 40, suppl., p. 156.

Hacker, B., and Liou, J. G., ed., 1998, When continents collide: Geodynamics and geochemistry of UHP rocks: Dordrecht, Kluwer Acad. Publ.

Hacker, B. R., and Peacock, S. M., 1995, Creation, preservation, and exhumation of UHPM rocks, *in* Coleman, R. G., and Wang, X., eds., 1995, Ultrahigh-pressure metamorphism: New York, Cambridge University Press, p. 159–181.

Hacker, B. R., Ratschbacher, L., Webb, L., Ireland, T., Walker, D., and Dong, S.W., 1998, U/Pb zircon ages constrain the architecture of the ultrahigh-pressure Qinling-Dabie orogen, China: Earth Planet Sci. Lett., v. 161, p. 215–230.

Jahn, B.-M., Cornichet, J., and Cong, B., 1995, Crustal evolution of the Qinling-Dabie orogen: Isotopic and geochemical constraints from coesite-bearing eclogites of the Su-Lu and Dabie terranes, China: Chinese Sci. Bull., v. 40, (suppl.), p. 116–119.

Jian, P., Ye, B., and Li, Z., 1994, Geochronology, P-T-t path, and tectonic implications of the eclogite from the Dabie Mountains, *in* Chen, H., ed., 1994, Collection on isotope geochemistry: Hangzhou, Zhejiang University Press, p. 205–213 (in Chinese).

Li, S., et al., 1993, Collision of the North China and Yangtze blocks and formation of coesite-bearing eclogites, timing and processes: Chem. Geol., v. 109, p. 89–111.

Li, S., Jagoutz, E., Xiao, Y., Ge, N., and Chen, Y., 1996, Chronology of ultrahigh-pressure metamorphism in the Dabie Mountains and Su-Lu terrane: I. Sm-Nd isotope system: Sci. in China, ser. D, v. 39, p. 597–608 (in Chinese).

Li, S., Jagoutz, E., Zhang Z., Lo, Q.-H., Chen, W., and Chen, Y. , 1998, Geochemical and geochronological constraints on the tectonic outline of the Dabie Mountains, central China: A continental-microcontinental collision model: Continental Dymanics, v. 5, p. 211–234.

Li, S., Nie, Y., Zheng, S., and Liu, D., 1997, Interaction between subducted continental crust and upper mantle: I. Major and trace elements of the syn-collision mafic-ultramafic rocks: Sci. in China, ser. D., v. 27, p. 466–493 (in Chinese).

Liou, J. G., Cong, B., and Maruyama, S., ed., 1998, Geodynamics for high- and ultrahigh-pressure metamorphism: The Island Arc, v. 7, no. 1/2 (spec. iss.), p. 1–200.

Liou, J. G., Wang, Q., Zhang, R. Y., Zhai, M., and Cong, B., 1995, Ultrahigh-P metamorphic rocks and their associated lithologies from the Dabie Mountains, central China: A field trip guide to the 3rd International Eclogite Field Symposium: Chinese Sci. Bull., v. 40 (suppl.), p. 1–40.

Liou, J. G., and Zhang, R. Y., 1996, Occurrences of intergranular coesite in ultrahigh-P rocks from the Sulu region, eastern China: Implications for lack of fluid during exhumation: Amer. Mineral., v. 81, p. 1217–1221.

______, 1998, Petrogenesis of an ultrahigh-pressure garnet-bearing ultramafic body from Maowu, Dabie Mountains, east-central China: The Island Arc, v. 7, p. 115–134.

Liu, X., 1998, Eclogites and their related rocks in the Shuanghe area, Dabie Mountains: Their petrography, metamorphism, and tectonic implications: Unpubl. master's thesis, Inst. Geol., Chinese Acad. Sci., 93 p. (in Chinese).

Maruyama, S., Liou, J. G., and Zhang, R., 1994, Tectonic evolution of the ultrahigh-pressure and high-pressure metamorphic belts from central China: The Island Arc, v. 3, p. 112–121.

Mysen, B. O., ed., 1987, Magmatic processes: Physicochemical principles: The Geochem. Soc., Spec. Publ., no. 1.

Okay, A. I., Xu, S., and Sengör, A. M. C., 1989, Coesite from the Dabie Shan eclogites, central China: Eur. Jour. Mineral., v. 1, p. 595–598.

Rolfo, F., Compagnoni, R., Xu, S.-T., and Jiang, L.-L., 1997, First occurrence of talc+kyanite association in a meta-aplite of the UHPM zone of Dabie Shan, China [abs.]: Terra Nova, Abs. suppl. 1, p. 30.

Schmidt, M. W., 1996, Experimental constraints on recycling of potassium from subducted oceanic crust: Science, v. 272, p. 1927–1930.

Schreyer, W., 1988, Experimental studies on metamorphism of crustal rocks under mantle pressure: Mineral. Mag., v. 52, p. 1–26.

______, 1995, Ultradeep metamorphic rocks: The retrospective viewpoint: Jour. Geophys. Res., v. 100, p. 8353–8366.

Wallis, S. R., Ishiwatari, A., Hirajima, T., Ye, K., Guo, J., Nakamura, D., Kato, T., Zhai, M., Enami, M., Cong, B., and Banno, S., 1997, Occurrence and field relationships of ultrahigh-pressure metagranitoid and coesite eclogite in the Su-Lu terrane, eastern China: Jour. Geol. Soc. Lond., v. 154, p. 45–54.

Wallis, S. R., Nakamura, D., and Hirajima, T., 1998, Rapid exhumation of the Su Lu UHP terrain [abs.]: Abs. Amer. Geophys. Union Fall Meeting, no. 14244.

Wang, Q., and Cong, B., 1999, Exhumation of UHP terranes: A case study from the Dabie Mountains, Eastern China: INT. GEOL. REV., v. 41, in press.

Wang, X., Liou, J. G., and Mao, H. K., 1989, Coesite-bearing eclogites from the Dabie Mountains in central China: Geology, v. 17, p. 1085–1088.

Xiao, Y., Li, S., Jagoutz, E., et al., 1995, P-T-t path for coesite-bearing peridotite-eclogite association in Bixiling, Dabie Mountains, China: Chinese Sci. Bull., v. 40, suppl., p. 156.

Xu, P., Sun, R., Liu, F., Wang, Q., and Cong, B., 1999, Subduction and breakoff of the Yangtze continental block evidenced by seismic tomography: Chinese Sci. Bull., in press.

Xu, S., Jiang, L., Liu, Y., and Zhang, Y., 1992, Tectonic framework and evolution of the Dabie Mountains in Anhui, eastern China: Acta Geol. Sinica, v. 66, p. 1-14 (in Chinese).

Xu, S., Liu, Y., Jiang, L., Su, W., and Ji, S., 1994, Tectonic framework and evolution of the Dabie Mountains: Beijing, Science Press, 175 p. (in Chinese).

Xu, Z., 1987, Etude tectonique et microtectonique de la chine paleozoique et triasique des Qintings (Chine): Unpubl. Ph.D. thesis, Univ. Sci. Tech. Languedoc, Montpellier.

Yamamoto, K., and Akimoto, S.-I., 1977, The system MgO-SiO_2-H_2O at high pressures and a temperature-stability field for hydroxyl-chondrodite, hydroxyl-clinohumite, and 10 Å phase: Amer. Jour. Sci., v. 277, p. 288–312.

Yang, J., Godard, G., Kienast, J. R., Lu, Y., and Sun, J., 1993, Ultrahigh-pressure (60 kbar) magnesite-bearing garnet peridotites from northeastern Jiangsu, China: Jour. Geol., v. 101, p. 541–554.

Yui, T. F., Rumble, D., and Lo, C. H., 1995, Unusually low ^{18}O ultrahigh-pressure metamorphic rocks from Su-Lu terrane, China: Geochim. et Cosmochim. Acta, v. 59, p. 2859–2864.

Zhang, Q., Mao, B., Liu, R., Zhao, D., Fan, Q., Li, Q., and Li, X., 1995, A relic slice of continental lithospheric mantle: Sci. in China, ser. B, v. 25, p. 867–873 (in Chinese).

Zhang, R. Y., Liou, J. G., and Cong, B., 1994, Petrogenesis of garnet-bearing ultramafic rocks and associated eclogites in the Su-Lu ultrahigh-P metamorphic terrane, eastern China: Jour. Metamor. Geol., v. 12, p. 169–186.

______, 1995, Talc-, magnesite-, and Ti-clinohumite-bearing ultrahigh-pressure meta-mafic and ultramafic complex in the Dabie Mountains, China: Jour. Petrol., v. 36, p. 1011–1037.

Zhang, R. Y., Rumble, D., III, Liou, J. G., and Wang, Q. C., 1998, Low $\delta^{18}O$, ultrahigh-P garnet-bearing mafic and ultramafic rocks from Dabie Shan, China: Chem. Geol., v. 150, p. 161–170.

Zhang, Z., You, Z., Han, Y., and Song, L., 1995, Petrology, metamorphic process and genesis of the Dabie-Sulu eclogite belt, east-central China: Acta Geol. Sinica, v. 69, p. 306–325 (in Chinese).

Zheng, Y. F., 1997, Oxygen and carbon isotope anomalies in the ultrahigh pressure metamorphic rocks of the Dabie-Sulu terranes: Implications for geodynamics: Episodes, v. 20, p. 104–108.

Zheng, Y. F., Fu, B., Cong, B. L., and Li, S. G., 1996, Extreme ^{18}O depletion in eclogite from the Su-Lu terrane in east China: Eur. Jour. Mineral., v. 8, p. 317–323.

The Sulu UHP Terrane: A Review of the Petrology and Structural Geology

SIMON WALLIS,

Department of Geology and Mineralogy, Kyoto University, Kyoto, 606-8107 Japan

MASAKI ENAMI,

Department of Earth and Planetary Sciences, Nagoya University, Nagoya, 464-8602 Japan

AND SHOHEI BANNO

Department of Geology and Mineralogy, Kyoto University, Kyoto, 606-8107 Japan

Abstract

Petrological and isotopic evidence suggests that the protolith of the Sulu ultrahigh-pressure (UHP) terrane was Precambrian continental crust consisting of granite, granodiorite, gabbro, marble, and basic dikes, with local granulite-facies assemblages. Around 220 Ma this unit of continental rocks was buried to depths up of ~120 km within the mantle. Structures formed during exhumation suggest highly mobile behavior of acidic rocks, even under conditions of very low water activity. Petrological studies show that the Sulu terrane underwent isothermal decompression, which implies relatively rapid exhumation, and suggests that the role of melting during exhumation may have been underestimated. The later stages of exhumation are associated with NW-SE–directed tectonic transport and the formation of at least one major normal detachment.

Introduction

THE QINLING-DABIE-SULU orogenic belt formed in the Triassic as a result of the collision between the Yangtze and North China cratons (Fig. 1). Petrological studies over the last 10 years have shown that coesite-eclogite, characteristic of ultrahigh-pressure (UHP) metamorphism, occurs widely throughout the eastern part of the orogen in the Dabie and Sulu terranes. Together the Sulu and Dabie terranes represent the most extensive UHP metamorphic belt in the world. The only UHP metamorphic belt of comparable dimensions is the Bohemian massif in central Europe (e.g., O'Brien and Carswell, 1993). A large amount of data are now available regarding the metamorphic and structural evolution of the Dabie terrane (e.g., Coleman and Wang, 1995; Hacker et al., 1995; Cong and Wang, 1996), but there is much less comparable information about the associated Sulu terrane. In this contribution we summarize the main geological features of the Sulu terrane, concentrating on the results of petrological and structural studies. Reviews of the geochemical and isotopic studies in the Sulu terrane are presented in Jahn (1998) and Rumble (1998).

Division of the Shangdong Peninsula and the Sulu Terrane

The UHP Sulu terrane is located on the Shangdong Peninsula between Beijing and Shanghai (Fig. 1), extending south into Jiangsu province. Many workers have treated the gneisses of this area as having a common geological history. The results of recent studies have shown, however, that the UHP metamorphic region and associated metamorphic rocks are restricted to one part of the peninsula, as discussed in the following section.

The Shangdong Peninsula can be divided into northwestern and southeastern areas, with a boundary along the Yantai-Qindao-Wulian (YQW) fault (Fig. 1). The entirety of the region has been intruded by Cretaceous granitic plutons, and has well-developed Mesozoic and more recent sedimentary sequences. However, the earlier, pre-Jurassic geological development of the two areas to the northwest and southeast of the YQW fault is strikingly different.

The basement rocks of the northwest area are characterized by amphibolite-granulite schists and gneisses, locally overlain by low-grade to unmetamorphosed rocks of the Upper Proterozoic Sinian system. No eclogite has yet been found in this area,

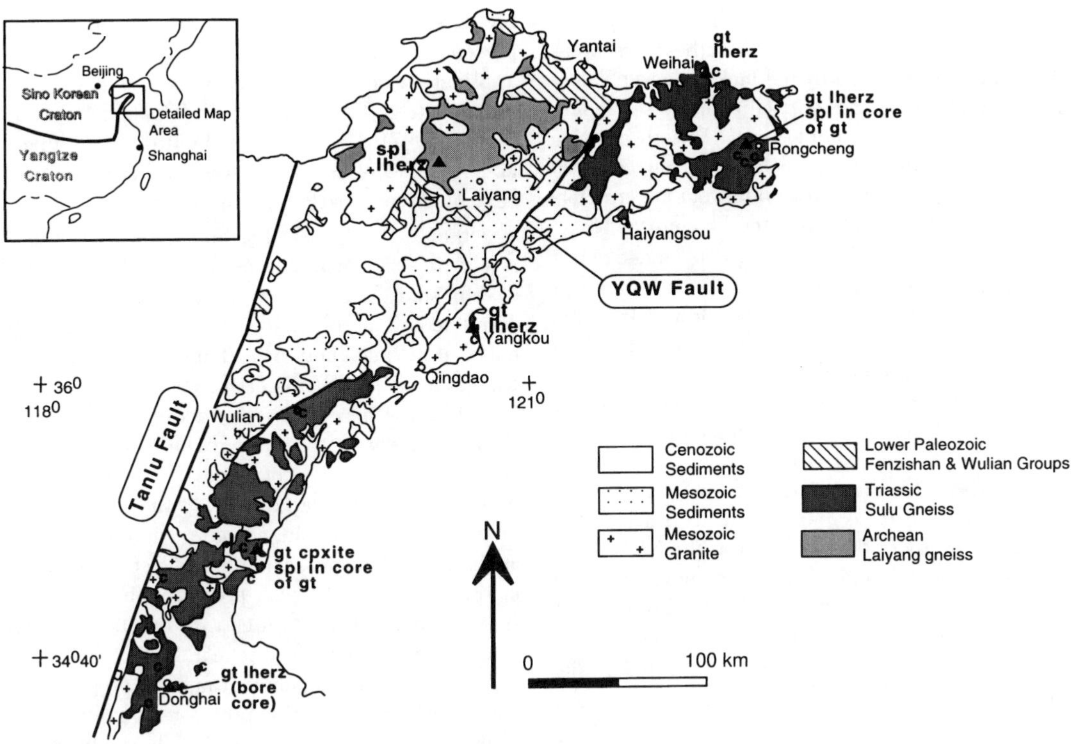

FIG. 1. Geological map of the Shangdong Peninsula, showing the known ultramafic occurrences (▲) and localities where coesite has been reported (c). The main place names mentioned in the text also are shown.

although rare spinel lherzolite exists (Ishizaka et al., 1994). Rb-Sr whole-rock and chemical Th-U total Pb dating of high-grade gneisses, including hornblende-granulite and migmatitic gneisses, yield mid-Proterozoic (1500 to 1800 Ma) ages (Enami, Suzuki et al., 1993; Ishizaka et al., 1994). Similar ages were obtained by Sm-Nd dating of granulite in this area (Li et al., 1997), and these are compatible with the age of the North China craton (e.g., Wang 1986).

In contrast, the southeastern area is composed predominantly of acidic tonalitic and granitic gneiss recrystallized in the amphibolite facies that locally contains lenses of eclogite-facies rocks. Many studies have shown that these eclogite lenses commonly preserve evidence for UHP metamorphism (e.g., Enami and Zang, 1990; Hirajima et al., 1990; Yang, 1991; Wang et al., 1993). Garnet peridotite also is found throughout the area (Fig. 1), as bodies with dimensions of up to several kilometers (Yang, 1991; Yang et al., 1993; Zhang et al., 1994; Hiramatsu et al., 1995). Both the eclogite-facies metamorphism and the garnet peridotite bodies are only found in the southeastern area.

A variety of methods have been used to determine the ages of formation and metamorphism of rocks in the southeast area. U-Pb dating of zircon suggests protolith ages for the eclogite and acidic gneiss of ~700 to 800 Ma and an age of UHP metamorphism of ~218 Ma (Ames et al., 1996). A protolith age of ~700 Ma also was suggested by Ishizaka et al. (1994), on the basis of whole-rock Rb-Sr dating of acidic gneiss. Jahn (1998) gave the age of UHP metamorphism as ~220 Ma, based on Sm-Nd dating of eclogite. He also proposed a period of metasomatism at ~1700 Ma.

The isotopic data from the southeast area are clearly distinct from those associated with the North China craton. Ames et al. (1996) suggested that they are best interpreted as representing the formation of a region of crust in a rift zone at the end of the Proterozoic on the northern margin of the Yangtze craton. This occurred after the main consolidation of the Yangtze craton at ~850 Ma (e.g., Wang 1986).

The difference in geological development between the northwestern and southeastern areas also can be recognized by the lack of Lower Paleozoic to Proterozoic (Sinian) sedimentary sequences southeast of the YQW fault (Fig 1). This observation is compatible with the age data, which suggest that the entire southeastern area underwent major deformation and metamorphism in the Phanerozoic.

Unfortunately, without the aid of advanced petrological and isotopic methods, it is not easy to recognize the differences between the northwestern and southeastern parts of the Sulu region. The most commonly used division of this region is based on geographical location, with an E-W boundary at the latitude of Qingdao. Gneisses to the north and south of this line are referred to as Jiaodong and Jiaonan, respectively. This division has no known geological significance. In this contribution, we refer to the Mesozoic gneiss to the southeast of the YQW fault as the Sulu gneiss and following the suggestions of Liu et al. (1998), we refer to the Precambrian gneiss to the northwest of the YQW fault as the Laiyang gneiss. The UHP Sulu terrane is restricted to the area southeast of the YQW fault.

Recognition of Ultrahigh-Pressure Metamorphism

UHP metamorphism can be defined as recrystallization in the coesite stability field of crustal rocks originally formed at, or close to, the Earth's surface. UHP metamorphic rocks, therefore, have a geological history that is distinct from the rocks derived from the mantle, such as garnet peridotite. The mantle-derived rocks of the Sulu terrane will be discussed in a separate section.

The key mineral for recognizing UHP conditions, coesite, was first reported in China from the Dabie terrane (Okay et al., 1989; Wang et al., 1989). UHP metamorphism of the Sulu terrane was later confirmed by the discovery of coesite (Hirajima et al., 1990) and quartz pseudomorphs after coesite in garnet (Yang and Smith, 1989; Enami and Zang, 1990). Coesite subsequently has been found in minerals other than garnet, such as jadeite, omphacite, dolomite, epidote, and kyanite (e.g., Schertl and Okay, 1994; Wang et al., 1995; Su et al., 1996). Other minerals suggesting very high peak pressures of metamorphism have also been reported, such as Mg-staurolite (Enami and Zang, 1988), nyböite (Hirajima et al., 1992), and Al-titanite (Zhang, Hirajama et al., 1995; Carswell et al., 1996). The ultrahigh-pressure event can be recognized not only by the presence of appropriate minerals, but also by the application of geothermobarometry to coexisting minerals (e.g., Carswell et al., 1997; Nakamura and Banno, 1997). There can now be little doubt that a large number of eclogite bodies in the Sulu gneiss have undergone UHP metamorphism.

Other rock types in the Sulu gneiss only very rarely preserve evidence for UHP metamorphism. Careful study, however, has revealed that both metagranitoid and marble in the region also have experienced UHP metamorphism (Hirajima et al., 1993; Kato et al., 1997; Wallis et al., 1997; Ogasawara et al., 1998). Whereas the true extent of the UHP metamorphism in the Sulu region remains to be determined, the weight of evidence very much supports the idea that large parts of the Sulu gneiss, including metagranitoid, eclogite, and other associated minor lithologies, experienced a common metamorphic history with peak pressures in the coesite stability field. This implies that crustal material became deeply buried in the mantle. One important constraint on any models that may be proposed for the burial and exhumation processes is given by the peak metamorphic conditions, which are discussed below.

Peak Metamorphic Conditions

High-pressure polymorphs of minerals are good indicators of minimum pressures for P_{max}. At temperatures of ~700°C, the presence of coesite implies depths in excess of 90 km. At similar temperatures and depths in excess of 120 km, diamond becomes stable. The only report of diamond from the Sulu region is from the Donghai area by Xu et al. (1992). This discovery, however, has yet to be verified independently. Nakamura and Banno (1997) used geothermobarometry to analyze kyanite eclogite from the Rongcheng area. The results show considerable scatter, probably indicating post-peak re-equilibration. The maximum values are compatible with local recrystallization in the diamond stability field, but in general lie in the graphite stability field. In addition, Enami and Nagasaki (1999) applied phengite geobarometry to the Junan kyanite eclogite, following Carswell et al. (1997), and obtained P-T values that fall within the coesite-graphite field. With the presently available data, therefore, we suggest that 120 km is a reasonable value for the maximum depth of burial for the Sulu UHP metamorphic rocks. Such a depth is compatible with the local development of diamond in suitable rock types.

The temperature of eclogite usually is measured by the Fe-Mg partitioning between Cpx and Gt using the parameter $K_D = (Fe^{2+}/Mg)Grt/(Fe^{2+}/Mg)Cpx$. This is one of the experimentally and theoretically most extensively studied geothermometers and, in general, can be used with a high degree of confidence that the results are meaningful. This thermometer is, however, sensitive to the estimation of the Fe^{3+}/Fe^{2+} ratio. At the pressure of UHP metamorphism, the amount of $Al_2Mg_{-1}Si_{-1}$ in clinopyroxene is usually considered to be small, and it is then safe to assume that Fe^{3+} = Na-Al. However, if the Al_2MgSi component is not negligible, the estimation of Fe^{3+} by pyroxene stoichiometry becomes quite sensitive to the analytical accuracy of all the relevant elements in the pyroxene. There is no single best method for estimating Fe^{3+} in pyroxene from microprobe analyses, and the reliability of the temperature estimates could be improved if there were independent measurements of the Fe^{3+}/Fe^{2+} ratio. With only microprobe analyses, it is possible to be internally consistent in the estimates, but one can never be really certain of the real value.

The difficulty of estimation of peak temperature not only reflects the aforementioned problems, but also the sample heterogeneity itself. Concentric growth zoning and retrograde zoning as a result of resorption or exchange equilibrium is well known. Lasaga et al. (1977) made the important point that merely the existence of a flat central part to a compositional profile is no guarantee that any particular stage of equilibrium crystallization is preserved. Mass-balance considerations show that symplectite formation is an open-system reaction (e.g., Obata, 1980). Furthermore, Nakamura and Banno (1997) showed that symplectite formation affects the core of minerals even when there is no clear microstructural evidence for this. The most important observation is that the core of partially symplectized omphacite is depleted in Na_2O. It follows that even the compositions of mineral cores may not represent equilibrium at the time of their formation. Banno et al. (in review) have shown that in the granulitized coesite eclogite of Weihai, omphacite inclusions in garnet commonly are zoned, with a rimward decrease of $Al_2Mg_{-1}Si_{-1}$ and $NaAlMg_{-1}Ca_{-1}$ components—i.e., toward the lower-pressure pyroxene of the eclogite system.

A rather different problem is raised by Hirajima (1996), who suggested that KD increases with increasing jadeite component. However, the problem with estimating Fe^{3+}/Fe^{2+}, as mentioned above, is also relevant in this case. Reliable formation temperatures are difficult to obtain in general, and particularly in UHP rocks where the applicability of geothermobarometery commonly used at lower pressures has yet to be fully verified. It is therefore desirable to have well-constrained estimates from independent methods in the same region. We have emphasized some of the difficulties in estimating metamorphic temperatures. There is, however, general agreement using a variety of geothermometers that peak temperatures were ~800°C in the north of the terrane around Weihai, and somewhat lower in the south around Donghai. A compilation of maximum T estimates in the Sulu terrane is presented in Zhang and Liou (1998).

Pre-UHP Mineralogy: Metagranitoid and Metagabbro

The dominant lithology of the Sulu gneiss is a granitic to tonalitic garnet-bearing gneiss that has been pervasively deformed and recrystallized, although there also are local occurrences of dolomitic marble and paraschist. It is therefore difficult to say much definite about the protolith of the UHP region. One important result derives from geochemical and isotopic studies of Jahn (1998), who show the basic rocks now found as eclogite have continental and not oceanic affinities. Further evidence about the protoliths of the UHP region comes from the Yangkou area, where an outcrop of several hundred square meters contains well-developed coesite eclogite associated with undeformed basic and acidic rocks (Wallis et al., 1997). Geochemistry and crosscutting relationships between acidic and basic rocks suggest that the unit originated as a single igneous complex. The igneous complex is strongly but heterogeneously deformed, and the original textures of both gabbro and granodiorite bodies are locally preserved (Hirajima et al., 1993; Wallis et al., 1997; Zhang and Liou, 1997). The metagranodiorite contains potassium feldspar, plagioclase, quartz, biotite, clinopyroxene, and orthopyroxene. This may be an original igneous assemblage, but the clinopyroxene composition and presence of orthopyroxene also are compatible with an origin as a charnockite.

The reason for the unusually good preservation of the original igneous textures is probably low water activity, which can act to increase the shear strength of rocks and impede metamorphic reactions. Low water activity is shown by the following observations. (1) There was very limited migration of elements, such that composition of garnet aggregates developed between plagioclase and augite, now rimmed by omphacite, is grossular-rich (X_{ca} = 0.57) at the contact with augite, and is grossular-poor at the contact with altered biotite (X_{Ca} = 0.11). (2) Coesite occurs not only as inclusions but also as a matrix mineral (Liou and Zhang, 1996; Ye et al., 1996; Wallis et al., 1997; Zhang and Liou, 1997). The coesite-quartz transition is very rapid in the presence of even a small amount of water (e.g., Hacker and Peacock, 1995). The presence of matrix coesite, therefore, is good evidence supporting very low water activity.

The Yangkou unit also preserves a variety of deformation features, most of which developed before the regional amphibolite-facies metamorphism (Wallis et al., 1997). The presence of axial planar cleavage defined by coesite, kyanite, and phengite in eclogite suggest that some of the folding took place under UHP conditions.

Sulu gneiss that preserves original granitic textures is very rare. However, another possible location is Chenlianzipo (location 35°04'13" N, 118°53'06" E), where strongly deformed xenolith-bearing granite is found. Elongate K-feldspar and plagioclase define a strong linear fabric with only a weak schistosity. Under the microscope, this rock contains a garnet corona texture, suggesting that this grantioid has been metamorphosed, and is not, therefore, one of the post-orogenic Mesozoic granitoid plutons.

Another indication of the protoliths to the Sulu gneiss is seen in the eastern part of the Sulu terrane around Haiyangsuo (Ye and Hirajima, 1996). In this region, three different metamorphic stages can be recognized. A protolith of garnet + brown hornblende granulite is cut by narrow shear zones along which transitional eclogite mineral assemblages are developed (max. Xjd = 0.24 + albite + hornblende + kyanite + zoisite). Garnet develops along the boundary between plagioclase and mafic phases, and omphacite/diopside develops around the granulite clinopyroxene and is lower in $Al_2Mg_{-1}Si_{-1}$ and slightly higher in Xjd than the host pyroxene. The eclogitic rocks developed in the sheared part of the granulite block. This eclogitic rock underwent metamorphism under conditions of ~10 to 18 kbar and 560 to 680°C. They are therefore distinct from the UHP eclogite and may be correlated with the cold eclogite belt of the Dabie region. The eclogitic assemblages are then overprinted by amphibolite-facies metamorphism and associated deformation.

In summary, scant direct evidence exists regarding the original protoliths of the Sulu region prior to UHP metamorphism. Locally preserved original textures and isotopic evidence suggests that the Sulu gneiss originally consisted of continental basement containing granite, granodiorite, gabbro, and numerous basic intrusive bodies in the form of dikes or sills. At least some of the basic rocks and perhaps some of the granitic material underwent granulite-facies metamorphism before the later Triassic high-pressure metamorphism.

Prograde P-T Path

Until recently there has been very little control on the prograde path followed by the UHP rocks. However, some critical minerals were reported to occur as prograde inclusions in garnet and omphacite, as follows: (1) Kfs (Yang and Smith, 1989; Enami and Zang, 1990; Enami, Zang et al., 1993; Zhang et al., 1994; Yang et al., 1998); (2) Pl (Enami, Zang et al., 1993); (3) Bar + Pg + Mag + Qtz + Zo + Rt (Zhang, Hirajama et al., 1995); (4) Hbl + Chl + Zo + Grt; (5) Prg + St + Pg (Enami and Nagasaki, 1999); (6) Mrg + Pg, where Kfs is potassium feldspar, Pl is plagioclase, Bar is barroisite, Pg is paragonite, Mag is magnetite, Qtz is quartz, Zo is zoisite, Rt is rutile, Hbl is hornblende, Chl is chlorite, Gt is Fe-rich garnet, Prg is pargasite, St is staurolite, and Mrg is margarite.

These assemblages suggest that the early part of the metamorphic history passed through the epidote-amphibolite to amphibolite facies. It is unnecessary to assume that all the eclogites followed the same P-T path, and the presence of pargasite may represent relatively low pressure, high-temperature conditions within the epidote amphibolite–amphibolite facies when compared to barroisite-bearing samples.

Decompression P-T Path

The retrograde metamorphism of the Sulu terrane can be constrained by two well-defined stages. The peak conditions are at ~800°C and 30 kbar, and these are overprinted by epidote-amphibolite facies metamorphism that produced the dominant assemblages of the Sulu gneiss. The P-T conditions of the epidote-amphibolite stage are ~500°C and <10 kbar

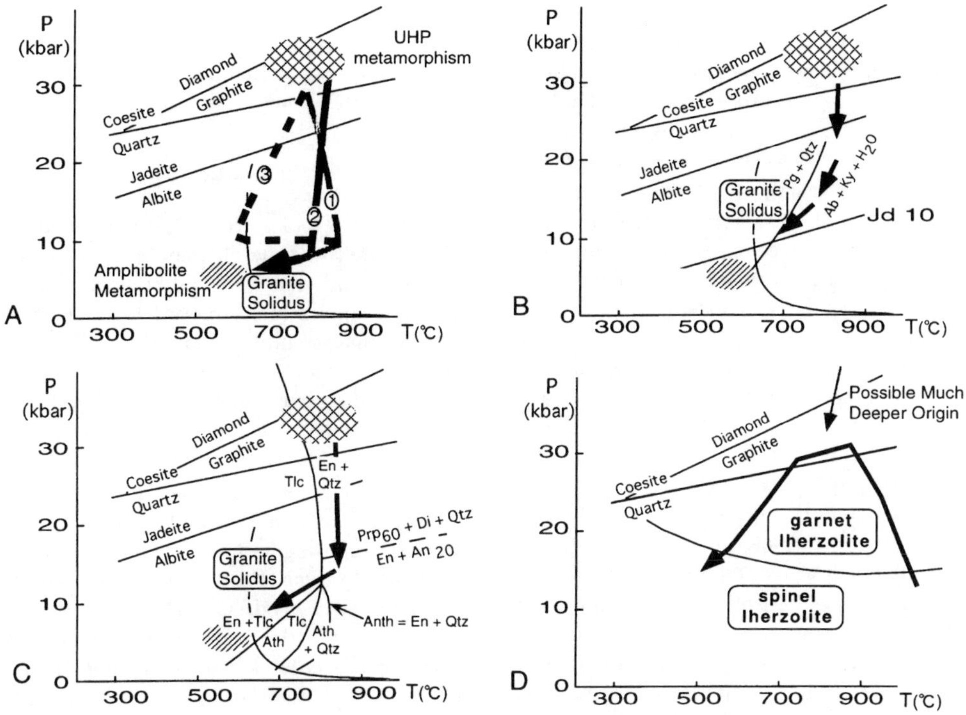

FIG. 2. A. P-T paths for the Weihai eclogite suggested by (1) Wang et al. (1993); (2) Zhang, Liou, et al. (1995); and (3) Harley and Carswell (1995). B. P-T path for the Rongcheng area (Enami, Zang et al., 1993). C. P-T path for the Rongcheng area (Nakamura and Hirajima, in review). D. P-T path for ultramafic rock (Hiramatsu and Hirajima, 1995).

(e.g., Wang et al., 1993). However, it should be noted that the replacement of eclogite minerals by hornblende symplectite does not necessarily indicate re-equilibration in the amphibolite facies. In the absence of quartz, amphibole + plagioclase can be stable up to 900°C under water-excess conditions (Ernst and Liu, 1998).

A comparison of the P-T conditions of the two dominant stages of re-equilibration reveals both cooling and decompression, but does not give any constraint on the path followed between the two stages. Determining the shape of the decompression path in P-T space has proved difficult, but now through a variety of studies it is becoming clear that the early decompression stage of the Sulu UHP terrane is characterized by approximately isothermal decompression. The reasons for this conclusion are outlined below.

One of the best-documented decompression paths is that of the Weihai coesite eclogite (Fig. 2A). In this area, coesite-bearing eclogite is replaced by granulite, hornblende granulite, and amphibolite. A number of researchers have estimated the peak temperature during the eclogite stage by using Fe-Mg partitioning between garnet and omphacite (Wang et al., 1993; Zhang, Liou et al., 1995; Banno et al., in prep). These workers also estimated the temperature of the granulite stage using two-pyroxene geothermometers (e.g., Wood and Banno, 1973; Wells, 1977; Carlson and Lindsley, 1988). The temperature of the formation of the amphibole symplectite also was estimated using amphibole geothermometry (Graham and Powell, 1984; Holland and Blundy, 1994). In the same locality, Zhang, Liou et al. (1995) recognized the presence of neoblastic garnet, which they related to the granulite stage that overprints the eclogite-facies assemblages. These workers also used garnet-clinopyroxene geothermometry to estimate the associated temperature. The recognition of two phases of garnet growth makes it possible to compare temperatures at different stages during the metamorphism of one rock using a single and, therefore, internally consistent geothermometer. The P-T paths for the Weihai area determined by Wang et al.

(1993) and Zhang, Liou et al. (1995) are shown in Figure 2. In contrast to Zhang, Liou et al. (1995), Banno et al. (in review) consider that the neoblasts represent the low-pressure eclogite assemblage. This is evidence against the suggestion of Harley and Carswell (1995) that there was a phase of cooling prior to granulite-facies metamorphism.

Enami, Zang et al. (1993) presented further data supporting the idea of decompression with relatively minor cooling throughout most of the Sulu region. The critical observation of this study is the presence of albite around kyanite, implying that decompression took place on the high-temperature side, and therefore had a steeper P-T path than the slope of the reaction curve paragonite + quartz = albite + kyanite + H_2O (Fig 2B). One potential difficulty with this line of argumentation is that the location of the reaction curve in P-T space is dependent on the activity of water. A low water activity would allow a decompression path with a shallower gradient.

A convincing case for isothermal decompression in the Rongcheng area is made by Nakamura and Hirajima (in review). These workers recognized local overprinting of the eclogite-facies assemblages by granulite-facies metamorphism, with the development of orthopyroxene. These researchers used three independent geothermometers (garnet-clinopyroxene for the eclogite facies, orthopyroxene-clinopyroxene and garnet-biotite for the granulite facies) to show that decompression of at least 10 kbar (roughly 30 km) is associated with the same temperature within the errors of the methods (Fig. 2C). Two of the applied geothermometers are independent of any assumption about the activity of water.

Kato et al. (1997) also proposed isothermal decompression based on overprinting reactions in metadolomite. The peak temperature estimate of this work remains rather uncertain; however, the general shape of the P-T path seems well established and agrees with the conclusions of the studies above.

In summary, details of P-T paths, in particular the estimated peak temperature, are dependent on the approach taken in the application of geothermobarometers. However, where good evidence has been gathered, it all indicates that a large part of the Sulu UHP terrane underwent roughly isothermal decompression from depths of ~100 km to mid-crustal levels of ~20 to 30 km. One of the important conclusions from this work is that during exhumation, the Sulu gneiss crossed the granite solidus and partial melting would have been very likely (Fig. 2). The role of partial melting in the development of UHP terranes is one that may be an area for fruitful research in the future.

Ultramafic Rocks

Ultramafic rocks are widespread in the Sulu UHP terrane. In general, they have small dimensions but locally reach several kilometers in size—e.g., the Hujialin area (Yang, 1991; Zhang et al., 1994; Hiramatsu and Hirajima, 1995). The P-T path followed by these ultramafic rocks is distinct from eclogite of the same region and can, therefore, help furnish further constraints on possible processes acting at depth in continental collisional (Fig. 2D).

Garnet lherzolite and garnet clinopyroxenite are relatively common in the Sulu terrane and represent an origin from depths in excess of 100 km. Hiramatsu and Hirajima (1995) suggested that the garnet clinopyroxenite originated at relatively shallow depths in the mantle, based on the observation that garnet locally contains both spinel and Al-rich cpx as inclusions. Hiramatsu et al. (1995) also considered that the Cr_2O_3-rich portion of garnet grains in peridotite represents the remnant of a precursor spinel formed under upper-mantle conditions. Pressures of ultramafic rocks are generally estimated using the garnet-orthopyroxene geobarometer. The Al_2O_3 content of orthopyroxene is as low as 0.11 wt%. Yang et al. (1993) used the Al zoning of orthopyroxene to suggest that the equilibrium pressure of a garnet harzburgite from Donghai was ~60 kbar. Hiramatsu et al. (1995) also showed that conventional geobarometry yields equilibrium pressures for garnet peridotite of Chijiardian ranging from 45 to 68 kbar. The relatively large uncertainty is partly a result of the calibration of the barometer, but also to the presence of minor elements such as Cr, Ca, and Na in amounts comparable with Al. The low Al in orthopyroxene makes it difficult to estimate the activity of the $MgAl_2SiO_6$ component.

Unconfirmed reports of diamond in heavy mineral separates from garnet lherzolite at Zhimafang would be compatible with an origin of these rocks at depths exceeding 120 km. Recently, there also have been reports of the discovery of P2/c pyroxene (Furukawa and Ogasawara, 1998; Zhang et al., 1998), which may have transformed from high-pressure clinohypersthene, which is stable at pressures higher than 60 kbar. However, despite suggestions of origins at even greater depths, there is still no evi-

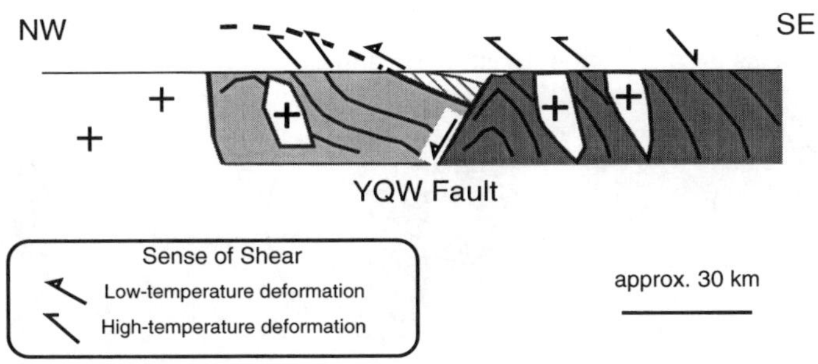

FIG. 3. Generalized NW-SE cross section through the northern Shangdong Peninsula. The true shape of the granite batholiths at depth are unknown. The YQW fault is a major normal fault forming the boundary between the Laiyang and Sulu gneisses. The patterns are the same as for Figure 1.

dence P_{max} was as great as 100 kbar. If this were the case, it would be expected that the Al/Si ratio of the associated garnet should significantly deviate from 2/3—i.e., from the stoichiometric formula of $R_3Al_2Si_3O_{12}$ (Akaogi and Akimoto, 1977; Irifune, 1987).

It is significant that none of the ultramafic bodies in the Sulu region show ophiolitic affinities. There is, therefore, no evidence for the preservation of oceanic crust in the Sulu terrane.

Regional Structure and Kinematics

In order to understand the present distribution of rock types including the UHP units, it is essential to investigate the associated structures. Where these structures can be related to the metamorphic history, they can also be used to help unravel the early (pre-Jurassic) tectonic evolution of the region, which is the prime interest of the present study. The pre-Jurassic structures have been affected to some extent by later deformation. The clearest example is the presence of a series of faults trending NNW-SSW. These faults displace the Mesozoic granitic batholiths, showing that the associated movements are relatively young (BGMR, 1991). Map and field relationships show that the granitic bodies cross-cut the regional foliation and cause only minor warping. This is an important observation because it implies that the granitoids were intruded passively, and that their emplacement has not affected the original structural grain of the Sulu gneiss. In the southern part of the Sulu gneiss, the Mesozoic granite bodies have locally experienced high-strain deformation, and the distinction between gneisses on the one hand and deformed Mesozoic granite on the other becomes very difficult.

Kinematic studies in metamorphic belts are useful in tectonic reconstructions because they can help define movement directions and flow patterns. However, in high-temperature metamorphic belts such as the Sulu and Laiyang gneisses of the present study, kinematic data can be difficult to obtain because of the speed at which recrystallization takes place. If ambient temperatures are still high when deformation stops, both the microstructures and mesostructures formed during deformation will be rapidly altered, if not obliterated. In the present study, the best-preserved sense-of-shear indicators were asymmetric K-feldspar porphyroclasts and deformed vein sets.

We have studied the structures of units on both sides of the YQW fault. Below we first summarize the main structural features of the Precambrian Laiyang gneiss and the Sulu UHP terrane and then look at some particular tectonic problems that can be addressed using these data.

Mesoscopic Structures

Laiyang gneiss

The Laiyang gneiss northwest of the YQW fault consists of high-grade gneisses of variable composition, overlain by a series of low-grade metamorphic sediments, the Fezishan Group (Fig. 3). The regional gneissosity is locally associated with a well-developed linear fabric. On average, the gneissosity trends E-W, although there is considerable scatter (Fig. 4A). At least two distinct phases of penetrative deformation can be defined by the relationship to high-temperature metamorphism. Locally, post-metamorphic strike-slip faults show sinistral displacement.

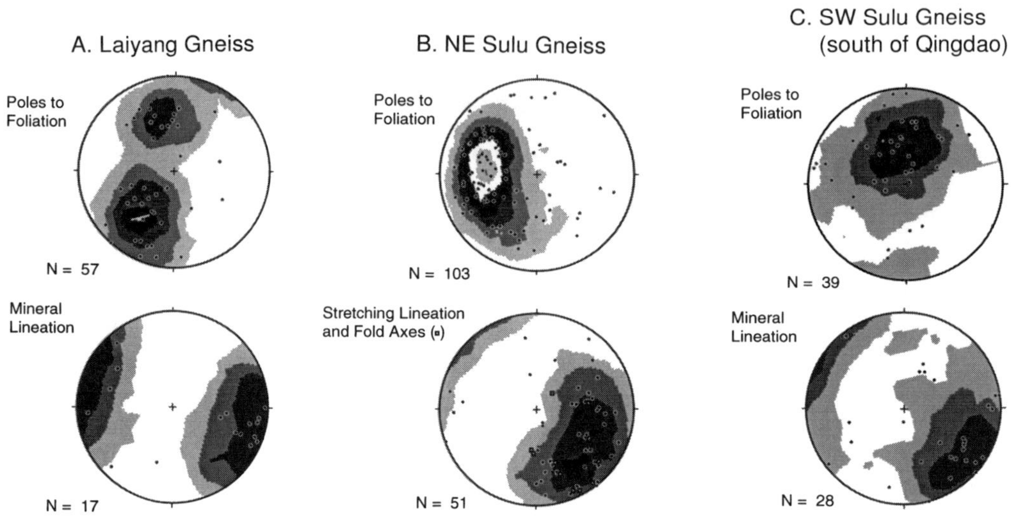

FIG. 4. Mesoscopic orientation data for the Laiyang gneiss (A), and northern (B) and southern Sulu gneiss (C). All plots are equal area, with Kamb contours at the 2σ level drawn with the use of R. Allmendinger's STEREONET program. Foliation is given as poles. True north was taken as 7° east of magnetic north.

The relationship between the dominant deformation and metamorphism is indicated by several features. Coarse-grained patches of quartz + feldspar + mica leucosome are locally developed in the necks of boudins of relatively competent basic material surrounded by acidic gneiss (Fig. 5A). The patches of leucosome can be interpreted as former melt, and these are locally wrapped around by the foliation in the acidic gneiss, showing that the foliation was forming under high-temperature conditions. Similar coarse-grained minerals also are locally found in synkinematic veins. There is a transition from undeformed veins to tightly folded equivalents, with the fold axial planes approximately parallel to the main foliation. These structural relationships also suggest the development of the foliation under temperatures sufficiently high to cause partial melting. A further general indication of high-temperature deformation is the presence of ductiley deformed feldspar.

The high-temperature deformation is locally associated with a strong mineral lineation trending roughly WNW-ESE (Fig. 4A), roughly perpendicular to the trend of the metamorphic belt. The associated sense of shear can be determined from the geometry of 1 to 10 cm–scale shear bands (Fig. 5B) and the distribution of deformed veins. These kinematic indicators provide a consistent top-to-the-west sense of shear.

In addition to the above high-temperature structures, there also is evidence for ductile deformation at much lower temperatures (greenschist facies). The low temperature is recognized by the brittle behavior of minerals such as feldspar and lack of evidence for post-deformational recovery. This deformation is most clearly developed along the tectonic contact between the Fenzishan metasedimentary units and underlying gneiss (Fig. 3). This contact is mapped as an unconformity, but is associated with the development of blastomylonites with a strong E-W stretching lineation. Kinematic indicators such as shear bands and asymmetric pressure shadows give a top-to-the-west sense of shear. The similarity of the kinematics of this zone with that of the high-temperature deformation in the underlying gneisses suggests that deformation may be related to the same tectonic event, but at different structural levels.

Elsewhere in this region, there is only minor evidence for ductile, low-temperature deformation. In contrast to the above examples, the stretching lineation is dominantly oriented N-S. Both top-to-the-north and top-to-the-south senses of shear can be recognized.

Sulu gneiss

The Sulu gneiss region has a well-developed foliation, and locally also mineral lineation. The pres-

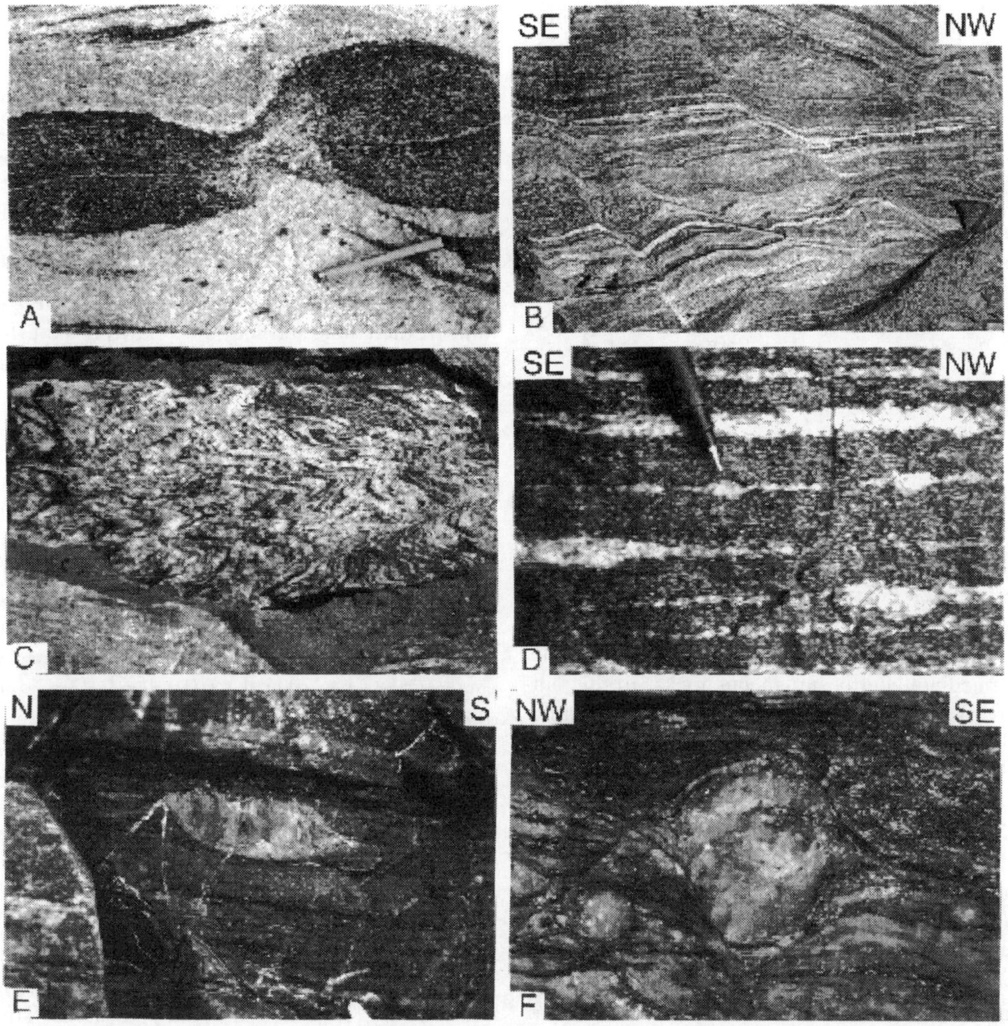

FIG. 5. A. Boudinaged basic layers in acidic gneiss. Coarse-grained leucosome indicative of partial melting is formed in the boudin necks. B. Boundin necks localized along shear bands; these structures can be used as a sense-of-shear indicator. C. Ductiley deformed feldspar, indicating relatively high-temperature deformation of the Sulu gneiss. D. Melt formed along the axial planes to folds in the Sulu gneiss from the Weihai region. E. Potassium feldspar porphyroclast with asymmetric recrystallized tails that can be used to determine the sense of shear in the Sulu gneiss (σ-clast). F. Same as in E, but the clast (size roughly 5 cm) has rotated to form a δ-clast.

ence of ductilely deformed feldspar indicates deformation at relatively high temperatures (Fig. 5C). In the northern part of the region around Weihai, features suggest that temperatures during deformation were high enough to cause partial melting. This includes the presence of melt formed along the axial planes of folds (Fig. 5D) and minor dikes penetrating former eclogite bodies with the foliation wrapping around them. In the Haiyangsou region, evidence exists for syndeformational intrusion of acidic pegmatites.

Eclogite pods within the dominant siliceous gneiss are commonly foliated, and in many cases folded. The foliation of the eclogite pods is, however, not always concordant with the surrounding gneiss, and is locally clearly crosscut by shear zones formed in the amphibolite facies. The dominant foliation of the Sulu gneiss, therefore, formed after the UHP conditions, and is related to syn- to post-amphibolite–facies metamorphism. This implies that most of the structural data collected reflect events at relatively shallow levels in the crust rather than the early part of the exhumation history.

The main foliation of the Sulu gneiss in general strikes NE-SW and dips to the southeast, although this changes close to the western margin to a northwest dip (Fig. 3). The associated stretching lineation locally is well developed and is oriented on average NW-SE, although there is considerable scatter (Fig. 4B). To the west, a top-to-the-NW sense of shear is dominant, but this changes to a region with both top-to-the-SE and -NE senses of shear in the east (Figs. 5E and 5F). Ductile deformation at relatively low temperatures is locally present in the east, and shows similar kinematics to the high-temperature deformation.

The Yangkou unit to the northeast of Qingdao preserves a set of particularly interesting structures. The Yangkou unit is composed of a series of relatively small bodies of chemically related acidic, intermediate, and basic igneous rocks. In this outcrop, recrystallization associated with the amphibolite-facies metamorphism clearly overgrows an earlier deformational fabric. Close study of these microstructures shows that this is a composite fabric developed during UHP and later conditions. Studies of this outcrop can then tell us something about the ways that various rock types behave mechanically during deformation in the root zones of mountain belts. A structural analysis of the fabric, including quantitative estimates of strain and vorticity (Makino, 1999), shows that the deformation resulted from a complex type of non-plane strain, non-coaxial flow that formed under relatively high temperatures. Perhaps the most important point of these studies is that deformation in the Yangkou unit during exhumation was highly complex, showing multiple overprinting relationships and suggesting a highly mobile behavior of the rocks, indicating a low effective viscosity. This is all the more surprising because petrological features, such as the presence of coesite in the matrix, clearly indicate that the water activity in the outcrop was very low. Low water activity increases the shear strength of minerals and the acidic gneisses of the Yangkou unit are, therefore, likely to have been stronger than the rest of the Sulu gneiss. The implication is, therefore, that the whole of the Sulu gneiss behaved as a highly mobile fluid during exhumation.

Tectonic Issues

In this section, we briefly mention some ideas that have been proposed regarding the tectonic evolution of the Sulu terrane, and show how the data discussed above can be used to test some of the predictions of models or to clarify certain problems.

What is the form of the UHP units at a large scale?

Many researchers have suggested that the UHP unit in the Dabie terrane is a relatively thin, flat-lying sheet that was extruded along a former subduction zone (e.g., Maruyama et al., 1994; Xue et al., 1996). In the present study, we find local evidence for units that may represent higher structural units overlying the UHP domain, but we do not recognize tectonic windows or other structural features that would indicate the presence of a regional UHP nappe. The foliation and mineral lineations of the earliest recognizable deformation commonly are steeply inclined, suggesting that vertical movements were more dominant than were horizontal ones. We suggest, therefore, that the UHP domain represents the core of a former collisional belt rather than a nappe. Proposed drilling in the Donghai region may help resolve this conflict of views. We predict that beneath the UHP rocks there will be more UHP rocks, and not a low-P footwall to a UHP nappe.

Did the Tanlu fault play a significant role in the tectonic evolution of the Dabie-Sulu terranes?

It has been suggested that the Tanlu fault originated in the Triassic and acted as a major lateral ramp displacing the orogenic front several hundred kilometers (Okay and Sengör, 1992). The alternative view is that the Tanlu fault is a relatively late-stage structure unrelated to the metamorphic history, and simply cross-cuts a pre-existing set of features. These possibilities can be tested with kinematic data. In the first case, the synmetamorphic movement directions should be constrained by the orientation of the fault, and should be roughly parallel to its trend. In the second case, any relationship between the two would be coincidental. The mineral lineation recorded in the Sulu terrane trends NW-SE and matches closely with that in the Dabie terrane (e.g., Hacker et al., 1995). This orientation shows that the Tanlu fault cross-cuts the tectonic fabrics of the Dabie-Sulu terranes and therefore is a post-orogenic feature.

Did the Triassic orogen affect both the Laiyang and Sulu gneiss regions, or was the Laiyang gneiss almost completely unaffected?

Despite the evidence for very different geological histories northwest and southeast of the YQW fault, the two regions do not show significant differences in mesoscopic structural data. The kinematic data also are very similar, although in the west there is a consistent top-to-the-west sense of shear, whereas in the east, both senses are recognized. These similarities may be purely coincidence, but we suggest that it is possible that much of the deformation seen in the Laiyang gneiss is related to the Triassic Sulu orogeny. More detailed radiometric dating should make it possible to resolve this problem.

How does quartz-rich gneiss behave mechanically at depths of 100 km?

It is difficult to obtain reliable information about processes occurring at depth in root zones of mountain belts, largely because deformation and metamorphism developed in these regions are easily overprinted. However, in the Yangkou outcrop, some of these early structures are clearly preserved. Structural analyses combined with petrological studies suggest that in this region, acidic gneiss was highly mobile and flowed around the more rigid eclogitic blocks. These observations are all the more significant because of the very low associated water activity, and the implied relatively strong behavior of the unit. This implies, therefore, that appealing to tectonic overpressure to create the usually high metamorphic pressures of UHP metamorphism or suggesting that eclogite blocks can somehow be squeezed out from between two rigid plates is unworkable in this region.

Conclusions

Petrological and isotopic evidence suggests that the Sulu UHP terrane originally formed part of the northern fringe of the Yangtze craton. The main rock types were granite, granodiorite, gabbro, and basic dikes, which were at least locally metamorphosed in the granulite facies. At ~220 Ma, this unit of continental rocks was buried to depths up to ~120 km, where they were surrounded by mantle rocks. The prograde P-T path implies a geothermal gradient of ~10°C /km. Structures formed during exhumation suggest highly mobile behavior of acidic rocks, even under conditions of very low water activity, implying that models requiring that eclogite be squeezed out from between rigid slabs are inapplicable in this region. Petrological studies show that the Sulu terrane underwent regional isothermal decompression during the initial stages of exhumation. These P-T paths have yet to be properly accounted for, but imply relatively rapid exhumation, and suggest that the role of melting during exhumation may be more important than previously realized. The later stages of exhumation are associated with NW-SE–directed tectonic transport, and the formation of at least one major normal detachment along the YQW fault. The Tanlu fault is a relatively late-stage feature, unrelated to the early phase of orogenesis in the Sulu terrane.

Acknowledgments

Takao Hirajima, Gary Ernst, and Juhn Liou provided useful comments on an earlier draft of this paper.

REFERENCES

Akaogi, M., and Akimoto, S., 1977, Pyroxene-garnet solid-solution equilibria in the system $Mg_4Si_4O_{12}$-$Mg_3Al_1Si_3O_{12}$ and $Fe_4Si_4O_{12}$-$Fe_3Al_1Si_3O_{12}$ at high pressure and temperatures: Phys. Earth Planet. Inter., v. 15, p. 90–106.

Ames, L., Xhou, G., and Xiong, B., 1996, Geochronology and isotopic character of ultrahigh-pressure metamorphism with implications for collision on the Sino-Korean and Yangtze cratons, central China: Tectonics, v. 15, p. 472–489.

Banno, S., Enami, M., Hirajima, T., Ishiwatari, A., and Wang, Q., in review, Decompression P-T path of coesite eclogite to granulite from Weihai, eastern China: Lithos.

BGMR (Bureau of Geology and Mineral Resources of Shangdong Province), 1991, Regional geology of Shangdong Province: Beijing, Geol. Publ. House, Geol. Mem. Ser. 1, no. 26 (in Chinese with English abstract).

Carlson, W. D., and Lindsley, D. H., 1988, Thermochemistry of pyroxenes on the join $Mg_2Si_2O_6$-$CaMgSi_2O_6$: Amer. Mineral., v. 73, p. 242–252.

Carswell, D. A., O'Brien, P. J., Wilson, R. N., and Zhai, M., 1997, Thermobarometry of phengite-bearing eclogites in the Dabie Mountains of central China: Jour. Metamor. Geol., v. 15, p. 239–252.

Carswell, D. A., Wilson, R. N., and Zhai, M., 1996, Ultra-high pressure aluminous titanites in carbonate-bearing eclogites at Shuanghe in Dabieshan, central China: Mineral. Mag., v. 60, p. 461–471.

Coleman, R. G., and Wang, X., 1995, Overview of the geology and tectonics of UHPM, *in* Coleman, R. G., and Wang, X., eds., Ultrahigh-pressure metamorphism: New York, Cambridge University Press, p. 1–32.

Cong, B., and Wang, Q., 1996, A review of research on UHPM rocks in the Dabieshan-Sulu region, *in* Cong, B., ed., Ultrahigh-pressure metamorphic rocks in the Dabieshan-Sulu region of China: Petrology and structural geology: Beijing, Science Press, p. 1–7.

Enami, M., and Nagasaki, A., 1999, Prograde P-T path of kyanite eclogites from Junan in the Su-Lu ultrahigh-pressure province, eastern China: The Island Arc, v. 8, in press.

Enami, M., Suzuki, K., Zhai, M., and Zheng, X., 1993, Th-U-total Pb isochron ages of Jiaodong and Jiaonan metamorphic rocks in the Shandong Peninsula, eastern China: The Island Arc, v. 2, p. 104–113.

Enami, M., and Zang, Q., 1988, Magnesian staurolite in garnet-corundum rocks and eclogite from the Donghai district, Jiangsu province, east China: Amer. Mineral., v. 73, p. 48–56.

_______, 1990, Quartz pseudomorphs after coesite in eclogites from Shandong province, east China: Amer. Mineral., v. 75, p. 381–386.

Enami, M., Zang, Q., and Yin, Y., 1993, High-pressure eclogites in northern Jiangsu–southern Shandong province, eastern China: Jour. Metamor. Geol., v. 11, p. 589–603.

Ernst, W. G., and Liu, J., 1998, Experimental phase-equilibrium study of Al- and Ti-contents of calcic amphibole in MORB—A semiquantitative thermobarometer: Amer. Mineral., v. 83, p. 952–969.

Furukawa, N., and Ogasawara, Y., 1998, New occurrence of clinohypersthene(?) in garnet-bearing ultramafic rock from Donghai in the Su-Lu UHP terrane, eastern China: EOS (Trans. Amer. Geophys. Union), v. 79, p. 990.

Graham, C. M., and Powell, R., 1984, A garnet-hornblende geothermometer: Calibration, testing, and application to the Pelona schist, southern California: Jour. Metamor. Geol., v. 2, p. 13–31.

Hacker, B. R., and Peacock, S. M., 1995, Creation, preservation, and exhumation of UHPM rocks, *in* Coleman, R. G., and Wang, X., eds., Ultrahigh pressure metamorphism: New York, Cambridge University Press, p. 159–181.

Hacker, B. R., Ratschbacher, L., Webb, L., and Dong, S., 1995, What brought them up? Exhumation of the Dabie Shan ultrahigh pressure rocks: Geology, v. 23, p. 743–746.

Harley, S. L., and Carswell, D. A., 1995, Ultradeep crustal metamorphism: A prospective view: Jour. Geophys. Res., v. 100, p. 8367–8380.

Hirajima, T., 1996, Effect of jadeite content on the garnet-clinopyroxene geothermometer for an ultrahigh-pressure eclogite complex: Proc. Japan Acad., v. 72, ser. B., p. 208–213.

Hirajima, T., Ishiwatari, A., Cong, B., Zhang, R., Banno, S., and Nozaka, T., 1990, Coesite from Mengzhong eclogite at Donghai country, northeastern Jiangsu province, China: Mineral. Mag., v. 54, p. 579–583.

Hirajima, T., Wallis, S. R., Zhai, M., and Ye, K., 1993, Eclogitized metagranitoid from the Su-Lu ultrahigh pressure (UHP) province, eastern China: Proc. Japan Acad., v. 68, p. 249–254.

Hirajima, T., Zhang, R., Li, J., and Cong, B., 1992, Petrology of the nyböite-bearing eclogite in the Donghai area, Jiangsu province, eastern China: Mineral. Mag., v. 56, p. 37–46.

Hiramatsu, N., Banno, S., Hirajima, T., and Cong, B., 1995, Ultrahigh pressure garnet lherzolite from Chijiadian, Rongcheng country, in the Su-Lu region of eastern China: The Island Arc, v. 4, p. 324–333.

Hiramatsu, N., and Hirajima, T., 1995, Petrology of the Hujialin garnet clinopyroxenite in the Su-Lu ultrahigh-pressure province, eastern China: The Island Arc, v. 4, p. 310–323.

Holland, T., and Blundy, J., 1994, Non-ideal interactions in calcic amphiboles and their bearing on amphibole-plagioclase thermometry: Contrib. Mineral. Petrol., v. 116, p. 433–447.

Irifune, T., 1987, An experimental investigation of the pyroxene-garnet transformation in a pyrolite composition and its bearing on the constitution of the mantle: Phys. Earth Planet. Inter., v. 45, p. 324–336.

Ishizaka, K., Hirajima, T., and Zheng, X., 1994, Rb-Sr dating for the Jiaodong gneiss of the Su-Lu ultrahigh pressure province, eastern China: The Island Arc, v. 3, p. 232–241.

Jahn, B.-M., 1998, Geochemical and isotopic characteristics of UHP eclogites and ultramafic rocks of the Dabie orogen: Implications for continental subduction and collisional tectonics, *in* Hacker, B., and Liou, J. G., eds., When continents collide: Geodynamics and geochemistry of ultrahigh-pressure rocks: Petrology and structural geology: Dordrecht, Kluwer Acad. Publ., p. 203–239.

Kato, T., Enami, M., and Zhai, M., 1997, The ultrahigh-pressure (UHP) marble and eclogite in the Su-Lu UHP terrane, eastern China: Jour. Metamor. Geol., v. 15, p. 169–182.

Lasaga, A. C., Richardson, S. M., and Holland, H. D., 1977, The mathematics of cation diffusion and exchange between silicate minerals during retrograde metamorphism, *in* Saxena, S. K., and Bhattacharji, S. D., eds., Energetics of geodynamic process: New York, Springer, p. 353–388.

Li, Y., Zhai, M., Liu, W., and Guo, J., 1997, Sm-Nd geochronology of the high-pressure basic granulite in Laixi, eastern Shandong: Scientia Geol. Sinica, v. 32, p. 283–290 (in Chinese with English abstract).

Liou, J. G., and Zhang, R. Y., 1996, Occurrence of intergranular coesite in ultrahigh-P rocks from the Sulu region, eastern China: Implications for lack of fluid during exhumation: Amer. Mineral., v. 81, p. 1217–1221.

Liu, W., Zhai, M., and Li, Y., 1998, Metamorphism of the high-pressure basic granulites in Laixi, eastern Shandong, China: Acta Petrol. Sinica, v. 14, p. 449–459 (in Chinese with English abstract).

Makino, T., 1999, Deformation during exhumation of UHP metamorphic rocks: A case study of the Yangkou Unit, E. China: Unpubl. master's thesis, Kyoto University (in Japanese).

Maruyama, S., Liou, J. G., and Zhang, R., 1994, Tectonic evolution of the ultrahigh-pressure and high-pressure metamorphic belts from central China: Island Arc, v. 3, p. 112–121.

Nakamura, D., and Banno, S., 1997, Thermodynamic modeling of sodic pyroxene solid solution and its application in a garnet-omphacite-kyanite-coesite geothermobarometer for UHP metamorphic rocks: Contrib. Mineral. Petrol., v. 130, p. 93–102.

Nakamura, D., and Hirajima, T., in review, Granulite facies overprinting of ultrahigh pressure metamorphic rocks, northeast Su-Lu region, eastern China: Jour. Petrol.

Obata, M., 1980, The Ronda peridotite; garnet-, spinel-, and plagioclase-lherzolite facies and the P-T trajectories of a high-temperature mantle intrusion: Jour. Petrol., v. 21, p. 533–572.

O'Brien, J., and Carswell, D. A., 1993, Tectonometamorphic evolution of the Bohemian massif: Evidence from high pressure metamorphic rocks: Geol. Rundsch., v. 82, p. 531–555.

Ogasawara, Y., Zhang, R. Y., and Liou, J. G., 1998, Petrogenesis of dolomitic marbles from Rongcheng in the Sulu UHP metamorphic terrane, eastern China: The Island Arc, v. 7, p. 82–97.

Okay, A. I., and Sengör, A. M. C., 1992, Evidence for intracontinental thrust-related exhumation of the ultrahigh-pressure rocks in China: Geology, v. 20, p. 411–414.

Okay, A. I., Xu, S., and Sengör, A. M. C., 1989, Coesite from the Dabie Shan eclogites, central China: Eur. Jour. Mineral., v. 1, p. 595–598.

Rumble, D., 1998, Stable isotope geochemistry of ultrahigh pressure rocks, *in* Hacker, B., and Liou, J. G., eds., When continents collide: Geodynamics and geochemistry of ultrahigh-pressure rocks: Petrology and structural geology: Dordrecht, Kluwer Acad. Publ., p. 241–259.

Schertl, H.-P., and Okay, A. I., 1994, A coesite inclusion in dolomite in Dabie Shan, China: Petrological and rheological significance: Eur. Jour. Mineral., v. 6, p. 995–1000.

Su, W., Xu, S., Jiang, L., and Liu, Y., 1996, Coesite from quartz-jadeitite in the Dabie Mountains, eastern China: Mineral. Mag., v. 60, p. 659–662.

Wallis, S. R., Ishiwatari, A., Hirajima, T., Ye, K., Guo, J., Nakamura, D., Kato, T., Zhai, M., Enami, M., Cong, B., and Banno, S., 1997, Occurrence and field relationships of UHP meta-granitoid and coesite eclogite in the Su-Lu terrane, Eastern China: Jour. Geol. Soc. Lond., v. 154, p. 45–54.

Wang, H., 1986, Geotectonic development of China, *in* Yang, Z., Yuqi, C., and Wang, H., eds., The geology of China: Oxford Monog. Geol. Geophys., no. 3, p. 237–273.

Wang, Q., Ishiwatari, A., Zhao, Z. Y., Hirajima, T., Hiramatsu, N., Enami, M., Zhai, M., Li, J., and Cong, B., 1993, Coesite-bearing granulite retrograded from eclogite in Weihai, eastern China: Eur. Jour. Mineral., v. 5, p. 141–152.

Wang, X., Liou, J. G., and Mao, H. K., 1989, Coesite-bearing eclogites from the Dabie Mountains in central China: Geology, v. 17, p. 1085–1088.

Wang, X., Zhang, R., and Liou, J. G., 1995, UHPM terrane in east central China, *in* Coleman, R. G., and Wang, X., eds., Ultrahigh-pressure metamorphism: New York, Cambridge University Press, p. 356–390.

Wells, P. R. A., 1977, Pyroxene thermometry in simple and complex systems: Contrib. Mineral. Petrol., v. 62, p. 129–139.

Wood, J. B., and Banno, S., 1973, Garnet-orthopyroxene and orthopyroxene-clinopyroxene relationships in simple and complex systems: Contrib. Mineral. Petrol., v. 42, p. 109–124.

Xu, S., Okay, A. I., Ji, S., Sengor, A. M. C., Su, W., Liu, Y., and Jiang, L., 1992, Diamond from the Dabie Shan metamorphic rocks and its implication for tectonic setting: Science, v. 256, p. 80–82.

Xue, F., Rowley, D. B., and Baker, J., 1996, Refolded syn-ultrahigh pressure thrust sheets in the south Dabie complex, China: Field evidence and tectonic significance: Geology, v. 24, p. 455–458.

Yang, J., 1991, Eclogites, garnet pyroxenites and related ultrabasics in Shandong and north Jiangsu of east China: Beijing, Geol. Publ. House, 99 p. (in Chinese with English abstract).

Yang, J., Godard, G., Kienast, J.-R., Lu, Y., and Sun, J., 1993, Ultrahigh-pressure (60 kbar) magnesite-bearing garnet peridotites from northeastern Jiangsu, China: Jour. Geol., v. 101, p. 541–554.

Yang, J., Godard, G., and Smith, C., 1998, K-feldspar-bearing coesite pseudomorphs in an eclogite from Lanshantou (Eastern China): Eur. Jour. Mineral., v. 10, p. 969–985.

Yang, J., and Smith, D. C., 1989, Evidence for a former sanidine-coesite-eclogite at Lanshantou, Eastern China, and the recognition of the Chinese "Su-Lu coesite-eclogite province," East China [abs.]: Terra Nova Abs., v. 1, p. 26.

Ye, K., and Hirajima, T., 1996, Kinetic process of eclogitization of granulites from Haiyangsuo, southeastern Shangdong Peninsula [abs.]: Abs. 30th Int. Geol. Cong., Beijing, v. 2, p. 596.

Ye, K., Hirajima, T., Ishiwatari, A., Guo, J., and Zhai, M., 1996, Significance of interstitial coesite in eclogite at Yangkou, Qingdao city, eastern China: Chinese Sci. Bull., v. 41, p. 1407–1408 (in Chinese).

Zhang, R. Y., Hirajima, T., Banno, S., Cong, B., and Liou, J. G., 1995, Petrology of ultrahigh-pressure rocks from the southern Su-Lu region, eastern China: Jour. Metamor. Geol., v. 13, p. 659–675.

Zhang, R. Y., and Liou, J. G., 1997, Partial transformation of gabbro to coesite-bearing eclogite from Yangkou, the Sulu terrane, eastern China: Jour. Metamor. Geol., v. 15, p. 183–202.

______, 1998, Ultrahigh pressure metamorphism of the Sulu Terane, Eastern China: A prospective view: Continental Dynamics, v. 3, p. 32–53.

Zhang, R. Y., Liou, J. G., and Cong, B., 1994, Petrogenesis of garnet-bearing ultramafic rocks and associated eclogites in the Su-Lu ultrahigh-P metamorphic terrane, eastern China: Jour. Metamor. Geol., v. 12, p. 169–186.

Zhang, R. Y., Liou, J. G., and Ernst, W. G., 1995, Ultrahigh-pressure metamorphism and decompression P-T paths of eclogites and country rocks from Weihai, eastern China: The Island Arc, v. 4, p. 293–309.

Zhang, R. Y., Liou, J. G., Shau, Y. H., and Lo, C. H., 1998, Occurrence of clinoenstatite in ultradeep (> 200 km) garnet pyroxenite from the Sulu UHP terrane, eastern China: EOS (Trans. Amer. Geophys. Union), v. 79, p. 971.

MINERALOGY AND GEOCHEMISTRY OF ULTRAHIGH-PRESSURE ROCKS

Major Rock-Forming Minerals in UHP Metamorphic Rocks

MASAKI ENAMI

Department of Earth and Planetary Sciences, Graduate School of Science, Nagoya University, Nagoya 464-8602, Japan

AND SHOHEI BANNO

Department of Geology and Mineralogy, Kyoto University, Kyoto 606-8502, Japan

Abstract

This paper reviews chemical characteristics of common minerals in ultrahigh-pressure (UHP) metamorphic rocks. Garnet in UHP metamorphic rocks belongs to the almandine-pyrope-grossular series and has a wide compositional range in X_{alm} (up to 0.73), X_{prp} (0.98), and X_{grs} (0.92). Garnet with pyroxene exsolution microstructures reported from the Western Gneiss Region, Norway, shows relicts of majoritic garnet formed under P = 6 to 7 GPa. Most clinopyroxenes are Na-Ca groups and their X_{jd} ranges from 0.0 to 0.89. Members of the clinoenstatite-clinohypersthene ($P2_1/c$) were reported from peridotites in UHP terranes, and are considered to have been originally the high-pressure polymorph (C2/c) formed under P > 7 to 8 GPa. The presence of sodium in garnet and of potassium in clinopyroxene characterizes UHP equilibrium. Attention is drawn to minerals of the epidote group as strontium and REE containers, to amphibole as a sensor of metamorphic fluid composition, and to phengite as a geobarometer.

Introduction

THIS PAPER SUMMARIZES the chemistries of major ultrahigh-pressure (UHP) minerals, mainly on the basis of data from the Sulu-Dabie province of east-central China, one of the best-surveyed UHP provinces in the world. Major UHP rock types in the Sulu-Dabie province are eclogites, paraschists, and metagranitoids. The eclogites are subdivided into G-, P-, and M-eclogites based on the enclosing country rocks: orthogneiss, garnet peridotite, and marble, respectively. The UHP minerals cover wide compositional ranges, mainly because of differences in bulk composition of the host rocks. In this paper, chemical data of minerals are grouped into the above five categories.

Garnet

Garnet is a grossular-rich member of the almandine-pyrope series with low MnO content (usually less than 1 wt%) (Fig. 1).[1] Most garnets show prograde zoning with increasing X_{prp} and decreasing X_{sps} and X_{grs} toward the rim (Wang et al., 1992). Some garnets from Yangzhuang in the Junan area show complex zoning analogous to the "crack-filled zoning" reported from an eclogite in the Great Caucasus in Russia by Perchuk and Philippot (1997) (see also Enami and Nagasaki, 1999). The mosaic core of the "crack-filled zoning" may record that garnet had once dissolved at an early stage of prograde metamorphism. Retrograde zonation with decreasing X_{prp} and X_{alm} and increasing X_{grs} rarely occurs in garnet of G-type eclogite (Enami et al., 1993). Multiple stages of garnet formation also are well documented in UHP eclogites from the Zermatt-Saas zone, Western Alps (Reinecke, 1998)

Garnets in G-eclogites, which are the most common type, exhibit a wide compositional range of $X_{alm}(0.23–0.59)X_{prp}(0.11–0.50)X_{grs}(0.08–0.44)X_{sps}(0.0–0.03)$. M-eclogite has a CaO-rich bulk-rock composition (up to 25 wt%) (Kato et al., 1997) and their garnets are distinctly higher in X_{grs} (0.53 ± 0.07) than those in the other rock types. Garnets in P-eclogite and paraschist are rich in X_{prp}(0.37 ± 0.09) and poor in X_{grs} (0.20 ± 0.08), respectively. Garnets in metagranitoid have variable chemical compositions controlled by the compositions of the local domains in which they occur. The

[1] The data plotted on Figures 1, 4, and 6 are from Carswell et al. (1997), Cong et al. (1995), Enami and Nagasaki (1999), Enami and Zang (1988, 1990), Enami et al. (1993), Enami (unpubl. data), Hirajima et al. (1992, 1993), Kato et al. (1997), Liou et al. (1997), Okay (1993, 1994, 1995), Wang and Liou (1993), Wang (1997), Wu et al. (1998), Xu et al. (1979), Yang (1991), Ye and Xu (1992), Zhang and Liou (1994, 1997), and Zhang et al. (1992, 1994, 1995).

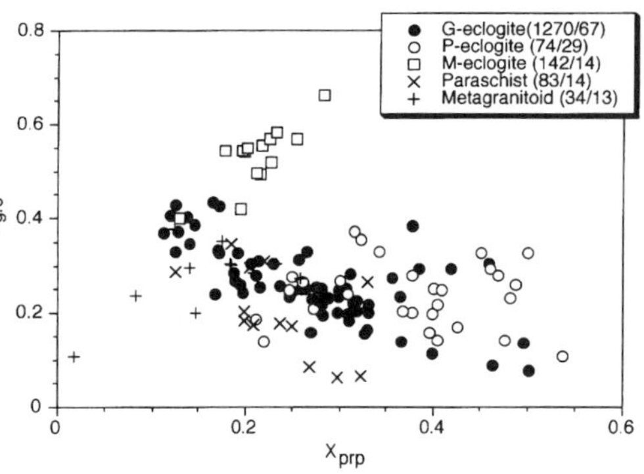

FIG. 1. Average compositions of garnet in UHP metamorphic rocks from the Sulu-Dabie belt, China. Numbers in parentheses in the legend indicate the number of analyzed points (numerator) and samples (denominator).

garnets have developed along grain boundaries between igneous plagioclase and mafic phases, and show a systematic increase of X_{alm} (0.44 to 0.73) and decrease of X_{grs} (0.55 to 0.10) from the plagioclase side to the mafic phases side (Hirajima et al., 1993; Zhang and Liou, 1997). Pyrope (X_{prp} = 0.98) and grossular (X_{grs} = 0.92) are common in UHP metamorphic rocks such as pyrope-coesite (e.g., Dora-Maira massif; Schertl et al., 1991) and calc-silicates (e.g., Münchberg gneiss complex; Klemd et al., 1994).

The kyanite + omphacitic pyroxene-bearing assemblage characterizes UHP environments in which paragonite is unstable (Holland, 1979). Garnet that coexists with clinopyroxene, kyanite, and a SiO_2 phase has maximum and minimum X_{grs} in common eclogite and grospydite, respectively (Banno, 1966, 1967). For example, garnets in kyanite-bearing and kyanite-free eclogites collected from an outcrop in Junan of the Sulu province contains X_{grs} of 0.23 to 0.25 (8 samples) and 0.21 to 0.23 (3 samples) at the rims. The kyanite-bearing eclogite assemblages thus lead to the following geothermobarometer (Nakamura and Banno, 1997):

$$\underset{\text{garnet}}{Ca_3Al_2Si_3O_{12} + Mg_3Al_2Si_3O_{12}} + SiO_2 = \underset{\text{kyanite}}{2Al_2SiO_5} + \underset{\text{diopside}}{3CaMgSi_2O_6}. \quad (1)$$

Garnet with pyroxene lamellae was reported from the Western Gneiss Region, Norway (van Roermund and Drury, 1998). This garnet might be a relict majoritic phase, and the amount of exsolved pyroxene, 3.5 to 4.6 vol%, indicates a pressure of ~6 to 7 GPa.

Sodium is an important minor element in garnet of kimberlite and related rocks, and the following three main mechanisms have been proposed for Na incorporation in garnet: $^{[8]}Na^{[6]}P(Ca, Mg, Fe)_{-1}Si_{-1}$ (Thompson, 1975); $^{[8]}Na^{[6]}Ti(Ca, Mg, Fe)_{-1}Al_{-1}$ (Ringwood and Major, 1971); and $^{[8]}Na^{[6]}Si(Ca, Mg, Fe)_{-1}Al_{-1}$ (Gasparik, 1989). Kato et al. (1997) documented that: (1) garnet in M-eclogite from Rongcheng of the Sulu province characteristically contains Na_2O of up to 0.2 wt%; (2) there is a strong positive correlation between Na and Ti contents (Fig. 2); and (3) Na concentrates in the Ca-rich part of zoned crystals. Shatsky et al. (1995) reported calcic garnet (X_{grs} = 0.51) with Na_2O = 0.3 wt% from eclogite of the Kokchetav Massif. These data imply that Na incorporation in garnet of UHP metamorphic rocks is mainly controlled by the following exchange reaction:

$$Ca_3Al_2Si_3O_{12} + NaTiAlSiO_6 = NaCa_2AlTiSi_3O_{12} + CaAl_2SiO_6, \quad (2)$$
$$\text{i.e., } CaAl = NaTi.$$

Ono and Yasuda (1996) and Okamoto and Maruyama (1998) experimentally documented the potential for Na substitution into garnet as a geobarometer. Low Na content of natural UHP metamorphic rocks (usually less than 0.05 wt%) and its strong compositional dependence, however, imply that much more detailed petrological data are needed before we discuss pressure conditions using garnet chemistry.

Chondrite-normalized REE distribution patterns of garnets are similar among eclogites from the Sulu-Dabie province showing: (1) a distinct enrichment in Sm and heavier REE; and (2), a slight decrease toward Lu (Fig. 3). This REE distribution contrasts markedly from the general REE pattern of garnet in eclogite with distinct HREE enrichment (Griffin and Brueckner, 1985; Bernard-Griffiths et al., 1991; Messiga et al., 1995).

Clinopyroxene

UHP Na-Ca clinopyroxene in eclogite belongs to the augite-jadeite series, and its X_{acm} is usually less than 0.1 (Fig. 4). It is homogeneous, and the X_{jd} variation in individual samples is usually less than 0.02. Most clinopyroxenes in G-eclogites are omphacitic with end-member compositions of

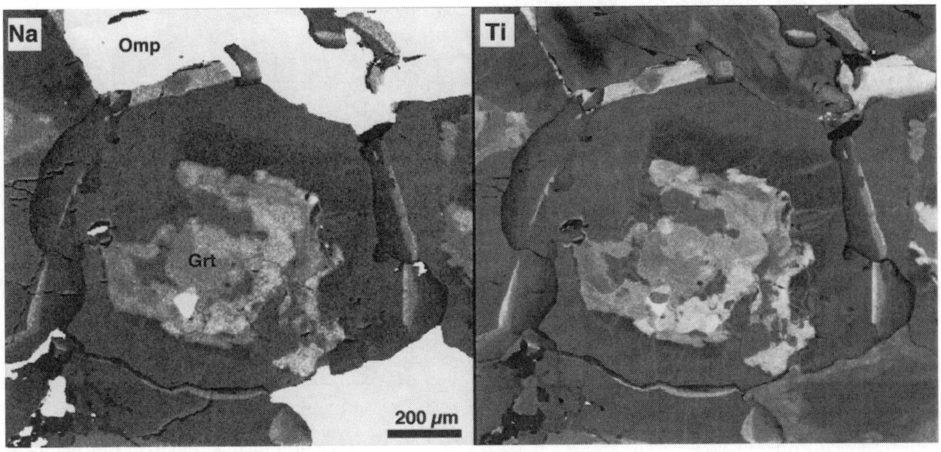

FIG. 2. Element X-ray mapping images of garnet in an M-eclogite from Rongcheng in Sulu province, eastern China (Kato, unpubl. data). Abbreviations: Grt = garnet; Omp = omphacite.

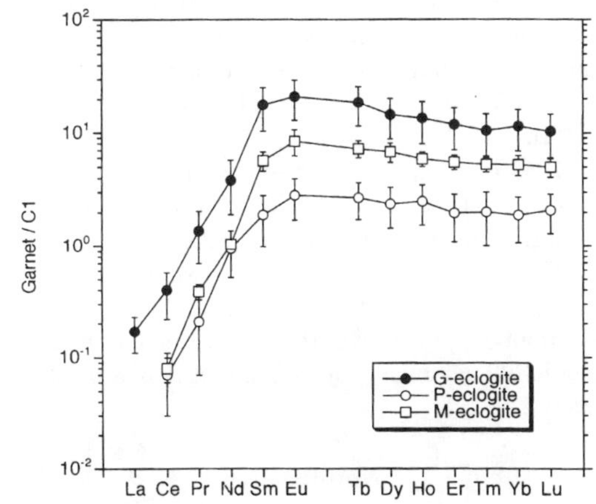

FIG. 3. Chondrite-normalized REE distribution patterns of garnet in eclogites from Sulu province, China (Enami, Kimura, and Yoshida, unpubl. data).

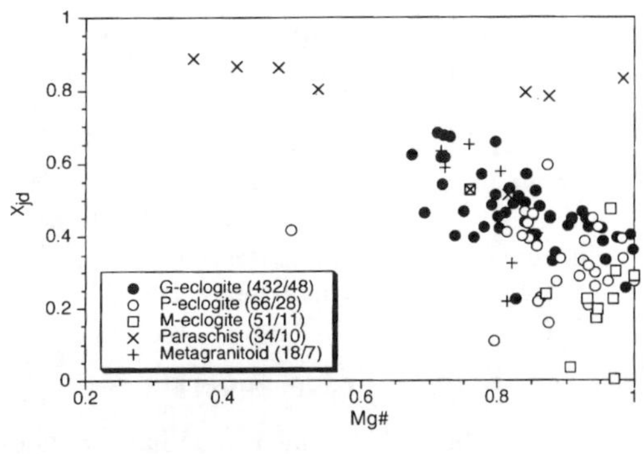

FIG. 4. Average compositions of clinopyroxene in UHP metamorphic rocks from the Sulu-Dabie belt, China. Numbers in parentheses in the legend indicate the number of analyzed points (numerator) and samples (denominator).

X_{aug}(21–72)X_{jd}(23–68)X_{acm}(0–22) and Mg# (0.67–1.0). Clinopyroxenes in P- and M-eclogites tend to have low X_{jd} (0.24 ± 0.16 and 0.34 ± 0.11) and high Mg# (0.93 ± 0.07 and 0.88 ± 0.09). A distinctly low Mg# composition was reported for clinopyroxene in a P-eclogite from the Junan area of the Sulu province (Zhang et al., 1992). Clinopyroxene in paraschist is typically jadeite-rich (up to 0.89 X_{jd}) and has a variable Mg# value (0.36 to 0.98). Such high X_{jd} apparently is caused by the unique bulk chemistry of the protolith (Wang et al., 1995).

Potassium incorporation in clinopyroxene is discussed on the basis of high-pressure experimental work (Shimizu, 1971; Okamoto and Maruyama, 1998) and petrological study of natural samples. Becker and Altherr (1992) reported clinopyroxene with K-rich feldspar exsolution in calc-silicate marble from the Bohemian Massif and discussed possible crystallization of K-rich clinopyroxene at pressures above 3 to 4 GPa. Shatsky et al. (1995) described K-rich calcic clinopyroxene (up to 1.2 wt% K_2O and CaO = 22 to 34 wt%) from diamond-bearing pyroxene-carbonate rock from the Kokchetav massif. The potassium solubility of natural samples seems to be an order of magnitude higher than that experimentally proposed by Okamoto and Maruyama (1998) (~0.2 wt% at 1000°C and 7 GPa). Clinopyroxene analyzed by these authors is more sodic (X_{jd} = 0.19 to 0.34) than the Kokchetav sample. Potassium incorporation into clinopyroxene thus may be strongly controlled by the major-element composition of the $^{[6]}M(2)$ site. In the Sulu-Dabie province, the K_2O content of UHP clinopyroxene is usually less than the analytical detection limit (0.02 wt% at 2σ level), and no one has positively documented the potassium solubility of clinopyroxene as a function of physical conditions.

Clinopyroxene in eclogite shows an asymmetric, bell-shaped REE pattern, which is common in typical eclogite (Griffin and Brueckner, 1985; Bernard-Griffiths et al., 1991; Messiga et al., 1995). The peak position of the REE pattern is scattered from Nd in G-eclogite to Eu in P-eclogite (Fig. 5).

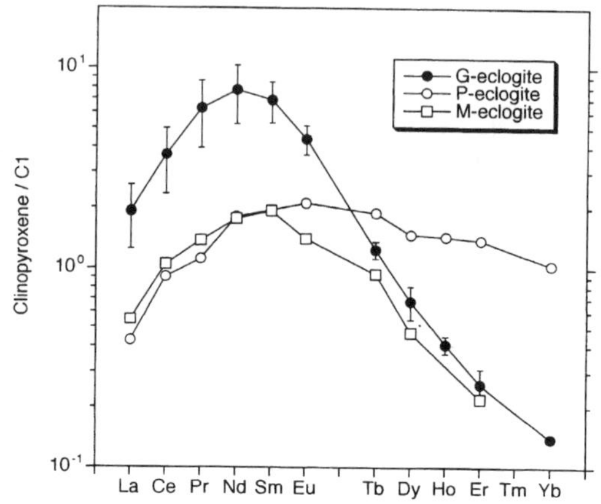

FIG. 5. Chondrite-normalized REE distribution patterns of clinopyroxene in eclogites from Sulu province, China (Enami, Kimura, and Yoshida, unpubl. data).

Members of the clinoenstatite-clinohypersthene series of low-pressure type (space group: $P2_1/c$) occur as lamellae in diopside and orthoenstatite, and as discrete crystals in garnet-bearing ultramafic rocks (Furukawa and Ogasawara, 1998; Green et al., 1998; Zhang et al., 1998). Originally they may have been the high-pressure polymorph (C2/c), showing equilibrium under P> 7 to 8 GPa (Pacalo and Gasparik, 1990; Angel and Hugh-Jones, 1994).

Epidote-Group Minerals

Orthorhombic zoisite and monoclinic epidote-allanite are major Ca-Al silicates of UHP metamorphic rocks. The Y_{Fe} $[=Fe^{3+}/(Fe^{3+} + Al)]$ value is 0.04 ± 0.01 in zoisite and 0.15 ± 0.04 in monoclinic phases of the Sulu UHP metamorphic rocks. Continuous solid solution exists between epidote and allanite, and total REE contents reach 12 wt% (La_2O_3 = 3.0 wt%, Ce_2O_3 = 6.0 wt%, Nd_2O_3 = 3.0 wt%). The REE content of zoisite is less than 1 wt% (La_2O_3< 0.2 wt%, Ce_2O_3< 0.3 wt%, Nd_2O_3< 0.2 wt%), and thus epidote-allanite is the main REE reservoir in UHP metamorphic rocks. Dissakisite-(Ce), the Mg-analogue of allanite-(Ce) (Grew et al., 1991), occurs in some UHP metamorphic rocks (Enami and Zang, 1988).

Nagasaki and Enami (1998) reported that epidote-group minerals in the Sulu UHP terrane are characteristically enriched in SrO (up to 3.2 wt%). Their SrO contents are one to four orders of magnitude higher than those in coexisting phases. More than 70% of bulk-rock SrO is estimated to be concentrated in the epidote-group minerals. Sr-bearing zoisite and epidote have been reported from other UHP metamorphic rocks (Getty and Selverstone, 1994).

Recent high-pressure experiments show that zoisite is stable up to 3 GPa at 650 to 700°C in basaltic and andesitic systems (Poli and Schmidt, 1995). Epidote-group minerals in subducted slabs transport SrO and REE from the crust to the mantle to depths approaching 100 km and may effectively modify the chemical and isotopic composition of the mantle.

Amphiboles

Experimental studies have shown that glaucophane-nyböite amphiboles are stable under coesite-stable conditions (Holland, 1988; Pawley, 1992). These high-pressure amphiboles have specific compositions (Chopin and Sobolev, 1995). The rarity of glaucophane and more common occurrence of nyböite may be ascribed to the following reaction:

$$\underset{\text{glaucophane}}{Na_2Mg_3Al_2Si_8O_{22}(OH)_2} + \underset{\text{jadeite}}{NaAlSi_2O_6} = \underset{\text{nyböite}}{NaNa_2MgAl_2(Al, Si_7)O_{22}(OH)_2} + 3SiO_2. \quad (3)$$

Nyböite has been described from the Western Gneiss Region, Norway (Ungaretti et al., 1981; Smith, 1988); the Sulu province, China (Smith et al., 1990; Hirajima et al., 1992); and the Dora-Maira

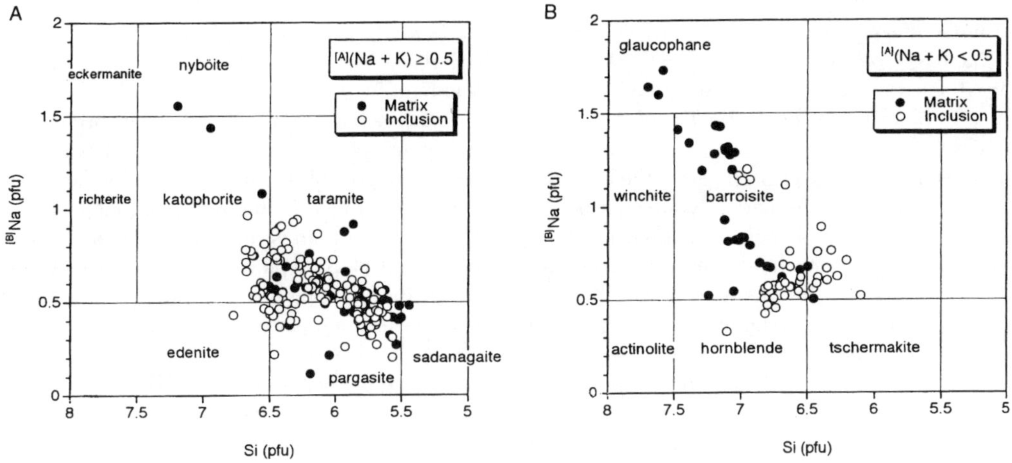

FIG. 6. Compositional variation of amphibole in G-eclogites from the Sulu-Dabie belt, China. A. $^{[A]}(Na + K) \geq 0.5$. B. $^{[A]}(Na + K) < 0.5$. Nomenclature follows Leake et al. (1997).

Complex, Italy (Chopin et al., 1991; Hirajima and Compagnoni, 1993). In all of these occurrences, nyböite is considered to be the first phase formed by hydration of UHP metamorphic rocks, and could have been stable under UHP conditions. In a multicomponent system, the stability of amphibole is considerably reduced, as shown in Figure 1 of Ernst and Liu (1998), who determined the stability of amphibole in MORB composition. It has been suggested, however, by Enami et al. (1993) that barroisite-pargasite that occurs as a matrix mineral in Sulu eclogite is a prograde mineral stable under the UHP environment because of a continuous increase of Al toward the crystal margin. Wang et al. (1992) also reported that katophorite is stable at the climax UHP stage in coesite-bearing eclogites. Discrete grains of amphibole in the rock matrix have a wide compositional range from pargasite, through taramite, to katophorite or barroisite, with Si = 6.4 ± 0.6 pfu, $^{[B]}Na = 0.74 \pm 0.37$ pfu and $^{[A]}(Na + K) = 0.63 \pm 0.27$ pfu (Fig. 6). Al_2O_3 content reaches 21.5 wt%.

Pre–UHP-stage amphiboles have been described as inclusions in garnet and omphacite. They have a slightly narrower compositional range than does the matrix amphibole: Si = 6.2 ± 0.4 pfu, $^{[B]}Na = 0.57 \pm 0.17$ pfu, and $^{[A]}(Na + K) = 0.66 \pm 0.18$ pfu. Nyböitic amphibole has not been reported as an inclusion phase. These relations suggest that stable assemblages include: (1) barroisite-pargasite + magnetite + rutile + quartz; (2) barroisite + zoisite + chlorite + relict garnet; and (3) pargasite + epidote + paragonite + margarite (Enami et al., 1993; Zhang et al., 1995). The formation of barroisite-taramite in basic rocks implies a rather high-pressure part of the epidote-amphibolite facies.

Amphiboles that occur as inclusions or matrix phases characteristically contain Cl up to 2.4 wt%; fluorine contents are less than 0.15 wt%. Halogen enrichment in the metamorphic fluid is documented from other HP-UHP terranes (Larsen et al., 1998; Philippot et al., 1998; Scambelluri et al., 1998). These data imply successive concentration of halogens into the metamorphic fluid during prograde dehydration.

Amphibole formed during decompression occurs as a pseudomorph after garnet and as symplectite around clinopyroxene. The former is pargasite-taramite with lower $^{[B]}Na$ than the UHP amphibole: Si = 6.0 ± 0.3 pfu, $^{[B]}Na = 0.39 \pm 0.23$ pfu, and $^{[A]}(Na + K) = 0.77 \pm 0.14$ pfu. Symplectitic amphibole has a wider compositional range from pargasite-taramite to hornblende: Si = 6.4 ± 0.5 pfu, $^{[B]}Na = 0.40 \pm 0.24$ pfu, and $^{[A]}(Na + K) = 0.68 \pm 0.25$ pfu. Both chlorine and fluorine contents are usually less than 0.2 wt%, indicating infiltration of halogen-poor fluid during the decompression stage.

Phengite

Coarse-grained phengitic mica in UHP metamorphic rocks is usually considered to be a UHP phase.

In the Sulu province, phengite is a common constituent in G-eclogite and paraschist. Silicon and X_{Na} [Na/(Na + K + Ba)] values are 3.40 ± 0.09 pfu and 0.09 ± 0.04 in G-eclogite and 3.45 ± 0.07 pfu and 0.05 ± 0.02 in paraschists. BaO contents reach 1.0 wt%.

Equilibria involving phengite are quite sensitive to pressure, but also are markedly affected by the activity of H_2O. Three phengite geobarometers have been proposed. Massonne and Schreyer (1989) presented isopleths of Si of phengite for the limiting assemblages phengite + kyanite + talc + SiO_2 and phengite + phlogopite + talc + SiO_2 for 500 < T (°C) < 800, 1 < P (GPa) < 3.5. Massonne (1995) extracted 6 Margules parameters of phengite from Massonne and Schreyer (1986, 1987, 1989), and applied it to the garnet + omphacite + phengite assemblage.

Carswell et al. (1997) and Enami and Nagasaki (1999) employed the solid-solution model of Holland and Powell (1990), and applied this to the eclogitic assemblage using the following equilibria:

$$\underset{\text{diopside}}{6CaMgSi_2O_6} + \underset{\text{muscovite}}{3KAl_3Si_3O_{10}(OH)_2} = \underset{\text{grossular}}{2Ca_3Al_2Si_3O_{12}} + \underset{\text{pyrope}}{Mg_3Al_2Si_3O_{12}} + \underset{\text{celadonite}}{3KMgAl_2Si_4O_{10}(OH)_2}, \qquad (4)$$

and

$$\underset{\text{diopside}}{3CaMgSi_2O_6} + \underset{\text{muscovite}}{3KAl_3Si_3O_{10}(OH)_2} + 2SiO_2 = \underset{\text{grossular}}{Ca_3Al_2Si_3O_{12}} + \underset{\text{kyanite}}{2Al_2SiO_5} + \underset{\text{celadonite}}{3KMgAl_2Si_4O_{10}(OH)_2}. \qquad (5)$$

They considered that differences among the calculated pressure values for different combinations of thermodynamic models for clinopyroxene and garnet are 0.5 GPa or less. Carswell et al. (1997) used the Mg and Al end members because the mole fractions of Mg end members are less affected by the Fe^{3+}/Fe^{2+} estimation of microprobe analysis than are the Fe^{2+} and Fe^{3+} end members.

Conclusions

1. Garnet and clinopyroxene in UHP metamorphic rocks have wide compositional ranges controlled by the bulk chemistries of protoliths: X_{alm} (0.23–0.73), X_{prp} (0.11–0.98), X_{grs} (0.08–0.92), and X_{sps} (0.0–0.1) for garnet and X_{jd} (0.0–0.89) for clinopyroxene.

2. Increasing pressure promotes incorporation of sodium into garnet and potassium into Ca-Na pyroxene. The degree of their concentrations, however, is strongly controlled by the bulk-rock major-element composition, and thus much more detailed experimental and petrographical data are needed before we can discuss pressure conditions quantitatively using these minor elements in garnet and Ca-Na pyroxene.

3. Recent investigations of UHP rocks produced the discoveries of (i) garnet containing pyroxene lamellae, and (ii) members of the clinoenstatite-clinohypersthene series in orogenic peridotites. The pyroxene lamellae-bearing garnet might represent recrystallized majoritic garnet. The monoclinic Mg-Fe pyroxenes are considered to have been originally the high-pressure polymorph, C2/c. They reflect equilibrium under unusually high pressure conditions, more than 6 to 8 GPa.

4. A member of the epidote group is a main Ca-Al silicate in UHP metamorphic rocks, and is an important container of Sr and REE in the subducted slab. Some amphiboles such as nyböite might be stable under coesite-stable conditions, implying potential use to constrain chemistry of the UHP metamorphic fluid. Phengite is a major K- and Ba-rich phase in UHP metamorphic rocks. Solid-solid equilibria involving phengite are useful geobarometers for UHP metamorphic rocks.

Acknowledgments

We would like to thank T. Kato (Nagoya University), J. Kimura (Shimane University), and T. Yoshida (Tohoku University) for permitting us to cite their unpublished data. The comments of W. G. Ernst (Stanford University), S. R. Wallis (Kyoto University), and T. Hirajima (Kyoto University) improved the manuscript.

REFERENCES

Angel, R. J., and Hugh-Jones, D. A., 1994, Equations of state and thermodynamic properties of enstatite pyroxenes: Jour. Geophys. Res., v. 99, p. 19,777–19,783.

Banno, S., 1966, Eclogite and eclogite facies: Japan. Jour. Geol. Geog., v. 37, p. 105–122.

______, 1967, Effect of jadeite component on the paragenesis of eclogitic rocks: Earth Planet. Sci. Lett., v. 2, p. 249–254.

Becker, H., and Altherr, R., 1992, Evidence from ultrahigh-pressure marbles for recycling of sediments into the mantle: Nature, v. 358, p. 745–748.

Bernard-Griffiths, J., Peucat, J. J., and Menot, R. P., 1991, Isotopic (Rb-Sr, U-Pb, and Sm-Nd) and trace-element geochemistry of eclogites from the Pan-African Belt: A case study of REE fractionation during high-grade metamorphism: Lithos, v. 27, p. 43–57.

Carswell, D. A., O'Brien, P. J., Wilson, R. N., and Zhai, M., 1997, Thermobarometry of phengite-bearing eclogites in the Dabie Mountains of central China: Jour. Metamor. Geol., v. 15, p. 239–252.

Chopin, C., Henry, C., and Michard, A., 1991, Geology and petrology of the coesite-bearing terrain, Dora Maira massif, Western Alps: Eur. Jour. Mineral., v. 3, p. 263–291.

Chopin, C., and Sobolev, N. V., 1995, Principal mineralogic indicators of UHP in crustal rocks, *in* Coleman, R. G., and Wang, X., eds., Ultrahigh-pressure metamorphism: New York, Cambridge University Press, p. 96–131.

Cong, B., Zhai, M., Carswell, D. A., Willson, R. N., Wang, Q., and Zhao, Z., 1995, Petrogenesis of ultrahigh-pressure rocks and their country rocks at Shuanghe in Dabieshan, Central China: Eur. Jour. Mineral., v. 7, p. 119–138.

Enami, M., and Nagasaki, A., 1999, Prograde P-T path of kyanite eclogites from Junan in the Sulu ultrahigh-pressure province, eastern China: The Island Arc, v. 8, in press.

Enami, M., and Zang, Q., 1988, Magnesian staurolite in garnet-corundum rocks and eclogite from the Donghai district, Jiangsu province, east China: Amer. Mineral., v. 73, p. 48–56.

______, 1990, Quartz pseudomorphs after coesite in eclogites from Shandong province, east China: Amer. Mineral., v. 75, p. 381–386.

Enami, M., Zang, Q., and Yin, Y., 1993, High-pressure eclogites in northern Jiangsu–southern Shandong province, eastern China: Jour. Metamor. Geol., v. 11, p. 589–603.

Ernst, W. G., and Liu, J., 1998, Experimental phase-equilibrium study of Al- and Ti-contents of calcic amphibole in MORB—a semiquantitative thermobarometer: Amer. Mineral., v. 83, p. 952–969.

Furukawa, N., and Ogasawara, Y., 1998, New occurrence of clinohypersthene(?) in garnet-bearing ultramafic rock from Donghai in the Su-Lu UHP terrane, eastern China [abs.]: EOS (Trans. Amer. Geophys. Union), v. 79, p. 990.

Gasparik, T., 1989, Transformation of enstatite-diopside-jadeite pyroxenes to garnet: Contrib. Mineral. Petrol., v. 102, p. 389–405.

Getty, S. R., and Selverstone, J., 1994, Stable isotopic and trace element evidence for restricted fluid migration in 2 GPa eclogites: Jour. Metamor. Geol., v. 12, p. 747–760.

Green, H. W., II, Bozhilov, K. N., and Dobrzhinetskaya, L., 1998, Alpe Arami: More pieces of the puzzle [abs.]: EOS (Trans. Amer. Geophys. Union), v. 79, p. 971.

Grew, E. S., Essene, E. J., Peacor, D. R., Su, S. C., and Asami, M., 1991, Dissakisite-(Ce), a new member of the epidote group and the Mg analogue of allanite-(Ce), from Antarctica: Amer. Mineral., v. 76, p. 1990–1997.

Griffin, W. L., and Brueckner, H. K., 1985, REE, Rb-Sr, and Sm-Nd studies of Norwegian eclogites: Chem. Geol., v. 52, p. 249–271.

Hirajima, T., and Compagnoni, R., 1993, Petrology of a ferro-nyböite–bearing jadeite-quartz/coesite-almandine-phengite fels from the Brossasco-Isaca unit, southern Doria-Maira massif, western Alps, Italy: Eur. Jour. Mineral., v. 5, p. 943–955.

Hirajima, T., Wallis, S. R., Zhai, M., and Ye, K., 1993, Eclogitized metagranitoid from the Sulu ultra-high pressure (UHP) province, eastern China: Proc. Japan Acad., v. 68, p. 249–254.

Hirajima, T., Zhang, R., Li, J., and Cong, B., 1992, Petrology of the nyböite-bearing eclogite in the Donghai area, Jiangsu province, eastern China: Mineral. Mag., v. 56, p. 37–46.

Holland, T. J. B., 1979, Experimental determination of the reaction paragonite = jadeite + kyanite + water, and internally consistent thermodynamic data for part of the system Na_2O-Al_2O_3-SiO_2-H_2O, with applications to eclogites and blueschists: Contrib. Mineral. Petrol., v. 68, p. 293–301.

______, 1988, Preliminary phase relations involving glaucophane and applications to high pressure petrology: New heat capacity and thermodynamic data: Contrib. Mineral. Petrol., v. 99, p. 134–142.

Holland, T. J. B., and Powell, R., 1990, An enlarged and updated internally consistent thermodynamic dataset with uncertainties and correlations: The system K_2O-Na_2O-CaO-MgO-MnO-FeO-Fe_2O_3-Al_2O_3-TiO_2-SiO_2-C-H_2-O_2: Jour. Metamor. Petrol., v. 8, p. 89–124.

Kato, T., Enami, M., and Zhai, M., 1997, The ultrahigh-pressure (UHP) marble and eclogite in the Sulu UHP terrane, eastern China: Jour. Metamor. Geol., v. 15, p. 169–182.

Klemd, R., Matthes, S., and Schüssler, U., 1994, Reaction textures and fluid behavior in very high pressure calc-silicate rocks of the Münchberg gneiss complex, Bavaria, Germany: Jour. Metamor. Geol., v. 12, p. 735–745.

Larsen, R. B., Eide, E. A., and Burke, E. A. J., 1998, Evolution of metamorphic volatiles during exhumation of microdiamond-bearing granulites in the Western Gneiss Region, Norway: Contrib. Mineral. Petrol., v. 133, p. 106–121.

Leake, B. E., Woolley, A. R., Arps, C. E. S., Birch, W. D., Gilbert, M. C., Grice, J. D., Hawthorne, F. C., Kato, A., Kisch, H. J., Krivovichev, V. G., Linthout, K., Laird, J., Mandarino, J., Maresch, W. V., Nickel, E. H., Schumacher, J. C., Smith, D. C., Stephenson, N. C. N., Ungaretti, L., Whittaker, E. J. W., and Youzhi, G., 1997, Nomenclature of amphiboles: Report of the subcommittee on amphiboles of the International Mineralogical Association Commission on new minerals and mineral names: Mineral. Mag., v. 61, p. 295–321.

Liou, J. G., Zhang, R. Y., and Jahn, B.-M., 1997, Petrology, geochemistry, and isotope data on a ultrahigh-pressure jadeitite quartet from Shuanghe, Dabie Mountains, east-central China: Lithos, v. 41, p. 59–78.

Massonne, H.-J., 1995, Experimental and petrogenetic study of UHPM, *in* Coleman, R. G., and Wang, X., Ultrahigh-pressure metamorphism: New York, Cambridge University Press, p. 33–95.

Massonne, H.-J., and Schreyer, W., 1986, High-pressure syntheses and X-ray properties of white micas in the system K_2O-MgO-Al_2O_3-SiO_2-H_2O: Neues Jahrb. für Mineral. Abh., v. 153, p. 177–215.

______, 1987, Phengite geobarometry based on the limiting assemblage with K-feldspar, phlogopite and quartz: Contrib. Mineral. Petrol., v. 96, p. 212–224.

______, 1989, Stability field of the high-pressure assemblage talc + phengite and two new phengite barometers: Eur. Jour. Mineral., v. 1, p. 391–410.

Messiga, B., Tribuzio, R., Bottazzi, P., and Ottolini, L., 1995, An ion microprobe study on trace element composition of clinopyroxenes from blueschist and eclogitized Fe-Ti-gabbros, Ligurian Alps, northwestern Italy: Some petrologic considerations: Geochim. et Cosmochim. Acta, v. 59, p. 59–75.

Nagasaki, A., and Enami, M., 1998, Sr-bearing zoisite and epidote in ultra high-pressure (UHP) metamorphic rocks from the Sulu province, eastern China: An important Sr-reservoir under UHP conditions: Amer. Mineral., v. 83, p. 240–247.

Nakamura, D., and Banno, S., 1997, Thermodynamic modeling of sodic pyroxene solid solution and its application in a garnet-omphacite-kyanite-coesite geothermobarometer for UHP metamorphic rocks: Contrib. Mineral. Petrol., v. 130, p. 93–102.

Okamoto, K., and Maruyama, S., 1998, Multi-anvil re-equilibration experiments of a Dabie Shan ultrahigh-pressure eclogite within the diamond-stability fields: The Island Arc, v. 7, p. 52–69.

Okay, A. I., 1993, Petrology of a diamond and coesite-bearing metamorphic terrain: Dabie Shan, China: Eur. Jour. Mineral., v. 5, p. 569–675.

______, 1994, Sapphirine and Ti-clinohumite in ultra-high-pressure garnet-pyroxenite and eclogite from Dabie Shan, China: Contrib. Mineral. Petrol., v. 116, p. 145–155.

______, 1995, Paragonite eclogites from Dabie Shan, China: Re-equilibration during exhumation?: Jour. Metamor. Geol., v. 13, p. 449–460.

Ono, S., and Yasuda, A., 1996, Compositional change of majoritic garnet in a MORB composition from 7 to 17 GPa and 1400 to 1600°C: Phys. Earth Planet. Inter., v. 96, p. 171–179.

Pacalo, R. E. G., and Gasparik, T., 1990, Reversals of the orthoenstatite-clinoenstatite transition at high pressures and high temperatures: Jour. Geophys. Res., v. 95, p. 15,853–15,858.

Pawley, A. R., 1992, Experimental study of the compositions and stabilities of synthetic nyböite and nyböite-glaucophane amphiboles: Eur. Jour. Mineral., v. 4, p. 171–192.

Perchuk, A., and Philippot, P., 1997, Rapid cooling and exhumation of eclogitic rocks from the Great Caucasus, Russia: Jour. Metamor. Geol., v. 15, p. 299–310.

Philippot, P., Arrinier, P., and Scambelluri, M., 1998, Chlorine cycling during subduction of altered oceanic crust: Earth Planet. Sci. Lett., v. 161, p. 33–44.

Poli, S., and Schmidt, M. W., 1995, H_2O transport and release in subduction zones: Experimental constraints on basaltic and andesitic systems: Jour. Geophys. Res., v. 100, p. 22,299–22,314.

Reinecke, T., 1998, Prograde high- to ultrahigh-pressure metamorphism and exhumation of oceanic sediments at Lago di Cignana, Zermatt-Saas Zone, western Alps: Lithos, v. 42, p. 147–189.

Ringwood, A. E., and Major, A., 1971, Synthesis of majorite and other high pressure garnets and perovskites: Earth Planet. Sci. Lett., v. 12, p. 411–418.

Scambelluri, M., Pennacchioni, G., and Philippot, P., 1998, Salt-rich aqueous fluids formed during eclogitization of metabasites in the Alpine continental crust (Austroalpine Mt. Emilius unit, Italian western Alps): Lithos, v. 43, p. 151–167.

Schertl, H.-P., Schreyer, W., and Chopin, C., 1991, The pyrope-coesite rocks and their country rocks at Parigi, Dora Maira massif, Western Alps: Detailed petrography, mineral chemistry and PT-path: Contrib. Mineral. Petrol., v. 108, p. 1–21.

Shatsky, V. S., Sobolev, N. V., and Vavilov, M. A., 1995, Diamond-bearing metamorphic rocks of the Kokchetav massif (northern Kazakhstan), *in* Coleman, R. G., and Wang, X., eds., Ultrahigh-pressure metamorphism: New York, Cambridge University Press, p. 427–455.

Shimizu, N., 1971, Potassium contents of synthetic clinopyroxenes at high pressures and temperatures: Earth Planet. Sci. Lett., v. 11, p. 374–380.

Smith, D. C., 1988, A review of the peculiar mineralogy of the "Norwegian-eclogite province," with crystal-chemical, petrological, geochemical and geodynamical notes and an extensive bibliography, *in* Smith, D. C., ed., Eclogites and eclogite-facies rocks: Amsterdam, Elsevier, p. 1–206.

Smith, D. C., Yang, J., Oberti, R., and Previde-Massara, E., 1990, A new locality of nyböite and taramite, the Jianchang eclogite pod in the "Chinese Sulu coesite-eclogite province," compared with the nyböite- and taramite-bearing Liset eclogite pod in the "Norwegian coesite-eclogite province" [abs.]: 15th Int. Mineral. Assoc. Meeting, Beijing, Abs., p. 889–890.

Thompson, R. N., 1975, Is upper-mantle phosphorus contained in sodic garnet?: Earth Planet. Sci. Lett., v. 26, p. 417–424.

Ungaretti, L., Smith, D. C., and Rossi, G., 1981, Crystal-chemistry by x-ray structure refinement and electron microprobe analysis of a series of sodic-calcic to alkali-amphiboles from the Nybö eclogite pod, Norway: Bull. Minéral., v. 104, p. 400–412.

Van Roermund, H. L. M., and Drury, M. R., 1998, An ultra-deep (d > 200 km) orogenic peridotite body in Western Norway [abs.]: EOS (Trans. Amer. Geophys. Union), v. 79, p. 971.

Wang, S., 1997, Eclogites' chemical classification: Scientia Geol. Sinica, v. 32, p. 275–282 (in Chinese with English abstract).

Wang, X., and Liou, J. G., 1993, Ultra-high-pressure metamorphism of carbonate rocks in the Dabie Mountains, central China: Jour. Metamor. Geol., v. 11, p. 575–588.

Wang, X., Liou, J. G., and Maruyama, S., 1992, Coesite-bearing eclogites from the Dabie Mountains, central China: Petrogenesis, P-T paths, and implications to tectonics: Jour. Geol., v. 100, p. 231–250.

Wang, X., Zhang, R. Y., and Liou, J. G., 1995, Ultrahigh-pressure metamorphic terrane in eastern central China, *in* Coleman, R. G., and Wang, X., eds., Ultra-high-pressure metamorphism: New York, Cambridge University Press, p. 356–390.

Wu, W., Xu, S., Jiang, L., Liu, Y., and Su, W., 1998, Quartz-jadeitite in ultrahigh-pressure metamorphic belt in the Dabie Mountains, eastern China: Acta Petrol. Sinica, v. 14, p. 60–70 (in Chinese with English abstract).

Xu, H., Wang, W., and Go, J., 1979, Deep-seated eclogites in Junan and Rizhao, Shandong province, China: Seismol. Geol., v. 1, p. 57–66 (in Chinese with English abstract).

Yang, J., 1991, Eclogites, garnet pyroxenites and related ultrabasics in Shandong and north Jiangsu of east China: Beijing, Geol. Publ. House, 99 p. (in Chinese with English abstract).

Ye, K., and Xu, P., 1992, Petrogenesis of eclogites and related peridotitic rocks in Datuan area of Rongcheng country, Shandong Province: Acta Petrol. Sinica, v. 8, p. 27–39 (in Chinese with English abstract).

Zhang, R. Y., Hirajima, T., Banno, S., Cong, B., and Liou, J. G., 1995, Petrology of ultrahigh-pressure rocks from the southern Sulu region, eastern China: Jour. Metamor. Geol., v. 13, p. 659–675.

Zhang, R. Y., and Liou, J. G., 1994, Coesite-bearing eclogite in Henan Province, central China: Detailed petrography, glaucophane stability, and P-T path: Eur. Jour. Mineral., v. 6, p. 217–233.

______, 1997, Partial transformation of gabbro to coesite-bearing eclogite from Yangkou, the Sulu terrane, eastern China: Jour. Metamor. Geol., v. 15, p. 183–202.

Zhang, R. Y., Liou, J. G., and Cong, B., 1994, Petrogenesis of garnet-bearing ultramafic rocks and associated eclogites in the Sulu ultrahigh-P metamorphic terrane, eastern China: Jour. Metamor. Geol., v. 12, p. 169–186.

Zhang, R. Y., Liou, J. G., Shau, Y. H., and Lo, C. H., 1998, Occurrence of clinoenstatite in ultradeep (> 200 km) garnet pyroxenite from the Sulu UHP terrane, eastern China [abs.]: EOS (Trans. Amer. Geophys. Union), v. 79, p. 971.

Zhang, X., Chen, Y., and Wang, G., 1992, Characteristics of eclogites in Junan county, southeastern Shandong: Acta Petrol. Sinica, v. 8, p. 40–49 (in Chinese with English abstract).

Exsolution Lamellae in Minerals from Ultrahigh-Pressure Rocks

R. Y. ZHANG AND J. G. LIOU

Department of Geological and Environmental Sciences, Stanford University, Stanford, California 94305

Abstract

Exsolution microstructures have recently been identified in ultrahigh-pressure (UHP) ultramafic and eclogitic minerals from several orogenic belts, including the Western Gneiss Region, Norway; Alpe Arami, Switzerland; the Bohemian massif, central Europe; the Kokchetav massif, Kazakhstan; the Dabie-Sulu UHP terrane, east-central China; and in central Sulawesi, Indonesia. They include ilmenite rods and magnetite lamellae in olivine; K-feldspar, Mg-Al-Cr titanomagnetite and/or ilmenite, quartz, and garnet rods in clinopyroxene; pyroxene and rutile lamellae in garnet; and lamellae of monazite and an unknown phase in apatite. Although the mechanisms for exsolution in some UHP minerals are poorly known, most exsolution microstructures, such as K-feldspar and quartz in clinopyroxene and pyroxene in garnet, may have resulted from nearly isothermal decomposition of very high pressure precursor phases at mantle depths. Detailed characterization and experimental investigations regarding exsolution microstructures of these and other minerals in UHP rocks will improve our understanding of mantle processes and continental subduction.

Introduction

SINCE ULTRAHIGH-PRESSURE (UHP) metamorphism of supracrustal rocks was recognized (Chopin, 1984; Smith, 1984; Sobolev and Shatsky, 1990), exsolution microstructures in UHP minerals such as olivine, pyroxene, and garnet from both eclogitic and ultramafic rocks have attracted much attention (Smith, 1988; Dobrzhinetskaya et al., 1996; Hacker et al., 1997; Zhang and Liou, 1998; Zhang et al., 1999; van Roermund et al., in press). A summary description of several exsolution minerals in UHP rocks, mainly from the Dabie-Sulu terrane, was made by Liou et al. (1998). Lamellae-bearing minerals preserve information on the physicochemical conditions of the precursor homogeneous host as well as on the inferred exhumation P-T path. Lamellae of UHP minerals may have exsolved during nearly isothermal decompression from a depth of >100 km to the Earth's surface. The mechanism is significantly different from that producing common exsolution lamellae in primary igneous minerals with falling temperature; hence, previous studies on the nature of solid solutions and formation of exsolution lamellae have been restricted to the effect of temperature decreases at constant pressure.

The extent of solid solution for some UHP phases at mantle pressures and mechanisms for mineral exsolution are poorly known. Several exsolution features of quartz lamellae in clinopyroxene, garnet in orthopyroxene, and rutile rods in garnet were identified by Smith (1988) in UHP rocks from the Western Gneiss Region (WGR). Other exsolution lamellae in magnesiostaurolite from the Dora-Maira massif were reported by Chopin and Schertl (1999). Thus far, descriptions of such lamellae are of a reconnaissance nature. In this paper, we summarize several recognized exsolution microstructures in olivine, pyroxene, garnet, and apatite from UHP eclogitic and ultramafic rocks. Where experimental data are available, they are used to support their extremely high-P origins. We hope that this review will lead mineral physicists and experimental geochemists to pursue systematic studies of compositions and structures of host mineral and lamellae pairs and of stabilities of mantle minerals under UHP conditions. Such investigations will help bridge the gap between crustal petrogenesis and mantle processes of concern to many Earth scientists.

Olivine

The most common exsolution features in olivine from UHP garnet peridotites are ilmenite lamellae or rods; magnetite lamellae are minor. Similar exsolution lamellae also occur in clinohumite. Micron-sized ilmenite rods in olivine have been identified in garnet peridotites from several UHP terranes, including the Alpe Arami garnet peridotite massif (Dobrzhinetskaya et al., 1996), the Dabie-Sulu UHP terrane (Hacker et al., 1997; Zhang and Liou, 1998), and central Sulawesi, Indonesia (Kadarusman and

Parkinson, in press). Magnetite lamellae in olivine from UHP garnet peridotites have only been reported in the Dabie-Sulu terrane (Zhang et al., 1999). The characteristics of these exsolution microstructures and speculations concerning their origins are described below.

Ilmenite rods in olivine

Alpe Arami. The Alpe Arami garnet peridotite consists of garnet, diopside, enstatite, and two generations of olivine. The olivine exhibits considerable amounts of $FeTiO_3$ and chromite exsolution rods. $FeTiO_3$ rods constitute 1 to 3 vol% of the olivine, and range from 2 to 20 μm in length and from 0.2 to 4 μm in width, oriented parallel to $[010]_{Ol}$. Examination by transmission electron microscope (TEM) and selected-area electron diffraction shows that $FeTiO_3$ rods display four crystal structures—from a previously unknown orthorhombic structure through two structural intermediates to ilmenite. A topotaxial relationship exists between ilmenite rods and olivine: $[0001]_{Ilm}//[100]_{Ol}$, $[11\bar{2}0]_{Ilm}//[010]_{Ol}$, and $[01\bar{1}0]_{Ilm}//[001]_{Ol}$ (Dobrzhinetskaya et al., 1996).

The occurrence of $FeTiO_3$ rods in olivine from the Alpe Arami garnet peridotite was used to suggest that the body was derived from depths of >300 km (Dobrzhinetskaya et al., 1996). Their evidence includes a distinct lattice preferred orientation of olivine, three previously unknown crystal structures of $FeTiO_3$ rods, and recalculated high TiO_2 contents (7000 to 20,000 ppm). Subsequently, they indicated that these precipitates are older than the Alpine deformation structures and incompatible with derivation from decomposition of titanoclinohumite (Green et al., 1997). Recent experiments on titanium dissolution and precipitation in olivine at high–P-T conditions suggested that TiO_2 solubility in olivine increases nonlinearly with increasing pressure: 1500 K, 0.4 wt% at 8 GPa, 0.5 wt% at 10 GPa, and >1 wt% at 12 GPa. A similar trend was found at 1600 and 1650 K at 8, 10, and 12 GPa (Dobrzhinetskaya et al., 1998).

Dabie-Sulu terrane. Ilmenite rods or lamellae in olivine have been found in many garnet peridotite bodies in the Sulu-Dabie terrane (Zhang et al., in press). Except for those in the Chijiadian lherzolite (Hacker et al., 1997), others have been reported with only photomicrographs or mentioned in several publications (Zhang and Liou, 1998). One example is shown in Figure 1A.

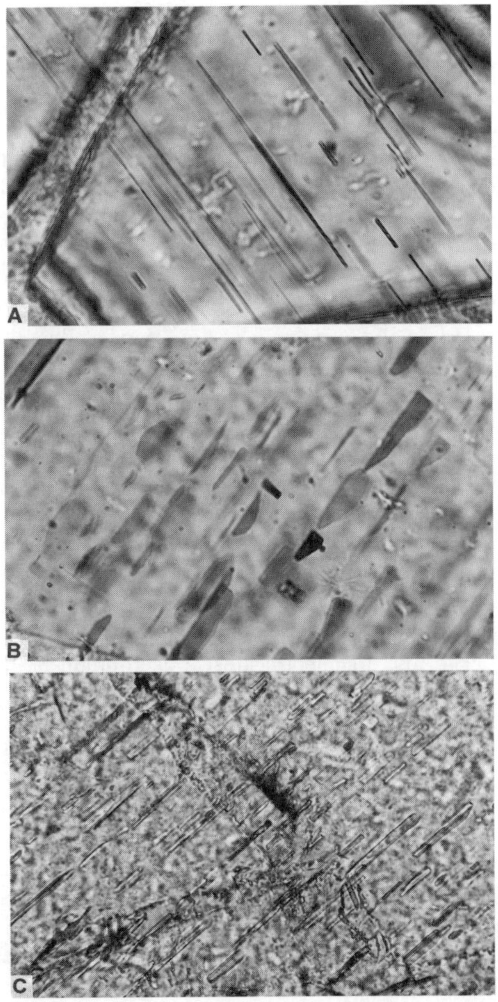

FIG. 1. Photomicrographs showing exsolution microstructures. A. Ilmenite rods in olivine from the Zhimafang lherzolite; width of view = 0.2 mm, crossed polars. B. Magnetite lamellae (lighter tones) and Cr-bearing magnetite (darker tones) lamellae in olivine from a Maowu harzburgite, Dabieshan; width of view = 0.2 mm, plane light. C. Quartz rods in omphacite from the Rongcheng coesite-bearing eclogite within gneiss; width of view = 0.32 mm, plane light.

The Chijiadian lherzolite of Rongcheng experienced UHP metamorphism at 820 to 900°C, >4 to 6 GPa based on Al partitioning between enstatite and garnet (Zhang et al., 1994; Hiramatsu et al., 1995) or at ~4.4 GPa based on the distribution of Ca between olivine and diopside (Hacker et al., 1997). $FeTiO_3$ rods with the composition $Fe_{0.82}Mg_{0.15}Mn_{0.03}TiO_3$ are nearly square prisms 0.2 to 0.5 μm across, whereas the third dimension is typically ~20 μm; they are smaller and more magnesian than the Alpe Arami iron titanates ($Fe_{0.94}Mg_{0.06}TiO_3$. TEM images indicate that rods in the Chijiadian olivines also have long axes parallel to [010] of the host olivine crystals (Hacker et al., 1997). The bulk TiO_2 content of the Chijiadian olivine ranges from 100 ± 20 ppm in rod-free portions to 269 ± 188 ppm (rod-bearing portions). Values less than ~300 ppm for TiO_2 contents were obtained independently by Nakajima and Ogasawara (1997) for olivines from the same lherzolite.

Origin. Interpretation of the origin of ilmenite exsolved from olivines is paradoxical (e.g., Hacker et al., 1997; Ulmer and Trommsdorff, 1997). Since the electrifying report by Dobrzhinetskaya et al. (1996), at least three other independent groups have begun working on such assemblages. Meanwhile, experimental solubility data obtained for TiO_2 in olivine under UHP conditions by Ulmer and Trommsdorff (1997), Okamoto et al. (1997), Dobrzhinetskaya and Green (1997), and Dobrzhinetskaya et al. (1998) are mutually contradictory.

The first argument against a deep origin is the measurement of the TiO_2 content in olivine. Detailed analyses of Alpe Arami olivine revealed bulk olivine TiO_2 contents of ~220 ppm (Hacker et al., 1997). Risold and Trommsdorff (1996) also gave a low TiO_2 content of <400 ppm for olivines from both the Alpe Arami and Cima di Gagnone in the Central Alps and concluded that the ilmenite rods formed as exsolution from Ti-rich olivine during cooling and decompression from ~3 GPa, 800°C. Green and Dobrzhinetskaya (1997) reported that their samples contain olivines of two generations: only the older one has high TiO_2 contents. The second argument regards the crystal structure of $FeTiO_3$. The electron diffraction patterns of Dobrzhinetskaya et al. (1996) were subsequently explained by dynamical diffraction between overlapping olivine and ilmenite structures (Hacker et al., 1997). Furthermore, Risold et al. (1997) indicated that the orthorhombic structure (β-ilmenite) reported by Dobrzhinetskaya et al. (1996) could be indexed as a trigonal structure based on selected-area electron diffraction.

The third argument regards experimental results on the solubility of TiO_2 in olivine as a function of P; such a relationship has not been confirmed by other groups. Ulmer and Trommsdorff (1997) used synthetic magnesio-ilmenite (X_{mg} = 0.6) and an oxide mix of olivine (X_{mg} = 0.9) in the proportions 5:95, respectively (+ brucite as the source of H_2O), as starting materials and ran experiments at 1 to 10 GPa and 1200 to 1600°C. These P-T conditions cross the reaction $Mg_2SiO_4 + TiO_2 = MgSiO_3 + MgTiO_3$, thus leading to the paragenesis Ol + Rt + Ilm at 1 to 6 GPa and Ol + En + Ilm at 9 GPa. Coarse-grained, homogeneous olivine crystals >150 μm in size were produced. Olivine contains a constant amount of TiO_2—0.135 ± 0.01 wt% at 1400°C and 0.105 ± 0.01 wt% at 1200°C; extrapolating these values to 800°C results in a TiO_2 content of olivine saturated with ilmenite and/or rutile on the order of 0.04 to 0.06 wt%, corresponding to the amount measured in natural Alpe Arami specimens by Risold et al. (1997). These experiments demonstrate the lack of P dependence of TiO_2 in α-olivine; rather, the TiO_2 content of olivine increases with increasing T. Okamoto et al. (1997) conducted a few reconnaissance experiments at 6, 8, and 10 GPa at 1200°C and 13 GPa at 1300°C using 90 wt% San Carlos olivine + 10 wt% rutile. Recrystallized olivines contain ~0.25 wt% TiO_2, and Mg-ilmenite formed in 13 GPa products. One experiment at 15 GPa yielded β-olivine containing ~0.15 wt% TiO_2. Although higher Ti solubilities in α-olivine were reported than values determined by Ulmer and Trommsdorff (1997), they are far less than those of Green et al. (1997). In other words, the solubility of TiO_2 in olivine and the origin of $FeTiO_3$ rods exsolved from olivine need to be investigated further.

Magnetite lamellae in olivine

Magnetite lamellae or intergrowths of magnetite + another silicate phase in olivine from some mafic-ultramafic complexes and basalts have been described (Goode, 1974; Ambler and Ashley, 1977; Moseley, 1984). They have been interpreted as oxidation of olivine or as decomposition products of Fe^{3+}-bearing olivine during cooling. Magnetite exsolution lamellae in olivine and Ti-clinohumite were identified from Dabieshan UHP ultramafic rocks several years ago (Zhang, Hacker et al., 1997a; Jin et al., 1998); details are reported by Zhang et al. (1999).

Light-brown magnetite lamellae without any accompanying silicate phases are common in olivine cores and mantles in a garnet harzburgite from Maowu, Dabieshan, and vary from a trace to 1.5 vol%. Most lamellae are 3 to 75 μm long, 2 to 30 μm wide, and 0.5 to 2 μm thick (Fig. 1B). The host olivine has orthorhombic cell parameters: a = 6.007(5) Å, b = 10.249(8) Å, c = 4.757(2) Å, V = 239.42(18)

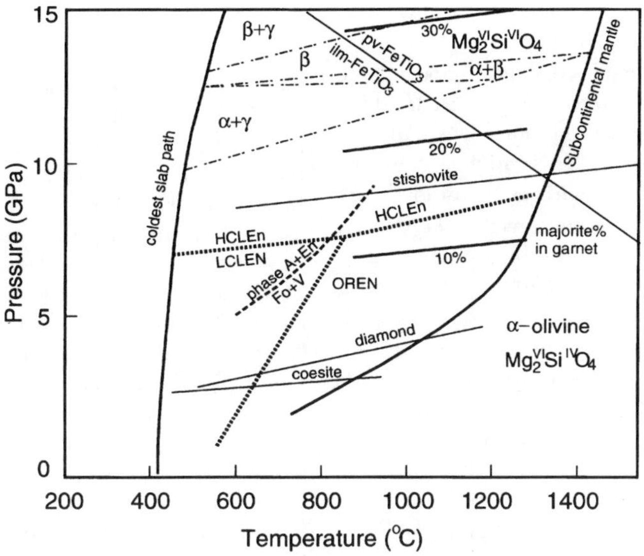

FIG. 2. P-T stability fields are shown for olivine (Akaogi et al., 1989), ilmenite (Mehta et al., 1994), quartz-coesite-stishovite (Akaogi et al., 1995), diamond (Bundy, 1980), and enstatite (Angel and Hugh-Jones, 1994). The coldest subduction zone and subcontinental mantle geotherms are after Helffrich et al. (1989) and Mercier and Nicolas (1975). The 10, 20, and 30 percent isopleths of majorite in pyrope are after Gasparik (1990). The polymorphs of enstatite ($MgSiO_3$)—orthoenstatite (OREN), low-P clinoenstatite (LCLEN), and high-P clinoenstatite (HCLEN)—are from Angel and Hugh-Jones (1994). Reaction of phase A + EN = Ol + V is also shown (Pawley and Wood, 1996).

Å^3. The magnetite is cubic: a = 8.411(2) Å and V = 594.95(38) Å^3. A topotaxial relationship exists between olivine and magnetite: $[220]_{Mag} \parallel [200]_{Ol}$, $[111]_{Mag} \parallel [3\bar{3}\bar{1}]_{Ol}$, $[11\bar{1}]_{Mag} \parallel [331]_{Ol}$, and $[242]_{Mag} \parallel [2\bar{2}0]_{Ol}$.

Oxidation of olivine (Haggerty and Baker, 1967; Champness, 1970; Putnis, 1979; Rietmeuer, 1996) or decomposition of Fe^{3+}-bearing high-T olivine (Moseley, 1984) invariably requires additional silica or silicate phases associated with magnetite; a topotaxial relationship between magnetite lamellae and the host olivine may not occur. The lack of additional silica or silicate phases and the observed topotaxial relations described above for the Dabieshan case rule out these hypotheses to explain the magnetite exsolution in olivine.

Two new hypotheses were employed to explain such exsolution microstructures: (1) derived from a possible precursor phase A [$Mg_7Si_2(OH)_6$]; and (2) exsolution of magnetite from a very high P olivine phase, a spinel solid solution [Fe_3O_4-$(Fe,Mg)_2SiO_4$]. The Maowu harzburgite was metamorphosed at ~750°C and 6 GPa (Liou and Zhang, 1998); this estimate is very close to the experimentally calibrated equilibrium points at 720 to 740°C at 7 GPa and 800 to 820°C at 8 GPa for the reaction (Fig. 2):

$$\underset{\text{phase A}}{Mg_7Si_2(OH)_6} + \underset{\text{enstatite}}{3MgSiO_3} = \underset{\text{forsterite}}{5Mg_2SiO_4} + \underset{\text{vapor}}{3H_2O}$$

(Luth, 1995a, 1995b; Pawley and Wood, 1996). If peak metamorphism of the Maowu harzburgite occurred at pressures slightly higher than the above estimates, the paragenesis of phase A + enstatite would be expected. During decompression, the dehydration reaction of phase A + enstatite to olivine would occur, and the phase A would expel Fe, Cr, and other impurities without formation of additional accompanying silicates (Ulmer, pers. commun., 1998). Although phase A is a potential precursor to the Maowu olivine, suggesting a high-P origin at ~200 km, this hypothesis does not satisfactorily explain the process of magnetite exsolution from olivine, as the observed topotaxial relationship is not consistent with the dehydration reaction mentioned above.

Zhang et al. (1999) suggested that the exsolution of magnetite occurred from a very HP olivine phase as a spinel solid solution: Fe_3O_4-$(Fe,Mg)_2SiO_4$. Olivine from the Dabie harzburgite contains 1 to 1.5 vol% magnetite; this implies that the original olivine could have contained 1.0 wt% Fe_2O_3. Such a high Fe_2O_3 content is difficult to accommodate in the forsterite structure at low P. If the original "olivine" was actually wadsleyite with a distorted-spinel structure, it could readily accommodate Fe_3O_4 as a spinel solid solution along the binary join Fe_3O_4-$(Fe,Mg)_2SiO_4$. Ross et al. (1992) synthesized a spinelloid phase, $Fe_{5.2}Si_{0.8}O_8$ ($Fa_{40}Mt_{60}$), at 1200°C, 7 GPa. More recent experiments in the system Fe_2SiO_4-Fe_3O_4 by Woodland and Angel (1998) yielded a new spinelloid phase ($Fe_{2.45}Si_{0.55}O_4$) with the wadsleyite structure, suggesting that significant amounts of Fe^{3+} can be incorporated into a wadsleyite-type structure through the substitution of $2Fe^{3+} = Fe^{2+} + Si^{4+}$. The difference in the stability of Fe^{3+}-substituted wadsleyite (5.0 to 6.0 GPa at 1100 to 1200°C) and $(Mg,Fe)_2SiO_4$ wadsleyite (i.e., 12 GPa at 1200°C) suggests that the addition of Fe^{3+} could stabilize $(Mg,Fe)_2SiO_4$ wadsleyites to lower P compared with the Fe^{3+}-free system (Woodland and Angel, 1998).

This hypothesis leads us to suggest that the host olivine may have had the wadsleyite structure; if so, the Maowu ultramafic body was subducted to depths greater than 200 km. Such a suggestion requires further independent documentation of the occurrences of other UHP minerals, including majoritic garnet and high-P clinoenstatite.

Clinopyroxene

Clinopyroxene is a major phase that can accommodate large cations in ultramafic rocks and shows various exsolution microstructures. This includes exsolved K-feldspar, ilmenite, Mg-Al-Cr titanomagnetite, and garnet in diopside, and quartz in omphacite and diopside. The mechanism of exsolution of these phases in clinopyroxene may be complicated. It is well known that solubilities of K_2O and SiO_2 in clinopyroxene are pressure dependent; in fact, the K_2O content of Ca-Na pyroxenes has been considered to be a potential barometer for UHP rocks (e.g., Schmidt, 1996; Harlow, 1997; Okamoto and Maruyama, 1998). However, the solubilities of other excess components as a function of pressure for exsolution of garnet and other phases are poorly known.

K-feldspar lamellae in diopside

K-feldspar lamellae in diopside were found in diamond-bearing dolomitic marble in the Kokchetav massif, Kazakhstan, and it has been suggested that primary Cpx contained K_2O up to ~1.5 wt% (Sobolev and Shatsky, 1990; Shatsky et al., 1995; Zhang, Liou et al., 1997). The diopsides of dolomitic marble display variable K_2O contents and amounts of exsolved K-feldspar lamellae. Some diopside grains lack "coarse" exsolved K-feldspar lamellae and contain 0.17 wt% K_2O, whereas other grains exhibit "coarse" K-feldspar lamellae (see Fig. 4A of Zhang, Liou et al., 1997) and range in K_2O content from 0.2 to 0.7 wt%, with an average of 0.47 wt%. The amount of the K-feldspar lamellae in some diopside grains may be as much as ~7%. Assuming end-member composition for the exsolved K-feldspar lamellae, primary diopside should contain 1.56 wt% K_2O (Zhang, Liou et al., 1997). Similar exsolution structures have been reported in Cpx inclusions from eclogitic garnets in kimberlites (Harlow and Veblen, 1991).

In order to determine K_2O solubility in clinopyroxene at high pressures, a few experiments have been performed (Ryabchikov and Ganeyev, 1990; Doroshev et al., 1992; Luth, 1992, 1995; Harlow, 1997; Okamoto and Maruyama, 1998; Wang and Takahashi, 1999). Clinopyroxene containing 3 wt% K_2O has been synthesized in the $KAlSi_2O_6$-$CaMgSi_2O_6$ system (Luth, 1995). The maximum K_2O content is 4.7 wt% in Cpx ($Di_{38}Ko_{39}Kcpx_{22}En_1$) formed from a 50% Di + 50% Ko ($NaCrSi_2O_6$) mixture at 10 GPa, 1400°C (here Kcpx = $KCrSi_2O_6$). The K uptake is proportional to the P as well as the Cpx composition, but not the T (Harlow, 1997). Okamoto and Maruyama (1998) calibrated the K_2O content of clinopyroxene as a potential geobarometer using Dabie coesite-bearing eclogite (UHP 303), or the eclogite + 1.1 wt% K_2O (UHK 303), for reequilibration experiments at P = 2.6 to 7 GPa and T = 900 to 1200°C. All run products consisted of garnet, sodic augite, with or without a SiO_2 phase, and phengite. The stable Ca-Na pyroxenes are sodic augite rather than omphacite at T > 900°C and P = 3 to 7 GPa. The K_2O content of Cpx increases from 0.07 to 0.23 wt% with increasing P from 3 to 7 GPa. A similar pressure dependence of the K_2O content of clinopyroxene is observed at 1000°C; it increases from below the detection limit at 3 GPa through 0.13 wt% at 5 GPa to 0.36 wt% at 7 GPa. These experiments suggest

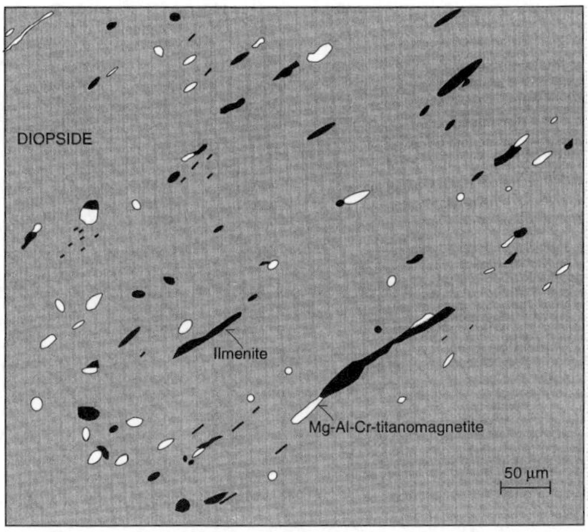

FIG. 3. Schematic sketch of ilmenite + Mg-Al-Cr titanomagnetite rods in diopside from a Sulu garnet olivine clinopyroxenite, based on a backscattered electron image.

that K_2O solubility increases with increasing pressure in clinopyroxene.

Mg-ilmenite and Mg-Al-Cr titanomagnetite

Ilmenite ± Mg-Al-Cr titanomagnetite rods commonly occur in relatively coarse grained diopside rather than in the fine-grained (0.3 to 0.5 mm) diopside matrix in the Rizhao clinopyroxenite of the Sulu terrane (Zhang and Liou, 1998; Zhang et al., in press). The oriented ilmenite rods are >5 vol%, 5 to 30 μm in width, and 25 to 200 μm in length, whereas the Mg-Al-Cr titanomagnetite rods are closely associated with ilmenite but are shorter than the ilmenite rods (Fig. 3). The ilmenite contains >2 wt% MgO, and Mg-Al-Cr titanomagnetite contains a considerable amount of MgO (1.5 wt%), TiO_2 (3.5 wt%), Al_2O_3 (2.2 wt%), and Cr_2O_3 (8.2 wt%) (Zhang et al., in press). Based on the volumes and compositions of the exsolved phase and host, the original diopside would have contained >1.0 wt% TiO_2; such Ti-rich diopsides have been experimentally synthesized (e.g., I. MacGregor, pers. comm., 1997).

Quartz rods in omphacite and clinopyroxene

Exsolved needles of quartz occur in clinopyroxene, suggesting that the original clinopyroxene was supersilicic. Earlier, Smith and Cheeney (1980) reported oriented quartz needles in sodic augite. Quartz needles of variable size have been identified in eclogitic omphacite from several UHP terranes including the WGR, Norway (Smith, 1988), the Bohemian massif (Bankun-Czubaow, 1992), the Munchberg massif (Gayk et al., 1995), and the Dabie-Sulu UHP terrane (Zhang et al., 1995; Liou and Zhang, 1998; Tsai, 1998; Zhang and Liou, 1998). Smith (1988) suggested that the existence of supersilicic Cpx requires high P-T conditions. Oriented quartz lamellae in omphacite are widely developed in the Rongcheng eclogite within gneisses of the Sulu region. The oriented quartz lamellae are 2 to 6 μm in width and 12 to 85 μm in length and are elongate along the *c*-axis of omphacite (Fig. 1C); they appear to have a topotaxial relation with the host phase.

Several experiments have demonstrated that high pressure is the most important factor required to stabilize supersilicic clinopyroxene. For example, Mao (1971) documented that: (1) jadeite-CaTsch clinopyroxene can contain up to 7.5 wt% excess SiO_2 at P up to 4.0 GPa at 1100 to 1700°C; and (2) the incorporation of excess SiO_2 in clinopyroxene is highly P-sensitive. A similar conclusion was confirmed for diopside-CaTsch ($CaAl_2SiO_6$) pyroxene by Khanukhova et al. (1976) and Wood and Henderson (1978) at 2.5 to 3.5 GPa, 1200 to 1450°C, and for diopside-Ca-eskola ($Ca_{0.5}AlSi_2O_6$) clinopyroxene by Zharikov et al. (1984) at 3.5 to 7 GPa, 1200°C. In a synthesis experiment for silica-rich Na-Mg pyroxene at 10 to 15 GPa and 1600°C, Angel et al. (1988) showed that the incorporation of excess silica could stabilize the Na-Mg pyroxene possibly up to 20 GPa, and that UHP Na-Mg clinopyroxene contains both tetrahedral- and octahedral-coordinated silicon. In summary, topotaxial intergrowths of quartz needles and omphacite in eclogitic rocks are best explained by quartz exsolution from a former supersilicic clinopyroxene stabilized at UBP conditions.

Clinoenstatite exsolution in diopside

So far, five polymorphs of (Mg,Fe)-pyroxene are known: orthoenstatite (space group *Pbca*), protoenstatite (*Pbcn*), low-P clinoenstatite ($P2_1/c$), high-P clinoenstatite (*C2/c*), and high-T clinoenstatite (*C2/c*) (Gasparik, 1990; Pacalo and Gasparik, 1990). Enstatite with orthorhombic structure has long been thought to be the stable polymorph of $(MgFe)_2Si_2O_6$ in the Earth's upper mantle. Several recent experiments indicate that orthoenstatite transforms to high-P *C2/c* clinoenstatite at P > 7 GPa (Fig. 2), corresponding to mantle depths of 200 to 300 km (Angel et al., 1992; Hugh-Jones et al., 1994; Woodland and Angel, 1997).

Clinoenstatite ($P2_1/c$) as exsolution lamellae in diopside recently has been found in the Alpe Arami garnet peridotite (Bozhilov et al., 1999); the low-P clinoenstatite lamellae are up to several microns long and tens to hundreds of nanometers wide. They are subparallel to (802) of the diopside oriented at about 22° with (100) planes of both host and exsolution phases. The lamellae contain nanometer-size antiphase domains with displacements at their boundaries of $1/2(a+b)$. This characteristic has been interpreted as a solid-state phase transition from initially high-P clinoenstatite ($C2/c$) at a minimum depth of ~250 km (Bozhilov et al., 1999). The clinoenstatite contains a considerable amount of $CaSiO_3$ and has a composition close to that of pigeonite. Experimentally determined phase relations concerning orthoenstatite and high-P and low-P clinoenstatite are restricted to end-member $FeSiO_3$ and $MgSiO_3$ compositions; the effect of minor Ca contents on phase boundaries has not been considered (Angel et al., 1992; Hugh-Jones et al., 1994; Woodland and Angel, 1997).

It should be mentioned that intergrowths of both orthoenstatite and clinoenstatite in magnesite-garnet pyroxenite from China have been described recently (Zhang, Shau et al., in review). The garnet pyroxenite is associated with coesite-bearing eclogite and consists of garnet, "enstatite," magnesite, and titanomagnetite. Petrochemical data suggest that this and associated rocks were recrystallized at P > 2.8 to 3.6 GPa and 750 to 850°C. High-resolution TEM study reveals that the enstatite (En_{85}) consists of clinoenstatite ($P2_1/c$) lamellae and orthoenstatite (*Pbca*); clinoenstatite lamellae and orthoenstatite host have nearly identical compositions. The intergrowth texture can be clearly seen on the grains of an *a-c* section denoted as the [010] zone. A few lamellae are very closely spaced and add up to 0.2 μm in total width, but most lamellae range from tens to hundreds of Å in width. In addition to the majority of the 18 Å for orthoenstatite and 9 Å lamellae for clinoenstatite, periodicities of 36 Å and 72 Å also are observed. All clinoenstatite lamellae have even numbers of 9 Å fringes without twins; the intensities of (2n,0,0) reflections are much stronger than (2n+1,0,0) reflections in electron-diffraction patterns.

It should be emphasized here that formation of the clinoenstatite lamellae in orthoenstatite is different from the exsolution process of a solid solution phase—i.e., an initially homogeneous solid solution phase separates into two distinct phases during cooling. The clinoenstatite lamellar textures can be treated as different polytypes of enstatite, with 9 Å, 18 Å, 36 Å, and 72 Å periodicities (Buseck and Iijima, 1975). Zhang et al., (in review) proposed that the clinoenstatite lamellae within orthoenstatite may have formed by three possible mechanisms: (1) it formed at conditions near the P-T boundary of ortho- and low-P clinoenstatite at the peak metamorphic stage; (2) it inverted from orthoenstatite as a result of deformation at shallow depths; or (3) it may have formed by a displacive transformation from a high-P clinoenstatite during decompression. On the basis of petrological information and TEM analyses, the last hypothesis seems to be the most reasonable. If so, some of the Yangtze supracrustal materials must have been subducted to mantle depths ≥200 km prior to Triassic continent collision.

Garnet exsolution in pyroxene

Exsolution of garnet in pyroxene was first recognized by Beck (1907) in eclogite from South African kimberlite pipes and has been described in a few UHP rocks (e.g., Reiche and Bautsch, 1985; Smith, 1988; Yang, 1991; Zhang et al., 1994; Zhang, Liou et al., in press). Exsolution of garnet from clinopyroxene is generally lamellar, although lens- and bleb-shaped garnets also are present. Exsolution lamellae of garnet ± ilmenite occur in coarse-grained diopside, but not in fine-grained diopside of garnet clinopyroxenites from the Sulu UHP terrane. Lamellae range from 0.01 to 0.1 mm in width and may be up to >10 vol% (Figs. 4A and 4B). The garnet is rich in pyrope and grossular (Zhang et al., 1994).

Garnet exsolution also has been found in clinopyroxene in eclogite xenoliths in kimberlite from South Africa and Yakutia, Russia (Jerde et al., 1993); in orthopyroxene in Grt-orthopyroxenite associated with an eclogite pod in the WGR, Norway (Smith, 1988); and in eclogite from the Gilsberg, Saxony Granulite Complex (Reiche and Bautsch, 1985). The Gilsberg eclogite occurs as an inclusion in serpentinite and shows oriented, coherent garnet exsolution lamellae in orthopyroxene. The garnet lamellae have widths of up to ~5 μm and orientation of $\{100\}_{Grt}//(100)_{Opx}$ and $<110>_{Grt}//[001]_{Opx}$. Exsolution of garnet from pyroxene has been suggested as having formed at high pressure and temperature, but the conditions permitting such exsolutions are obscure (Reiche and Bautsch, 1985).

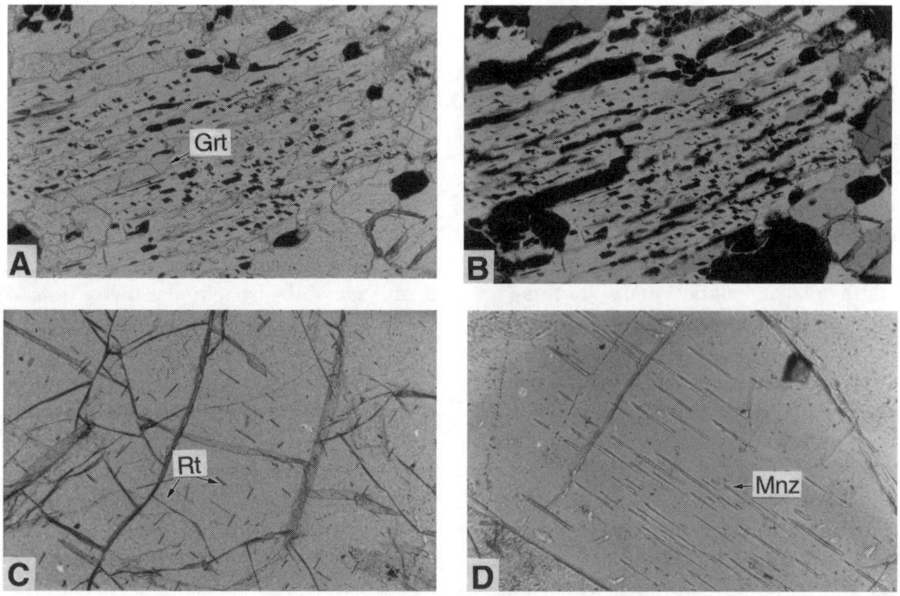

FIG. 4. Photomicrographs showing exsolution textures identified in garnet pyroxenites from the Dabie-Sulu UHP terrane. All are plane light except B (crossed polars). A and B. Lamellae-like exsolution of ilmenite (dark) and garnet in diopside from the Rizhao garnet clinopyroxenite in the Sulu region; width of view = 1 mm. C. Rutile needles in garnet from the Yangkou garnet clinopyroxenite; width of view = 1 mm. D. Monazite rods in apatite from the Maowu garnet clinopyroxenite; width of view = 0.39 mm.

Garnet

Pyroxene exsolution in garnet

Garnet is a principal upper-mantle phase hosting Al. Clinopyroxene exsolution lamellae in pyropic garnet in mantle xenoliths within diamond-bearing kimberlite pipes were first described by Haggerty and Sautter (1990) and Sautter et al. (1991). Pyropic garnet shows large concentrations of exsolved clinopyroxene along three directions, indicating solid solution of pyroxene in garnet (majorite) derived from mantle depths >300 km. The solid solution of pyroxene and garnet can be expressed as $^{VIII}M_3{}^{VI}(Al_{2-2n}M_nSi_n)^{IV}Si_3O^{12}$ ($0 \geq n \leq 1$) (Irifune, 1987), where M = Mg, Fe, and Ca; 2Al (+Cr)$^{3+}$ iron in the octahedral sites of garnet are replaced by $M^{2+}Si^{4+}$, and this phase may have tetragonal symmetry (Heineman et al., 1997). The majorite component in garnet, as shown in Figure 2, increases with increasing pressure.

Similar majoritic garnets (or supersilicic garnets) have been recently discovered in the Ugelvik and Raudhaugene garnet peridotite bodies from Otrøy, in the Western Gneiss Region of Norway (van Roermund et al., in press). Two distinct garnets occur in the peridotite: (1) matrix garnets ≥4 mm in diameter; and (2) huge isolated single garnet crystals up to 4 cm and polycrystalline garnet nodules up to 10 cm in size. The core of polycrystalline garnet in some larger garnet nodules contains two-pyroxene exsolution needles; interstitial Opx (also thought to be exsolution) occurs at grain boundaries. The spindle-shaped pyroxene needles are parallel to $<111>_{Grt}$, whereas the octagonal prismatic sections are parallel to $<110>_{Grt}$. Garnet is rich in pyrope (Prp_{77}); equilibrium P-T conditions for the coexistence of exsolved pyroxene and garnet are 805 ± 40°C and 3.2 ± 0.2 GPa. The maximum content of exsolved pyroxenes in garnet is 4.6 vol%, including 1 vol% intracrystalline and 3.6 vol% intercrystalline pyroxene exsolution; the recalculated composition of the majoritic host corresponds to a higher pressure of 6.4 GPa. This implies that the exsolution microstructure in garnet records much higher pressure than does the mineral chemistry; these mantle rocks may have been exhumed to the Earth's surface from depths of 185 to 200 km (van Roermund et al., in press).

Exsolved needles of rutile in garnet

Exsolved rutile needles in garnet were found in both eclogite and garnet pyroxenite in the Sulu UHP terrane (Ye et al., 1997; Zhang and Liou, 1998) and in eclogite from the WGR, Norway (Smith, 1988). Eclogite and/or garnet pyroxenite from Yangkou and Rizhao contain both porphyroblastic garnets 1 to 2.5 mm in size and fine-grained matrix garnets 0.1 to 0.3 mm in diameter. Rutile needles occur only in the cores of porphyroblastic garnets and show three orientations. These rutile needles are 0.01 to 0.2 mm by <0.02 to 0.03 mm, some up to 0.4 mm in length (Fig. 4C). The mechanism of rutile exsolution in garnet has not been investigated.

Apatite

Apatite occurs as a common minor phase in UHP eclogites and ultramafic rocks. In some apatite-bearing garnet pyroxenites with high REE concentrations from the Dabie and Kokchetav massifs, monazite was identified as an accessory rare-earth phosphate with the formula $REE^{3+}(PO_4)^{3-}$ (e.g., Zhang, Hacker et al., 1997; Liou and Zhang, 1998). Monazite shows three modes of occurrence in the Dabie pyroxenite: (1) elliptical grains as a matrix phase ranging from 0.1 to 0.5 mm in size; (2) fine-grained inclusions in garnet, clinopyroxene, and orthopyroxene; and (3) exsolved lamellae in apatite. Monazite lamellae 1 to 3 μm in width and 10 to 130 μm long occur in the core of apatite and parallel to the *c*-axis of the host phase (Fig. 4D); it seems to bear topotaxial relations with the host apatite, resulting from exsolution. The composition (wt%) of monazite lamellae in apatite is: Ce_2O_3, 26.8–30.3; La_2O_3, 26.8–30.3; Nd_2O_3, 16.3–17.6; Pr_2O_3, 4.1–4.3; Sm_2O_3, 2.1–2.8; Gd_2O_3, 3.01–3.8; CaO, 2.9–6.8; and P_2O_5, 30.5–31.0; with negligible ThO_2 (0.2–0.6) and SiO_2 (0.1).

For most rocks, small amounts of REE are included in apatites under UHP conditions, inasmuch as apatites are major sinks of REE elements. Similar to other trace phases described above, monazite lamellae may have been exsolved from host apatite during decompression. The size, abundance, and composition of monazite lamellae depend on the bulk-rock REE content. For extremely high REE–bearing rocks, such as those from the Dabie Shan, discrete monazite grains occur as a matrix mineral, as inclusions in omphacite and garnet, and as monazite lamellae in apatites. Many eclogitic apatites contain extremely fine exsolution lamellae of unknown phases in addition to monazite lamellae. These have been recognized in the Dabie-Sulu terrane (Carswell et al., 1996; Zhang and Liou, 1998) and the WGR, Norway (Smith, 1988).

Conclusions

1. Some exsolution microstructures in UHP minerals described above have also been reported in mantle xenoliths from kimberlite or from kimberlite itself. These include pyroxene lamellae in garnet (Haggerty and Sautter, 1990; Sautter et al., 1991), Mg-Al-Cr titanomagnetite exsolved from picroilmenite (Haggerty, 1991), garnet exsolved from pyroxene (Reiche and Bautsch, 1985; Jerde et al., 1993), and abundant, coarse, lamellar ilmenite in clinopyroxene (e.g., Dawson and Reid, 1970; Ringwood and Lovering, 1970; Boyd and Nixon, 1973). The intergrowth of ilmenite and clinopyroxene may have resulted from breakdown of a UHP Cpx (Ringwood and Lovering, 1970; Boyd and Nixon, 1973) or by eutectoid crystallization from kimberlitic magma (e.g., Wyatt, 1977), whereas the pyroxene lamellae in garnet are considered to represent majoritic garnet (e.g., Haggerty and Sautter, 1990). Generally speaking, these exsolution structures in kimberlitic rocks and xenoliths formed under very high P-T conditions at mantle depths >150 km. The similarity between exsolution microstructures in kimberlites and those in UHP rocks implies that most observed intergrowths in UHP minerals also have very high pressure precursors formed under deep-mantle conditions.

2. Mineral exsolution from breakdown of a solid-solution phase is a subsolidus reaction. A principal characteristic of UHP solid-solution phases, such as majoritic garnet and supersilicic clinopyroxene, is the substitution of sixfold coordinated Si for Al, Fe, and Mg in octahedral sites. For example, the substitution of $[2Al(+Cr)]^{VI}$ by $[M^{2+}Si^{4+}]^{VI}$, where M = Mg and Fe, occurs in majoritic garnet and $[Mg^{2+}Na^{1+}Al^{3+}]^{VI-VIII}$ by $[1.5Si^{4+}]^{VI}$ occurs in supersilicic clinopyroxene. Similarly, some large cations, such as K, are considered too large to enter the M2 site of clinopyroxene at low pressures. However, experimental studies (e.g., Harlow, 1997; Wang and Takahashi, 1999) concluded that clinopyroxene can accommodate a considerable amount of K_2O in octahedral sites and garnet can carry an important amount of Na_2O at mantle pressures. The results are consistent with the occurrence of K-bearing diopside in diamond-bearing rocks and higher Na con-

tents in garnets from diamond inclusions. Therefore, compositional and crystal-chemical features of minerals at very high pressures need to be investigated in order to understand the origin of mineral exsolutions, such as garnet lamellae from clinopyroxene, in UHP rocks.

3. Thus far, TEM studies on exsolution microstructures in UHP minerals are few in number and preliminary in detail; exsolution mechanisms are still unknown. The exsolved lamellae in many UHP minerals may have been produced during nearly isothermal decompression, in contrast to exsolution with falling temperatures for primary igneous minerals. Recent studies of mineral exsolution and polymorphs have provided a new window, allowing us to identify high-pressure precursors. Documentation of compositions, crystal structures, and stabilities of host and lamellae mineral pairs probably will lead to better constraints regarding the processes of mantle dynamics and continental subduction.

Acknowlegments

We thank W. G. Ernst for his constructive review and many scientists for their contributions to the topic covered in this study. The research was supported by NSF EAR 95-06468 and EAR 9814468.

REFERENCES

Akaogi, M., Ito, M., and Navrotsky, A., 1989, Olivine-modified spinel-spinel transitions in the system Mg_2SiO_4-Fe_2SiO_4 calorimetric measurements, thermochemical calculation, and geophysical application: Jour. Geophys. Res., v. 94, p. 15,671–15,685.

Akaogi, M., Yusa, H., Shiraishi, K., and Suzuki, T., 1995, Thermodynamic properties of a-quartz, coesite, and stishovite and equilibrium phase relations at high pressure and high temperature: Jour. Geophys. Res., v. 100, p. 22,337–22,347.

Ambler, E. P., and Ashley, P. M., 1977, Vermicular orthopyroxene-magnetite symplectites from the Wateranga layered mafic intrusion, Queensland, Australia: Lithos, v. 10, p. 163–172.

Angel, R. J., and Hugh-Jones, D. A., 1994, Equation of state and thermodynamic properties of enstatite pyroxenes: Jour. Geophys. Res., v. 99, p. 19,777–19,783.

Angel, R. J., Chopelas, A., and Ross, N. E., 1992, Stability of high-density clinoenstatite at upper-mantle pressures: Nature, v. 358, p. 322–324.

Angel, R. J., Gasparik, T., Ross, N. L., Finger, L. W., Prewitt, C. T., and Hazen, R. M., 1988, A silica-rich pyroxene phase with six-coordinated silicon: Nature, v. 335, p. 156–158.

Bakun-Czubarow, N., 1992, Quartz pseudomorphs after coesite and quartz exsolutions in eclogitic omphacites of the Zlote Mountains in the Sudetes (SW Poland): Archiwum Mineral., v. 48, p. 3–25.

Beck, R., 1907, Untersuchungen über einige südafrikanische Diamantlagerstätten: Zeit. Deutsch. Geol. Ges., v. 59, p. 275–307.

Blanchard, M. B., and Cunningham, G. C., 1974, Artificial meteor ablation studies: Olivine: Jour. Geophys. Res., v. 79, p. 3973–3990.

Bozhilov, K. N., Green H. W., and Dobrzhinetskaya, L., 1998, Clinoenstatite in the Alpe Arami Peridotite: Additional evidence of very high pressure: Science, v. 284, p. 128–132.

Boyd, F. R., and Nixon, P. H., 1973, Origin of the ilmenite-silicate nodules in kimberlites from Lesotho and South Africa, *in* Nixon, P. H., ed., Lesotho kimberlites: Lesotho Nat. Dvlpt. Corp., p. 254–268.

Bundy, F. R., 1980, The P, T phase and reaction diagram for elemental carbon: Jour. Geophys. Res., v. 85, p. 6930–6936.

Buseck, P. R., and Iijima, S., 1975, High resolution electron microscopy of enstatite II: Geology application: Amer. Mineral., v. 60, p. 771–784.

Carswell, D. A., Wilson, R. N., and Zhai, M., 1996, Ultra-high pressure aluminous titanites in carbonate-bearing eclogites at Shuanghe in Dabieshan, central China: Mineral. Mag., v. 60, p. 461–471.

Champness, P. E., 1970, Nucleation and growth of iron oxides in olivine $(Mg,Fe)_2SiO_4$: Mineral. Mag., v. 37, p. 790–800.

Chopin, C., 1984, Coesite and pure pyrope in high-grade blueschists of the western Alps: A first record and some consequences: Contrib. Mineral. Petrol., v. 86, p. 107–118.

Chopin, C., and Schertl, H.-P., 1999, The UHP unit in the Dora-Maira massif, Western Alps: INT. GEOL. REV., v. 41, p. 765–780.

Dawson, J. B., and Reid, A. M., 1970, A pyroxene-ilmenite intergrowth from the Monastery mine, South Africa: Contrib. Mineral. Petrol., v. 26, p. 296–301.

Dobrzhinetskaya, L., and Green, H. W., 1997, TiO_2-solubility in olivine and composition of coexisting titanate at very high P & T: Preliminary results [abs.]: EOS (Trans. Amer. Geophys. Union), v. 78, p. F786.

Dobrzhinetskaya, L., Green, H. W., and Bozbilov, K. N., 1998, Titanium and precipitation in olivine at high pressure and temperature [abs.]: EOS (Trans. Amer. Geophys. Union), v. 79, p. F972.

Dobrzhinetskaya, L., Green, H. W., and Wang, S., 1996, Alpe Arami: A peridotite massif from depths of more than 300 kilometers: Science, v. 271, p. 1841–1846.

Doroshev, A. M., Sobolev, N. V., and Brey, G., 1992, Experimental evidence of high-pressure origin of the potassium-bearing pyroxenes: 29th Int. Geol. Cong., v. 2, p. 602.

Gasparik, T., 1990, Phase relations in the transition zone: Jour. Geophys. Res., v. 95, p. 15,751–15,769.

Gayk, T., Kleinschrodt, R., Langosch, A., and Seidel, E., 1995, Quartz exsolution in clinopyroxene of high-pressure granulite from the Munchberg Massif: Eur. Jour. Mineral., v. 7, p. 1217–1220.

Goode, A. D. T., 1974, Oxidation of natural olivines: Nature, v. 248, p. 500–501.

Green, H. W., II, and Dobrzhinetskaya, L., 1997, Response (Determining the origin of ultrahigh-pressure lherzolites, by Hacker et al., 1997): Science, v. 278, p. 704–707.

Green, H. W., II, Dobrzhinetskaya, L., Riggs, E. M., and Jin, Z. M., 1997, Alpe Arami: A peridotite massif from the mantle transition zone?: Tectonophys., v. 279, p. 1–21.

Hacker, B. R., Sharp, T., Zhang, R. Y., Liou, J. G., and Hervig, R. L., 1997, Determining the origin of ultrahigh-pressure lherzolites: Science, v. 278, p. 702–704.

Haggerty, S. E., 1991, Oxide textures—a mini-atlas, *in* Lindsley, D. H., ed., Oxide minerals: Petrologic and magnetic significance: Rev. Mineral., v. 25, p. 129–219.

Haggerty, S. E., and Baker, I., 1967, The alteration of olivine in basaltic and associated lavas. Part 1. High-temperature alteration: Contrib. Mineral. Petrol., v. 16, p. 233–257.

Haggerty, S. E., and Sautter, V., 1990, Ultradeep (greater than 300 kilometers) ultramafic upper mantle xenoliths: Science, v. 248, p. 993–996.

Harlow, G. E., 1997, K in clinopyroxene at high pressure and temperature: An experimental study: Amer. Mineral., v. 82, p. 259–269.

Harlow, G. E., and Veblen, D. R., 1991, Potassium in clinopyroxene inclusion from diamonds: Science, v. 251, p. 652–655.

Heineman, S., Sharp, T. G., Seifert, F., and Rubie, R., 1997, The cubic-tetragonal phase transition in the system majorite-pyrope and garnet symmetry in the Earth's transition zone: Phys. Chem. Mineral., v. 24, p. 206–221.

Helffrich, G. R., Stein, S., and Wood, B. J., 1989, Subduction zone thermal structure and mineralogy and their relationship to seismic wave reflections and conversions at the slab/mantle interface: Jour. Geophys. Res., v. 94, p. 753–756.

Hiramatsu, N., Banno, S., Hirajama, T., and Cong, B., 1995, Ultrahigh-pressure garnet lherzolite from Chijiadian, Rongcheng county, in the Su-Lu region of eastern China: The Island Arc, v. 4, p. 324–333.

Hugh-Jones, D. A., Woodland, A. B., and Angel, R. J., 1994, The structure of high-pressure C2/c ferrosilite and crystal chemistry of high-pressure C2/c pyroxenes: Amer. Mineral., v. 79, p. 1032–1042.

Irifune, T., 1987, An experimental investigation of the pyroxene-garnet transformation in a pyrolite composition and its bearing on the constitution of the mantle: Phys. Earth Planet. Inter., v. 45, p. 324–336.

Jerde, E. A., Taylor, L. A., Crozaz, C., and Sobolev, N. V., 1993, Exsolution of garnet within clinopyroxene of mantle eclogite: Major- and trace-element chemistry: Contrib. Mineral. Petrol., v. 114, p. 148–159.

Jin, Z. M., Jin, S. Y. Gao, S., and Zhao, W. X., 1998, Is the depth of formation of ultrahigh-P rock from the Dabie Mountains limited to 100–150 km?: Discovery of Ti-Cr magnetite needle and its significance for dynamics: Chinese Sci. Bull., v. 43, p. 767–771 (in Chinese).

Kadarusman, A., and Parkinson, C. D., in press, Petrology and P-T evolution of garnet peridotites from central Sulawesi, Indonesia: Jour. Metamor. Geol.

Khanukhova, L. T., Zharikov, V. A., Ishbulatov, R. A., and Litvin, Y. A., 1976, Excess silica in solid solutions of high-pressure clinopyroxenes as shown by experimental study of the system $CaMgSi_2O_6$-$CaAl_2SiO_2$ at 35 kilobars and 1200°C: Doklady Earth Sci. Sect., v. 229, p. 170–172.

Liou, J. G., and Zhang, R. Y., 1998, Petrogenesis of an ultrahigh-P garnet-bearing ultramafic body from Maowu, the Dabie Mountains, central China: The Island Arc, v. 7, p. 115–134.

Liou, J. G., Zhang, R. Y., Ernst, W. G., Rumble, D., and Maruyama, S., 1998, High-pressure minerals from deeply subducted metamorphic rocks, *in* Hemley, R. J., ed., Ultrahigh-pressure mineralogy: Physics and chemistry of the Earth's deep interior: Rev. Mineral., v. 37, p. 33–96.

Luth, R. W., 1992, Potassium in clinopyroxene at high pressure: Experimental constraints [abs.]: EOS (Trans. Amer. Geophys. Union), v. 73, p. 608.

______, 1995a, Potassium in clinopyroxene at high pressure [abs.]: EOS (Trans. Amer. Geophys. Union), v. 76, p. F711.

______, 1995b, Is phase A relevant to the Earth's mantle?: Geochim. et Cosmochim. Acta., v. 59, p. 679–682.

Mao, H. K., 1971, The system jadeite ($NaAlSi_2O_6$)-anorthite ($CaAl_2Si_2O_8$) at high pressures: Carnegie Inst. Yearbook, v. 69, p. 163–168.

Mehta, A. K., Leinenweber, K., Navrotsky, A., and Akaogi, M., 1994, Calorimetric study of high pressure polymorphism in $FeTiO_3$; stability of perovskite phase: Phys. Chem. Minerals, v. 21, p. 207–212.

Mercier, J.-C. C., and Nicolas, A., 1975, Textures and fabrics of upper-mantle peridotites as illustrated by xenoliths from basalts: Jour. Petrol., v. 16, p. 454–487.

Moseley, D., 1984, Symplectic exsolution in olivine: Amer. Mineral., v. 69, p. 139–153.

Nakajima, Y., and Ogasawara, Y., 1997, Garnet lherzolite with olivine-Ti mineral intergrowth from Rongcheng in the Sulu UHP terrane, eastern China: EOS (Trans. Amer. Geophys. Union), v. 78, p. F787.

Okamoto, K., and Maruyama, S., 1998, Multi-anvil re-equilibration experiments of a Dabie Shan ultrahigh-pressure eclogite within the diamond-stability fields: The Island Arc, v. 7, p. 52–69.

Okamoto, K., Maruyama, S., and Liou, J. G., 1997, Experimental study of TiO_2 content in olivine under ultrahigh-pressure conditions: EOS (Trans. Amer. Geophys. Union), v. 78, p. F761.

Pacalo, R. E. G., and Gasparik, T., 1990, Reversals of the orthoenstatite-clinoenstatite transition at high pressures and high temperature: Jour. Geophys. Res., v. 95, p. 15,853–15,858.

Pawley, A. R., and Wood, B. J., 1996, The low-pressure stability of phase A, $Mg_7Si_2O_8(OH)_6$: Contrib. Mineral. Petrol., v. 124, p. 90–97.

Putnis, A., 1979, Electron petrography of high-temperature oxidation in olivine from the Rhum layered intrusion: Mineral. Mag., v. 43, p. 293–296.

Reiche, M., and Bautsch, H.-J., 1985, Electron microscopical study of garnet exsolution in orthopyroxene: Phys. Chem. Minerals, v. 12, p. 29–33.

Rietmeuer, F. J. M., 1996, Cellular precipitates of iron oxide in olivine in a stratospheric interplanetary dust particle: Mineral. Mag., v. 60, p. 877–885.

Ringwood, A. E., and Lovering, J. F., 1970, Significance of pyroxene-ilmenite intergrowths among kimberlite xenoliths: Earth Planet. Sci. Lett., v. 7, p. 371–375.

Risold, A. C., and Trommsdorff, V., 1996, Alpe Arami and Cima di Gagnone garnet peridotites: Observations contradicting the hypothesis of ultradeep origin: EOS (Trans. Amer. Geophys. Union), v. 77, p. F761.

Risold, A. C., Trommsdorff, V., Reusser, E., and Ulmer, P., 1997, Genesis of $FeTiO_3$ inclusions in garnet peridotites from the central Alps: Terra Abs., v. 9, p. 28–29.

Ross, C. R., II, Armbruster, T., and Canil, D., 1992, Crystal structure refinement of a spinelloid in the system Fe_3O_4-Fe_2SiO_4: Amer. Mineral., v. 77, p. 507–511.

Ryabchikov, I. D., and Ganeyev, I. I., 1990, Isotopic substitution of potassium in clinopyroxene at high pressures: Geokhimiya, v. 1, p. 3–12.

Sautter, V., Haggerty, S. E., and Field, F., 1991, Ultradeep (>300 kilometers) ultramafic xenoliths: Petrological evidence from the transition zone: Science, v. 252, p. 827–830.

Schmidt, M. W., 1996, Experimental constraints on recycling of potassium from subducted oceanic crust: Science, v. 272, p. 1927–1930.

Shatsky, V. S., Sobolev, N. V., and Vavilov, A A., 1995, Diamond-bearing metamorphic rocks of the Kokchetav massif (northern Kazakhstan), *in* Coleman, R. G., and Wang, X., eds., Ultrahigh-pressure metamorphism: New York, Cambridge University Press, p. 427–455.

Smith, D. C., 1984, Coesite in clinopyroxene in the Caledonides and its implications for geodynamics: Nature, v. 310, p. 641–644.

______, 1988, A review of the peculiar mineralogy of the "Norwegian coesite-eclogite province," with crystal-chemical, petrological, geochemical, and geodynamical notes and an extensive bibliography, *in* Smith, D. C., ed., Eclogites and eclogite-facies rocks: Amsterdam, Elsevier, p. 1–206.

Smith, D. C., and Cheeney, R. F., 1980, Oriented needles of quartz in clinopyroxene: Evidence for exsolution of SiO_2 from a non-stoichiometric supersilicic "clinopyroxene" [abs.]: 26th Int. Geol. Cong., Paris, p. 145.

Sobolev, N. V., and Shatsky, V. S., 1990, Diamond inclusions in garnets from metamorphic rocks: Nature, v. 343, p. 742–746.

Tsai, C. H., 1998, Petrology and geochemistry of mafic-ultramafic rocks in the north Dabie Complex, central-eastern China: Unpubl. Ph.D. thesis, Stanford University.

Ulmer, P., and Trommsdoff, V., 1997, Titanium content of mantle olivine: An experimental study to 10 GPa [abs.]: Terra Abs., v. 9, p. 39.

van Roermund, H., Drury, M., Barnhoom, A., and Ronde, A., in press, Super-silicic garnet microstructures from an orogenic garnet peridotite: Evidence for an ultra-deep (>6 GPa) origin: Jour. Metamor. Geol.

Wang, W., and Takahashi, E., 1999, Subsolidus and melting experiments of a K-rich basaltic composition to 27 GPa: Implication for the behavior of potassium in the mantle: Amer. Mineral., v. 84, p. 357–361.

Wood, B. J., and Henderson, C. M. B., 1978, Compositions and unit-cell parameters of synthetic non-stoichiometric tschermakitic clinopyroxene: Amer. Mineral., v. 63, p. 66–72.

Woodland, A. B., and Angel, R. J., 1997, Reversal of the orthoferrosilite–high-P clinoferrosilite transition, a phase diagram for $FeSiO_3$, and implications for the mineralogy of the Earth's upper mantle: Eur. Jour. Mineral., v. 9, p. 245–254.

______, 1998, Crystal structure of a new spinelloid with the wadsleyite structure in the system Fe_3O_4-Fe_2SiO_4 and implications for the Earth's mantle: Amer. Mineral., v. 83, p. 404–408.

Wyatt, B. A., 1977, The melting and crystallization behavior of a natural clinopyroxene-ilmenite intergrowth: Contrib. Mineral. Petrol., v. 61, p. 1–9.

Yang, J., 1991, Eclogite and related ultramafic rocks in the Su-Lu region, eastern China: Beijing, Geol. Publ. House, p. 1–99 (in Chinese with English abstract).

Ye, K., Cong, B. L., and Ye, D. L., 1997, Exsolutions of clinopyroxene, rutile, and apatite from garnet in eclogite at Yangkou, Shandong peninsula, eastern China [abs.]: Terra Nova, v. 9, p. 41.

Zhang, R. Y., Hacker, B., Liou, J. G., 1997, Exsolution in ultrahigh Sulu (China) and Kokchetav (Kazakhstan) terranes [abs.]: Terra Nova, v. 9, p. 43.

Zhang, R. Y., Hirajima, T., Banno, S., Cong, B., and Liou, J. G., 1995, Petrology of ultrahigh-pressure rocks from the southern Sulu region, eastern China: Jour. Metamor. Geol., v., 13, p. 659–675.

Zhang, R. Y., and Liou, J. G., 1998, Ultrahigh-pressure metamorphism of the Sulu terrane, eastern China: A prospective view: Continental Dynamics, v. 3, p. 32–53.

Zhang, R. Y, Liou, J. G., Coleman, R. G., Ernst, W. G., Sobolev, N. V., and Shatsky, V. S., 1997, Metamorphic evolution of diamond-bearing and associated rocks from the Kokchetav massif, northern Kazakhstan: Jour. Metamor. Geol., v. 15, p. 479–496.

Zhang, R. Y., Liou, J. G., and Cong, B. L., 1994, Petrogenesis of garnet-bearing ultramafic rocks and associated eclogites in the Su-Lu ultrahigh-pressure metamorphic terrane, China: Jour. Metamor. Geol., v. 12, p. 169–186.

Zhang, R. Y., Liou, J. G., Yang, J. S., and Yui, Z.-P., in press, Petrochemical constraints for dual origin of garnet peridotites from the Dabie-Sulu UHP terrane, eastern-central China: Jour. Metamor. Geol.

Zhang, R. Y., Shau, Y. H., Liou, J. G., and Lo, C. H., in review, Occurrence of clinoenstatite in ultradeep (>200 kilometers) magnesite-bearing garnet pyroxenite in the Sulu UHP terrane, eastern China: Jour. Metamor. Geol.

Zhang, R. Y., Shu, J. F., Mao, H. K., and Liou, J. G., 1999, Magnetite lamellae in olivine and clinohumite from Dabie UHP ultramafic rocks, central China: Amer. Mineral., in press.

Zharikov, V. A., Ishbulatov, R. A., and Chudinovskikh, L. T., 1984, High-pressure clinopyroxene and the eclogite barrier: Sov. Geol. Geophys., v. 25, p. 53–61.

Summary of Research on Stable-Isotope Geochemistry of UHP Rocks, 1993 to 1998

DOUGLAS RUMBLE, III

Geophysical Laboratory, 5251 Broad Branch Road, NW, Washington, District of Columbia 20015-1305

Abstract

The purpose of this paper is to provide a guide to research results and published literature on stable-isotope geochemistry of UHP rocks for the period from 1993 to 1998. Although some articles on blueschist- and lower-eclogite-facies rocks are included, the predominant focus is on coesite- and diamond-bearing rocks of the eclogite facies. Attention is called to a number of studies of fluid inclusions in UHP rocks. Even though stable-isotope geochemistry may not play a direct role in fluid-inclusion research, the link between the two methods is too strongly complementary to be ignored. A number of references are made to works in press and to abstracts presented at the 1998 International Workshop on UHP Metamorphism and Exhumation, Stanford University. These informal references to abstracted research are made in order to alert the reader to ongoing projects that are likely to be published in the near future.

Reviews

THE REVIEW BY BEBOUT (1996) assessed constraints on the cycling of volatile components through convergent margins, including inputs of oceanic crust into subduction zones and outputs via magmatic degassing and forearc venting, with a discussion of the evidence for fluid production and migration obtained in stable-isotope geochemical studies. Although not specifically concerned with UHP metamorphism, Bebout (1996) provided a valuable introduction to the current literature on the stable-isotope geochemistry of high-pressure, blueschist-facies metamorphism. Pearson et al. (1995) presented a broad array of geochemical data, including C and O isotopes, on diamond-bearing orogenic ultramafic rocks. Rumble (1998) reviewed stable-isotope data bearing on fluid-rock interactions during suduction of continental crust into the coesite- and diamond-eclogite facies of the upper mantle with an emphasis on recent results from China. The foregoing three reviews thus span a broad range of convergent tectonic environments including subduction of both oceanic and continental crust as well as emplacement of mantle slivers into the crust.

China

A recurring theme of stable-isotope studies of Chinese UHP rocks, both in Dabieshan and Sulu, is the surprising preservation of pristine, protolith isotope characteristics despite intense deformation and extreme pressure (P) and temperature (T) conditions during continental collision. The relatively high $\delta^{13}C$ values of +5 to +6‰ found in Dabieshan marbles, for example, are generally held to have been established prior to metamorphism during sedimentation and diagenesis (Baker et al., 1997; Yui et al., 1997; Zheng et al., 1998a). But the big surprise was the discovery of unusually low, negative $\delta^{18}O$ and δD values, as low as −10‰ for eclogite garnet and −120 for phengite, respectively (Baker et al., 1994, 1997; Yui et al., 1994, 1995a, 1995b, 1997; Chen et al., 1995; Zheng et al., 1995, 1996; Zhang et al., 1998). Most workers regard the low $\delta^{18}O$ and δD values as recording an ancient, pre-subduction geothermal episode that altered surface rocks with heated meteoric water in a cold climate (Rumble, 1998; Rumble and Yui, 1999; Zheng et al., 1998b; Fu et al., 1999). The persistence of protolith isotope values, and especially the maintenance of differences in isotopic composition throughout the geodynamic cycle, denies the existence of a pervasive, freely circulating fluid phase during UHP metamorphism.

Oxygen-isotope geothermometry has proven relatively successful, in that high-temperature fractionations are found for a number of mineral pairs in Chinese eclogites and, moreover, instances of agreement between isotopic and chemical thermometry are not uncommon (Yui et al., 1995a, 1995b; Zheng et al., 1996, 1998b; Fu et al., 1999; Rumble and Yui, 1999). High-temperature thermometry invalidates the widespread presence of a grain-boundary fluid phase during exhumation because high-T fractionations are not preserved under wet conditions (Rumble and Yui, 1999). These findings are consistent with the occurrence of coesite in surface outcrops, as the mineral would have inverted to quartz in the presence of water.

High-temperature oxygen-isotope fractionations have been measured between not only anhydrous UHP minerals such as pyrope, omphacite, kyanite, and rutile, but also between hydrous UHP minerals like phengite and epidote (Yui et al., 1995a, 1995b; Zheng et al., 1996, 1998b; Fu et al., 1999; Rumble and Yui, 1999). The presence of phengite and epidote in oxygen-isotope equilibrium with UHP mineral assemblages demonstrates that subduction is effective in transporting mineral-bound water at least as deeply as the upper mantle.

Data published thus far show that unusually low $\delta^{18}O$ values in eclogites, gneisses, quartzites, and metagranitoids are distributed over tens of kilometers in length and hundreds of km^2 in area both in Dabieshan and in Sulu (Baker et al., 1997; Yui et al., 1997; Rumble, 1998; Zheng et al., 1998b; Fu et al., 1999; Rumble and Yui, 1999). Most researchers agree that these relationships (together with low δD values) identify an ancient, pre-collision geothermal system that has not been obliterated by subsequent deformation during subduction and exhumation. A consequence of the interpretation is that eclogites and their host gneisses all must have experienced UHP metamorphism, a concept termed in situ UHP metamorphism.

Europe

Studies of European HP and UHP metamorphism have led the way in terms of fluid inclusion research and also have contributed noteworthy stable-isotope discoveries. In general, there is congruence between fluid-inclusion and stable-isotope results in that both give evidence of local rock control of fluid compositions at length scales as small as centimeters. Journal articles with fluid-inclusion data are listed in the references for the convenience of the reader, but not discussed in detail herein.

There are a number of striking parallels between European and Chinese observations. High-temperature mineral-mineral oxygen-isotope fractionations are preserved (Sharp et al.,1993; Medaris et al., 1995). Protolith oxygen- and carbon-isotope compositions are little changed and heterogeneities in isotopic composition are present at scales of centimeters to meters (Nadeau et al., 1993; Philippot, 1993; Sharp et al., 1993; Fruh-Green, 1994; Getty and Selverstone, 1994; Barnicoat and Cartwright, 1997; Philippot et al., 1998). Significant correlations between low $\delta^{18}O$ values and high fluid salinities in contrast to high $\delta^{18}O$ and low salinity suggest that a record of seafloor hydrothermal alteration has persisted throughout subduction, metamorphism, and exhumation (Philippot et al., 1998; cf. Barnicoat and Cartwright, 1997). As is the case in China, data for European HP and UHP rocks invalidate the existence of a pervasive fluid freely infiltrating across lithologic boundaries.

Among the differences between European and Chinese UHP rocks, perhaps none is more striking than the prevalence of oceanic crustal protoliths in Europe and the predominance of continental crust in China. Despite such a big difference in the provenance of subducted rocks in the two areas, the metamorphic processes they experienced were similar.

Another contrast between Chinese and European HP and UHP rocks is the action of fluid infiltration along shear zones found in the latter area. Oxygen-isotope shifts within shear zones relative to wall rocks are attributed to fluid advection parallel to shearing (Fruh-Green, 1994). The remarkable eclogitized shear zones found crosscutting granulite-facies gneisses at Holsnoy, Norway, show both fluid-inclusion and stable-isotope evidence of channelized fluid flow (Mattey et al., 1994; Van Wick et al., 1996). One hesitates to ascribe the absence of such features in China to differences in geodynamic processes, however, because accidents of weathering and exposure may play a decisive role.

Russia and Kazakhstan

Recent data from Russia and Kazakhstan echo themes evident in Chinese and European studies. Protolith isotopic values have been measured despite UHP metamorphism in the diamond-eclogite facies. The Maksyutov complex of the southern Urals, containing possible graphite pseudomorphs after diamond, exhibits the classic $\delta^{13}C$ split between marbles with marine carbonate values (approx. 0‰) and reduced carbon with biogenic values of –21 to –40‰ (Leech and Ernst, 1998). Metamorphic diamonds from Kokchetav, northern Kazakhstan, are –10.5 to –10.2 $\delta^{13}C$ and +5.9 to +8.5 $\delta^{15}N$ (‰). Diamond's intermediate carbon-isotope values are believed to reflect mixing of marine carbonate with biogenic reduced carbon, whereas its nitrogen isotopes are typical of metasedimentary rocks (De Corte et al., 1998).

Acknowledgments

G. E. Bebout, L. G. Medaris, Pascal Philippot, Z. D. Sharp, and J. W. Valley kindly supplied comments and references.

REFERENCES[1]

Reviews

Bebout, G. E., 1996, Volatile transfer and recycling at convergent margins: Mass-balance and insights from high–P/T metamorphic rocks, *in* Bebout, G. E., Scholl, D. W., Kirby, S. H., and Platt, J. P., eds., Subduction top to bottom, Washington, DC: Amer. Geophys. Union, p. 179–194.

Pearson, D. G., Davies, G. R., and Nixon, P. H., 1995, Orogenic ultramafic rocks of UHP (diamond facies) origin, *in* Coleman, R. G., and Wang, X., eds., Ultrahigh pressure metamorphism: New York, Cambridge: Cambridge University Press, p. 456–510.

Rumble, D., 1998, Stable isotope geochemistry of ultrahigh-pressure rocks, *in* Hacker, B. R., and Liou, J. G., eds., When continents collide: Geodynamics and geochemistry of ultrahigh-pressure rocks: Dordrecht, Kluwer Acad. Publ., p. 241–259.

Europe

Alps

Barnicoat, A. C., and Cartwright, I., 1997, Focused fluid flow during subduction: Oxygen isotope data from high pressure ophiolites of the western Alps: Earth Planet. Sci. Lett., v. 132, p. 53–61.

Fruh-Green, G. L., 1994, Interdependence of deformation, fluid infiltration, and reaction progress recorded in eclogitic metagranitoids (Sesia Zone, western Alps): Jour. Metamor. Geol., v. 12, p. 327–343.

Getty, S. R., and Selverstone, J., 1994, Stable isotope and trace element evidence for restricted fluid migration in 2 GPa eclogites: Jour. Metamor. Geol., v. 12, p. 747–760.

Nadeau, S., Philippot, P., and Pineau, F., 1993, Fluid inclusion and mineral isotopic compositions (H-C-O) in eclogite rocks as tracers of local fluid migration during high pressure metamorphism: Earth Planet. Sci. Lett., v. 114, p. 431–448.

Philippot, P., 1993, Fluid-melt-rock interaction in mafic eclogites and coesite-bearing metasediments: Constraints on volatile recycling during subduction: Chem. Geol., v. 108, p. 93–112.

______, 1996, The composition of high-pressure fluids from the Alps: Natural observations vs. experimental constraints: Earth Sci. Frontiers, v. 3 p. 39–52.

Philippot, P., Agrinier, P., and Scambelluri, M., 1998, Chlorine cycling during subduction of altered oceanic crust: Earth Planet. Sci. Lett., v. 161, p. 33–44.

Philippot, P., Chevallier, P., Chopin, C., and Dubessy, J., 1995, Fluid composition and evolution in coesite-bearing rocks (Dora Maira massif, western Alps): Implications for element recycling during subduction: Contrib. Mineral. Petrol., v. 121, p. 29–44.

Scambelluri, M., Pennacchioni, G., and Philippot, P., 1998, Salt-rich aqueous fluids formed during eclogitisation of metabasites in the alpine continental crust (Austroalpine Mt. Emilius unit, Italian western Alps): Lithos, v. 43, p. 151–167.

Scambelluri, M., and Philippot, P., 2000, Deep cycling of fluids in subduction zones, *in* Fluid inclusion studies at the turn of the century: Eur. Mineral. Union, in press.

Scambelluri, M., Piccardo, G., Philippot, P., Robbiano, A., and Negretti, L., 1997, High-salinity brines from recycled seawater in deeply subducted serpentinite: Earth Planet. Sci. Lett., v. 148, p. 485–499.

Sharp, Z. D., Essene, E. J., and Hunziger, J. C., 1993, Stable isotope geochemistry and phase equilibria of coesite-bearing whiteschists, Dora Maira massif, western Alps: Contrib. Mineral. Petrol., v. 114, p. 1–12.

Bohemia

Medaris, L. G., Beard, B. L., Johnson, C. M., Valley, J. W., Spicuzza, M. J., Jelinek, E., and Misar, Z., 1995, Garnet pyroxenite and eclogite in the Bohemian Massif: Geochemical evidence for Variscan recycling of subducted lithosphere: Geol. Rundsch., v. 84, p. 489–505.

Norway

Mattey, D., Jackson, D. H., Harris, N. B. W., and Kelley, S., 1994, Isotopic constraints on fluid infiltration from an eclogite facies shear zone, Holsenoy, Norway: Jour. Metamor. Geol., v. 12, p. 311–325.

Van Wick, N., Valley, J. W., and Austrheim, H., 1996, Oxygen and carbon isotopic constraints on the development of eclogites, Holsnoy, Norway: Lithos, v. 38, p. 129–145.

Russia

Maksyutov

Leech, M. L., and Ernst, W. G., 1998, Graphite pseudomorphs after diamond? A carbon isotope and spectro-

[1] References on stable-isotope geochemistry of UHP rocks, 1993 to 1999.

scopic study of graphite cuboids from the Maksyutov complex, south Ural Mountains, Russia: Geochim. et Cosmochim. Acta, v. 62, p. 2143–2154.

Kazakhstan

Kokchetav

Cartigny, P., De Corte, K., Shatsky, V. S., De Paepe, P., Sobolev, N. V., and Javoy, M., in prep., An insight into ultrahigh pressure metamorphic fluids; a nitrogen and carbon isotope study of microdiamonds from the Kokchetav massif, Kazakhstan.

De Corte, K., Cartigny, P., Shatsky, V. S., De Paepe, P., Sobolev, N. V., and Javoy, M., in press, Characteristics of microdiamonds from UHPM rocks of the Kokchetav massif (Kazakhstan): Proc. 7th Int. Kimberlite Conf.

De Corte, K., Cartigny, P., Shatsky, V. S., Sobolev, N. V., and Javoy, M., in press, First evidence of inclusions in metamorphic microdiamonds from the Kokchetav massif, northern Kazakhstan: Geochim. et Cosmochim. Acta.

De Corte, K., Shatsky, V. S., Taylor, W., Cartigny, P., Korsakov, A., and Trautman, R., 1998, What in situ microdiamonds tell us about UHPM [abs.], *in* International Workshop on UHP Metamorphism and Exhumation: Stanford, CA, Stanford University, p. A67–A69.

China

Dabieshan

Baker, J., Matthews, A., Mattey, D., Rowley, D. B., and Xue, F., 1997, Fluid-rock interaction during ultra-high pressure metamorphism, Dabie Shan, China: Geochim. et Cosmochim. Acta, v. 61, p. 1685–1696.

Baker, J., Matthews, A., Mattey, D., Xue, F., and Rowley, D. B., 1994, Fluid infiltration during ultra-high pressure metamorphism, Dabie Shan, China [abs.], *in* First Workshop on UHP Metamorphism: Stanford, CA, Stanford University, p. A75.

Chen, J., Shieh, Y. N., and Sun, X., 1995, A preliminary study of oxygen-isotope geochemistry of eclogites from the Dabie ultra-high pressure metamorphic terrane: Chinese Sci. Bull., v. 40, p. 120–122.

Fu, B., Zheng, Y.-F., Wang, Z., Xiao, Y., Gong, B., and Li, S., 1999, Oxygen and hydrogen isotope geochemistry of gneisses associated with ultrahigh pressure eclogites at Shuanghe in the Dabie Mountains: Contrib. Mineral. Petrol., v. 134, p. 52–66.

Rumble, D., 1998, Stable isotope geochemistry of the Dabie-Sulu UHP terrains, China [abs.], *in* International Workshop on UHP Metamorphism and Exhumation: Stanford, CA, Stanford University, p. A36–A38.

Rumble, D., Wang, Q., and Zhang, R. Y., 1998, Stable isotope geochemistry of marbles from the coesite UHP terrains of Dabieshan and Sulu, China [abs.], *in* International Workshop on UHP Metamorphism and Exhumation: Stanford, CA, Stanford University, p. A-165.

______, 1998b, Stable isotope geochemistry of marbles from UHP terrains in Dabieshan and Sulu, China, in prep.

Yui, T. F., Rumble, D., Chen, C. H., and Lo, C. H., 1997, Stable isotope characteristics of eclogites from the ultra-high-pressure metamorphic terrain, east-central China: Chem. Geol., v. 137, p. 135–147.

Yui, T. F., Rumble, D., and Lo, C. H., 1994, Stable-isotope characteristics of the ultra-high pressure (UHP) metamorphic rocks of China [abs.], *in* First Workshop on UHP Metamorphism: Stanford, CA, Stanford University, p. A76–A77.

Zhang, R. Y., Rumble, D., Liou, J. G., and Wang, Q. C., 1998, Low $\delta^{18}O$, ultrahigh-P garnet-bearing mafic and ultramafic rocks from Dabie Shan, China: Chem. Geol., v. 150, p. 161–170.

Zheng, Y. F., 1998, Oxygen and carbon isotope anomalies in the ultrahigh pressure metamorphic rocks of the Dabie-Sulu terranes: Implications for geodynamics: Episodes, v. 20, p. 104–108.

Zheng, Y. F., Fu, B., Gong, B., and Wang, Z.-R., 1998a, Carbon isotope anomaly in marbles associated with eclogites from the Dabie Mountains in China: Jour. Geol., v. 106, p. 97–104.

Zheng, Y. F., Fu, B., Li, Y., Xiao, Y., and Li, S., 1998b, Oxygen and hydrogen isotope geochemistry of ultrahigh-pressure eclogites from the Dabie Mountains and the Sulu terrane: Earth Planet. Sci. Let., v. 155, p. 113–129.

Sulu

Jahn, B. M., Cornichet, J., Cong, B., and Yui, T. F., 1996, Ultrahigh ε_{Nd} eclogites from an UHP metamorphic terrane of China: Chem. Geol., v. 127, p. 61–79.

Rumble, D., 1998, Stable isotope geochemistry of the Dabie-Sulu UHP terrains, China [abs.], *in* International Workshop on UHP Metamorphism and Exhumation: Stanford, CA, Stanford University, p. A36–A38.

Rumble, D., Wang, Q., and Zhang, R. Y., 1998, Stable isotope geochemistry of marbles from the coesite UHP terrains of Dabieshan and Sulu, China [abs.], *in* International Workshop on UHP Metamorphism and Exhumation: Stanford, CA, Stanford University, p. A165.

______, 1998b, Stable isotope geochemistry of marbles from UHP terrains in Dabieshan and Sulu, China, in prep.

Rumble, D., and Yui, T. F., 1998, The Qinglongshan oxygen and hydrogen isotope anomaly near Donghai in Jiangsu province, China: Geochim. et Cosmochim. Acta, in press.

Yui, T. F., Rumble, D., and Lo, C. H., 1994, Stable-isotope characteristics of the ultra-high pressure (UHP) metamorphic rocks of China [abs.], *in* First Workshop on UHP Metamorphism: Stanford, CA, Stanford University, p. A76–A77.

______, 1995, Unusually low $\delta^{18}O$ UHP metamorphic rocks from the Sulu terrain, eastern China: Chinese Sci. Bull., v. 40 (suppl.), p. 125.

______, 1995, Unusually low $\delta^{18}O$ ultrahigh-pressure metamorphic rocks from the Sulu terrain, eastern China: Geochim. et Cosmochim. Acta, v. 59, p. 2859–2864.

Zheng, Y. F., 1998, Oxygen and carbon isotope anomalies in the ultrahigh pressure metamorphic rocks of the Dabie-Sulu terranes: Implications for geodynamics: Episodes, v. 20, p. 104–108.

Zheng, Y. F., Fu, B., Gong, B., and Li, S., 1995, Unusually light oxygen in eclogite from Qinglongshan in the Sulu terrain: Chinese Sci. Bull., v. 40 (suppl.), p. 129–130.

______, 1996, Extreme ^{18}O depletion in eclogite from the Su-Lu terrane in east China: Eur. Jour. Mineral. v. 8, p. 317–323.

Zheng, Y. F., Fu, B., Li, Y., Xiao, Y., and Li, S., 1998b, Oxygen and hydrogen isotope geochemistry of ultra-high-pressure eclogites from the Dabie Mountains and the Sulu terrane: Earth Planet. Sci. Lett., v. 155, p. 113–129.

Sm/Nd, Rb/Sr, and $^{40}Ar/^{39}Ar$ Isotopic Systematics of the Ultrahigh-Pressure Metamorphic Rocks in the Dabie-Sulu Belt, Central China: A Retrospective View

SHUGUANG LI,

Department of Earth and Space Sciences, University of Science and Technology of China, Hefei, Anhui 230026, China

EMIL JAGOUTZ,

Department of Cosmochemistry, Max-Planck-Institut für Chemie, D55020 Mainz, Germany

CHING-HUA LO,

Department of Geology, National Taiwan University, Taipei 106

YIZHI CHEN, QIULI LI, AND YILIN XIAO

Department of Earth and Space Sciences, University of Science and Technology of China, Hefei, Anhui 230026, China

Abstract

Because of a complicated metamorphic history, the isotopic systematics of the ultrahigh-pressure (UHP) metamorphic rocks in the Dabie-Sulu belt, east China, appear to be rather different from what were expected. Depending on the degree of retrograde metamorphism and on the retentivity of isotopes, the radiogenic isotopic systematics in the UHP metamorphic rocks yielded a wide range of radiometric ages. Some of these ages are geologically meaningful, but others may not be. In some fine-grained UHP metamorphic rocks, Sm/Nd isotopic systematics appear to be in equilibrium among the UHP phases, showing the best estimate for the age of peak metamorphism at 226 ± 3 Ma. On the other hand, retrograde overprinting often makes the interpretation of isotopic data more difficult. It is common to find that the Sm/Nd and Rb/Sr isotopic systematics among the UHP phases and retrograde phases are not in equilibrium. Regression of isotopic data involving both UHP and retrograde minerals in isotopic correlation diagrams often yields geologically meaningless ages. Although $^{40}Ar/^{39}Ar$ dating of UHP metamorphic rocks has been reported not to be very helpful in establishing the thermal history because of the presence of excess argon, a good correlation between excess argon and rock type in the Dabie-Sulu belt would provide a criterion in identifying the possible sources of excess argon. By taking all the possible effects into consideration, a T-t path with two rapid cooling stages for UHP metamorphic rocks from the Dabie-Sulu belt can be postulated. An initial rapid cooling stage in the period from 226 to 219 Ma may have resulted from rapid exhumation of UHP metamorphic rocks immediately after the peak metamorphism. The second rapid cooling stage, from 450°C to 300°C, may have been caused by the exhumation of the entire terrane, including UHP metamorphic units and their host gneisses, during the period from 180 to 167 Ma.

Introduction

ONE OF THE MOST important achievements in the study of continental geodynamics during the past 15 years is the discovery of ultrahigh-pressure (UHP) minerals such as coesite and diamond in eclogite and metapelite within orogenic belts (Chopin, 1984; Smith, 1984; Okay et al., 1989; Wang et al., 1989; Sobolev and Shatsky, 1990; Xu et al., 1992). It suggests that the continental crust can be subducted to mantle depths during continent-continent collision. Yet scientists are still pondering the causes of the exhumation of the UHP metamorphic rocks from mantle depths. How were UHP minerals preserved under high-pressure and high-temperature conditions during the long history of subduction and exhumation? Answers to these questions are necessarily constrained by precise thermochronological data. Unfortunately, previous thermochronological results for the Dabie-Sulu belt appear to be quite controversial, mainly because of the complicated metamorphic history.

The E-W–trending Qinling-Dabie orogenic belt was formed through the collision between the North China Block and the South China Block. Geologically, the belt is truncated and separated by two major elements—the Nayang basin and the Tanlu fault. The belt therefore can be subdivided into three sections—the Qinling terrane (west), the Dabie terrane (middle), and the Sulu terrane (east) (Fig. 1). It is generally believed that the Sulu terrane is the eastern extension of the Dabie terrane and was displaced ~500 km by the left-lateral movement of the Tanlu fault. UHP rocks occur widely in the Dabie-Sulu terranes, and it has been proposed that the Dabie-Sulu UHP belt may be the largest UHP terrane on the Earth's surface. In addition to the occurrence of UHP rocks in the Dabie-Sulu terrane, a blueschist belt also crops out along the northern margin of the South China block.

Since the first Sm/Nd age of coesite-bearing eclogite from the Dabie terrane was published in 1992 (Li et al., 1992), there have been quite a few Sm/Nd mineral isochron and U-Pb discordia ages published for the coesite-bearing eclogites and UHP gneisses in the Dabie-Sulu terrane. The published Sm/Nd ages—in the range from 209 to 246 Ma (Fig. 1A) (Li et al., 1992, 1994, 1996, 1999; Li, Xiao et al., 1993; Okay and Sengör, 1993; Chavagnac and Jahn, 1996)—are in general agreement with the published U/Pb lower intercept ages, in the range from 202 to 238 Ma (Fig. 1B) (Ames et al., 1993, 1996; Li, Chen et al., 1993; Li et al., 1997; Rowley et al., 1997; Hacker et al., 1998). This agreement indicates that the ultrahigh-pressure metamorphism in the Dabie-Sulu belt is Triassic in age. Furthermore, previous $^{40}Ar/^{39}Ar$ dating of blueschists found in the northern margin of the South China block showed ages of 222 to 245 Ma (Fig. 1B). This may further confirm that the suturing of the North China and the South China blocks may indeed have occurred in the Triassic (Li, Xiao et al., 1993; Eide et al., 1994). The reported U/Pb upper-intercept ages, however, were much older, at 600 to 800 Ma, which would suggest that the protoliths of most of the UHP rocks in the Dabie-Sulu belt may have been late Proterozoic in age (Fig. 1B) (Ames et al., 1993, 1996; Li, Chen et al., 1993; Li et al., 1997; Rowley et al., 1997; Hacker et al., 1998).

Although all the reported Sm/Nd and U/Pb ages seem to be consistent in suggesting a Triassic age for the time of a UHP event for the Dabie-Sulu belt, it should be noted that the age range shown by these isotopic systematics is much larger than the analyti-

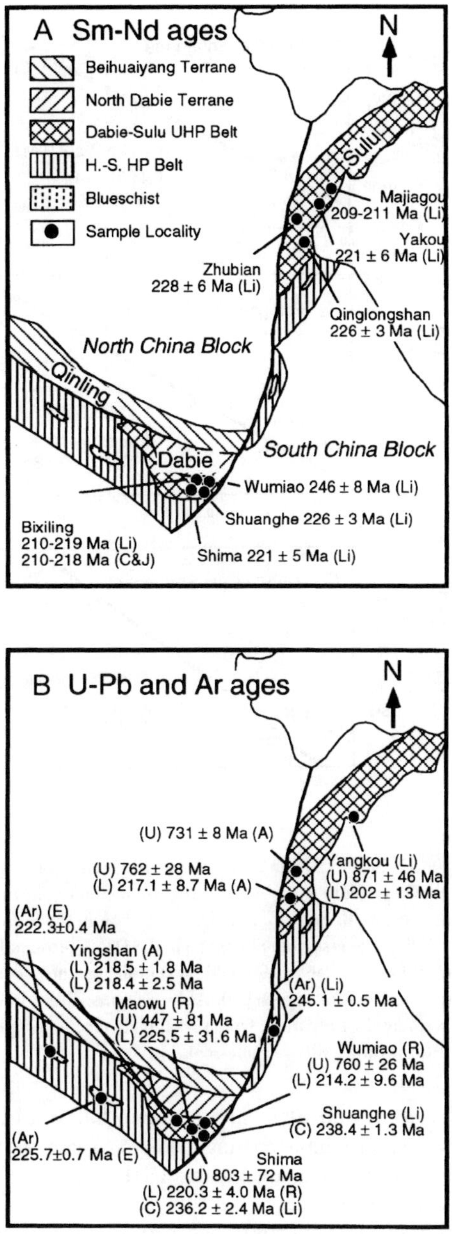

FIG. 1. A. Sketch map of the Dabie-Sulu UHP belt, with published Sm/Nd ages for coesite-bearing eclogites. Data sources: C & J = Chavagnac and Jahn, 1996; Li = Li, Xiao et al., 1993; Li et al., 1994, 1996. B. Sketch map with published U/Pb zircon ages for the UHP metamorphic rocks and $^{40}Ar/^{39}Ar$ ages of blueschists. Data sources: A = Ames et al., 1996; E = Eide et al., 1994; L = Li, Xiao et al., 1993; Li et al., 1997; R = Rowley et al., 1997. Symbols: U = U/Pb upper intercept age; L = lower intercept age; Ar = $^{40}Ar/^{39}Ar$ age; C = concordant age.

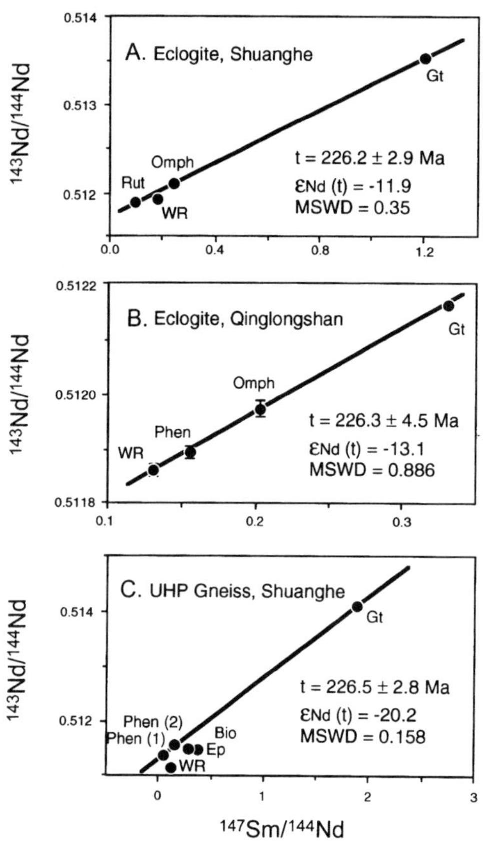

FIG. 2. Sm/Nd isochron plots for UHP metamorphic rocks in the Dabie-Sulu belt. A. Eclogite from Shuanghe in the Dabie terrane. B. Eclogite from Qinglongshan in the Sulu terrane. C. UHP gneiss from Shuanghe in the Dabie terrane. Data sources: Li et al. (1994, 1999). MSWD indicates the goodness of fit (least-squares regression). Abbreviations: Bio = biotite; Ep = epidote; Gt = garnet; Omph = omphacite; Phen = phengite; Rut = rutile; WR = whole-rock.

cal errors, and also larger than what would be expected if a simple cooling history is assumed. Isotopic equilibrium among the UHP phases at the peak of metamorphism and the possible effects of retrograde metamorphic overprinting are the two most likely mechanisms responsible for such a dispersion of isotopic ages. Without further clarification of this age dispersion, it may be difficult to employ those geochronological data in precisely constraining the age of the UHP event and in revealing the cooling history of the UHP rocks in the Dabie-Sulu belt. In order to resolve this age dispersion, the present study carefully examines the Sm/Nd, Rb/Sr, and K/Ar isotopic systematics in a variety of rock types in the Dabie-Sulu belt. It is hoped that the present results will shed some light on the thermal evolution of the UHP metamorphic rocks and assist in revealing the tectonic evolution of the Dabie-Sulu belt.

Isotopic Systematics in UHP Metamorphic Rocks

Sm/Nd systematics

There is general consensus that high metamorphic temperatures can induce a rapid exchange rate for isotopic species among metamorphic phases. However, according to rate equations, the degree of reaction (equilibration) not only is a function of temperature but also is dependent on grain size. Fine-grained minerals therefore more easily reach isotopic equilibrium than do coarse-grained minerals, if all other conditions are the same. Indeed, it has been reported that the Sm/Nd isotopic systematics are not in equilibrium between garnet and clinopyroxene for some coarse-grained eclogites and eclogitic xenoliths in kimberlite (Jagoutz, 1988, 1994; Thöni and Jagoutz, 1992). In addition, retrograde overprinting also may cause isotopic disequilibrium among mineral phases. It would be interesting to examine the possible effects of these factors on isotopic systematics of the UHP metamorphic rocks in the Dabie-Sulu belt.

As reported by many authors, the UHP rocks of the Dabie-Sulu belt were first metamorphosed at temperature conditions of ~600 to 900°C during the peak of metamorphism and were then subjected to a retrograde metamorphism at temperatures of ~ 480 to 800°C (see Liou et al, 1995, for a review). Presumably, these high metamorphic temperatures would have resulted in an isotopic equilibration among UHP phases and among the retrograde phases as well. Indeed, Sm/Nd isotopic compositions of UHP phases (such as garnet, omphacite, rutile, and phengite) in some fine-grained eclogites from the Shuanghe and Qinglongshan areas define isochrons with reasonable goodness of fit in the range from 0.35 to 0.886 (Figs. 2A and 2B) (Li et al., 1994, 1999). The Sm/Nd isochron ages inferred from these isochron plots are not only in agreement with each other (226.2 ± 2.9 Ma and 226.3 ± 4.5 Ma), but also agree well with reported U/Pb zircon ages of 225.5 +3/–6 Ma and 225 ± 4 Ma for UHP metamorphic rocks in the Dabie terrane (Rowley et al., 1997; Hacker et al., 1998). This concordance may therefore suggest that the well-defined Sm/Nd isochron ages could be considered as the best estimate for the age of peak metamorphism in the Dabie-Sulu belt. However, it also was determined that the Sm/Nd isotopic compositions of mineral phases in retrograded UHP metamorphic rocks in the Dabie-Sulu belt usually do not form a good linear array in the isochron plot. As shown in Figure 2C, the Sm/Nd isotopic compositions of the retrograde phases (i.e., biotite and epidote in a UHP gneiss from the Shuanghe area) are located below the linear array defined by garnet and phengite in the same rock sample. It is not surprising to find a lower $^{143}Nd/^{144}Nd$ value by retrograde phases, because retrograde metamorphism occurred much later than peak metamorphism. However, it is interesting to find that the linear array defined by garnet and phengite indicates an isochron age of 226.5 ± 2.8 Ma, which is in perfect agreement with the Sm/Nd isochron and the U/Pb zircon ages mentioned above. These would demonstrate that the Sm/Nd isotopic systematics of UHP phases may have remained closed and scarcely modified by later retrograde metamorphism, although the temperature conditions of retrograde metamorphism found in the UHP rocks of the Dabie-Sulu belt are as high as ~480 to 800°C. This clearly demonstrates that the Sm/Nd isotopic compositions of UHP phases are not in equilibrium with those of retrograde phases in the same sample.

Accordingly, any linear array defined by the UHP phases and retrograde phases/whole-rock should be considered as an "errorchron" or a "mixing line" instead of an "isochron." As a result, depending on the Sm/Nd compositions of the retrograde phases, such an errorchron would give an abnormal age either older or younger than the true one. As shown in Figure 2C, the UHP gneiss from the Shuanghe area contains retrograde epidote and biotite; both minerals are low in Sm content. The linear array defined by garnet and retrograde phases would yield an abnormally old age. A similar case can be found in a retrograde eclogite sample from the Wuhe area. The omphacite in this sample has been replaced by amphibole, which results in an abnormally old garnet–whole-rock Sm/Nd isochron age (246 ± 8 Ma) (Okay and Sengör, 1993). On the other hand, if the retrograde phase is rich in Sm, a younger Sm/Nd errorchron age would be expected. As an example, the coarse-grained eclogites and garnet-peridotites from Bixiling in the Dabie terrane are often found to contain second-generation garnets (pink colored) that exhibit higher $^{143}Nd/^{144}Nd$ values than do the earlier-generation garnets (light pink in color). Regression of the data for the retrograde garnet and other UHP phases yields a younger age (210 ± 1.5 Ma) with respect to that for peak metamorphism (i.e., sample DX36A-1, discussed in Li et al., 1996). It is suspected that similar young Sm/Nd isochron ages (210 to 218 Ma) obtained by Chavagnac and Jahn (1996) for the eclogite and garnet-peridotite samples from the same area might have resulted from such an unequilibrated Sm/Nd isotopic system among the UHP and retrograde phases.

In summary, although the temperature conditions of retrograde metamorphism in the UHP rocks in the Dabie-Sulu belt may have been as high as ~480 to 800°C, the Sm/Nd isotopic systematics of UHP phases may have remained closed and may not have been re-equilibrated with that of retrograde phases during retrograde metamorphism. Depending on the Sm/Nd compositions of retrograde phases, the Sm/Nd isochron plot for the retrograded UHP rocks may yield abnormal ages that are either older or younger than the true age. Some abnormally young and old Sm/Nd isochron ages reported by previous studies may have resulted from such nonequilibrated isotopic systems in retrograded UHP rocks. After careful examination, highly consistent Sm/Nd isochron ages of ~226 Ma can be obtained for fine-grained and unretrograded samples and are considered to be the best estimate for the age of ultrahigh-pressure metamorphism in the Dabie-Sulu belt.

$^{40}Ar/^{39}Ar$ systematics

In general, the $^{40}Ar/^{39}Ar$ technique is widely accepted as one of the most powerful dating tools in revealing the thermal history of metamorphic rocks (McDougall and Harrison, 1988). However, it may be difficult to obtain reasonable $^{40}Ar/^{39}Ar$ ages for the UHP metamorphic rocks because in many cases the $^{40}Ar/^{39}Ar$ ages are much older than the Sm/Nd and U/Pb ages in the same sample (Mattauer et al., 1991; Monié and Chopin, 1991; Tilton et al., 1991; Li et al., 1992; Ames et al., 1993; Li, Xiao et al., 1993). One of the major drawbacks of $^{40}Ar/^{39}Ar$ dating for UHP metamorphic rocks is the presence of excess argon (Li et al., 1994; Arnaud and Kelly, 1995; Hacker and Wang, 1995; Ruffet et al., 1995; Scaillet, 1996). Yet scientists are still pondering whether the excess argon was trapped or inherited. How did the excess argon reside in UHP phases? And from where did the excess argon come? Many

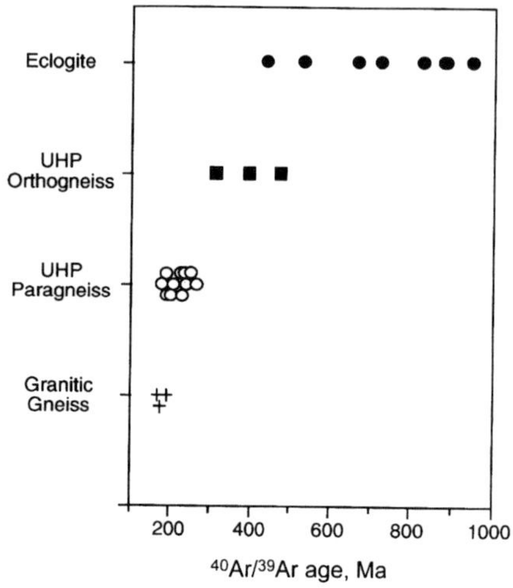

FIG. 3. Diagram showing the correlation between the $^{40}Ar/^{39}Ar$ apparent ages of micas from UHP metamorphic rocks and their country rocks in the Shuanghe area (cited from Li and Lo, 1999).

studies have shown that almost all minerals with excess argon in the UHP rocks yield perfect flat profiles for age spectra, indicating a homogeneous distribution of argon throughout the mineral grain (Li et al., 1994; Ruffet et al., 1995; Scaillet, 1996). This would suggest that the excess argon probably was incorporated into the minerals before and/or during the closure of the isotopic system and probably was residing either in structural defects (dislocations and vacancies) or in vacant sites (Ruffet et al., 1995; Scaillet, 1996, 1998). Ruffet et al. (1995) suspected that the excess argon residing in phengite in the UHP terrane probably was conveyed by a fluid phase enriched in ^{40}Ar through the degassing of adjacent minerals or rocks during UHP metamorphism. This postulation has been confirmed by the results of in-situ $^{40}Ar/^{39}Ar$ laserprobe analyses on UHP rocks from the Dora-Maira nappe (Scaillet 1996). On the basis of the distribution pattern of $^{40}Ar/^{39}Ar$ ages in a single mica grain, Scaillet (1996) further suggested that the incorporation of excess argon in micas may have been controlled by very low lattice and grain-boundary diffusion under dry conditions.

In order to further reveal the source of excess argon and the scope of effects from excess argon, a systematic $^{40}Ar/^{39}Ar$ dating survey was deployed by Li and Lo (1999) on UHP metamorphic rocks in the Shuanghe area. The results showed apparent $^{40}Ar/^{39}Ar$ ages of 431 to 943 Ma for micas in eclogites, 309 to 471 Ma in interlayering UHP orthogneisses, 187 to 266 Ma in UHP paragneisses, and 171 to 173 Ma in granitic country rocks (Fig. 3). The $^{40}Ar/^{39}Ar$ ages for micas in eclogites and UHP gneisses are much older than the Sm/Nd and U/Pb ages mentioned earlier, indicating that these abnormally old $^{40}Ar/^{39}Ar$ ages may have resulted from the presence of excess argon. This clearly indicates that $^{40}Ar/^{39}Ar$ dating may not be helpful in elucidating the thermal history of UHP rocks in the Dabie-Sulu belt. However, it is interesting to find that micas from the granitic country rocks at the same outcrop yield $^{40}Ar/^{39}Ar$ ages of 170 ± 1 Ma and 177 ± 1 Ma (Li and Lo, 1999), which are consistent with their Rb/Sr ages of 171 ± 2 Ma and 173 ± 3 Ma (Li et al., 1999). Since these ages are reasonable in view of the cooling history of the Dabie UHP rocks, as discussed below, they demonstrate that excess argon may not be present in the country rocks. On the other hand, substantial amounts of excess argon were residing in the UHP rocks and were not expelled from the rock body during retrograde metamorphism.

The difference of $^{40}Ar/^{39}Ar$ ages among the interlayering UHP eclogites and gneisses from the same outcrop also suggests that the UHP metamorphic rocks failed to equilibrate Ar isotopes with each other on a centimeter to meter scale. Certainly, there may be a number of mechanisms—for example, the density of schistosity, which is considered to be a potential conduit for the release of excess argon—controlling the amount of excess argon residing in the rock body. The fact that excess argon was widely distributed in the UHP rocks but was not present in the country rocks may suggest that excess ^{40}Ar has been derived internally from the UHP rocks, through the degassing of early-generation minerals during the UHP and/or retrograde metamorphism. However, the excess argon may not have been expelled from the rock body before the closure of the isotopic system. Thus, it is concluded that although $^{40}Ar/^{39}Ar$ dating of UHP rocks may not be helpful in elucidating the thermal history because of the presence of excess argon, useful information for regional tectonics can still be obtained from the $^{40}Ar/^{39}Ar$ dating of the country rocks.

Rb/Sr systematics

Similar to the case involving the Sm/Nd isotopic system, Rb/Sr isotopic dating has been carried out through the isochron-plot technique, which assumes

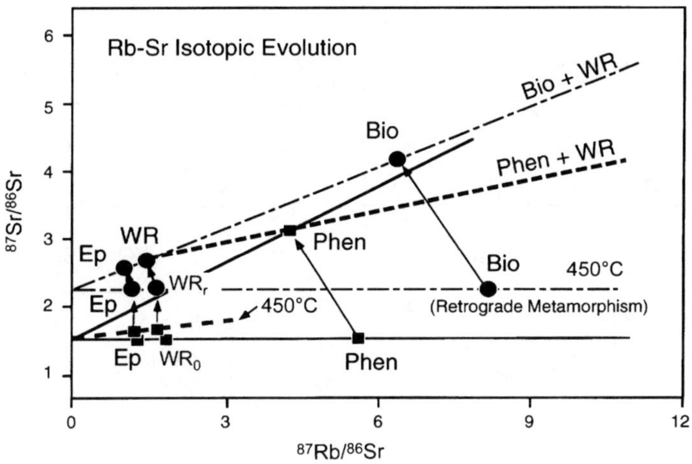

FIG. 4. Schematic diagram showing the evolution of Rb/Sr isotopic systematics in the UHP and retrograde phases (from Li et al., 1999). Abbreviations: WR_0 = whole-rock before retrograde metamorphism; WR_r = whole-rock during the retrograde metamorphism; WR = whole-rock after retrograde metamorphism; Bio = biotite; Ep = epidote; Phen = phengite.

that all minerals have remained in isotopic equilibrium since they formed. However, as discussed earlier, many phases in the UHP rocks are not in isotopic equilibrium with each other because of retrograde metamorphism. In such a case, careful examination of petrographic characteristics and isotopic systematics is required. If UHP metamorphic rocks experienced significant retrograde overprinting, the earlier-generation UHP minerals (e.g., garnet and phengite) might have remained isotopically closed while the later-generation minerals (e.g., biotite and epidote) grew during the retrograde metamorphism (Fig. 4). As a result of the time lag, these two sets of minerals would have grown in two different $^{87}Sr/^{86}Sr$ isotopic systems and the later-generation phases would have a higher initial $^{87}Sr/^{86}Sr$ value. As time passed, these two Rb/Sr isotopic systems would evolve independently (shown as the solid and dashed [Bio + WR and 450°C] lines in Fig. 4, respectively). In this case, regressions of data for these two separate isotopic systems will yield two different isochron ages that may closely represent the cooling ages after metamorphic events. However, any linear regression involving two generation phases (shown as the dotted line [Phen + WR] in Fig. 4) should be considered a "mixing line" or an "errorchron" rather than as an "isochron." The age implied by the errorchron regression is geologically meaningless. It could be either older or younger than the two ages of metamorphic events, depending on the isotopic compositions of the retrograde phases.

In the Dabie-Sulu belt, some eclogitic rocks are not significantly altered by the retrograde metamorphism and still exhibit UHP mineral assemblages, such as garnet-omphacite-phengite. Their Rb/Sr isochron plots yield consistent ages of ~214 to 224 Ma (Fig. 5) (Li et al., 1994; Chavagnac and Jahn, 1996), which are reasonable for rocks cooling through the closure temperature of phengite Rb/Sr isotopic systematics (~500°C). On the other hand, some UHP rocks have been significantly overprinted by retrograde effects and have been reported to exhibit much younger Rb/Sr isochron ages (169 to 181 Ma) (Li et al., 1994, 1999; Chavagnac and Jahn, 1996). These younger Rb/Sr isochron ages obviously result froom retrograde metamorphism because the Rb/Sr ratios of retrograde biotites are much higher than those of other minerals in the same rock. If this is the case, the reported ages could approximate the cooling ages of the UHP rocks after the retrograde metamorphic event.

In order to further clarify this problem, the Rb/Sr isotopic compositions of the UHP phases, as well as those of the retrograde phases, from a UHP gneiss (HT-1-1) from the Shuanghe area are shown in Figure 6 as an example. It is clear that the data points for UHP phases (i.e., garnet and phengite), retrograde phase (i.e., biotite), and whole-rock cannot define a statistically meaningful isochron. Alterna-

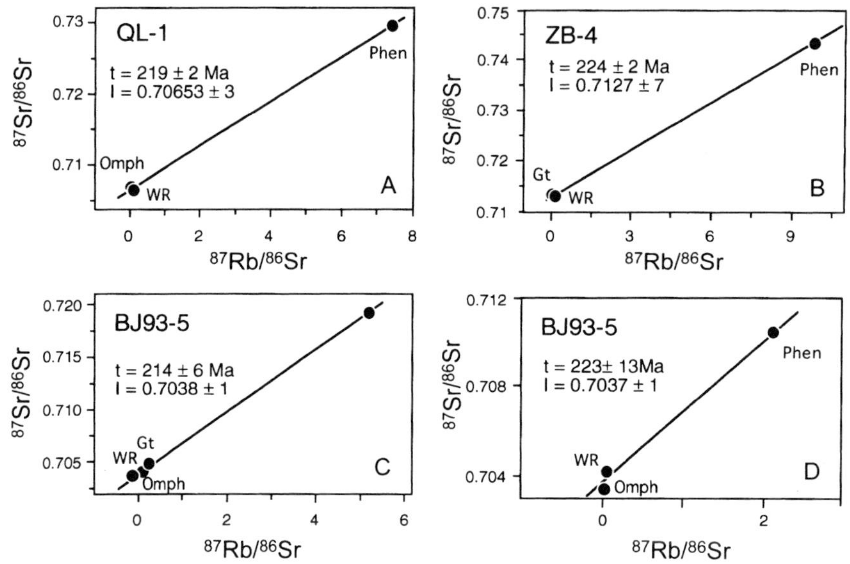

FIG. 5. Rb/Sr isochrons for unretrograded eclogites from Qinglongshan (QL-1) (A) and Zhubian (ZB-4) (B) in the Sulu terrane (from Li et al., 1994). C and D. Rb/Sr isochrons for unretrograded eclogites from Bixiling (BJ93-5, 7) in the Dabie terrane (from Chavagnac and Jahn, 1996). Abbreviations: Ep = epidote; Gt = garnet; Phen = phengite; WR = whole-rock.

tively, the linear array defined by garnet and phengite yields an isochron age of 218.5 ± 3.3 Ma, which is in perfect agreement with that for the unretrograded UHP rocks shown in Figure 5. On the other hand, the data for the retrograde phase (biotite) and whole-rock form another isochron with an age of 173.9 ± 1.7 Ma, which agrees with those reported for the retrograded eclogites in the Dabie-Sulu belt. Since biotite is high in Rb/Sr ratio, this isochron age could be considered to indicate the time the rock cooled through the closure temperature of biotite, which is ~300°C. However, if considering the mixing line formed by phengite and the whole-rock, a geologically meaningless errorchron age of 130.1 ± 4.1 Ma would be calculated. This demonstrates that during retrograde metamorphism, the Rb/Sr isotopic systematics of the UHP phases may not have been significantly altered, and the isotopic systematics of the retrograde phases may have formed an isotopic system independent from that defined by the UHP phases.

In addition to the minerals discussed above, it is also interesting to see how the Rb/Sr isotopic systematics of amphibole evolved during the retrograde metamorphism, because amphibole was found to be abundant in retrograded UHP rocks in the Dabie-Sulu belt; amphibole also is one of the minerals with a high Rb/Sr ratio, and hence is suitable for Rb/Sr dating. Unfortunately, amphibole from the Dabie-Sulu belt has seldom been dated by the Rb/Sr method. The only case was reported by Li et al. (1999) for a retrograded eclogite collected from the Shuanghe area. In this sample, a well-defined isochron formed by retrograde phases—including amphibole, titanite, and whole-rock—yielded an age of 167 ± 5 Ma (Li et al., 1999). This age appears to be in perfect agreement with the published Rb/Sr isochron ages (169 to 181 Ma) defined by retrograde phases in the Dabie-Sulu belt (Li et al., 1994, 1999; Chavagnac and Jahn, 1996). This age could be considered as the time when the sample cooled through the closure temperature of the Rb/Sr isotope systematics in amphibole, because the Rb/Sr ratio of amphibole is much higher than that of titanite and the whole-rock. Unfortunately, the closure temperature of the Rb/Sr isotopic systematics in amphibole has not yet been well established. However, the fair agreement between the Rb/Sr isochron age defined by amphibole + titanite + whole-rock and the $^{40}Ar/^{39}Ar$ age for amphibole in the same sample may suggest that the closure temperature for the Rb/Sr system in amphibole might be equal to or slightly lower than the Ar closure temperature of amphibole (i.e., ~450 ± 50°C) (Li et al., 1999). Thus, the aforemen-

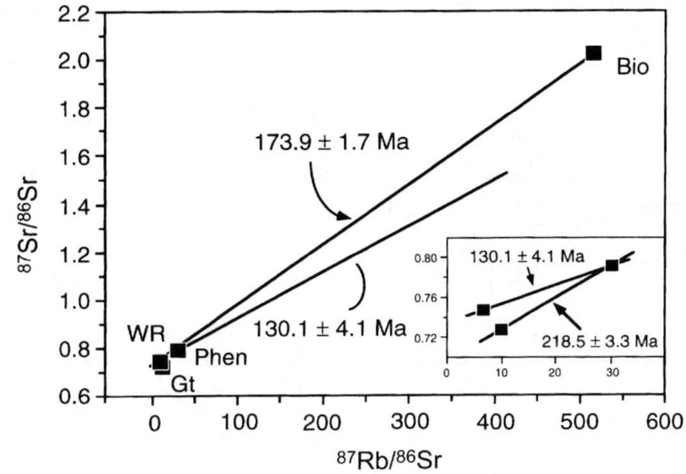

FIG. 6. Rb/Sr isotopic diagram for HT-1-1 UHP gneiss from Shuanghe (from Li et al., 1999). See Figures 2, 4, and 5 for abbreviations.

tioned Rb/Sr ages of 167 ± 5 Ma would indicate the time of the retrograded UHP rock cooling through a temperature of ~450 ± 50°C.

Tectonic Implications

Timing of ultrahigh-pressure metamorphism

As discussed above, because of the complicated metamorphic history, any interpretation of geochronological data for the UHP rocks must be cautious and should take such factors as closed/open system, closure temperature, and isotopic equilibration of minerals into consideration. In the Dabie-Sulu belt, the retrograde metamorphism has induced the formation of retrograde phases in UHP rocks. The Sm/Nd and Rb/Sr isotopic systematics of these retrograde phases equilibrated with each other and remained isotopically closed during the subsequent cooling. On the other hand, Sm/Nd and Rb/Sr isotopic systematics of the UHP phases seem to have not been significantly affected by the retrograde metamorphism and have preserved a closed system during the retrograde metamorphism. It is therefore possible to obtain an age constraint for the UHP metamorphism from these immobile isotopic systematics (Sm/Nd and Rb/Sr) in the UHP phases with high closure temperatures. As demonstrated above, the Sm/Nd isotopic systematics of UHP phases (i.e., garnet-phengite-omphacite-rutile) in three UHP rocks in the Dabie-Sulu belt yield well-defined isochrons with a narrow age range of 226.2 to 226.5 Ma (Fig. 2). These Sm/Nd isochron ages also agree well with published U/Pb zircon ages for the UHP rocks in the same belt. Thus, it is suggested that the best age estimate for the peak ultrahigh-pressure metamorphism in the Dabie-Sulu belt might be 226 ± 3 Ma.

T-t path for UHP rocks in the Dabie-Sulu belt

As demonstrated above, the Rb/Sr isotopic systematics of the UHP phases in the Dabie-Sulu belt seem to have not been significantly altered by retrograde metamorphism, and the Rb/Sr isochron ages would effectively record the time of subsequent cooling through closure temperatures. Considering the Rb/Sr isochron ages with their corresponding closure temperatures, a T-t path for UHP rocks from the Dabie-Sulu belt can be obtained, as shown in Figure 7. This T-t path shows that after peak metamorphism, the UHP rocks in the Dabie-Sulu belt experienced rapid cooling to ~500°C at a rate of ~40°C/Ma. The UHP rocks then went through a nearly isothermal stage from ~219 Ma to ~175 Ma. Starting from ~175 Ma, the rocks cooled rapidly again to ~300°C at a rate of >50°C/Ma. The first rapid cooling would probably reflect the initial exhumation of the UHP rocks from mantle depths to a middle crustal level characterized by pressures of ~6 to 8 kbar. It is interesting to note that the cooling path of the UHP rocks below ~500°C is similar to that recorded in the country rocks after 220 Ma, as reported by many thermochronological results based

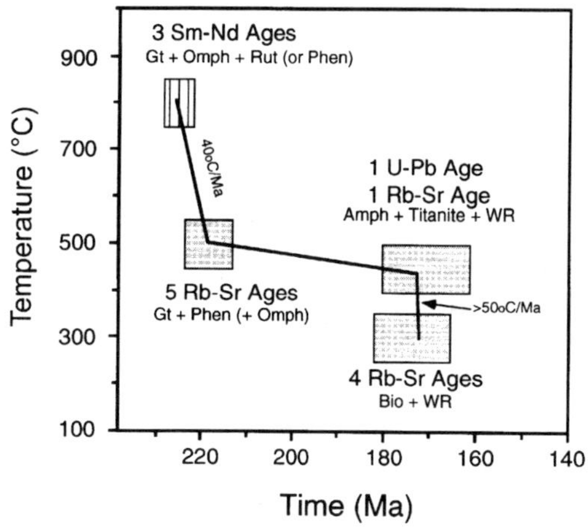

FIG. 7. T-t path of the UHP rocks in the Dabie-Sulu belt. Data sources: Li et al. (1994, 1999), Chavagnac and Jahn (1996), Li, H. (pers. commun.). Abbreviations: Amph = amphibole; Bio = biotite; Gt = garnet; Phen = phengite; Omph = omphacite; Rut = rutile; WR = whole-rock.

on Rb/Sr and Ar/Ar dating studies (Chen et al., 1992, 1995; Li et al., 1995; Chavagnac and Jahn, 1998). This suggests that both the UHP rocks and their country rocks may have exhumed together from the middle-crustal level to the upper-crustal level. Change in the cooling rate during exhumation, shown as a two-stage rapid cooling curve in Figure 7, is not uncommon in UHP terranes (Michard et al., 1993; Gebauer et al., 1997; Grasemann et al., 1998). It could simply reflect the change of thermal gradient during the exhumation of UHP rocks from depths at a constant rate in a compression regime (Chemenda et al., 1995; Ernst and Liou, 1995; Grasemann et al., 1998), or it may indeed indicate a thermal upwelling resulting from the breakoff of a subducted slab (Davies and von Blanckenburg, 1995; Jahn et al., 1999). Because the early rapid cooling record was only found in UHP rocks in the Dabie-Sulu belt, it is plausible that the two rapid cooling stages were caused by two different tectonic mechanisms. The earlier rapid cooling may reflect the exhumation of UHP rocks from mantle depths to the middle-crustal level by compression tectonics during the subduction and/or collision, whereas the later rapid cooling may reflect the exhumation of the entire subducted continental crust from middle-crustal levels to the surface via extension tectonics as a result of the breakoff of a subducted slab.

Acknowledgments

This study was financially supported by National Natural Science Foundation of China (Nos. 49132040, 49573190, 49794012) and partly supported by the Max-Planck-Institut für Chemie, Germany. The authors thank Dr. Tzen-Fu Yui for helpful discussion and comments.

REFERENCES

Ames, L., Tilton, G. R., and Zhou, G., 1993, Timing of collision of the Sino-Korean and Yangtze cratons: U-Pb zircon dating of coesite-bearing eclogites: Geology, v. 21, p. 339–342.

Ames, L., Zhou, G., and Xiong, B., 1996, Geochronology and isotopic character of ultrahigh-pressure metamorphism with implications for collision of the Sino-Korean and Yangtze cratons, central China: Tectonics, v. 15, no. 2, p. 472–489.

Arnaud, N. O., and Kelly, S., 1995, Evidence for excess Ar during high pressure metamorphism in the Dora-Maira (western Alps, Italy), using an ultraviolet laser ablation microprobe $^{40}Ar/^{39}Ar$ technique: Contrib. Mineral. Petrol., v. 121, p. 1–11.

Chavagnac, V., and Jahn, B., 1996, Coesite-bearing eclogites from the Bixiling complex, Dabie Mountains, China: Sm/Nd ages, geochemical characteristics, and tectonic implications: Chem. Geol., v. 133, p. 29–51.

______, 1998, Geochronological evidence for the in situ tectonic relationship in the Dabie UHP metamorphic terrane, Central China: Mineral. Mag., v. 62A, p. 312–313.

Chemenda, A. I., Mattauer, M., and Bokun, A. N., 1996, Continental subduction and a mechanism for exhumation of high-pressure metamorphic rocks: New modeling and field data from Oman: Earth Planet. Sci. Lett., v. 143, p. 173–182.

Chemenda, A. I., Mattauer, M., Malavieille, J., and Bokun, A. N., 1995, A mechanism for syncollisional rock exhumation and associated normal faulting: Results from physical modeling: Earth Planet. Sci. Lett., v. 132, p. 225–232.

Chen, W., Harrison, T., Heizerler, M. T., Liu, R., Ma, B., and Li, J., 1992, The cooling histories of the mélange zone of N. Jiangsu and S. Shandong provinces: Evidence from multiple diffusion domain $^{40}Ar/^{39}Ar$ thermal geochronology: Acta Petrol. Sinica, v. 8, p. 1–17 (in Chinese with English abstract).

Chen, J., Xie, Z., Liu, S., Li, X, and Foland, K. A., 1995, Cooling age of the Dabie orogen, China, determined by $^{40}Ar/^{39}Ar$ and fission-track techniques: Science in China (ser. B), v. 38, p. 749–757.

Chopin, C., 1984, Coesite and pure pyrope in high-grade blueschists of the Western Alps: A first record and some consequences: Contrib. Mineral. Petrol., v. 86, p. 107–118.

Davies, J. H., and von Blanckenburg, F., 1995, Slab breakoff : A model of lithosphere detachment and its test in the magmatism and deformation of collisional orogens: Earth Planet. Sci. Lett., v. 129, p. 85–102.

Eide, L., McWilliams, M. O., and Liou, J. G., 1994, $^{40}Ar/^{39}Ar$ geochronologic constraints on the exhumation of HP-UHP metamorphic rocks in east-central China: Geology, v. 22, p. 601–604.

Ernst, W. G., and Liou, J. G., 1995, Contrasting plate-tectonic styles of the Qinling-Dabie-Sulu and Franciscan metamorphic belts: Geology, v. 23, p. 353–356.

Gebauer, D., Schertl, H.-P., Brix, M., and Schreyer, W., 1997, 35 Ma old ultrahigh-pressure metamorphism and evidence for very rapid exhumation in the Dora Maira Massif, Western Alps: Lithos, v. 41, p. 5–24.

Grasemann, B., Ratschbacher, L., and Hacker, B. R., 1998, Exhumation of ultrahigh-pressure rocks: Thermal boundary conditions and cooling history, *in* Hacker, B. R., and Liou, J. G., eds., When continents collide: Geodynamics and geochemistry of ultrahigh-pressure rocks: Dordrecht, Kluwer Acad. Publ., p. 117–139.

Hacker, B. R., and Wang, Q., 1995 , Ar/Ar geochronology of ultrahigh-pressure metamorphism in central China: Tectonics, v. 14, p. 994–1006.

Hacker, B. R., Ratschbacher, L., Webb, L., Ireland, T., Walker, D., and Dong, S., 1998, U/Pb zircon ages constrain the architecture of the ultrahigh-pressure Qinling-Dabie Orogen, China: Earth Planet. Sci. Lett., v. 161, p. 215–230.

Jahn, B., Wu, F., Lo, C.-H., and Tsai, C.-H., 1999, Mafic-ultramafic intrusions of the Northern Dabie Complex, Central China: Geochemical and isotopic evidence for post-collisional crust-mantle interaction: Chem. Geol., in press.

Jagoutz, E., 1988, Nd and Sr systematics in an eclogite xenolith from Tanzania: Evidence for frozen mineral equilibrium in the continental lithosphere: Geochim. et Cosmochim. Acta, v. 52, p. 1285–1293.

______, 1994, Isotopic systematics of metamorphic rocks [abs.], U. S. Geol. Surv. Circ. 1107, Abs. for ICOG-8, Berkeley, p. 156.

Li, S., Chen, Y., Ge, N., Liu, D., Zhang, Z., Zhang, Q., and Zhao, D., 1993, U-Pb zircon ages of eclogite and gneiss from the Qingdao area—evidence for the later Proterozoic magmatism in the Qiaonan Group: Chinese Sci. Bull., v. 38, no. 19, p. 1773–1777.

Li, Q., Chen, W., Ma, B., Wang, Q., and Sun, M., 1995, Thermal evolution history after collision of North China plate with Yangtze plate: Seismol. Geol., v. 17, p. 193–203.

Li, S., Jagoutz, E., Chen, Y., and Li, Q., 1999, Sm-Nd and Rb-Sr isotopic chronology of ultrahigh-pressure metamorphic rocks of Dabie Mountains, Central China: Evidence for two stages of rapid uplift cooling: Geochim. et Cosmochim. Acta, submitted.

Li, S., Jagoutz, E., Xiao, Y., Ge, N., and Chen, Y., 1996, Chronology of ultrahigh-pressure metamorphism in the Dabie mountains and Su-Lu terrane: I. Sm-Nd isotope system: Science in China (ser. D), v. 39, p. 597–609.

Li, S., Li, H., Chen, Y., Xiao, Y., and Liu, D., 1997, Chronology of ultrahigh-pressure metamorphism in the Dabie mountains and Su-Lu terrane: II. U-Pb isotope system of zircons: Science in China (ser. D), v. 40, p. 200–206.

Li, S., Liu, D., Chen, Y., and Ge, N., 1992, The Sm/Nd isotopic age of coesite-bearing eclogite from the Southern Dabie Mountains: Chinese Sci. Bull., v. 37, p. 1638–1641.

Li, S., and Lo, C.-H., 1999, Inherited origin of excess argon in micas from UHPM rocks in the Dabie-Sulu belt: Chem. Geol., submitted.

Li, S., Wang, S., Chen, Y., Zhou, H., Zhang, Z., Liu, D., and Qiu, J., 1994, Excess argon in phengite from eclogite: Evidence from dating of eclogite minerals by Sm-Nd, Rb-Sr and $^{40}Ar/^{39}Ar$ methods: Chem. Geol., v. 112, p. 343–350.

Li, S., Xiao, Y., Liu, D., Chen, Y., Ge, N., Zhang, Z., Sun, S-S., Cong, B., Zhang, R., Hart, S. R., and Wang, S., 1993, Collision of the North China and Yangtze blocks and formation of coesite-bearing eclogites: Timing and processes: Chem. Geol., v. 109, p. 89–111.

Liou, J. G., Wang, Q., Zhai, M., Zhang, R. Y., and Cong, B., 1995, Ultrahigh-P metamorphic rocks and their associated lithologies from the Dabie Mountains, Central China: A field trip guide to the 3rd International Eclogite Field Symposium: Chinese Sci. Bull., v. 40 (suppl.), p. 1–71.

Mattauer, M., Matte, P., Maluski, H., Xu, Z., Zhang, Q., and Wang, Y., 1991, Paleozoic and Triassic plate boundary between North and South China—New structural and radiometric data on the Dabie-Shan (Eastern China): Comptes Rendus Acad. Sci. Paris, v. 312, p. 1227–1233.

McDougall, I., and Harrison, T. M., 1988, Geochronology and thermochronology by the $^{40}Ar/^{39}Ar$ method: Oxford, UK, Oxford University Press, p. 110.

Michard, A., Chopin, C., and Henry, C., 1993, Compression versus extension in the exhumation of the Dora-Maira coesite-bearing unit, Western Alps, Italy: Tectonophys., v. 221, p. 173–193.

Monié, P., and Chopin, C., 1991, $^{40}Ar/^{39}Ar$ dating in coesite-bearing and associated units of the Dora Maira massif, Western Alps: Eur. Jour. Mineral., v. 3, p. 239–269.

Okay, A. I., and Sengör, A. M. C., 1993, Tectonics of an ultrahigh-pressure metamorphic terrane: the Dabie Shan/Tongbai Shan orogen, China: Tectonics, v. 12, p. 1320–1334.

Okay, A. I., Xu, S., and Sengör, A. M. C., 1989, Coesite from the Dabie Shan eclogites, central China: Eur. Jour. Mineral., v. 1, p. 595–598.

Rowley, D. B., Xue, F., Tucker, R. D., Peng, Z. X., Baker, J., and Davis, A., 1997, Ages of ultrahigh pressure metamorphism and protolith orthogneisses from the eastern Dabie Shan: U/Pb zircon geochronology: Earth Planet. Sci. Lett., v. 151, p. 191–203.

Ruffet, G., Feraud, G., Balevre, M., and Kienast, J.-R., 1995, Plateau ages and excess argon in phengites: An$^{40}Ar/^{39}Ar$ laser probe study of Alpine micas (Sesia Zone, Western Alps, northern Italy): Chem. Geol., v. 121, p. 327–343.

Scaillet, S., 1996, Excess ^{40}Ar transport scale and mechanism in high-pressure phengites: A case study from an eclogitized metabasite of the Dora-Maira nappe, Western Alps: Geochim. et Cosmochim. Acta, v. 60, p. 1075–1090.

______, 1998, K-Ar ($^{40}Ar/^{39}Ar$) geochronology of ultrahigh pressure rocks, *in* Hacker, B. R., and Liou, J. G., eds., When continents collide: Geodynamics and geochemistry of ultrahigh-pressure rocks: Dordrecht, Kluwer Acad. Publ., p. 161–202.

Smith, D. C., 1984, Coesite in clinopyroxene in the Caledonides and its implications for geodynamics: Nature, v. 310, p. 641–644.

Sobolev, N. V., and Shatsky, V. S., 1990, Diamond inclusions in garnets from metamorphic rocks: Nature, v. 343, p. 742–746.

Thöni, M., and Jagoutz, E., 1992, Some new aspects of dating eclogites in orogenic belts: Sm-Nd, Rb-Sr and Pb-Pb isotopic results from the Austroalpine, Savalpine, and Koralpe type-locality: Geochim. et Cosmochim. Acta, v. 56, p. 347–368.

Tilton, G. R., Schreyer, W., and Schertl, H.-P, 1991, Pb-Sr-Nd isotopic behavior of deeply subducted crustal rocks from the Dora Maira Massif, Western Alps, Italy, II. What is the age of the ultrahigh-pressure metamorphism?: Contrib. Mineral. Petrol., v. 108, p. 22–23.

Wang, X., Liou, J. G., and Mao, H. J., 1989, Coesite-bearing eclogites from the Dabie Mountains in central China: Geology, v. 17, p. 1085–1088.

Xu, S., Okay, A. I., Ji, S., Sengör, A. M. C., Su, W., Liu, Y., and Jiang, L., 1992, Diamond from the Dabie Shan metamorphic rocks and its implication for tectonic setting: Science, v. 256, p. 80–82.

Sm-Nd Isotope Tracer Study of UHP Metamorphic Rocks: Implications for Continental Subduction and Collisional Tectonics

BOR-MING JAHN

Géosciences Rennes, Université de Rennes 1, Campus de Beaulieu, 35042 Rennes Cedex, France

Abstract

Sm-Nd isotope tracer techniques are powerful tools in identification of the protolith nature of UHP and HP rocks and can be used to constrain modeling of tectonic processes of continental collision. UHP rocks may have diverse origins, and not all of them carry the same significance for subduction of continental blocks. In this paper, Sm-Nd isotopic data are compiled for UHP and HP rocks, mostly represented by eclogites and garnet peridotites, from the Alpine, Hercynian (Variscan), and Caledonian belts of western Europe; the Pan-African belts of northern Africa; and the Ross belt of Antarctica. These data then are compared with the isotopic characteristics of the UHP rocks from the Dabie orogen of central China. Except for the coesite-bearing quartzitic metasediments of Dora-Maira (Western Alps), which are clearly of continental origin, all HP and UHP rocks (eclogites and ultramafic rocks) from the Alpine, Hercynian, and Pan-African belts have oceanic affinities with the characteristic positive $\varepsilon_{Nd}(T)$ values (= metamorphic initial $^{143}Nd/^{144}Nd$ ratios). They represent segments of oceanic lithosphere that were subducted, underwent eclogite-facies metamorphism, and later were tectonically transported into orogenic zones during continental collisions. By contrast, the majority of UHP rocks from the European Caledonide and the Dabie orogen have negative $\varepsilon_{Nd}(T)$ values, indicating continental affinity. This suggests that these mafic and ultramafic rocks have had a long crustal residence time and that their UHP metamorphism is indicative of subduction of ancient and cold continental blocks, as represented by some Precambrian gneiss terranes containing mafic components including greenschists, amphibolites, or basic granulites.

In the Dabie orogen, none of the UHP eclogites analyzed thus far have shown oceanic affinity; thus they do not represent subducted Tethys Ocean crust. The preservation of ultrahigh $\varepsilon_{Nd}(0)$ values (+170 to +260) in eclogites of very low Nd concentrations (average 0.5 ppm) from the Weihai region and of the extraordinarily low $\delta^{18}O$ in many eclogites and gneisses, the general absence of syntectonic granites in the Dabie Shan, and the available age data obtained by different techniques all point to a rapid rate of exhumation and the absence of a pervasive aqueous fluid phase during the entire process of subduction and exhumation of the Dabie UHP terrane.

Introduction

ULTRAHIGH-PRESSURE (UHP) metamorphic rocks provide important clues to an understanding of the subduction of continental blocks to mantle depths and processes of their exhumation. UHP mineral assemblages were first discovered in coesite-bearing metasedimentary rocks of the Dora-Maira massif, Western Alps (Chopin, 1984) and in eclogitic rocks of the Norwegian Caledonides (Smith, 1984). Subsequently, it was found that the majority of UHP assemblages are recorded in mafic and ultramafic rocks, including eclogites, garnet peridotites, and garnet pyroxenites, and less commonly in metapelites, marbles, and granitic gneisses (e.g., Wang and Liou, 1991, 1993; Cong et al., 1995; Cong, 1996; Liou et al., 1996; see also Carswell and Zhang, 1999; Compagnoni and Rolfo, 1999). UHP metamorphic conditions (600 to 900°C, ≥ 22 kbar) correspond to that of a cold upper mantle. Rocks equilibrated in this condition could come from subducted oceanic or continental blocks or from the upper mantle itself.

One important conclusion obtained from the recent UHP studies is the confirmation of the subduction of "light" continental blocks to mantle depths. The number of occurrences of UHP rocks appears to be increasing, but it is still debated whether all of them bear the same tectonic significance. Nevertheless, such a concept of continental subduction, not predicted by plate tectonics, came from UHP rocks of clear continental origin, such as the coesite-bearing metapelitic rocks of the Dora-Maira massif of the Western Alps (Chopin, 1984; Schreyer, 1988, 1995; Tilton et al., 1989, 1991), the Kokchetav massif in Kazakhstan (Shatsky et al., 1995, 1999), and the Dabie Shan (Liou et al., 1997); from metacarbonates (Liou et al., 1995; Ogasawara et al., 1995, 1998); and from eclogites of proven continental origin, but not from eclogites of oceanic-crustal derivation (Griffin and Brueckner, 1980, 1985; Liou and Zhang, 1998; Jahn, 1998). Moreover, evidence is also obtained from garnet peridotites that have preserved low P-T mineral assemblages of crustal residence prior to UHP metamorphism (Yang and Jahn, 1999), but not directly from the upper mantle, which were somehow emplaced into orogenic belts by tectonic processes during continental collision.

Isotope tracer studies have made important contributions to the understanding of the processes involved in deep burial and exhumation of continental rocks. For instance, the very low oxygen-isotope ratios of UHP rocks (eclogites and country rocks) from Sulu and Dabie Shan have been taken to imply that these rocks were once exposed at the Earth's surface, and the preservation of such low ratios indicates that the later, deep subduction and subsequent exhumation have not re-equilibrated the oxygen isotopes (Yui et al., 1995, 1997; Zheng et al., 1996, 1998; Baker et al., 1997; Rumble, 1998). On the other hand, quite spectacular ultrahigh ε_{Nd} values of ~+170 to +260 have been determined for six retrograded eclogites from Weihai (Sulu region) (Jahn et al., 1996). These are the highest values ever measured in terrestrial rocks (not minerals). They require a very long time lapse for any reasonable Sm/Nd systems to develop, and such values allowed us to estimate a rather precise protolith age of ~1.7 Ga (Jahn et al., 1996). The preservation of extraordinarily low $\delta^{18}O$ and extremely high ε_{Nd} values in UHP rocks probably indicates that a pervasive fluid phase was absent, preventing the re-equilibration of oxygen isotopes; moreover, the temperature was not sufficiently high to reset Nd isotopes during the entire process of subduction and exhumation of these continental rocks. This provides strong constraints to any tectonic modeling on the exhumation of UHP terranes. Furthermore, isotope tracer studies should provide information on the behavior of various elements and isotopes in rocks that have undergone UHP metamorphism. At present, such behavior is not well understood.

The main purpose of this paper is to document whole-rock Sm-Nd isotopic data for UHP and HP rocks worldwide, which will be used to decipher their pre-metamorphic history, as well as the nature and age limit of the protoliths of UHP rocks, and to discriminate the rocks that carry indicators of continental subduction from the other categories of ultrahigh- and high-pressure rocks. The subject of Sm-Nd geochronology will not be covered here, as this is treated in another paper by Li et al. (1999). Although generally important for studies of crustal evolution, an Sr-isotope tracer will not be included in the present review because the Rb-Sr systems in most UHP and HP metamorphic rocks have been shown to be highly disturbed, often leading to high but "unsupported" $^{87}Sr/^{86}Sr$ ratios (Griffin and Brueckner, 1985; Tilton et al., 1989; Ames et al., 1996; Jahn, 1998). Pb-isotope studies on the metapelites of Dora-Maira provided an important constraint to the age of their protolith (Tilton et al., 1989, 1991). Lu-Hf isotope analyses provided new age information for eclogites from Sesia and Monviso and the metapelite of Dora-Maira, all of the Western Alps (Duchêne et al., 1997). However, because of limited analyses, Pb and Hf isotopes will not be reviewed in this paper.

Selection and Representation of Data

Ultrahigh-pressure, high-temperature eclogite nodules in kimberlites emplaced in stable continental cratons may reveal important information concerning the age and evolution of the subcontinental lithosphere, the timing of large-scale and early mantle differentiation, and the perpetual controversy surrounding their ultimate origin (Jagoutz, 1988; Shervais et al., 1988; Richardson et al., 1990; Smith et al., 1991; Jerde et al., 1993; Snyder et al., 1993; Viljoen et al., 1996). However, these rocks are not related to orogenic zones and their "exhumation" process by explosive kimberlite volcanism distinguishes them from the orogenic UHP rocks; hence, they will be excluded from the present discussion.

In this review, only whole-rock Sm-Nd isotopic data are selected for evaluation of the protolith characteristics of UHP rocks, particularly concerning eclogites. The preferred use of whole-rock data is to minimize the problem of possible isotope disequilibrium between individual phases, which may give false information on the bulk-rock characteristics and result in rotated isochrons (e.g., Thöni and Jagoutz, 1992; Bowtell et al., 1994). However,

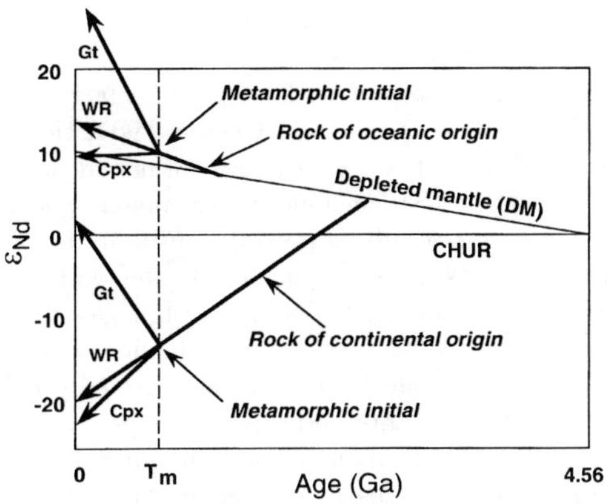

FIG. 1. Schematic diagram showing Nd-isotope evolution for two reservoirs of different Sm/Nd ratios. At the time of metamorphism (T_m), rocks of oceanic and continental origins would have substantially different Nd-isotopic compositions. They might be metamorphosed at the same time, but their metamorphic initial ratios can be used to deciper the nature of the protoliths.

because of the fact that many eclogites are coarse grained and often show metamorphic banding with concentration of individual phases of garnet, omphacite, or epidote, truly representative whole-rock compositions are difficult to obtain, mainly because of the comparatively small sizes of common samples. Nevertheless, the general conclusions reached in the present analysis probably remain valid, allowing the above uncertainty in the sample representativity.

Most of the data compiled here represent eclogites and garnet peridotites. The celebrated UHP metasediments from Dora-Maira are included in the Alpine chain (Appendix 1), but those from the Kokchetav Massif are not because the data have not been formally published (Shatsky et al., 1999). In the case where no whole-rock data are reported, the data for Cpx (or Omp) are used, because the Nd-isotopic compositions and Sm/Nd ratios of Cpx are often found to be closely similar to that of whole rocks; hence their metamorphic initial Nd ratios can be estimated.

During the course of the data compilation, it was discovered that most protolith ages of UHP and HP metamorphic rocks have not been determined with certainty. Thus, all the "initial" $^{143}Nd/^{144}Nd$ ratios, expressed as $\varepsilon_{Nd}(T)$ values, are calculated relative to their metamorphic ages. Together with additional age information and geochemical data, these metamorphic initials can be used to evaluate the nature of the protoliths. Figure 1 illustrates the Nd-isotopic evolution in two hypothetical protoliths of oceanic and continental origins and in their individual metamorphic phases. Both rocks could have the same metamorphic age (T_m), but their distinct metamorphic $\varepsilon_{Nd}(T)$ values indicate highly contrasting protolith natures. The protolith age of rock 1 (continental origin) could be estimated by its depleted mantle model age (T_{DM}), whereas that of rock 2 (oceanic origin) is difficult to determine because its $^{147}Sm/^{144}Nd$ ratio is closely similar to that of the depleted mantle reservoir (DM).

The Sm-Nd isotopic data of UHP and HP rocks (mainly eclogites and garnet peridotites) from six orogenic belts will be presented individually in separate diagrams. $^{143}Nd/^{144}Nd$ ratios are expressed as ε_{Nd} values, and by definition, for the present-day values:

$$\varepsilon_{Nd}(0) = [(^{143}Nd/^{144}Nd)_{sample}/0.512638 - 1] \times 10000.$$

The initial Nd-isotopic ratios are similarly expressed as $\varepsilon_{Nd}(T)$ values, which are approximated by:

$$\varepsilon_{Nd}(T) = \varepsilon_{Nd}(0) - Q \times f_{Sm/Nd} \times T$$

in which Q is a time constant (25.1 Ga^{-1}), $f_{Sm/Nd} = [(^{147}Sm/^{144}Nd)_{sample} / 0.1967] - 1$, and T = age (in Ga).

A typical diagram is shown in Figure 2. The $^{147}Sm/^{144}Nd$ ratio (abscissa) is roughly equivalent to the chemical nature of rocks. Ratios higher than 0.20 (or chondritic value of 0.1967) imply light-rare-earth (LREE) depletion, whereas lower ratios indicate LREE enrichment. In each set of data for an individual orogenic belt, two companion plots of $\varepsilon_{Nd}(0)$ versus $^{147}Sm/^{144}Nd$ and $\varepsilon_{Nd}(T)$ versus $^{147}Sm/^{144}Nd$ are presented. The $\varepsilon_{Nd}(T)$ values in most cases are calculated on the basis of the metamorphic ages; hence they are "metamorphic initial ratios." The time interval (ΔT) between the protolith formation and metamorphism are not known in most cases. True $\varepsilon_{Nd}(T)$ for the protoliths requires further correction based on ΔT values. The magnitudes of such correction, depending on individual $^{147}Sm/^{144}Nd$ ratios, are shown for ΔT = 500 Ma and 1000 Ma in Figure 2. The magnitude increases when $^{147}Sm/^{144}Nd$ ratios are farther from the chondritic value of ~0.1967, and little or no correction is needed when they are close to this value.

Common terrestrial rocks have a rather restricted range of $^{147}Sm/^{144}Nd$ ratios between 0.10 (e.g., granite) to 0.25 (e.g., MORB). Rocks having ratios ≥ 0.25 are unusual, but they are often observed in UHP eclogites and garnet peridotites. An interpreta-

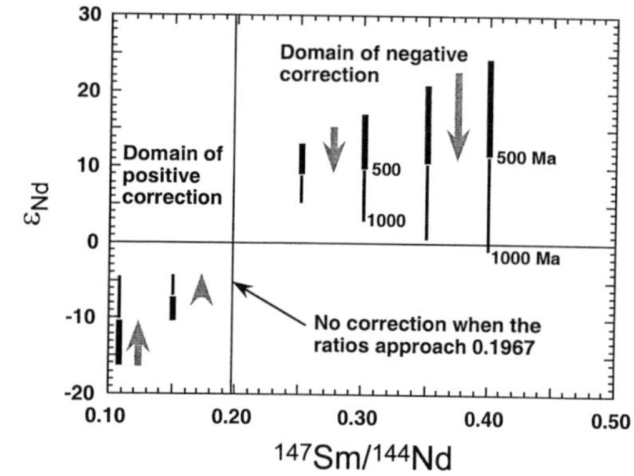

FIG. 2. ε_{Nd} versus $^{147}Sm/^{144}Nd$ diagram showing the magnitude of ε_{Nd} correction from a metamorphic initial to a true magmatic initial value. The correction is related to both the $^{147}Sm/^{144}Nd$ ratio in the rock and the time interval (ΔT) between protolith formation and metamorphism: upward or positive correction is needed for rocks with $^{147}Sm/^{144}Nd < 0.1967$ (chondritic value) and downward or negative correction for those > 0.1967. The magnitude for $\Delta T = 500$ and 1000 Ma is shown for reference.

tion is that these samples do not represent the bulk rocks; rather, they are biased samples probably from portions enriched in garnet. Indeed, as mentioned earlier, truly representative whole-rock samples are not always obtainable from coarsely recrystallized UHP rocks.

UHP Rock Assemblages and Implications for Continental Subduction

Eclogites are volumetrically minor in all UHP terranes, but they are the principal rock type that bears evidence of UHP metamorphism (Smith, 1984; Wang and Liou, 1991; Okay, 1994; Cong, 1996; Liou et al., 1996). If the eclogites were produced by metamorphism of a subducted oceanic lithosphere, as in the Hercynian and Alpine orogenic chains to be shown later, their contact relation with associated granitic gneisses must be tectonic. Thus, even if they show UHP parageneses, it is not necessary that their country gneisses also have undergone the same UHP metamorphism. However, if the UHP eclogites can be identified as part of the ancient continental crust, then their occurrence strongly implies deep subduction of the continental block. Thus, the long-standing controversy over the tectonic relationship—"in-situ" versus "foreign"—between eclogites and country gneisses very much hinges on the identification of oceanic or continental affinity of the protoliths.

UHP metamorphic rocks of sedimentary origin, such as those of Dora-Maira (Chopin, 1984; Tilton et al., 1989, 1991), are clearly of continental derivation, so their UHP metamorphism unambiguously indicates deep subduction of continental crust. However, it is not so clear for eclogites and ultramafic rocks with UHP mineral assemblages. Figure 3 illustrates the possible types of mafic-ultramafic rocks formed under UHP conditions. Broadly, eclogites could have three origins. Those produced by metamorphism of subducted oceanic crust and mantle eclogite enclaves in kimberlites do not imply continental subduction, whereas eclogites formed from the mafic components (greenschists, amphibolites, basic granulites) of an old and stable continent would persuasively argue for subduction of a continental block. In the Dabie orogen, eclogites of possible ocean-floor affinity have been reported (Zhai and Cong, 1996), but they appear to be quantitatively insignificant and remain to be verified. Eclogites of mantle origin have not been identified in the Sulu or Dabie terranes.

The implication of ultramafic rocks is even more controversial. Orogenic peridotites are of mantle origin, and thus have recorded mantle pressures or UHP conditions. Their occurrence in gneiss terranes does not necessarily indicate that they have been metamorphosed at UHP conditions, nor have the gneiss terranes necessarily undergone deep subduction. Ultramafic rocks of layered intrusions are exceptions. They were originally crystallized under crustal conditions, so UHP parageneses would imply deep subduction of these "crustal" rocks. In some cases, mantle blocks could have been emplaced by an earlier tectonic process into the continental crust, followed by subduction to UHP conditions together with the continental crust. Such a process is difficult to identify if the rocks have chemical and isotopic compositions comparable to that of the upper mantle. This kind of rock may be represented by the Rizhao clinopyroxenites (Hiramatsu and Hirajima, 1995). However, if mantle peridotites have recorded a history of crustal contamination as revealed by Nd-Sr-O isotopic analyses and/or petrological evidence, then their UHP assemblages would imply deep continental subduction. Such rocks may be represented by the garnet peridotites of Donghai in the southern Sulu region (Yang and Jahn, 1999).

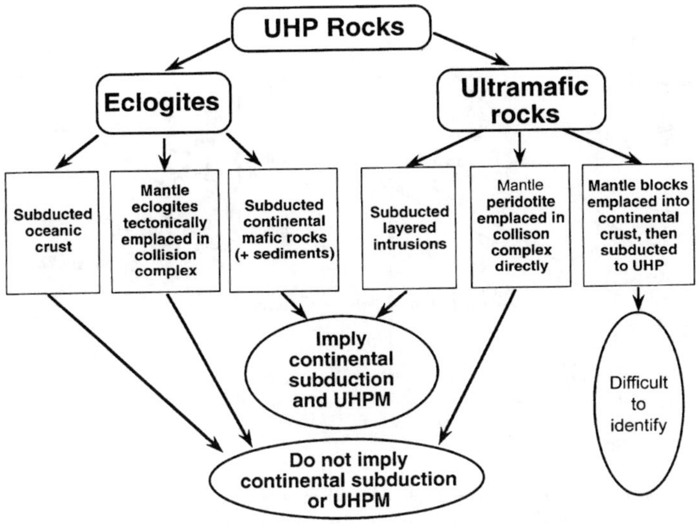

FIG. 3. Tectonic implications of a variety of UHP rocks. Most can be identified in the Dabie orogen by means of geochemical and isotope tracer studies. Only subducted continental mafic rocks, including some eclogites possibly transformed from para-amphibolites, and subducted layered mafic-ultramafic intrusions clearly indicate subduction of continental crust. Eclogites and peridotites of oceanic affinity and eclogites from kimberlite pipes, although bearing UHP parageneses, do not provide evidence for continental subduction.

It is known that P-T estimates for coexisting eclogite and garnet peridotite often provide incoherent information about the conditions of metamorphism (e.g., Liou et al., 1996). Likewise, geochronological studies of eclogites and garnet peridotites also often give discordant information concerning the time of UHP metamorphism (e.g., Griffin and Brueckner, 1985; Mearns, 1986; Jamtveit et al., 1991; Brueckner et al., 1996; Yang and Jahn, 1999). Coincidentally, isotopic studies of basalts and orogenic peridotites, such as the Ronda massif, often provide conflicting insights regarding the isotopic and compositional heterogeneity of the upper mantle and the genesis of magmas (Reisberg et al., 1989).

Sm-Nd Concentration Data of UHP Eclogites and Ultramafic Rocks

Before Nd-isotopic characteristics are presented, it is useful to first review the Sm and Nd concentration data of UHP and HP rocks. These concentration data and the Sm/Nd ratios may be useful in deciphering their protolith nature. However, as mentioned earlier, truly representative whole-rock samples are not always obtained for coarse-grained eclogites, so their analyses must be used with care. The concentration data are presented in the Appendix tables 1 to 6.

In the UHP terranes of central China, eclogites have been conveniently classified into three types on the basis of their field relationship with the country rocks (Wang et al., 1990; Zhang et al., 1995; Jahn, 1998): Type I = enclaves in granitic gneisses (including both ortho- and para-gneisses); Type II = enclaves in or interlayers with marbles or metaclastic sediments (e.g., garnet-mica schists); and Type III = enclaves and interlayers with ultramafic rocks. However, such a classification cannot be entirely adopted to describe the occurrences of eclogites from other orogenic belts. For example, the Piemonte zone of the Alps contains eclogites of classic ophiolitic origin (Barnicoat and Fry, 1986; Cliff et al., 1998), but an ophiolite complex has not been clearly identified in either the Dabie Shan or Sulu region. Moreover, eclogites from the Alpine external massifs (Aiguilles Rouges, Belledonne, Argentera) occur as intercalations with amphibolites or within a leptinite/amphibolite complex (Paquette et al., 1989), but such an occurrence is not present in the Dabie Shan. Eclogites of the Oztal Complex in the Tirolian Alps occur as relics within a much larger body of metabasic rocks; these in turn are surrounded by granitic gneisses. Consequently, the

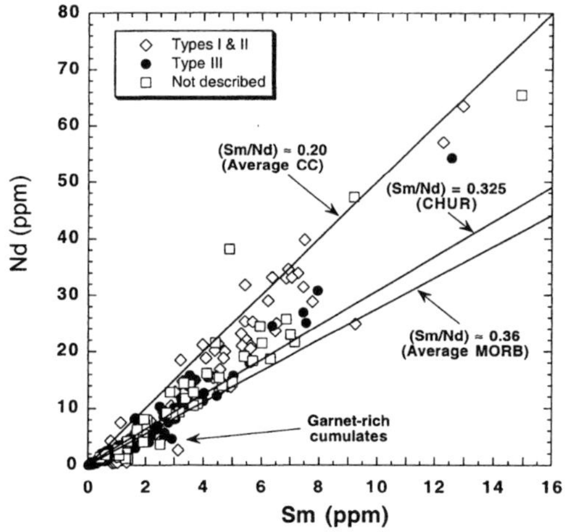

FIG. 4. Nd versus Sm concentration plot for eclogites and ultramafic rocks (garnet pyroxenites and garnet peridotites) from the world's UHP and HP terranes. The majority of the rocks have Nd < 40 ppm and Sm < 8 ppm. Sm/Nd ratios fall in the field bracketed by 0.2 (for alkali basalts or average continental crust, CC) and 0.36 (for average N-MORB). However, a number of low-Sm (or low-Nd) rocks show high Sm/Nd ratios, which probably indicate that they are not common magmatic rocks, but represent cumulates or garnet-enriched portions produced by metamorphic differentiation. Data sources: see Appendix tables.

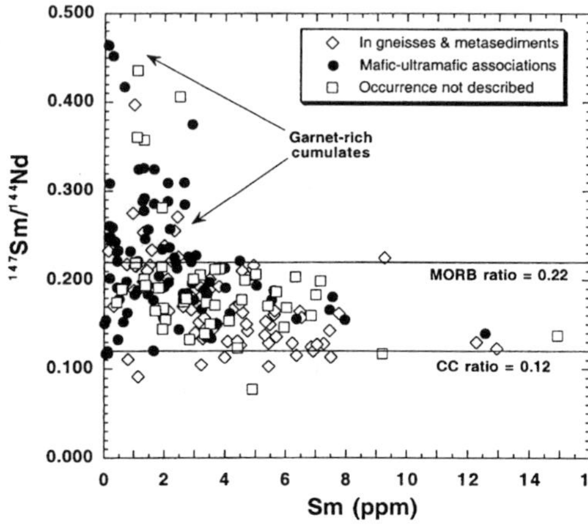

FIG. 5. $^{147}Sm/^{144}Nd$ versus Sm concentration plot. More than 20% of the rocks have $^{147}Sm/^{144}Nd$ ratios higher than the N-MORB value of 0.22, most of which are rocks of mafic-ultramafic associations. Data sources: see Appendix tables.

occurrence types used in the Dabie orogen will only be applied to other orogenic belts when clear and comparable field relations are described; the rest will be grouped as "not described" in Figures 4 and 5.

The eclogites are mainly of basaltic composition, but they show a wide range of major- and trace-element abundances, suggesting their multiple origins and heterogeneous sources, and in some cases, the results of metamorphic segregation (e.g., Cong, 1996; Jahn, 1998). Figures 4 and 5 show the ranges of Sm and Nd concentrations and their ratios for 223 analyses of eclogites and associated ultramafic rocks. Naturally, Sm and Nd are positively correlated. Sm concentrations range from less than 1 to 15 ppm, but most rocks contain 8 ppm or less (grand average = 3.1 ± 2.5, 1σ; Fig. 4). Nd ranges from less than 1 to 65 ppm, and most have 30 ppm or less (grand average = 11.3 ± 11.4, 1σ). The Sm/Nd ratios for the majority of eclogites lie between 0.20 (for alkali basalts, continental basalts, or average continental crust) and 0.36 (for depleted N-MORB). Also noticeable is that many rocks of low Sm and Nd concentrations have Sm/Nd ratios higher than depleted N-MORB values. This suggests that the protoliths of these eclogites probably are not common basaltic rocks, but they could represent garnet-rich cumulate portions of the metamorphic assemblages or simply pyroxene cumulates.

Similarly, Figure 5 illustrates the relation of the $^{147}Sm/^{144}Nd$ ratio, which is roughly equal to 0.6 × (Sm/Nd), to Sm concentration. Again, more than 20% of the rocks have $^{147}Sm/^{144}Nd$ ratios higher than the N-MORB value of 0.22. The majority are rocks of mafic-ultramafic associations. In conclusion, Type I and II eclogites appear generally to have higher Sm and Nd concentrations but lower Sm/Nd ratios than Type III rocks.

Sm-Nd Isotopic Characteristics of the World's UHP and HP Eclogites

Sm-Nd isotopic data on whole-rock samples are presented in Appendix tables 1 to 6. In this paper, all Nd-isotopic ratios are adjusted relative to the La Jolla Nd standard of 0.511860. Analyses of individual phases (e.g., clinopyroxene and garnet) are obviously vital to studying the ages of various events (peak metamorphism, cooling, reheating, etc.) and to understanding the isotopic evolution and petrogenesis of eclogites, but this aspect is reviewed by Li et al. (1999) and is not discussed further here. It is understood that petrogenetic interpretation based only on whole-rock analyses may lead to erroneous conclusions if post-magmatic and post-metamorphic modification of chemical and isotopic compositions has taken place. However, whole-rock analyses provide the most direct means of protolith identification. Some authors have tried to use clinopyroxene and garnet analyses to reconstitute whole-rock compositions (e.g., Shervais et al., 1988; Jerde et al., 1993), but the validity of this approach hinges on the assumption that the bulk rocks contain only two phases and both phases are free of inclusions. This is not the case for most UHP eclogites, so this approach is not adopted in the present study. In all ε_{Nd} versus Sm/Nd diagrams, only "global" features are examined and discussed. Individual aberrant data points that require additional explanations will be downplayed.

REE are generally considered to be less mobile than Rb and Sr in metamorphic processes. Thus, Sm-Nd isotope tracers may provide more accurate information about the nature of the eclogite protoliths. That is not to say that REE always are immobile, as the open-system behavior of REE has been reported for a number of cases. Among them the metasomatic effect on the protolith of the Weihai eclogites, which led to an extreme Sm/Nd fractionation, is a prominent example (Jahn et al., 1996). Figures 6 through 9 show Sm-Nd isotopic data for eclogites and associated garnet peridotites or pyroxenites from the European, Pan-African, and Antarctic Ross orogenic belts. In the present compilation, rocks occurring in a given orogenic belt are grouped together, regardless of their protolith or metamorphic ages. For example, some eclogites from the Alpine chain are clearly of pre-Alpine ages; many eclogites belong to the Variscan evolution with only minor Alpine tectonic effects (e.g., Paquette et al., 1989). Similarly, some eclogites of mid-Proterozoic (?) protolith that underwent Grenvillian metamorphism at ~1.0 Ga occur in the Scottish Caledonides (e.g., Sanders et al., 1984). These rocks are grouped in the Caledonian belt. In addition, because the dominant majority of the data compiled herein represent those of eclogites and only a small minority belong to ultramafic rocks, the term "eclogites" will be used to mean both eclogites and ultramafic rocks unless specified otherwise in the following discussion.

In all figures, the present-day Nd isotopic ratios, $\varepsilon_{Nd}(0)$, as well as the age-corrected values, $\varepsilon_{Nd}(T)$, are shown for comparison. The principal trends are summarized below.

1. The most celebrated UHP rocks, jadeite-kyanite or garnet quartzites of Dora-Maira, have nega-

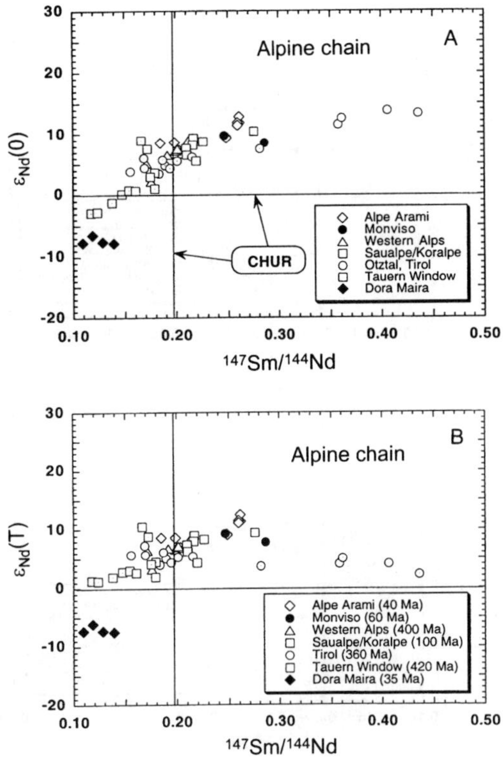

FIG. 6. $\varepsilon_{Nd}(0)$ versus $^{147}Sm/^{144}Nd$ (A) and $\varepsilon_{Nd}(T)$ versus $^{147}Sm/^{144}Nd$ (B) plots for eclogites and garnet peridotites from the Alpine belt. Except for the UHP metasediments of Dora-Maira, all rocks show positive $\varepsilon_{Nd}(T)$ values. Data sources: see Appendix 1.

tive $\varepsilon_{Nd}(T)$ values, –5 to –8 (Fig. 6). This is consistent with their ultimate derivation from middle Proterozoic continental crust, as estimated from their T_{DM} ages of 1.3 to 1.8 Ga (Tilton et al., 1989, 1991; see Appendix 1). Except for these metasediments, the entire population of eclogites from the Alpine chain have positive $\varepsilon_{Nd}(T)$ values (Fig. 6B), suggesting that all originated from subducted oceanic crust, or directly from the depleted upper mantle, such as the garnet peridotites of Alpe Arami (Becker, 1993). This conclusion is largely supported by available geochemical and petrological data (Paquette et al., 1989; Thöni and Jagoutz, 1992; Miller and Thöni, 1995; Dobrzhinetskaya et al., 1996; von Quadt et al., 1997).

2. In the Variscan or Hercynian belt, most eclogites possess positive $\varepsilon_{Nd}(T)$ values (Fig. 7B), suggesting their dominant oceanic derivation (Bernard-Griffiths and Cornichet, 1985; Bernard-Griffiths et

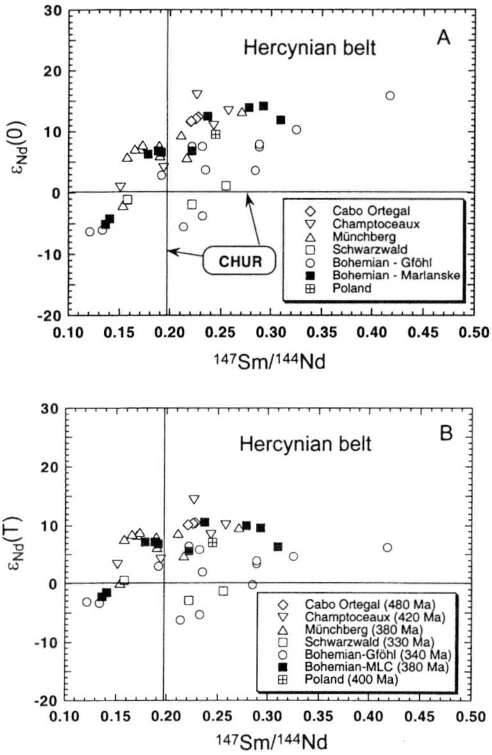

FIG. 7. $\varepsilon_{Nd}(0)$ versus $^{147}Sm/^{144}Nd$ (A) and $\varepsilon_{Nd}(T)$ versus $^{147}Sm/^{144}Nd$ (B) plots for eclogites and garnet peridotites from the Hercynian belt of western Europe. Note that most rocks show positive $\varepsilon_{Nd}(T)$ values, suggesting an oceanic affinity. Those having negative values have been interpreted as resulting from crustal contamination by reinjection of crustal rocks (oceanic clays) into the upper mantle. Data sources: see Appendix 2.

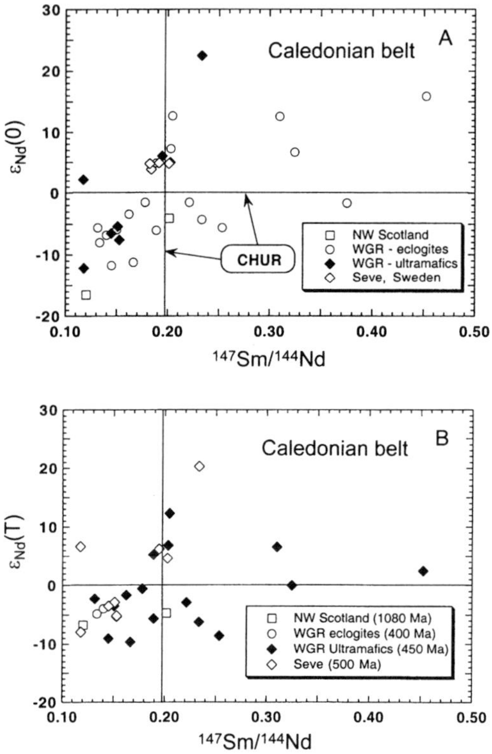

FIG. 8. $\varepsilon_{Nd}(0)$ versus $^{147}Sm/^{144}Nd$ (A) and $\varepsilon_{Nd}(T)$ versus $^{147}Sm/^{144}Nd$ (B) plots for eclogites and garnet peridotites from the European Caledonides. The majority of the rocks show negative $\varepsilon_{Nd}(T)$ values, in strong contrast to those of the Alpine and Hercynian chains. These rocks appear to have resided in the continental crust for a lengthy period of time before UHP metamorphism. Data sources: see Appendix 3.

al., 1985; Stosch and Lugmair, 1990; Beard et al, 1992, 1995; Medaris et al., 1995). One eclogite from Münchberg and two from Schwarzwald with low $\varepsilon_{Nd}(T)$ values of ~0 to –3 are probably the result of crustal contamination (Stosch and Lugmair, 1990; Kalt et al., 1994). A few eclogites from the Bohemian Massif also show negative $\varepsilon_{Nd}(T)$ values (–3 to –6) (Medaris et al., 1995), but such an isotopic signature was interpreted to result from the involvement of oceanic clay in the melting of subducted oceanic lithosphere, and the eclogites were considered to represent high-pressure cumulates of the "enriched" basaltic liquid (Medaris et al., 1995). In any case, the oceanic-crustal origin of the eclogites from the Hercynian belt is clearly demonstrated, and the negative $\varepsilon_{Nd}(T)$ values do not imply the deep subduction of continental crust.

3. Eclogites from the European Caledonides have a much more complex evolution. Two eclogites from northwestern Scotland (Sanders et al., 1984) and over half of the UHP rocks (including about equal proportions of eclogites and garnet pyroxenites) (Griffin and Brueckner, 1980, 1985; Mearns, 1986; Mork and Mearns, 1986; Jamtveit et al., 1991) show negative $\varepsilon_{Nd}(T)$ values to as low as –10 (Figs. 8A and 8B). It appears that the Norwegian and Scottish eclogites have both oceanic and "continental" affinities.

The meaning of "continental" affinity merits further explanation. From the compositional point of view, all the analyzed rocks (eclogites, garnet pyroxenites, and garnet peridotites) are clearly of mantle origin. Some, if not all, were emplaced into the continental crust in Precambrian time(s) and became an

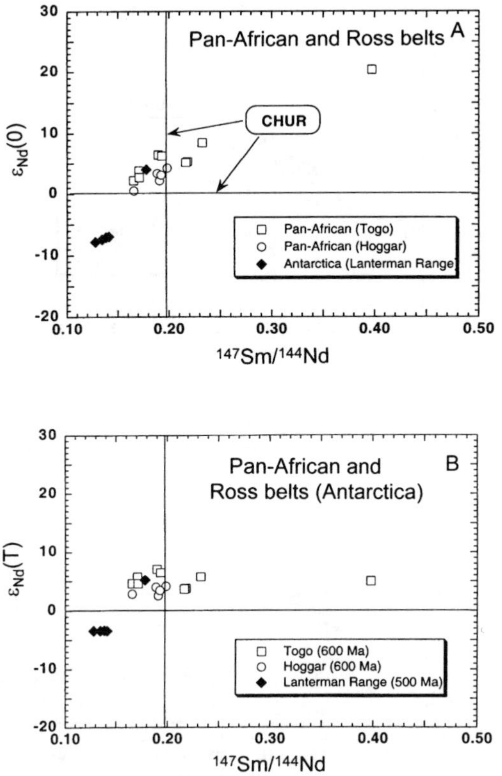

FIG. 9. $\varepsilon_{Nd}(0)$ versus $^{147}Sm/^{144}Nd$ (A) and $\varepsilon_{Nd}(T)$ versus $^{147}Sm/^{144}Nd$ (B) plots for eclogites from the Pan-African belt of northern Africa and the Ross orogenic belt of Antarctica. The Pan-African eclogites have an oceanic affinity with positive $\varepsilon_{Nd}(T)$ values; in contrast, those of Antarctica possess $\varepsilon_{Nd}(T)$ values of ca. –3.4; their protoliths were probably emplaced at 750 Ma as enriched-MORB or crustally contaminated basalts. Data sources: see Appendix 4.

integral part of the continental crust then. This represents the first stage of evolution. However, the process of the first-stage emplacement could not be precisely determined, as it is a seperate problem, reflecting a controversy over the styles of early Precambrian tectonics. This will not be further explored here.

We note that many eclogites have $^{147}Sm/^{144}Nd$ ratios lower than the chondritic values of 0.1967 (Fig. 8), suggesting that they were likely emplaced as LREE-enriched continental basalts, later metamorphosed to amphibolites. In this scenario, these rocks became an integral part of the continental crust. This further implies that an ancient crustal segment was subducted to mantle depths, and UHP metamorphic assemblages were produced during the Caledonian orogeny. In the Western Gneiss Region of Norway, eclogites recorded only the Caledonian thermal event of 410 to 440 Ma (Griffin and Brueckner, 1980, 1985; Mearns, 1986; Mork and Mearns, 1986), whereas garnet peridotites and garnet pyroxenites often yielded a Proterozoic event of ~1.7 Ga (Jamtveit et al., 1991). It is possible that the isotopic systems registered at 1.7 Ga in the garnet peridotites at mantle P-T conditions have not been erased or homogenized during the Caledonian event, whereas the newly formed garnets from basaltic protoliths record the Caledonian orogeny at 410 to 400 Ma. Thus, the evolution of the European Caledonides is clearly distinguished from that of the Alpine and Hercynian chains of western Europe and the Pan-African belt of North Africa (see below). Finally, eclogites from the Seve Nappe of Sweden have an oceanic affinity, judging from the Nd-isotope and geochemical compositions (T-MORB?) (Mork et al., 1988). These rocks occur as lenses enclosed within marble beds or quartzite schists.

4. The Pan-African eclogites from Togo have been interpreted as derived from continental rift and oceanic arc basalts, whereas those from Hoggar in Algeria probably originated from continental basalts (Bernard-Griffiths et al., 1991). These rocks have a narrow range of $\varepsilon_{Nd}(T)$ values (calculated at 600 Ma), from ~+3 to +7 (Figs. 9A and 9B). Their protolith ages are unknown, although a late Precambrian age of ~1 Ga was attributed to the Togo eclogites and an Archean age to the Hoggar eclogites (Bernard-Griffiths et al., 1991). Because of their positive $\varepsilon_{Nd}(T)$ values, it cannot be determined if these rocks have a "continental" affinity.

5. Except for one case, eclogites from the Lanterman Range of the Ross orogenic belt in Antarctica possess $\varepsilon_{Nd}(T)$ values of ca. –3.4 (Figs. 9A and 9B), and their protoliths were probably emplaced at 750 Ma as enriched-MORB or crustally contaminated basalts, as implied from the negative Nb-Ta anomalies in their spidergrams (Di Vincenzo et al., 1997).

Sm-Nd Isotopic Characteristics of UHP Eclogites from the Dabie Orogen

Figures 10 and 11 illustrate the Sm-Nd isotope characteristics of eclogites from the Sulu and Dabie terranes, respectively. Except for Weihai, Type I and Type II eclogites from both terranes had low $\varepsilon_{Nd}(T)$ values (–6 to –20) at the time of peak metamorphism or continental collision (~220 Ma). These values are

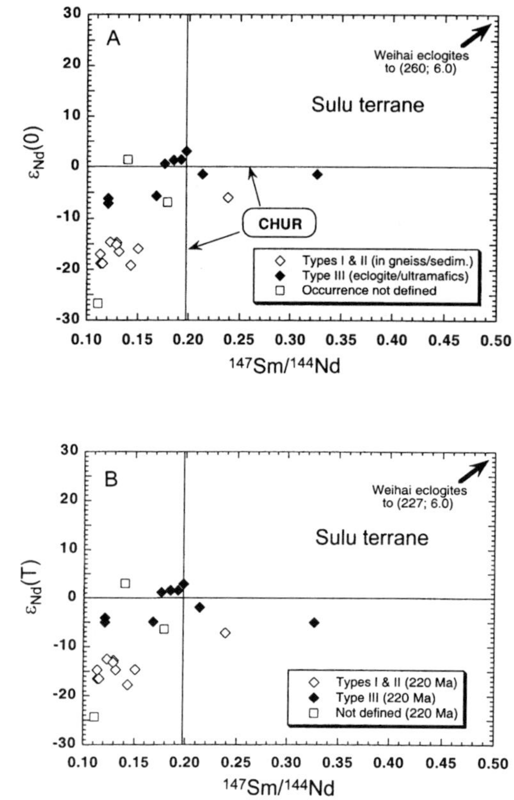

FIG. 10. $\varepsilon_{Nd}(0)$ versus $^{147}Sm/^{144}Nd$ (A) and $\varepsilon_{Nd}(T)$ versus $^{147}Sm/^{144}Nd$ (B) plots for eclogites from the Sulu terrane of the Dabie orogen, China. All Type I and II eclogites are characterized by negative $\varepsilon_{Nd}(T)$ values, suggesting continental affinities. The data of the Weihai eclogites are off scale and their ε_{Nd} and $^{147}Sm/^{144}Nd$ ratios are given in parentheses. Data sources: see Appendix 5.

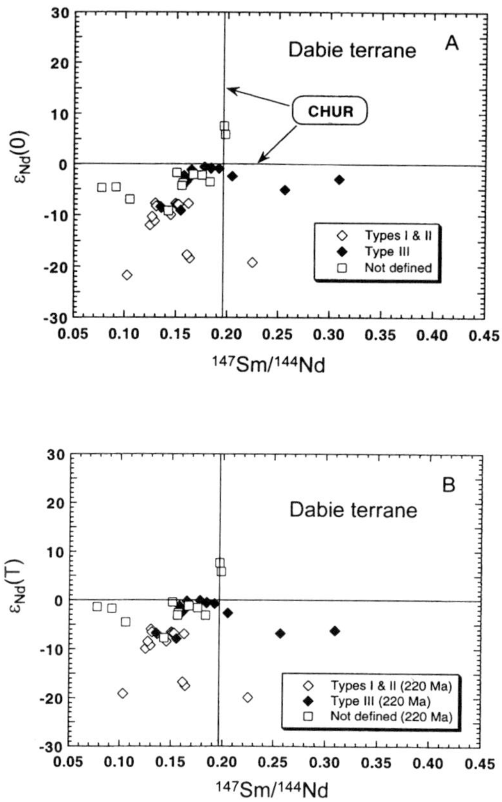

FIG. 11. $\varepsilon_{Nd}(0)$ versus $^{147}Sm/^{144}Nd$ (A) and $\varepsilon_{Nd}(T)$ vs $^{147}Sm/^{144}Nd$ (B) plots for eclogites from the Dabie terrane of the Dabie orogen, China. All Type I and II eclogites are characterized by negative $\varepsilon_{Nd}(T)$ values, suggesting continental affinities. Data sources: see Appendix 6.

among the lowest of the known eclogites and differ significantly from those of the Alpine and Hercynian chains in western Europe (Figs. 6 and 7). Moreover, most rocks have $^{147}Sm/^{144}Nd$ ratios less than the chondritic value, suggesting LREE-enriched geochemical characteristics. To some extent, only eclogites from the Caledonian belt (e.g., northwestern Scotland and Western Norway, Fig. 8) have Sm-Nd isotopic compositions similar to those of the Dabie orogen. The isotopic data and geochemical characteristics undoubtedly indicate that the eclogite protoliths of the Dabie orogen have "continental" affinities and could not be produced from the subducted Tethyan oceanic slab. The very low $\varepsilon_{Nd}(T)$ values further require that the protoliths of these eclogites were formed long before the Triassic collision, probably in mid-Proterozoic times. The Tethys Ocean, regardless of its petrological and geochemical nature, was too young to have generated such an isotopic evolution. That is, these eclogites cannot be taken to represent ophiolites or to mark the suture zone between the Sino-Korean and Yangtze blocks. It appears that, in addition to the UHP metasedimentary rocks of the Dora-Maira, the Dabie Shan (including Sulu) and the Western Gneiss Region (WGR) are two regions where coesite- and diamond-bearing eclogites and ultramafic rocks show clear evidence of subduction of ancient, cold continental crustal blocks to mantle depths.

Type III eclogite-ultramafic suites in the Sulu and Dabie terranes have two different tectonic origins. The first suite, represented by the Bixiling and Maowu complexes, consists of layered intrusions initially emplaced in crustal levels and later subjected to UHP metamorphism as a result of continental subduction (Okay, 1994; Zhang et al., 1995; Chavagnac and Jahn, 1996; Fan et al., 1996; Liou et al., 1997). The second suite consists of mantle rocks exhumed together with crustal UHP metamorphic rocks; they are considered here as "tectonic enclaves of mantle origin" within granitic gneisses. A probable example is the garnet pyroxenites of Rizhao. Even though these pyroxenites represent igneous cumulate rocks from a mantle-derived liquid, the magma differentiation probably occurred within the mantle.

The time of intrusion for the Bixiling complex has not been precisely determined. However, based on garnet Sm-Nd and phengite Rb-Sr mineral-isotopic systematics, Chavagnac and Jahn (1996) argued that the Bixiling complex was intruded shortly ($\leq$ 300 Ma) before the UHP metamorphism (~220 Ma). A recent zircon U-Pb dating of Type III eclogite from Maowu gave an upper intercept age of ~450 Ma, which was interpreted as the time of protolith basic magmatism (Rowley, 1997). It remains to be determined whether all the Type III eclogites in the Dabie orogen have similarly "young" protolith ages. The isotopic systematics of the Maowu layered intrusion ($\varepsilon_{Nd}(T)$ ca. –5 to –6, I_{Sr} ~0.707 to 0.708; unpubl. data) are distinguished from those of the Bixiling complex (Jahn, 1998). Upper-crustal contamination during magma-chamber processes is likely to have occurred in the Maowu intrusion (Jahn, Fan et al., 1999). However, the timing of this process cannot be easily determined. Identification of sapphirine along with clinochlore, enstatite, etc. as inclusions in garnet led Okay (1994) to conclude that the Maowu intrusion underwent an earlier granulite-facies metamorphism (730°, 4 kbar), probably in Precambrian time, which was followed by UHP metamorphism (740°, $\geq$ 40 kbar) during the Triassic continental collision. Because most Type III rock possesses $^{147}Sm/^{144}Nd$ ratios higher than 0.15, the calculated T_{DM} ages have no rigorous meaning. The Maowu rocks show an extreme variation in T_{DM}, from –6.2 Ga for garnet orthopyroxenite to +5.8 Ga for eclogite (Jahn, Fan et al., 1999).

The interpretation of mantle slices for Rizhao and Rongcheng garnet peridotites is not unchallenged. It is still possible that a pre-UHP, low-pressure event might be identified for these ultramafic rocks, which would imply their earlier emplacement under crustal conditions. Indeed, a subduction history of ultramafic rocks that once resided in the continental crust is generally difficult to identify. However, combining detailed petrographic, mineral-chemistry, and isotopic studies, a subduction history of garnet peridotites and garnet pyroxenites from WGR and Sulu have been convincingly demonstrated (Jamtveit et al., 1991; Yang and Jahn, 1999). In any case, the thermal gradients recorded in the present mineral assemblages of garnet pyroxenites and garnet lherzolites are significantly lower than that of the upper mantle (Yang et al., 1993; Zhang et al., 1994; Hiramatsu et al., 1995). Consequently, these ultramafic rocks implicitly have undergone UHP metamorphism, as have the associated eclogites.

Eclogite samples from the Weihai area in the easternmost Shandong Peninsula show the most spectacular Nd-isotopic compositions (off scale in Fig. 10). Their chemistry has been severely modified by metasomatism, and it preserves the world record whole-rock $\varepsilon_{Nd}(0)$ values, ranging from +260 to +174, with extreme Sm/Nd fractionation ($f_{Sm/Nd}$ = +6.1 and +4.6). The Sm-Nd isotopic compositions were used to calculate precise T_{DM} ages of 1.67 and 1.61 Ga, which constrain the minimum time for protolith formation to the middle Proterozoic (~1.7 Ga) (Jahn et al., 1996). Metasomatism was not related to the UHP metamorphism, because the small time interval of ~220 m.y. is not sufficient to develop such large isotopic growth (270 ε_{Nd} units).

Significance of Eclogites/Peridotites from the Alpine, Hercynian, and Pan-African Belts

Except for a few occurrences, eclogites and garnet peridotites from the Alpine, Hercynian, and Pan-African belts are characterized by depleted mantle or oceanic-crust Nd-isotopic signatures (Figs. 6, 7, and 9). Most of these eclogites record high- but not ultrahigh-pressure conditions (10 to 20 kbar; see Appendix tables). Evidently, they are not of the same tectonic significance as those from the Dabie and Caledonian UHP terranes. Eclogites from the Otztal of the Austrian Alps have recorded P-T conditions of 27 kbar and 730°C (Miller and Thöni, 1995). Such ultrahigh pressure probably represents the metamorphic condition of the subducted oceanic crust at ~360 Ma and is irrelevant to subduction of an older continental crust.

Garnet peridotites from all orogenic belts have invariably been shown to have recorded ultrahigh-pressure conditions (above the coesite-quartz transition, or ≥ 23 kbar) (e.g., Brueckner and Medaris, 1998) and their presure ranges are always higher than those of the associated eclogites. Garnet peridotites of the Alpe Arami have recorded pressure conditions of at least 35 to 40 kbar on the basis of thermobarometric studies (Ernst, 1977; Medaris and Carswell, 1990) or considerably higher (~100 kbar) if calculation is based on TiO_2 solubility in olivine (Dobrzhinetskaya et al., 1996). This latter pressure estimate has remained controversial, and the mode of emplacement for these peridotites is not clear.

The ultrahigh-pressure conditions recorded by garnet peridotites do not necessarily imply a subduction history, because all mantle peridotites must have been equilibrated in mantle P-T conditions prior to their final exposure to the surface. According to Brueckner (1998), garnet peridotites could have been formed in a subduction zone through transformation of the low-pressure mantle-wedge peridotite (spinel or plagioclase peridotite) into garnet peridotite by sinking into and being transported along with subducting continental crust to a greater depth of the garnet-stability field. Brueckner and Medaris (1998a, 1998b) used this model to interpret garnet peridotites of the Bohemian Massif and termed this kind of peridotite "subduction-zone peridotite" (Brueckner and Medaris, 1998b). Moreover, Brueckner and Medaris envisioned that garnet peridotites of the Norwegian Caledonides represent fragments of ancient, cold, static mantle that were collected by subducted continental crust. This contradicts observations that the garnet peridotites of the WGR and Sulu recorded low P-T conditions prior to UHP metamorphism (Jamtveit, 1984; Yang and Jahn, 1999). In other words, they had been tectonically emplaced into upper crustal levels and formed an integral part of the continental crust in Precambrian times. Occurrences of mafic and ultramafic layers or enclaves in amphibolite- or granulite-facies terranes are common in stable Precambrian cratons. If segments of such cold cratonic crust were subducted during continental collision, these mafic and ultramafic enclaves would be transformed into UHP eclogite and garnet peridotites. In this scenario, the tectonic relationship between eclogites and country gneisses must be "in situ." However, in a collision complex, we should expect no intact crustal field relationship to be preserved for a block of continental crust undergoing deep subduction and exhumation. Consequently, the debate of "in-situ" versus "foreign" relationships appears trivial and depends upon the scale or dimension in question.

Conclusions

The present review of Sm-Nd isotopic tracer studies leads to the following conclusions.

1. Eclogites from the Dabie Shan and Sulu UHP terranes are generally of basaltic composition, but show a wide range of major- and trace-element abundances, suggesting multiple origins and heterogeneous protoliths as a result of igneous differentiation and metamorphic segregation (Jahn, 1998). Type I and II eclogites show LREE-enriched patterns and highly negative $\varepsilon_{Nd}(T)$ values (–6 to –20). The geochemical and isotopic data indicate that protoliths of the eclogites resemble continental mafic rocks of Precambrian gneiss terranes, and they constitute an integral component of continental cratons. They are not obducted oceanic crust of the Tethys Ocean. Sm-Nd model-age data suggest that the eclogite protoliths were emplaced at different times, but mainly during the Proterozoic (Appendix tables 5 and 6).

2. Type III eclogites are genetically related to the associated ultramafic rocks (lherzolite, harzburgite, or pyroxenites). Their Sr-Nd isotopic signatures suggest that they were derived from differentiation of basaltic magmas in layered intrusions at crustal levels (e.g., Bixiling, Maowu), within the mantle domain (e.g., Rizhao garnet clinopyroxenite), or represent mantle slices emplaced as tectonic enclaves during exhumation (Rongcheng and Donghai garnet peridotites). The isotopic signatures of the tectonic enclaves indicate that some mantle segments have evolved in long-term enriched sources (Rongcheng), whereas others are from slightly depleted reservoirs (Rizhao). The UHP parageneses of layered intrusions have a connotation of continental subduction, whereas those of other ultramafic rocks do not directly imply continental subduction.

3. In addition to the Dora-Maira quartzites of the Western Alps, the UHP eclogites and garnet peridotites of the European Caledonides (Norway and northwestern Scotland) have a similar crustal and subduction history to those of Dabie Shan and Sulu. Protoliths of these rocks represent the mafic and ultramafic components of old and stable continental blocks. They are characterized by negative $\varepsilon_{Nd}(T)$ values (metamorphic initial values). By contrast, most of the eclogites and garnet peridotites from the Alpine, Hercynian, Pan-African, and Ross orogenic

belts were derived from protoliths of oceanic affinity. The relics of UHP conditions, if any, do not directly indicate subduction of continental crust to mantle depths during the orogenic processes.

4. Finally, in the Dabie orogen, the preservation of the ultrahigh $\varepsilon_{Nd}(0)$ values (+170 to +260) in eclogites of very low Nd concentrations (average 0.5 ppm) from the Weihai region (Jahn et al., 1996) and of the extraordinarily low $\delta^{18}O$ in many eclogites and gneisses (see Rumble, 1999), the general absence of syntectonic granites in the Dabie Shan, and the available age data obtained by different techniques all point to a rapid rate of exhumation and the absence of a pervasive aqueous fluid phase during the entire process of subduction and exhumation of the Dabie UHP terrane.

Acknowledgments

J. Bernard-Griffiths read an early draft and provided useful comments and suggestions. Gary Ernst pointed out some errors and improved the final version of this paper. The work is supported by "Intérieur de la Terre" of the Institut National des Sciences de l'Universe (INSU)-1997. This is INSU contribution No. 179.

REFERENCES

Ames, L., Zhou, G., and Xiong, B., 1996, Geochronology and isotopic character of high-pressure metamorphism with implications for collision of the Sino-Korean and Yangtze cratons, central China: Tectonics, v. 15, p. 472–489.

Baker, J., Matthews, A., Mattey, D., Rowley, D., and Xue, F., 1997, Fluid-rock interactions during ultra-high pressure metamorphism, Dabie Shan, China: Geochim. et Cosmochim. Acta, v. 61, p. 1685–1696.

Barnicoat, A. C., and Frey, N., 1986, High-pressure metamorphism in the Zermatt-Saas Fee ophiolite zone, Switzerland: Jour. Geol. Soc. Lond., v. 143, p. 607–618.

Beard, B. L., Medaris, L. G., Johnson, C. M., Brueckner, H. K., and Misar Z., 1992, Petrogenesis of Variscan high-temperature Group A eclogites from the Moldanubian zone of the Bohemian Massif, Czechoslovakia: Contrib. Mineral. Petrol., v. 111, p. 468–483.

Beard, B. L., Medaris, L. G., Jr., Johnson, C. M., Jelinek, E., Tonika, J., and Riciputi, L. R., 1995, Geochronology and geochemistry of eclogites from the Marianske Lazne Complex, Czech Republic: Implications for Variscan orogenesis: Geol. Rundsch., v. 84, p. 552–567.

Becker, H., 1993, Garnet peridotite and eclogite Sm-Nd mineral ages from the Lepontine dome (Swiss Alps): New evidence for Eocene high-pressure metamorphism in the central Alps: Geology, v. 21, p. 599–602.

Bernard-Griffiths, J., and Cornichet, J., 1985, Origin of eclogites from South Brittany, France: A Sm-Nd isotopic and REE study: Chem. Geol., v. 52, p. 185–201.

Bernard-Griffiths, J., Peucat, J. J., Cornichet, J., Ponce de Leon, M. I., and Ibarguchi, J. I. G., 1985, U-Pb, Nd isotopic, and REE geochemistry in eclogites from the Cabo Ortegal Complex, Galicia, Spain: An example of REE immobility conserving MORB-like patterns during high-grade metamorphism: Chem. Geol., v. 52, p. 217–225.

Bernard-Griffiths, J., Peucat J. J., and Menot, R. P., 1991, Isotopic (Rb-Sr, U-Pb, and Sm-Nd) and trace-element geochemistry of eclogites from the Pan-African Belt: A case study of REE fractionation during high-grade metamorphism: Lithos, v. 27, p. 43–57.

Bowtell, S. A., Cliff, R. A., and Barnicoat, A. C., 1994, Sm-Nd isotopic evidence on the age of eclogitization in the Zermatt-Saas ophiolite: Jour. Metamor. Geol., v. 12, p. 187–196.

Brueckner, H. K., 1998, Sinking intrusion model for the emplacement of garnet-bearing peridotites into continent collision orogens: Geology, v. 26, p. 631–634.

Brueckner, H. K., Blusztajn, J., and Bakun-Czubarow, N., 1996, Trace-element and Sm-Nd "age" zoning in garnets from peridotites of the Caledonian and Variscan Mountains and tectonic implications: Jour. Metamor. Geol., v. 14, p. 61–73.

Brueckner, H. K., and Medaris, Jr., L. G., 1997, A tale of two orogens—the contrasting P-T-t history and geochemical evolution of mantle in ultrahigh-pressure (UHP) metamorphic terranes of the Norwegian Caledonides and the Czech Variscides: Schweiz. Mineral. Petrog. Mitteil., v. 78, p. 293–307.

______, 1998, A general model for the intrusion and evolution of peridotites in high-pressure and ultrahigh-pressure metamorphic terranes, *in* International Workshop on UHP Metamorphism and Exhumation [abs.]: Stanford, CA, Stanford University. p. A-1–A-5.

Carswell, D. A., and Zhang, R. Y., 1999, Petrographic characteristics and metamorphic evolution of ultrahigh-pressure eclogites in plate-collision bodies: INT. GEOL. REV., v. 41, p. 781–797.

Chavagnac, V., and Jahn, B. M., 1996, Coesite-bearing eclogites from the Bixiling Complex, Dabie Mountains, China: Sm-Nd ages, geochemical characteristics and tectonic implications: Chem. Geol., v. 133, p. 29–51.

Chopin, C., 1984, Coesite and pure pyrope in high-grade blueschists of the western Alps: A first record and some consequences: Contrib. Mineral. Petrol., v. 86, p. 107–118.

Cliff, R. A., 1998, Early Tertiary eclogite facies metamorphism in the Monviso ophiolite: Jour. Metamor. Geol., v. 16, p. 447–455.

Compagnoni, R., and Rolfo, F., 1999, Characteristics of UHP pelites, gneisses, and other unusual rocks: INT. GEOL. REV., v. 41 p. 552–570.

Cong, B., ed., 1996, Ultrahigh-pressure metamorphic rocks in the Dabieshan-Sulu region of China: Beijing and Dordrecht, Sci. Press, Kluwer Acad. Publ., 224 p.

Cong, B., Zhai, M. G., Carswell, D. A., Wilson, R. N., Wang, Q. C., Zhao, Z., and Windley, B. F., 1995, Petrogenesis of ultrahigh-pressure rocks and their country rocks at Shuanghe in Dabieshan, central China: Eur. Jour. Mineral., v. 7, p. 119–138.

Di Vincenzo, G., Palmeri, R., Talarico, F., Andriessen, P. A. M., and Ricci, C. A., 1997, Petrology and geochronology of eclogites from the Lanterman Range, Antarctica: Jour. Petrol., v. 38, p. 1391–1417.

Dobrzhinestskaya, L., Green, H. W., and Wang, S., 1996, Alpe Arami: A peridotite massif from depths of more than 300 kilometers: Science, v. 271, p. 1841–1845.

Duchêne, S., Blichert-Toft, J., Luais, B., Télouk, P., Lardeaux, J. M., and Albarède, F., 1997, The Lu-Hf dating of garnets and the ages of the Alpine high-pressure metamorphism: Nature, v. 387, p. 586–589.

Ernst, W. G., 1977, Mineralogical study of eclogitic rocks from Alpe Arami, Lepontine Alps, southern Switzerland: Jour. Petrol., v. 18, p. 371–398.

Fan, Q. C., Liu, R. X., Ma, B. L., Zhao, D. S., and Zhang, Q., 1996, The protolith and ultrahigh-pressure metamorphism of the Maowu mafic-ultramafic rock block in Dabieshan Mountains: Acta Petrol. Sinica, v. 12, p. 29–47 (in Chinese with English abstract).

Griffin, W. L., and Brueckner, H. K., 1980, Caledonian Sm-Nd ages and a crustal origin for Norwegian eclogites: Nature, v. 285, p. 319–321.

______, 1985, REE, Rb-Sr and Sm-Nd studies of Norwegian eclogites: Chem. Geol, v. 52, p. 249–271.

Hiramatsu, N., Banno, S., Hirajima, T., and Cong, B., 1995, Ultrahigh-pressure garnet lherzolite from Chijiadian, Rongcheng County, in the Su-Lu region of eastern China: The Island Arc, v. 4, p. 324–333.

Hiramatsu, N., and Hirajima, T., 1995, Petrology of the Hujialin garnet clinopyroxenite in the Su-Lu ultrahigh pressure province, eastern China: The Island Arc, v. 4, p. 310–323.

Jagoutz, E., 1988, Nd and Sr systematics in an eclogite xenolith from Tanzania: Evidence for frozen mineral equilibria in the continental lithosphere: Geochim. et Cosmochim. Acta, v. 52, p. 1285–1293.

Jahn, B.-M., 1998, Geochemical and isotopic characteristics of UHP eclogites of the Dabie orogen: Implications for continental subduction and collisional tectonics, *in* Hacker, B., and Liou, J. G., eds., When continents collide: Geodynamics and geochemistry of ultrahigh-pressure rocks: Dordrecht, Kluwer Acad. Publ., p. 203–239.

Jahn, B.-M., Cornichet, J., Cong, B. L., and Yui, T. F., 1996, Ultrahigh-ε_{Nd} eclogites from an ultrahigh pressure metamorphic terrane of China: Chem. Geol., v. 127, p. 61–79.

Jahn, B.-M., Fan, Q. C., Yang, J. J., and Henin, O., 1999, Genesis of the Maowu pyroxenite-eclogite body from the UHP metamorphic terrrane of Dabieshan: Chemical and isitopic constraints and tectonic implications: Chem. Geol., submitted,

Jahn, B.-M., Wu, F. Y., Lo, C. H., and Tsai, C. H., 1999, Crust-mantle interaction induced by deep subduction of the continental crust: Geochemical and Sr-Nd isotopic evidence from post-collisional mafic-ultramafic intrusions of the Northern Dabie Complex, central China. Chem. Geol., v. 157, p. 119–146.

Jamtveit, B., 1984, High-P metamorphism and deformation of the Gurskebotn garnet peridotite: Norsk Geol. Tidsskrift, v. 64, p. 97–110.

Jamtveit, B., et al., 1991, Chronology of the high-pressure metamorphism of Norwegian garnet peridotites/pyroxenes: Jour. Metamor. Geol., v. 9, p. 125–139.

Jerde, E. A., Taylor, L. A., Crozaz, G., Sobolev, N. V., and Sobolev, V. N., 1993, Diamondiferous eclogites from Yakutia, Siberia: Evidence for a diversity of protoliths: Contrib. Mineral. Petrol., v. 114, p. 189–202.

Kalt, A., Hanel, M., Schleicher, H., and Kramm, U., 1994, Petrology and geochronology of eclogites from the Variscan Schwarzwald (F.R.G.): Contrib. Mineral. Petrol., v. 115, p. 287–302.

Li, S., Jagoutz, E., Lo, C. H., Chen, Y. Z., Li, Q. L., and Xiao, Y. L., 1999, Sm/Nd, Rb/Sr, and $^{40}Ar/^{39}Ar$ isotopic systematics of ultrahigh pressure metamorphic rocks from the Dabie-Sulu belt in central China: A retrospective view: INT. GEOL. REV., v. 41, in press.

Li, S., Xiao, Y., Liou, D., Chen, Y., Ge, N., Zhang, Z., Sun, S. S., Cong, B., Zhang, R., Hart, S., and Wang, S., 1993, Collision of the North China and Yangtze blocks and formation of coesite-bearing eclogites: Timing and processes: Chem. Geol., v. 109, p. 89–111.

Liou, J. G., and Zhang, R. Y., 1998, Petrogenesis of the ultrahigh-P garnet-bearing ultramafic body from Maowu, Dabie Mountains, Central China: The Island Arc, v. 7, p. 115–134.

Liou, J. G., Zhang, R. Y., and Ernst, W. G., 1995, Occurrence of some hydrous and carbonate phases in ultrahigh-P rocks from east-central China—implications for the role of fluids in deep subduction zones: The Island Arc, v. 4, p. 362–375.

Liou, J. G., Zhang, R. Y., and Jahn, B. M., 1997, Petrology, geochemistry, and isotope data on a ultrahigh-pressure jadeite quartzite from Shuanghe, Dabie Mountains, east-central China: Lithos, v. 41, p. 59–78.

Liou, J. G., Zhang, R. Y., Wang, X., Eide, E. A., Ernst, W. G., and Maruyama, S., 1996, Metamorphism and tectonics of high-pressure and ultrahigh-pressure belts in the Dabie-Sulu regions, China, *in* Yin, A., and Harrison, M., eds., The tectonic evolution of Asia: Cambridge, U.K., Cambridge University Press, p. 300–344.

Mearns, E. W., 1986, Sm-Nd ages for Norwegian garnet peridotite: Lithos, v. 19, p. 269–278.

Medaris, L. G., Jr., 1999, Garnet peridotites in Eurasian high-pressure and ultrahigh-pressure terranes: A diversity of origins and thermal histories: INT. GEOL. REV., v. 41, p. 798–814.

Medaris, L. G., Jr., Beard, B. L., Johnson, C. M., Valley, J. W., Spicuzza, M. J., Jelinek, E., and Misar, Z., 1995, Garnet pyroxenite and eclogite in the Bohemian Massif: Geochemical evidence for Variscan recycling of subducted lithosphere: Geol. Rundsch, v. 84, p. 489–505.

Medaris, L. G., and Carswell, D. A., 1990, Petrogenesis of Mg-Cr garnet-peridotites in European metamorphic belts, *in* Carswell, D. A., ed., Eclogite facies rocks: Glasgow, Blackie and Sons, p. 260–290.

Miller, C., and Thöni, M., 1995, Origin of eclogites from the Austroalpine Otztal basement (Tirol, Austria): Geochemistry and Sm-Nd vs. Rb-Sr isotope systematics: Chem. Geol., v. 122, p. 199–225.

Mork, M. B. E., and Mearns, E. W., 1986, Sm-Nd isotopic systematics of a gabbro-eclogite transition: Lithos, v. 19, p. 255–267.

Mork, M. B. E., Kellerud, K., and Stabel, A., 1988, Sm-Nd dating of Seve eclogites, Norrbotten, Sweden—Evidence for early Caledonian (505 Ma) subduction: Contrib. Mineral. Petrol., v. 99, p. 344–351.

Ogasawara, Y., Liou, J. G., and Zhang, R. Y., 1995, Petrogenetic grid for ultrahigh-P metamorphism in the model system $CaO-MgO-SiO_2-CO_2-H_2O$: The Island Arc, v. 4, p. 240–253.

Ogasawara, Y., Zhang, R. Y., and Liou, J. G., 1998, Petrogenesis of dolomitic marbles from Rongcheng in the Sulu UHP metamorphic terrane, eastern China: The Island Arc, v. 7, p. 82–97.

Okay, A. I., 1994, Sapphirine and Ti-clinohumite in ultrahigh-pressure garnet-pyroxenite and eclogite from Dabie Shan, China: Contrib Mineral, Petrol., v. 116, p. 145–155.

Okay, A. I., Sengör, A. M. C., and Satir, M., 1993, Tectonics of an ultrahigh-pressure metamorphic terrane: The Dabieshan/Tongbaishan Orogen, China: Tectonics, v. 12, p. 1320–1334.

Paquette, J. L., Menot, R. P., Peucat, J. J., 1989, REE, Sm-Nd, and U-Pb zircon study of eclogites from the Alpine External Massif (western Alps): Evidence for crustal contamination: Earth Planet. Sci. Lett., v. 96, p. 181–198.

Reisberg, L., Zindler, A., and Jagoutz, E., 1989, Further Sr and Nd isotopic results from peridotites of the Ronda ultramafic complex: Earth Planet. Sci. Lett., v. 96, p. 161–180.

Richardson, S. H., Erlank, A. J., Harris, J. W., Hart, S. R., 1990, Eclogitic diamonds of Proterozoic age from Cretaceous kimberlites: Nature, v. 346, p. 54–56.

Rowley, D. B., Xue, F., Tucker, R. D., Peng, Z. X., Baker, J., and Davis, A., 1997, Ages of ultrahigh pressure metamorphism and protolith orthogneisses from the eastern Dabie Shan: U/Pb zircon geochronology: Earth Planet. Sci. Lett., v. 151, p. 191–203.

Rumble, D., 1998, Stable isotope geochemistry of ultrahigh-pressure rocks, *in* Hacker, B. and Liou, J. G., eds., When continents collide: Geodynamics and geochemistry of ultrahigh-pressure rocks: Dordrecht, Klumer Acad. Publ., p. 241–259.

Rumble, D., III, Summary of research on stable-isotope geochemistry of UHP rocks, 1993 to 1998: INT GEOL. REV., v. 41, in press.

Sanders, I. S., van Calsteren, P. W. C., and Hawkesworth, C. J., 1984, A Grenville Sm-Nd age for the Glenelg eclogite in northwest Scotland: Nature, v. 312, p. 439–440.

Shatsky, V. S., Jagoutz, E., Sobolev, N. V., Kozmenko, O. A., Parkhomenko, V. S., and Troesch, M., 1999, Geochemistry and age of ultrahigh-pressure metamorphic rocks from the Kokchetav Massif (northern Kazakhstan). Contrib: Mineral. Petrol. (in press).

Shatsky, V. S., Sobolev, N. V., Vavilov, M. A., 1995, Diamond-bearing metamorphic rocks of the Kokchetav massif (northern Kazakhstan), *in* Coleman, R. G., and Wang, X., eds., Ultrahigh-pressure metamorphism: New York, Cambridge University Press, p. 427–455.

Shervais, J. W., Taylor, L. A., Lugmair, G. W., Clayton, R. N., Mayeda, T. K., and Korotev, R. L., 1988, Early Proterozoic oceanic crust and the evolution of subcontinental mantle: Eclogites and related rocks from southern Africa: Geol. Soc. Amer. Bull., v. 100, p. 411–423.

Schreyer, W., 1988, Subduction of continental crust to mantle depths: Petrological evidence: Episodes, v. 11, p. 97–104.

———, 1995, Ultradeep metamorphic rocks: The retrospective viewpoint: Jour. Geophys. Res., v. 100, p. 8353–8366.

Smith, C. B., Gurney, J. J., Harris, J. W., Otter, M. L., Kirkley, M. B., and Jagoutz, E., 1991, Neodynium and strontium isotope systematics of eclogite and websterite paragenesis inclusions from single diamonds, Finsch and Kimberley Pool, RSA: Geochim. et Cosmochim. Acta, v. 55, p. 2579–2590.

Smith, D. C., 1984, Coesite in clinopyroxene in the Caledonides and its implication for geodynamics: Nature, v. 310, p. 641–644.

Snyder, G. A., Jerde, E. A., Taylor, L. A., Halliday, A. N., Sobolev, V. N., and Sobolev, N. V., 1993, Nd and Sr isotopes from diamondiferous eclogites, Udachnaya kimberlite pipe, Yakutia, Siberia: Evidence of differentiation in the early Earth?: Earth Planet. Sci. Lett., v. 118, p. 91–100.

Stosch, H. G., and Lugmair, G. W., 1990, Geochemistry and evolution of MORB-type eclogites from the Münchberg Massif, southern Germany: Earth Planet. Sci. Lett., v. 99, p. 230–249.

Thöni, M., and Jagoutz, E., 1992, Some new aspects of dating eclogites in orogenic belts: Sm-Nd, Rb-Sr, and Pb-Pb isotopic results from the Austroalpine Saualpe and Koralpe type-locality (Carinthia/Styria, southeastern Austria): Geochim. et Cosmochim. Acta, v. 56, p. 347–368.

Tilton, G. R., Schreyer, W., and Schertl, H. P., 1989, Pb-Sr-Nd behavior of deeply subducted crustal rocks from the Dora Maira massif, western Alps, Italy: Geochim. et Cosmochim. Acta, v. 53, p. 1391–1400.

______, 1991, Pb-Sr-Nd isotopic behavior of deeply subducted crustal rocks from the Dora Maira massif, western Alps, Italy. II: What is the age of the ultrahigh-pressure metamorphism?: Contrib. Mineral. Petrol., v. 108, p. 22–33.

Viljoen, K. S., Smith, C. B., Sharp, Z. D., 1996, Stable and radiogenic isotope study of eclogite xenoliths from the Orapa kimberlite, Botswana: Chem. Geol., v. 131, p. 235–255.

Von Quadt, A., Günther, D., Frischknecht, R., Zimmermann, R., and Franz, G., 1997, The evolution of pre-Variscan eclogites of the Tauern Window (eastern Alps): A Sm/Nd, conventional, and laser ICP-MS zircon U-Pb study: Schweiz. Mineral. Petrog. Mitteil., v. 77, p. 265–279.

Wang, X., Jing, Y., Liou, J. G., Liang, W., Xia, M., and Maruyama, S., 1990, Field occurrences and petrology of eclogites from the Dabie Mountains, Anhui, central China: Lithos, v. 25, p. 119–131.

Wang, X. M., and Liou, J. G., 1991, Regional ultrahigh-pressure coesite-bearing eclogite terrane in central China: Evidence from country rocks, gneiss, marble, and metapelite: Geology, v. 19, p. 933–936.

______, 1993, Ultrahigh-pressure metamorphism of carbonate rocks in the Dabie Mountains, central China: Jour. Metamor. Geol, v. 11, p. 575–588.

Yang, J., 1991, Eclogite, garnet pyroxenites, and related ultrabasics in Shandong and north Jiangsu of east China: Beijing, Geol. Publ. House, p. 1–99 (in Chinese, with English abstract).

Yang, J., Godard, G., Kienast, J. R., Lu, Y., and Sun, J., 1993, Ultrahigh-pressure (60 kbar) magnesite-bearing garnet peridotites from northeastern Jiangsu, China: Jour. Geol., v. 101, p. 541–554.

Yang, J., and Jahn, B. M., 1999, Deep subduction of the Su-Lu garnet peridotite and its mineral chemical and isotopic consequences: Jour. Metamor. Geol., submitted.

Yui, T. F., Rumble, D., Chen, C. H., Lo, C. H., 1997, Stable isotope characteristics of eclogites from the ultrahigh-pressure metamorphic terrain, east-central China: Chem. Geol., v. 137, p. 135–147.

Yui, T. F., Rumble, D., and Lo, C. H., 1995, Unusually low ^{18}O ultra-high-pressure metamorphic rocks from the Sulu terrain, eastern China: Geochim et Cosmochim Acta, v. 59, p. 2859–2864.

Zhai, M. G., and Cong, B. L., 1996, Major- and trace-element geochemistry of eclogites and related rocks, *in* Cong, B., ed., Ultrahigh-pressure metamorphic rocks in the Dabieshan-Sulu region of China: Beijing and Dordrecht, Science Press and Kluwer Acad. Publ., p. 69–89.

Zhang, R. Y., Liou, J. G., and Cong B., 1994, Petrogenesis of garnet-bearing ultramafic rocks and associated eclogites in the Su-Lu ultrahigh-P metamorphic terrane, eastern China: Jour. Metamor. Geol., v. 12, p. 169–186.

______, 1995, Talc-, magnesite-, and Ti-clinohumite-bearing ultrahigh-pressure metamafic and ultramafic complex in the Dabie Mountains, China: Jour. Petrol., v. 36, p. 1011–1037.

Zheng, Y. F., Fu, B., Gong, B., and Li, S., 1996, Extreme ^{18}O depletion in eclogite from the Su-Lu terrane in East China: Eur. Jour. Mineral., v. 8, p. 317–323.

Zheng, Y. F., Fu, B., Li, Y., Zhao, Y., and Li, S., 1998, Oxygen and hydrogen isotope geochemistry of ultrahigh-pressure eclogites from the Dabie Mountains and the Sulu terrane: Earth Planet. Sci. Lett., v. 155, p. 113–129.

APPENDIX 1. Sm-Nd Isotopic Compositions of Eclogites and Garnet Peridotites from the Alpine Belt

Sample no.	Locality	Age, Ma (metamor.)	Rock type	Occurrence type	[Sm]	[Nd]	$^{147}Sm/^{144}Nd$	$^{143}Nd/^{144}Nd$[1]	± 2σm	f(Sm/Nd)[2]	εNd (0)	εNd (T1)	εNd (T2)	T_{DM},[3] Ga	Supplementary information: P, kbar	T, °C	Protolith nature	Protolith age, Ma
Dora Maira, Western Alps												35 Ma						
15592 WR	Dora Maira	≈35	Jd-ky quartzite		11.89	56.00	0.1284	0.512248	26	–0.35	–7.6	–7.3	—	1.61	≈ 30	700–800	Evaporitic sediments (Hi-Mg clay)	Proterozoic sediments
15589a WR1	Dora Maira	≈35	Jd-ky quartzite		5.12	31.88	0.0971	0.512254	7	–0.51	–7.5	–7.0	—	1.17	≈ 30	700–800		
15615a WR1	Dora Maira	≈35	Pyrope quartzite		6.19	26.90	0.1391	0.512240	12	–0.29	–7.8	–7.5	—	1.85	≈ 30	700–800		
15615a WR2	Dora Maira	≈35	Pyrope quartzite		8.84	48.84	0.1094	0.512243	18	–0.44	–7.7	–7.3	—	1.32	≈ 30	700–800		
15589a WR2	Dora Maira	≈35	Jd-ky quartzite		4.80	24.48	0.1187	0.512307	24	–0.40	–6.5	–6.1	—	1.35	≈ 30	700–800		
Cima Lunga nappe												40 Ma						
AA13-A (1)	Alpe Arami	40	Gt lherzolite	Ultra-mafic-mafic assoc.	0.218	0.532	0.2477	0.513124	10	0.26	9.5	9.2	—	–0.12	35–42	850–900	Mantle perid.	young (?)
AA13-A (2)	Alpe Arami	40	Gt lherzolite		0.151	0.368	0.2481	0.513117	13	0.26	9.3	9.1	—	–0.15	35–42	850–900	Mantle perid.	young (?)
AA13-B	Alpe Arami	40	Gt pyroxenite		0.742	2.258	0.1987	0.513079	12	0.01	8.6	8.6	—	0.72	35–42	850–900	Mantle perid.	young (?)
AA6 (1)	Alpe Arami	40	Gt lherzolite		0.158	0.366	0.2610	0.513241	11	0.33	11.8	11.4	—	0.29	35–42	850–900	Mantle perid.	young (?)
AA6 (2)	Alpe Arami	40	Gt lherzolite		0.194	0.450	0.2606	0.513294	11	0.33	12.8	12.5	—	0.47	35–42	850–900	Mantle perid.	young (?)
AA15	Alpe Arami	40	Eclogite		3.425	11.211	0.1847	0.513076	10	–0.06	8.5	8.6	—	0.39	35–42	850–900	Mantle eclogite	young (?)
CG5 (1)	Cima di Gagnone	40	Gt lherzolite		0.269	0.628	0.2590	0.513245	15	0.32	11.8	11.5	—	0.32	35–42	850–900	Mantle perid.	young (?)
CG5 (2)	Cima di Gagnone	40	Gt lherzolite		0.275	0.642	0.2590	0.513222	11	0.32	11.4	11.1	—	0.24	35–42	850–900	Mantle perid.	young (?)
Monviso, Western Alps 1												60 Ma						
53910 - Cpx	Monviso, W Alps	60	Eclogite	Ophiolite	1.360	3.340	0.2462	0.513133	24	0.25	9.7	9.3	—	–0.08	20	400–500	Ferrogabbro	?
54423 - Cpx	Monviso, W Alps	60	Eclogite	Ophiolite	1.650	3.490	0.2858	0.513073	18	0.45	8.5	7.8	—	–0.16	20	400–500	Ferrogabbro	?
Western Alps												400 Ma	500 Ma					
R7756	Aiguilles Rouges	≈400	Eclogite	Inter-calated with amphib. gneiss	7.15	21.72	0.1990	0.512982	18	0.01	6.7	6.6	6.6	1.74	>11	780	N-MORB	500 Ma
R7757	Aiguilles Rouges	≈400	Eclogite		3.18	9.32	0.2063	0.512980	21	0.05	6.7	6.2	6.1	3.46	>11	780	N-MORB	
R7758	Aiguilles Rouges	≈400	Eclogite		4.64	14.04	0.1998	0.513005	11	0.02	7.2	7.0	7.0	1.59	>11	780	N-MORB	

(table continues)

APPENDIX 1. (*Continued*)

Sample no.	Locality	Age, Ma (metamor.)	Rock type	Occurrence type	[Sm]	[Nd]	$^{147}Sm/^{144}Nd$	$^{143}Nd/^{144}Nd$[1]	± 2σm	f(Sm/Nd)[2]	εNd (0)	εNd (T1)	εNd (T2)	T_{DM},[3] Ga	Supplementary information: P, kbar	T, °C	Protolith nature	Protolith age, Ma
								Western Alps (*continued*)										
												400 Ma	500 Ma					
B7751	Belledonne Msf	≈ 400	Eclogite		3.80	10.81	0.2125	0.513088	35	0.08	8.8	8.0	7.8	7.82			N-MORB	
B7752	Belledonne Msf	≈400	Eclogite	In leptinite-	3.66	10.44	0.2119	0.513092	12	0.08	8.9	8.1	7.9	4.97			N-MORB	
B7754	Belledonne Msf	≈400	Eclogite	amphibolite	6.31	18.75	0.2035	0.513025	21	0.03	7.5	7.2	7.1	1.85			N-MORB	
B7755	Belledonne Msf	≈400	Eclogite	complex	5.00	14.64	0.2065	0.513016	30	0.05	7.4	6.9	6.7	2.81			N-MORB	
AJ8017	Argentera Msf	≈400	Eclogite		1.33	3.99	0.2015	0.513021	12	0.02	7.5	7.2	7.2	1.61	14 kb	700°C	N-MORB	
AJ17	Argentera Msf	≈400	Eclogite	In leptinite-	1.80	5.68	0.1916	0.512972	26	−0.03	6.5	6.8	6.8	1.23	14 kb	700°C	Cumulate	
AC12	Argentera Msf	≈400	Eclogite	amphibolite	2.63	9.05	0.1757	0.512747	8	−0.11	2.1	3.2	3.5	1.61	14 kb	700°C	N-MORB	
AC18	Argentera Msf	≈400	Eclogite	complex	5.41	19.14	0.1709	0.512899	9	−0.13	5.1	6.4	6.7	0.89	14 kb	700°C	E-MORB	
								Austrian Alps 1										
												100 Ma	275 Ma					
88T35	(Saualpe and	≈100	Eclogite	Enclaves in	1.53	4.27	0.2166	0.513113	22	0.10	9.3	9.0	8.6	−1.95	18	580–630	MORB	
86T09	Koralpe)	≈100	Eclogite	gneiss/	2.45	6.56	0.2258	0.513082	18	0.15	8.7	8.3	7.6	−0.86	18	580–630	MORB	
SK30	"	≈100	Eclogite	schist	1.06	2.97	0.2158	0.513060	22	0.10	8.2	8.0	7.6	−6.80	18	580–630	MORB	≈275 Ma
GE/a	"	≈100	Eclogite		4.84	13.96	0.2096	0.513027	20	0.07	7.6	7.4	7.1	4.52	18	580–630	MORB	
GE2	"	≈100	Eclogite		4.56	13.11	0.2103	0.512972	20	0.07	6.5	6.3	6.0	7.76	18	580–630	MORB	
88T32	"	≈100	Eclogite		0.95	2.09	0.2751	0.513169	20	0.40	10.4	9.4	7.6	0.05	18	580–630	MORB	
								Austrian Alps 2										
												360 Ma	520 Ma					
CM88/48	Otztal, Tirol	≈360	Eclogite	As relics in	0.59	1.88	0.1896	0.512883	11	−0.04	4.8	5.1	5.2	1.69	27	730	Olivine gabbro	530–520
86T42	Otztal, Tirol	≈360	Eclogite	much larger	2.03	7.91	0.1553	0.512834	8	−0.21	3.8	5.7	6.6	0.83	27	730	Olivine gabbro	530–520
OE/MO	Otztal, Tirol	≈360	Eclogite	body of	1.97	7.06	0.1684	0.512946	17	−0.14	6.0	7.3	7.9	0.69	27	730	Olivine gabbro	530–520
BE207	Otztal, Tirol	≈360	Eclogite	meta-	5.69	18.40	0.1870	0.512930	9	−0.05	5.7	6.1	6.3	1.26	27	730	T-MORB	530–520
CM88/38	Otztal, Tirol	≈360	Eclogite	basic rocks	6.02	21.51	0.1692	0.512863	8	−0.14	4.4	5.7	6.2	0.98	27	730	T-MORB	530–520
T694	Otztal, Tirol	≈360	Eclogite	surrounded	2.95	8.88	0.2012	0.512919	10	0.02	5.5	5.3	5.2	2.80	27	730	T-MORB	530–520
T698	Otztal, Tirol	≈360	Eclogite	by granitic gneisses	6.99	23.03	0.1835	0.512814	11	−0.07	3.4	4.0	4.3	1.69	27	730	T-MORB	530–520
OE318	Otztal, Tirol	≈360	Eclogite		2.49	3.70	0.4064	0.513343	25	1.07	13.8	4.1	−0.2	0.15	27	730	Ferro-gabbro	530–520
2329	Otztal, Tirol	≈360	Eclogite		1.31	2.21	0.3577	0.513225	7	0.82	11.5	4.1	0.8	0.08	27	730	Ferro-gabbro	530–520
CM9244	Otztal, Tirol	≈360	Eclogite		1.09	1.51	0.4359	0.513320	11	1.22	13.3	2.3	−2.6	0.12	27	730	Ferro-gabbro	530–520
OE455-1	Otztal, Tirol	≈360	Eclogite		1.39	1.20	0.7015	0.513995	10	2.57	26.5	3.3	−7.0	0.26	27	730	Ferro-gabbro	530–520

								Austrian Alps 2 (*continued*)										
												360 Ma	520 Ma					
OE455-2	Otztal, Tirol	≈360	Eclogite		1.30	1.20	0.6556	0.514026	12	2.33	27.1	6.0	−3.4	0.30	27	730	Ferro-gabbro	530–520
OE92T4	Otztal, Tirol	≈360	Eclogite		1.91	5.35	0.2153	0.512957	7	0.09	6.2	5.4	5.0	−20.18	27	730	Ferro-gabbro	530–520
OE92T5	Otztal, Tirol	≈360	Eclogite		1.37	4.27	0.1944	0.512858	14	-0.01	4.3	4.4	4.4	2.29	27	730	Ferro-gabbro	530–520
OE92T6	Otztal, Tirol	≈360	Eclogite		1.08	1.81	0.3610	0.513280	20	0.84	12.5	5.0	1.6	0.13	27	730	Ferro-gabbro	530–520
93T42 OE	Otztal, Tirol	≈360	Eclogite		1.91	4.11	0.2810	0.513024	7	0.43	7.5	3.7	1.9	−0.29	27	730	Ferro-gabbro	530–520
								Eastern Alps										
												420 Ma	488 Ma					
LT25-87	Tauern Window	420	Eclogite	Individual samples	1.04	2.87	0.2192	0.512921	6	0.11	5.5	4.3	4.1	−6.47	8–12	450–500	E-MORB (? or OIB (?)	488 ± 12 (zircon)
LT26C-87	(Lower Schist)	420	Eclogite	not defined	5.94	24.42	0.1471	0.512644	5	−0.25	0.1	2.8	3.2	1.16	(Lower schist cover)			
LT26B-87	Tauern Window	420	Eclogite		4.12	16.16	0.1541	0.512674	18	−0.22	0.7	3.0	3.4	1.22				
LT29-87	(Lower Schist)	420	Eclogite		14.92	65.53	0.1376	0.512572	20	−0.30	−1.3	1.9	2.4	1.16				
LT36B	Tauern Window	420	Eclogite		2.68	9.04	0.1793	0.512821	16	−0.09	3.6	4.5	4.7	1.45				
LT28	(Lower Schist)	420	Eclogite		3.39	11.72	0.1749	0.512787	12	−0.11	2.9	4.1	4.3	1.42				
LT36A	Tauern Window	420	Eclogite		6.84	25.78	0.1604	0.512669	6	−0.18	0.6	2.6	2.9	1.37				
FRO-6	(Lower Schist)	420	Eclogite		3.66	12.88	0.1718	0.513023	6	−0.13	7.5	8.8	9.1	0.46				
FRO-3	Tauern Window	420	Eclogite		1.70	6.19	0.1660	0.513094	8	−0.16	8.9	10.5	10.8	0.18				
E3	(Lower Schist)	420	Eclogite		9.20	47.31	0.1176	0.512487	6	−0.40	−2.9	1.3	2.0	1.05				
E1	Tauern Window	420	Eclogite		2.72	9.18	0.1792	0.512687	12	−0.09	1.0	1.9	2.0	2.04				
TW22-89	(Lower Schist)	420	Eclogite		4.41	21.59	0.1235	0.512497	11	−0.37	−2.8	1.2	1.8	1.10				
								Monviso, Western Alps 2										
												T = 49						
	Monviso, W. Alps	49 ± 1	Mafic eclog.	Ophiolite	7.55	25.17	0.1813	0.513012	22	−0.08	7.3	7.4		0.65	15	~450		

[1] $^{143}Nd/^{144}Nd$ ratios are reported relative to the La Jolla Nd standard = 0.511860 and have been corrected for mass fractionation relative to $^{146}Nd/^{144}Nd$ = 0.7219.

[2]CHUR (chondritic uniform reservoir): $^{147}Sm/^{144}Nd$ = 0.1967; $^{143}Nd/^{144}Nd$ = 0.512638

[3]DM (depleted mantle): $^{147}Sm/^{144}Nd$ = 0.2137; $^{143}Nd/^{144}Nd$ = 0.51315.

Sources: Dora-Maira (Tilton, et al., 1989, 1991); Cima Lunga nappe (Becker, 1993); Monviso, Western Alps 1 (Cliff, 1998); Western Alps (Paquette et al., 1989); Austrian Alps 1 (Thöni and Jagoutz, 1992); Austrian Alps 2 (Miller and Thöni, 1995); Eastern Alps (vonQuadt et al., 1997); Monviso, Western Alps 2 (Duchêne et al., 1997).

APPENDIX 2. Sm-Nd Isotopic Compositions of Eclogites and Garnet Peridotites from the Hercynian Belt

Sample no.	Locality	Age, Ma (metamor.)	Rock type	Occurrence type	[Sm]	[Nd]	$^{147}Sm/^{144}Nd$	$^{143}Nd/^{144}Nd$	± 2σm	f(Sm/Nd)	εNd (0)	εNd (T)	εNd (T)	T_{DM}, Ga	Supplementary information: P, kbar	T, °C	Protolith nature	Protolith age, Ma
Bernard-Griffiths et al., 1985																		
												480 Ma	600 Ma					
7103	Cabo Ortegal	480	Eclogite	Large mafic-	3.04	8.13	0.2276	0.513275	38	0.16	12.4	10.5	10.1	1.37	17	800	N-MORB	≈600
7104	Cabo Ortegal	480	Eclogite	ultramafic	2.33	6.29	0.2252	0.513259	41	0.14	12.1	10.4	9.9	1.44	17	800	N-MORB	≈ 600
7107	Cabo Ortegal	480	Eclogite	complex	2.37	6.32	0.2203	0.513231	23	0.12	11.6	10.1	9.8	1.87	17	800	N-MORB	≈ 600
7108	Cabo Ortegal	480	Eclogite	granitic gneiss	2.89	7.97	0.2207	0.513236	19	0.12	11.7	10.2	9.8	1.87	17	800	N-MORB	≈ 600
Bernard-Griffiths and Cornichet, 1985																		
												420 Ma						
5506	Champtoceaux	413 ± 16	Eclogite	Enclosed in	5.03	15.80	0.1938	0.512846	46	−0.01	4.1	4.2		2.32			T-MORB	2200
4566	H. des Essarts	436 ± 15	Eclogite	kinzigitic	2.79	7.49	0.2263	0.513456	70	0.15	16.0	14.4		3.68	16–20	700 ± 100	N-MORB	1300
5498	H. des Essarts	436 ± 15	Eclogite	gneiss	2.17	5.14	0.2572	0.513323	26	0.31	13.4	10.1		0.61	16–20	700 ± 100	N-MORB	1300
5499	H. des Essarts	436 ± 15	Eclogite	≈ ophiolite	0.37	0.94	0.2428	0.513201	66	0.23	11.0	8.5		0.27	16–20	700 ± 100	N-MORB	1300
5500	H. des Essarts	436 ± 15	Eclogite	complex	3.73	15.06	0.1509	0.512683	36	−0.23	0.9	3.3		1.13	16–20	700 ± 100	N-MORB	1300
Stosch and Lugmair, 1990																		
												380 Ma	480 Ma					
W1.60m	Münchberg msf.	380–395	Eclogite	As isolated	4.35	15.44	0.1701	0.513001	14	−0.14	7.1	8.4	8.7	0.52	15–20	660–680	MORB	480 ± 23
W4.00m	S. Germany	380–395	Eclogite	bodies or	6.47	23.70	0.1650	0.513001	14	−0.16	7.1	8.6	9.0	0.47	15–20	660–680	(both N- and	(WR isochron)
W15.2m	Münchberg msf.	380–395	Eclogite	intercalated	6.52	25.00	0.1576	0.512935	14	−0.20	5.8	7.7	8.2	0.59	15–20	660–680	E-MORB)	480 ± 23
W27.5m	S. Germany	380–395	Eclogite	with gneisses	4.47	15.64	0.1728	0.513036	14	−0.12	7.8	8.9	9.2	0.43	15–20	660–680		(WR isochron)
W33.7m	Münchberg msf.	380–395	Eclogite		1.44	4.11	0.2109	0.513121	14	0.07	9.4	8.7	8.6	1.59	15–20	660–680		480 ± 23
W36.0m	S. Germany	380–395	Eclogite	All samples	5.34	21.07	0.1533	0.512536	14	−0.22	−2.0	0.1	0.7	1.55	15–20	660–680	Crustally	(WR isochron)
W43.0m	Münchberg msf.	380–395	Eclogite	from a drill	2.43	5.42	0.2706	0.513318	14	0.38	13.3	9.7	8.7	0.45	15–20	660–680	contaminated	480 ± 23
W114.7m	S. Germany	380–395	Eclogite	core	3.48	11.11	0.1895	0.513033	14	−0.04	7.7	8.1	8.1	0.74	15–20	660–680	(both N- and	(WR isochron)
																	E-MORB)	
W205m	Münchberg msf.	380–395	Eclogite		4.94	13.80	0.2165	0.512931	14	0.10	5.7	4.8	4.5	−12.24	15–20	660–680		480 ± 23
S114	S. Germany	380–395	Eclogite		3.05	9.71	0.1902	0.512947	28	−0.03	6.0	6.3	6.4	1.32	15–20	660–680		(WR isochron)
Kalt et al., 1994																		
												330 Ma	460 Ma					
Lo-1	Schwarzwald	330 Ma	Eclogite	Enclaves in	2.33	5.52	0.2550	0.512695	7	0.30	1.1	−1.3	−2.3	−1.69	14–16	670–730	Not discussed	460 Ma
Z-48	Schwarzwald	330 Ma	Eclogite	gneisses	2.23	6.08	0.2215	0.512541	14	0.13	−1.9	−2.9	−3.3	−12.43	10 to 20	750	Not discussed	460 Ma
Alp-1	Schwarzwald	330 Ma	Eclogite	(not defined)	3.32	12.74	0.1577	0.512582	15	−0.20	−1.1	0.6	1.2	1.54	16	685 ± 45	Not discussed	460 Ma

								Beard et al., 1992[1]										
												340 Ma						
CZ 2E	Nové Dvory	342	Eclogite	Mafic and	2.66	5.65	0.2846	0.512823	19	0.45	3.6	–0.2		–0.71	22.5	985	High P ga-	370–360
ND 2	Nové Dvory	336	Eclogite	ultramafic	0.642	2.03	0.1917	0.512781	48	–0.03	2.8	3.0		2.54	22.5	985	cpx cumu-	370–360
CZ-14F	Biskupice	340	Eclogite	association;	1.270	2.660	0.2886	0.513019	35	0.47	7.4	3.4		–0.27	16	855	lates of	370–360
CZ-14I-1	Biskupice	340	Opx Eclogite	emplaced	1.930	4.970	0.2348	0.512826	38	0.19	3.7	2.0		–2.37	16	855	basaltic	370–360
CZ-6C	Becvary	340	Gt clinopyrox.	from mantle	0.418	1.090	0.2318	0.513020	47	0.18	7.5	5.9		–1.10	17	890	liquid	370–360
CZ-6D	Becvary	340	Eclogite	to cont. crust	0.664	0.961	0.4177	0.513455	81	1.12	15.9	6.3		0.23	17	890	formed in	370–360
CZ-6E	Becvary	340	Gt clinopyrox.		0.436	1.190	0.2215	0.513025	26	0.13	7.5	6.5		–2.47	17	890	the mantle	370–360
								Medaris et al., 1995[2]										
												340 Ma						
CZ-12A	Dobesovice	340	Kyanite eclog.	Mafic and	3.990	11.300	0.2135	0.512358	17	0.09	–5.5	–6.2		227	13–20	770–1190	High P ga-	370–360
CZ-12D	Dobesovice	340	Kyanite eclog.	ultramafic	0.872	2.270	0.2322	0.512447	31	0.18	–3.7	–5.3		–5.91	13–20	770–1190	cpx cumu-	370–360
UR-4	Uhrov	340	Eclogite	association;	2.110	4.420	0.2886	0.513044	24	0.47	7.9	3.9		–0.22	13–20	770–1190	lates of basal-	370–360
UR-3	Uhrov	340	Eclogite	emplaced	1.640	3.050	0.3251	0.513166	19	0.65	10.3	4.7		0.02	13–20	770–1190	tic liquid	370–360
CZ-4A	Nihov	340	Gt websterite	from mantle to cont. crust	1.630	8.120	0.1214	0.512314	32	–0.38	–6.3	–3.1		1.38	13–20	770–1190	formed in the mantle	370–360
CZ-4B	Nihov	340	Gt websterite		0.464	2.100	0.1336	0.512328	19	–0.32	–6.0	–3.3		1.56	13–20	770–1190		370–360
								Beard et al., 1995[3]										
												380 Ma	496 Ma					
ML-1A	Mnichov	380	Qtz eclogite	Ophiolite	5.60	18.00	0.1881	0.512986	21	–0.04	6.8	7.2	7.3	0.98	14–16.4	630–730	MORB	500 Ma(?)
ML-1B	Mnichov	380	Qtz eclogite	complex(?)	5.54	18.80	0.1782	0.512959	23	–0.09	6.3	7.2	7.4	0.82	14–16.4	630–730	MORB	500 Ma(?)
ML-2A	Louka	380	Ky-qtz eclogite	intercalated	4.02	12.70	0.1914	0.512977	23	–0.03	6.6	6.9	7.0	1.18	14–16.4	630–730	MORB	500 Ma(?)
ML-3A	Tisova	380	Qtz eclogite	with	4.47	12.20	0.2215	0.512989	14	0.13	6.8	5.6	5.3	–3.19	14–16.4	630–730	MORB	500 Ma(?)
ML-3B	Tisova	380	Ky-qtz eclogite	migmatitic	2.11	4.12	0.3096	0.513248	15	0.57	11.9	6.4	4.8	0.16	14–16.4	630–730	OIB?	500 Ma(?)
ML-3C	Tisova	380	Ky-qtz eclogite	gneisses and	3.44	14.80	0.1405	0.512422	13	–0.29	–4.2	–1.5	–0.7	1.51	14–16.4	630–730	OIB?	500 Ma(?)
ML-3C	Tisova	380	Ky-qtz eclogite	schists	3.38	15.00	0.1362	0.512377	20	–0.31	–5.1	–2.2	–1.3	1.52	14–16.4	630–730	OIB?	500 Ma(?)
ML-3D	Tisova	380	Ky eclogite		1.31	2.85	0.2779	0.513351	22	0.41	13.9	10.0	8.8	0.48	14–16.4	630–730	MORB	500 Ma(?)
ML-3D	Tisova	380	Ky eclogite		1.33	2.75	0.2924	0.513367	22	0.49	14.2	9.6	8.2	0.42	14–16.4	630–730	MORB	500 Ma(?)
ML-3E	Tisova	380	Ky eclogite		2.16	5.52	0.2366	0.513280	18	0.20	12.5	10.6	10.0	0.87	14–16.4	630–730	MORB	500 Ma(?)
								Brueckner et al., 1996										
												400 Ma						
1111	Bystrzyca Gorna (Poland)	402 ± 3 (Variscan)	Gt peridotite		0.22	0.54	0.2445	0.513125	27	0.24	9.5	7.1		–0.12	27 (core) 16 (rim)	1030 800	Mantle perid.	400 Ma (?)

[1]Bohemian massif, Moldanubian zone; Gföhl terrane + Kutnan Hora-Svratka complex.
[2]Bohenian massif, Gföhl terrane.
[3]Bohemian massif, Marianske Lazne Complex.

Appendix 3. Sm-Nd Isotopic Compositions of Eclogites and Garnet Peridotites from the Caledonian Belt

Sample no.	Locality	Age, Ma (metamor.)	Rock type	Occurrence type	[Sm]	[Nd]	$^{147}Sm/^{144}Nd$	$^{143}Nd/^{144}Nd$	± 2σm	f(Sm/Nd)	εNd (0)	εNd (T1)	εNd (T2)	T_{DM}, Ga	Supplementary information: P, kbar	T, °C	Protolith nature	Protolith age, Ma
								Sanders et al., 1984										
												1080 Ma	1700 Ma					
S215	NW Scotland	≈1080	Eclogite	Associated with marble	2.10	6.30	0.2015	0.512430	40	0.02	–4.1	–4.7	–5.1	8.78	>12	700°C	Basalt	≈ 1700(?)
S274	NW Scotland	≈1010	Eclogite		6.91	34.70	0.1204	0.511790	20	–0.39	–16.5	–6.7	0.0	2.21				unknown
								Mork and Mearns, 1986										
												400 Ma						
GE4a	W. Norway	400	Eclogite	Not described	3.33	14.40	0.1402	0.512290	5	–0.29	–6.8	–3.9		1.78	15–20	>700	Gabbro	unknown
GE7	W. Norway	400	Eclogite	Not described	2.84	12.90	0.1337	0.512226	11	–0.32	–8.0	–4.8		1.76	15–20	>700	Gabbro	
								Griffin and Brueckner, 1980, 1985										
												410 Ma						
K-6	WGR, Norway	410–440	Eclogite	In gneiss	4.70	18.90	0.1503	0.512335	28	–0.24	–5.9	–3.5		1.95	12 to 18	600–800	Disputable	≈ 1800
K-6 Cpx	WGR, Norway	410–440	Cpx		2.89	10.49	0.1666	0.512065	46	–0.15	–11.2	–9.6		3.48				
U-19	WGR, Norway	410–440	Eclogite	In gneiss	1.30	3.10	0.2535	0.512352	80	0.29	–5.6	–8.6		–3.09	12 to 18	600–800	Complex	≈ 1800
U-19 Cpx	WGR, Norway	410–440	Cpx		1.58	4.10	0.2335	0.512418	64	0.19	–4.3	–6.2		–5.77	12 to 18	600–800	lithology	≈ 1800
IB-51	WGR, Norway	410–440	Eclogite	In gneiss	4.40	20.20	0.1317	0.512353	50	–0.33	–5.6	–2.2		1.48	12 to 18	600–800	Complex	≈ 1800
TYS-1(B)	WGR, Norway	410–440	Eclogite or	Mafic-ultra-	2.91	4.68	0.3756	0.512557	30	0.91	–1.6	–10.9		–0.56	12 to 18	600–800	lithology	≈ 1800
TYS-1(C)	WGR, Norway	410–440	clino-pyroxenite	mafic assoc.	4.16	15.48	0.1623	0.512466	28	–0.17	–3.4	–1.6		2.02	12 to 18	600–800		≈ 1800
5/79 Cpx	WGR, Norway	410–440	Cpx	In gneiss	1.13	2.10	0.3248	0.512980	40	0.65	6.7	0.0		–0.23				
S-19 Cpx	WGR, Norway	410–440	Cpx	Gt websterite	2.98	10.13	0.1779	0.512560	36	–0.10	–1.5	–0.5		2.50				
1428 Cpx	WGR, Norway	410–440	Cpx	Eclog in marble	2.65	5.17	0.3101	0.513282	34	0.58	12.6	6.6		0.21				
N-16 Cpx	WGR, Norway	410–440	Cpx	Metagabbro	1.11	3.54	0.1889	0.512887	36	–0.04	4.9	5.3		1.61				
7/79A Cpx	WGR, Norway	410–440	Cpx	In gneiss	0.28	0.37	0.4524	0.513454	45	1.30	15.9	2.5		0.19				
T-96 Cpx	WGR, Norway	410–440	Cpx	Gt peridotite	1.82	5.37	0.2046	0.513288	46	0.04	12.7	12.3		–2.33				
TE-4 Cpx	WGR, Norway	410–440	Cpx	(?)	1.94	8.09	0.1452	0.512040	8	–0.26	–11.7	–9.0		2.46				
								Mearns, 1986										
												410 Ma	1700 Ma					
8067	Almklovdalen	408 ± 8	Eclogite	In gneiss	4.74	14.10	0.2032	0.513010	30	0.03	7.3	6.9	5.8	2.03				1700(?)
8065	WGR Norway	1700	Gt peridotite		0.13	0.65	0.1181	0.512020	20	–0.40	–12.1	–7.9	5.0	1.80	28		Metasomatized mantle	
7957	WGR Norway	1040	Gt websterite		0.64	2.53	0.1529	0.512247	12	–0.22	–7.6	–5.3	1.9	2.26			peridotite/pyroxenite	

								Jamtveit et al., 1991[1]									
												450 Ma	1700 Ma				
F12 (stage III)	Fjortoft	450	Gt pyroxenite	Mafic-ultra-mafic complx	0.184	0.550	0.2024	0.512894	13	0.03	5.0	4.7	3.7	3.44	23–33	710–900	
U95 (III)	Ugelvik	(stage III)	Gt pyroxenite		0.057	0.295	0.1177	0.512749	12	−0.40	2.2	6.7	19.3	0.64			
U539(III)	Raudhaugene		Gt pyroxenite		0.138	0.180	0.4639	0.518254	11	1.36	109.6	94.2	51.6	3.09	31–40	700–870	
T96 (II)	Kalskaret	1700	Gt pyroxenite		0.458	1.187	0.2333	0.513789	5	0.19	22.5	20.3	14.5	4.90	37–38	870–890	
T153 (II)	Kalskaret	(stage II)	Gt pyroxenite		1.453	4.518	0.1945	0.512953	7	−0.01	6.1	6.3	6.6	1.56	38	880	
8065 (II)	Almklavdalen		Gt peridotite		0.018	0.072	0.1511	0.512360	90	−0.23	−5.4	−2.8	4.5	1.92	40	900–930	
7957 (II)	Almklavdalen		Gt pyroxenite		0.640	2.530	0.1529	0.512247	12	−0.22	−7.6	−5.1	1.9	2.26			
E4	Eiksunddal	≈450	Fe-Ti gt perid.		2.479	10.339	0.1450	0.512305	6	−0.26	−6.5	−3.5	4.7	1.87			
8242 Cpx	Eiksunddal	≈450	Fe-Ti eclogite		3.323	10.623	0.1891	0.512330	4	−0.04	−6.0	−5.6	−4.4	5.02			
GB6	Gurskeboten	≈450	Super-Fe eclog.		1.125	3.072	0.2214	0.512561	5	0.13	−1.5	−2.9	−6.9	−12.17			
								Mork et al., 1988[2]									
												500 Ma					
TS2	Tsäkok	505 ± 18	Eclogite	Eclogite lenses in marbles	3.27	10.80	0.1838	0.512837	11	−0.07	3.9	4.7		1.59	12 to 15	500–630	T-MORB?
TS3	Tsäkok	503 ± 14	Eclogite		3.20	10.20	0.1915	0.512890	5	−0.03	4.9	5.2		1.78	12 to 15	500–630	T-MORB?
TS5	Tsäkok	503 ± 14	Eclogite		3.23	10.50	0.1824	0.512884	10	−0.07	4.8	5.7		1.29	12 to 15	500–630	T-MORB?
SE3	Grapesvarre	≈503	Eclogite	In qtz schist	3.50	10.60	0.2011	0.512885	5	0.02	4.8	4.5		3.18	19 ± 1	690–730	T-MORB?

[1] Western Gneiss Region Norway

[2] Seve Nappe Complex, Sweden.

APPENDIX 4. Sm-Nd Isotopic Compositions of Eclogites from the Pan-African Belt and the Ross Belt of Antarctica

Sample no.	Locality	Age, Ma (metamor.)	Rock type	Occurrence type	[Sm]	[Nd]	$^{147}Sm/^{144}Nd$	$^{143}Nd/^{144}Nd$	± 2σm	f(Sm/Nd)	εNd (0)	εNd (T)	εNd (T)	T_{DM}, Ga	Supplementary information: P, kbar	T, °C	Protolith nature	Protolith age, Ma
Bernard-Griffiths et al., 1991																		
												600 Ma	1200 Ma					
TGL10	Togo - Mt. Lato	600	Eclogite	Interbeded with meta-	3.32	10.58	0.1897	0.512972	20	–0.04	6.5	7.1	7.6	1.13	16–18	700–750	Continental rift basalt	≤ 1150
TGL11	Togo - Mt. Lato	600	Eclogite	sediments	1.00	1.52	0.3977	0.513685	25	1.02	20.4	5.0	–10.4	0.44	16–18	700–750	"	≤ 1150
TGL12	Togo - Mt. Lato	600	Eclogite		3.80	11.89	0.1932	0.512959	19	–0.02	6.3	6.5	6.8	1.42	16–18	700–750	"	≤ 1150
TGL13	Togo - Mt. Lato	600	Eclogite	"	2.62	9.28	0.1707	0.512832	26	–0.13	3.8	5.8	7.8	1.13	16–18	700–750	"	≤ 1150
TGK17	Togo - Kabié	600	Eclogite	As boudins in	5.63	20.57	0.1655	0.512752	19	–0.16	2.2	4.6	7.0	1.26	16–18	700–750	Oceanic arc	1300–1400
TGK18	Togo - Kabié	600	Eclogite	gneisses	0.32	1.13	0.1712	0.512774	24	–0.13	2.7	4.6	6.6	1.35	16–18	700–750	basalt	1300–1400
TGK19	Togo - Kabié	600	Eclogite	"	0.15	0.39	0.2325	0.513075	24	0.18	8.5	5.8	3.0	–0.61	16–18	700–750	"	1300–1400
TGK23	Togo - Kabié	600	Eclogite	"	0.76	2.11	0.2178	0.512916	20	0.11	5.4	3.8	2.2	–9.09	16–18	700–750	"	1300–1400
TGK24	Togo - Kabié	600	Eclogite	"	1.02	2.85	0.2164	0.512903	27	0.10	5.2	3.7	2.2	–14.85	16–18	700–750	"	1300–1400
												600 Ma	2600 Ma					
6906	Hoggar	609 ± 17	Eclogite	As boudins in	0.61	1.94	0.1910	0.512750	44	–0.03	2.2	2.6	4.1	2.67	13–15	750	Continental	≥ 2600
6908	Hoggar	609 ± 17	Eclogite	paragneisses	1.30	3.96	0.1985	0.512860	40	0.01	4.3	4.2	3.7	2.88	13–15	750	basalt	≥ 2600
6911	Hoggar	609 ± 17	Eclogite	"	1.00	3.20	0.1889	0.512814	26	–0.04	3.4	4.0	6.0	2.06	13–15	750	"	≥ 2600
7769	Hoggar	609 ± 17	Eclogite	"	1.42	4.46	0.1925	0.512799	34	–0.02	3.1	3.5	4.5	2.51	13–15	750	"	≥ 2600
7770	Hoggar	609 ± 17	Eclogite	"	2.15	7.85	0.1656	0.512664	23	–0.16	0.5	2.9	10.8	1.54	13–15	750	"	≥ 2600
Di Vincenzo et al., 1997[1]																		
												500 Ma	750 Ma					
TC13	Husky Pass	500	Well-preserved	In gneiss	5.31	23.20	0.1383	0.512280	6	–0.30	–7.0	–3.3	–1.4	1.75	15	850	E-MORB?	750(?)
TC13 - Cpx	Husky Pass	500	eclogite	In gneiss	3.02	12.91	0.1412	0.512284	6	–0.28	–6.9	–3.4	–1.6	1.82	15	850	or contam. Basalt?	750(?)
TC16	Husky Pass	500	Well-preserved	In gneiss	3.28	14.76	0.1345	0.512258	6	–0.32	–7.4	–3.4	–1.5	1.71	15	850	"	750(?)
TC16 - Cpx	Husky Pass	500	eclogite	In gneiss	7.04	33.23	0.1281	0.512237	6	–0.35	–7.8	–3.4	–1.3	1.62	15	850	"	750(?)
CP28		500	Retrogr. eclogite	(?)	4.54	15.40	0.1781	0.512845	7	–0.09	4.0	5.2	5.8	1.30	15	850	T-MORB	750(?)

[1]Lanterman Range, Antarctica.

APPENDIX 5. Sm-Nd Isotopic Compositions of Eclogites from the Sulu Terrane

Sample no.	Locality	Rock type	[Sm], ppm	[Nd], ppm	$^{147}Sm/^{144}Nd$	$^{143}Nd/^{144}Nd$	± 2σm	f(Sm/Nd)	εNd (0)	εNd (T), 200 Ma	T_{DM}, Ga	Reference
Occurrence type I: Enclaves in granitic gneisses												
CF92-103	Yangkou	Eclogite	7.501	39.84	0.1138	0.511673	5	−0.42	−18.8	−16.5	2.24	Jahn, 1998
CF92-105	Yangkou	Eclogite	6.354	33.20	0.1157	0.511675	3	−0.41	−18.8	−16.5	2.28	Jahn, 1998
CF92-106	Yangkou	Eclogite	3.983	21.21	0.1135	0.511764	4	−0.42	−17.0	−14.7	2.10	Jahn, 1998
CF92-107	Wehai	Eclogite	1.347	0.592	1.3802	0.525950	5	6.02	260	226.5	1.67	Jahn et al., 1996
CF92-108	Wehai	Eclogite	1.015	0.562	1.0937	0.522480	5	4.56	192	166.8	1.61	Jahn et al., 1996
HNG-1	Wehai	Eclogite	0.845	0.462	1.1097	0.521550	4	4.64	174	148.2	1.43	Jahn et al., 1996
HNG-2	Wehai	Eclogite	0.934	0.497	1.1398	0.521619	5	4.79	175	148.7	1.39	Jahn et al., 1996
HNG-3	Wehai	Eclogite	1.022	0.509	1.2188	0.524207	5	5.20	226	197.0	1.67	Jahn et al., 1996
CF92-109a	Wehai	Ga granulite	4.807	19.33	0.1503	0.511825	5	−0.24	−15.9	−14.6	3.16	Jahn et al., 1996
CF92-109b	Wehai	Ga granulite	11.525	52.94	0.1316	0.511792	4	−0.33	−16.5	−14.7	2.51	Jahn et al., 1996
CF92-114	Rongcheng	Eclogite	6.214	29.03	0.1294	0.511885	4	−0.34	−14.7	−12.8	2.28	Jahn, 1998
CF92-115	Rongcheng	Eclogite	1.997	5.065	0.2384	0.512328	4	0.21	−6.0	−7.2	−5.17	Jahn, 1998
CF92-129	Qinglongshan	Eclogite	12.948	63.57	0.1231	0.511891	3	−0.37	−14.6	−12.5	2.11	Jahn, 1998
QL-1	Qinglongshan	Eclogite	7.260	34.03	0.1290	0.511865	11	−0.34	−15.1	−13.2	2.30	Li et al., 1993
ZB-4	Zhubian	Eclogite	7.450	31.48	0.1431	0.511652	13	−0.27	−19.2	−17.7	3.21	Li et al., 1993
Occurrence type II: Enclaves in or interlayers with marbles or metapelites												
CF92-116	Rongcheng	Garnetite	3.135	2.670	0.7098	0.512790	3	2.61	3.0	−11.4	−0.11	Jahn, 1998
Occurrence type III: Enclaves in or interlayers with ultramafic rocks												
CF92-111	Rongcheng	Ga lherzolite	0.111	0.555	0.1210	0.512275	4	−0.38	−7.1	−5.0	1.44	Jahn, 1998
CF92-112	Rongcheng	Eclogite	1.650	8.26	0.1208	0.512321	3	−0.39	−6.2	−4.1	1.36	Jahn, 1998
CF92-120	Rizhao	Ga pyrox.	1.996	6.272	0.1924	0.512712	4	−0.02	1.4	1.6	3.12	Jahn, 1998
CF92-121	Rizhao	Ga pyrox.	2.132	6.519	0.1977	0.512790	5	0.01	3.0	2.9	3.41	Jahn, 1998
CF92-122	Rizhao	Ga pyrox.	2.741	8.941	0.1853	0.512704	5	−0.06	1.3	1.6	2.38	Jahn, 1998
CF92-123	Rizhao	Ga pyrox.	2.621	8.573	0.1848	0.512705	5	−0.06	1.3	1.6	2.34	Jahn, 1998
CF92-124	Rizhao	Ga pyrox.	2.641	9.072	0.1760	0.512669	4	−0.11	0.6	1.2	1.94	Jahn, 1998
JZA	Junan	Ga pyrox.	3.241	11.67	0.1679	0.512346	5	−0.15	−5.7	−4.9	2.66	Yang, 1991
JC313B	Junan	Eclogite	1.299	2.41	0.3264	0.512566	6	0.66	−1.4	−5.0	−0.79	Yang, 1991
H06	Rizhao	Ga pyrox.	2.416	6.85	0.2134	0.512564	8	0.08	−1.4	−1.9	151.33	Yang, 1991
Occurrence type not defined												
SL91-12	Rongcheng	Eclogite	0.500	1.69	0.1789	0.512282	8	−0.09	−6.9	−6.4	3.76	Ames et al., 1996
SL91-21	Junan	Eclogite	12.570	54.33	0.1399	0.512708	18	−0.29	1.4	3.0	0.91	Ames et al., 1996
SL91-23	Junan	Eclogite	0.790	4.30	0.1111	0.511268	10	−0.44	−26.7	−24.3	2.78	Ames et al., 1996

APPENDIX 6. Sm-Nd Isotopic Compositions of Eclogites and Ultramafic Rocks from the Dabie Terrane

Sample no.	Locality	Rock type	[Sm], ppm	[Nd], ppm	$^{147}Sm/^{144}Nd$	$^{143}Nd/^{144}Nd$	± 2σm	f(Sm/Nd)	εNd (0)	εNd (T), 200 Ma	T_{DM}, Ga	Reference
Occurrence type I: Enclaves in granitic gneisses												
BJ93-27a	Wumiao	Eclogite	5.48	22.10	0.1499	0.512240	20	–0.24	–7.8	–6.5	2.17	Jahn, 1998
BJ93-27c	Wumiao	Eclogite	12.27	57.06	0.1300	0.512242	7	–0.34	–7.7	–5.9	1.65	Jahn, 1998
BJ93-27d	Wumiao	Eclogite	7.77	28.91	0.1625	0.512237	7	–0.17	–7.8	–6.9	2.70	Jahn, 1998
BJ93-31	Shuanghe	Eclogite	4.57	16.89	0.1636	0.511695	7	–0.17	–18.4	–17.5	4.38	Jahn, 1998
BJ93-36	Shuanghe	Eclogite	3.72	14.89	0.1510	0.512229	7	–0.23	–8.0	–6.7	2.23	Jahn, 1998
BJ95-11	Shuanghe	Eclogite	5.62	21.08	0.1612	0.511729	10	–0.18	–17.7	–16.7	4.08	Jahn, 1998
BJ95-12	Shuanghe	Eclogite	3.62	15.06	0.1451	0.512133	10	–0.26	–9.9	–8.4	2.25	Jahn, 1998
BJ95-13	Shuanghe	Eclogite	3.15	12.47	0.1528	0.512226	6	–0.22	–8.0	–6.8	2.30	Jahn, 1998
BJ93-39	Tanshuao	Cold eclogite	6.84	33.10	0.1249	0.512029	5	–0.36	–11.9	–9.9	1.92	Jahn, 1998
BJ95-20	Zhujiachung	Cold eclogite	5.43	25.37	0.1293	0.512069	14	–0.34	–11.1	–9.2	1.95	Jahn, 1998
BJ95-21	Zhujiachung	Cold eclogite	4.50	21.41	0.1269	0.512108	8	–0.35	–10.3	–8.4	1.82	Jahn, 1998
BJ95-22	Huangzhen	Cold eclogite	5.69	25.28	0.1361	0.512198	8	–0.31	–8.6	–6.9	1.86	Jahn, 1998
252C	Shuanghe	eclogite	4.09	18.87	0.1310	0.512212	15	–0.33	–8.3	–6.5	1.73	Okay et al., 1993
Occurrence type II: Enclaves in or interlayers with marbles												
BJ93-15	Taolichung	Eclogite	9.26	24.91	0.2249	0.511656	4	0.14	–19.2	–19.9	–21.97	Jahn, 1998
230D	S of Wumaio	Eclogite	5.43	31.87	0.1030	0.511524	15	–0.48	–21.7	–19.1	2.23	Okay et al., 1993
Occurrence type III: Enclaves in or interlayers with ultramafic rocks												
BJ93-02	Bixiling	Amphibolite	6.84	25.82	0.1601	0.512457	8	–0.19	–3.5	–2.5	1.96	Chavagnac & Jahn, 1996
BJ93-03	Bixiling	Eclogite	1.96	7.51	0.1578	0.512520	4	–0.20	–2.3	–1.2	1.71	Chavagnac & Jahn, 1996
BJ93-04	Bixiling	Eclogite	1.63	5.55	0.1778	0.512608	6	–0.10	–0.6	–0.1	2.29	Chavagnac & Jahn, 1996
BJ93-05	Bixiling	Eclogite	1.45	4.75	0.1845	0.512596	4	–0.06	–0.8	–0.5	2.88	Chavagnac & Jahn, 1996
BJ93-06	Bixiling	Eclogite	1.03	3.37	0.1839	0.512590	4	–0.07	–0.9	–0.6	2.84	Chavagnac & Jahn, 1996
BJ93-07	Bixiling	Eclogite	1.84	6.74	0.1651	0.512584	4	–0.16	–1.1	–0.2	1.77	Chavagnac & Jahn, 1996
BJ93-08	Bixiling	Ga peridotite	0.77	2.85	0.1633	0.512500	5	–0.17	–2.7	–1.8	1.96	Chavagnac & Jahn, 1996
BJ93-09	Bixiling	Ga peridotite	0.43	1.36	0.1919	0.512594	17	–0.02	–0.9	–0.7	3.85	Chavagnac & Jahn, 1996
BJ93-20	Taolichung	Eclogite	1.46	3.44	0.2566	0.512378	7	0.30	–5.1	–6.8	–2.78	Jahn, 1998
BJ95-17	Maowu	Orthopyroxenite	0.17	0.33	0.3092	0.512484	23	0.57	–3.0	–6.2	–1.07	Jahn, 1998
BJ95-18	Maowu	Orthopyroxenite	0.06	0.23	0.1549	0.512174	13	–0.21	–9.1	–7.9	2.52	Jahn, 1998
BJ95-19	Maowu	2-pyroxenite	3.54	15.85	0.1352	0.512205	12	–0.31	–8.4	–6.7	1.83	Jahn, 1998
R-4	Raobazhai	Eclogite	4.96	14.63	0.2049	0.512515	15	–0.04	–2.4	–2.6	10.66	Li et al., 1993
Occurrence type not defined												
DB91-28	Dabie	Eclogite	3.39	11.20	0.1830	0.512460	14	–0.07	–3.5	–3.1	3.40	Ames et al., 1996
DB91-34	Dabie	Eclogite	4.74	20.08	0.1427	0.512168	10	–0.27	–9.2	–7.7	2.10	Ames et al., 1996

Wumiao	Dabie	Eclogite	3.22	18.51	0.1052	0.512277	12	–0.47	–7.0	–4.5	1.23	Ames et al., 1996
DB91-64	Dabie	Eclogite	1.14	7.51	0.0918	0.512401	12	–0.53	–4.6	–1.7	0.94	Ames et al., 1996
DB91-65	Dabie	Eclogite	0.42	1.45	0.1751	0.512526	19	–0.11	–2.2	–1.6	2.45	Ames et al., 1996
DB91-15a	Dabie	Eclogite	3.52	14.10	0.1509	0.512553	9	–0.23	–1.7	–0.4	1.45	Ames et al., 1996
DB91-61-2	Dabie	Eclogite	4.89	38.20	0.0774	0.512397	10	–0.61	–4.7	–1.4	0.84	Ames et al., 1996
Hu-105	Hong'an	Eclogite	3.49	10.65	0.1981	0.512940	8	0.01	5.9	5.9	2.05	Ames et al., 1996
Hu-117	Hong'an	Eclogite	6.36	24.55	0.1566	0.512445	8	–0.20	–3.8	–2.6	1.88	Ames et al., 1996
Hu-113	Hong'an	Eclogite	7.45	27.00	0.1668	0.512531	7	–0.15	–2.1	–1.2	2.01	Ames et al., 1996
Hu-115	Hong'an	Eclogite	2.91	8.95	0.1966	0.513030	9	0.00	7.6	7.7	1.07	Ames et al., 1996
Hu-121	Hong'an	Eclogite	7.95	30.87	0.1557	0.512420	8	–0.21	–4.3	–3.1	1.91	Ames et al., 1996

RECENT LITERATURE ON ULTRAHIGH-PRESSURE METAMORPHISM

References for Ultrahigh-Pressure Metamorphism and Tectonics Since 1984[1]

J. G. LIOU

Department of Geological and Environmental Sciences, Stanford University, Stanford, California 94305

AND CONG BOLIN

Institute of Geology, Chinese Academy of Sciences, P. O. Box 9825, Beijing 100029, China

BOOKS AND SPECIAL ISSUES

Bebout, R. E., Scholl, D. W., Kirby, S. H., and Platt, J., eds., 1996, Subduction: Top to bottom: Amer. Geophys. Union, Geophys. Monog. 96, 384 p.

Carswell, D. A., ed., 1990, Eclogite facies rocks: New York, Blackie, 396 p.

Coleman, R. G., and Wang, X., eds., 1995, Ultrahigh-pressure metamorphism: New York, Cambridge University Press, 528 p.

Cong, B., ed., 1996, Ultrahigh-pressure metamorphic rocks in the Dabieshan-Sulu Region of China: Beijing and Dordrecht, Science Press and Kluwer Acad. Publ., 224 p.

Dong, S., Cui, W., and Zhang, L., 1996, The Proterozoic glaucophane-schist belt and some eclogites of the North Yangtze Craton, Central China: Beijing, Science Press, 139 p.

Hacker, B., and Liou, J. G., eds., 1998, When continents collide: Geodynamics and geochemistry of ultrahigh-pressure rocks. Dordrecht, Kluwer Acad. Publ., 323 p.

Hemley, R. J., ed., 1998, Ultrahigh-pressure mineralogy: Physics and chemistry of the Earth's deep interior: Rev. Mineral., v. 37, 671 p.

Liou, J. G., Banno, S., and Ernst, W. G., eds., 1995, Ultrahigh-pressure metamorphism and tectonics: The Island Arc, v. 4, p. 233–383.

Liou, J. G., Cong, B., and Maruyama, S., eds., 1998, Geodynamics for high- and ultrahigh-pressure metamorphism: The Island Arc, v. 7, p. 1–200.

Smith, D. C., ed., 1988, Eclogites and eclogite-facies rocks: Amsterdam, Elsevier, Developments in Petrology, v. 12, 524 p.

Schreyer, W., and Stöckhert, B., eds., 1997, High pressure metamorphism in nature and experiment: Lithos, v. 41, p. 1–266.

You, Z., Han, Y., Rang, W., Zhang, Z., Wei, B., and Liu, R., 1996, The high-pressure and ultrahigh-pressure metamorphic belt in the East Qinling and Dabie Mountains, China: Wuhan, China University of Geoscience Press, 150 p.

GENERAL

Andre, L., Shatsky, V. S., De Corte, K., Sobolev, N. V., and Navez, J., in press, K-rich clinopyroxenes as mantle conveyers of crustal components: Nature.

Austrheim, H., 1991, Eclogite formation and dynamics of crustal roots under collision zones: Terra Nova, v. 3, p, 492–499.

______, 1998, Influence of fluid and deformation on metamorphism of the deep crust and consequences for the geodynamics of collision zones, *in* Hacker, B., and Liou, J. G., eds., When continents collide: Geodynamics and geochemistry of ultrahigh-pressure rocks: Dordrecht, Kluwer Acad. Publ., p. 297-323.

Austrheirn, H., Erambert, M., and Engvik, A. K., 1997, Processing of crust in the root of the Caledonian continental collision zone: The role of eclogitization: Tectonophys., v. 273, p. 129–153.

Banno, S., Wallis, S. R., and Hirajima, T., 1997, From the depths of continental collision belt: Discovery and significance of UHP metamorphic rocks: Kagaku, v. 67, p. 39–47.

Beaumont, C., Ellis, S., Hamilton, J., and Fullsack, P., 1996, Mechanical model for subduction-collision tectonics of Alpine-type compressional orogens: Geology, v. 24, p. 675–678.

Blythe, A. E., 1998, Active tectonics and ultrahigh-pressure rocks, *in* Hacker, B., and Liou, J. G., eds., When continents collide: Geodynamics and geochemistry of ultrahigh-pressure rocks: Dordrecht, Kluwer Acad. Publ., p. 141–160.

Brey, G. P., et al., 1986, Garnet-pyroxene equilibria in the system CaO-MgO-Al_2O_3-SiO_2, CMAS, prospects for simplified "T-independent" lherzolite barometry, and an eclogite barometer: Contrib. Mineral. Petrol., v. 92, p. 448–455.

Brey, G. P., and Kohler, T., 1990, Geothermobarometry in four lherzolites II. New thermobarometers, and practi-

[1] Excluded are theses, short notes, abstracts, and conference reports.

cal assessment of existing thermobarometers: Jour. Geol., v. 31, p. 1553–1378.

Bruckner, H. K., 1998, Sinking intrusion model for the introduction of garnet-bearing peridotites into continent collision orogens: Geology, v. 26, p. 631–634.

Brunet, F., Chopin, C., and Seifert, F., 1998, Phase relations in the $MgO-P_2O_5-H_2O$ system and the stability of phosphoellenbergerite: Petrological implications: Contrib. Mineral. Petrol., v. 131, p. 54–70.

Chemenda, A. I., Mattauer, M., and Bokun, A. N., 1996, Continental subduction and a mechanism for exhumation of high-pressure metamorphic rocks: New modeling and field data from Oman: Earth Planet. Sci. Lett., v. 143, p. 173–182.

Chopin, C., Klaska, R., Medenbach, O., and Dron, D., 1986, Ellenbergerite, a new high-pressure Mg-Al-(Ti,Zr)-silicate with a novel structure based on face-sharing octahedra: Contrib. Mineral. Petrol., v. 92, p. 316–321.

Chopin, C., and Sobolev, N. V., 1995, Principal mineralogical indicators of UHP in crustal rocks, *in* Coleman, R. G., and Wang, X., eds.., Ultrahigh-pressure metamorphism: New York, Cambridge University Press, p. 96–131.

Cloos, M., 1993, Lithospheric buoyancy and collisional orogenesis: Subduction of oceanic plateaus, continental margins, island arcs, spreading ridges, and seamounts: Geol. Soc. Amer. Bull., v. 105, p. 715–737.

Coleman, R. G., and Wang, X., 1995, Overview of the geology and tectonics of UHPM, *in* Coleman, R. G., and Wang, X. eds.., Ultrahigh-pressure metamorphism: New York, Cambridge University Press, p. 1–32.

Daniels, L. R. M., Gurney, J. J., and Harte, B., 1996, A crustal mineral in a mantle diamond: Nature, v. 379, p. 153–156.

Davies, J. H., and von Blanckenburg, F., 1995, Slab breakoff: A model of lithosphere detachment and its test in the magmatism and deformation of collisional orogens: Earth Planet. Sci. Lett., v. 129, p. 85–102.

______, 1998, Thermal controls on slab breakoff and the rise of high-pressure rocks during continental collisions, *in* Hacker, B., and Liou, J. G., eds., When continents collide: Geodynamics and geochemistry of ultrahigh-pressure rocks: Dordrecht, Kluwer Acad. Publ., p. 97–116.

Drury, M. R., and van Roermund, H. L. M., 1988, Metasomatic origin for Fe-Ti-rich multiphase inclusions in olivine: Geology, v. 16, p. 1035–1038.

Duchêne, S., Lardeaux, J.-M., and Albarède, F., 1997, Exhumation of eclogites: Insights from depth-time path analysis: Tectonophys., v. 280, p. 125–140.

Ernst, W. G., Liou, J. G., and Coleman, R. G., 1995, Comparative petrotectonic study of five Eurasian ultrahigh-pressure metamorphic complexes: INT. GEOL. REV., v. 37, p. 191–211.

Ernst, W. G., Liou, J. G., and Hacker, B. R., 1994, Petrotectonic significance of high- and ultrahigh-pressure metamorphic belts: Inferences for subduction-zone histories: INT. GEOL. REV., v. 36, p. 213–237.

Ernst, W. G., Maruyama, S., and Wallis, S., 1997, Buoyancy-driven, rapid exhumation of ultrahigh-pressure metamorphosed continental crust: Proceed. Nat. Acad. Sci., v. 94, p. 9532–9537.

Ernst, W. G., Mosenfelder, J. L., Leech, M. L., and Liu , J., 1998, H_2O recycling during continental collision: Phase equilibrium and kinetic considerations, *in* Hacker, B., and Liou, J. G., eds., When continents collide: Geodynamics and geochemistry of ultrahigh-pressure rocks: Dordrecht, Kluwer Acad. Publ., p. 275–296.

Ernst, W. G., and Peacock, S. M., 1996, A thermotectonic model for preservation of ultrahigh-pressure mineralogic relics in metamorphosed continental crust, *in* Bebout, G. E., et al., eds., Subduction: Top to bottom: Amer. Geophys. Union, Geophys. Monog., v. 96, p. 171–178.

Godard, G., and van Roermund, H. L. M., 1995, Deformation-induced clinopyroxene fabrics from eclogites: Jour. Struct. Geol., v. 17, p. 1425–1443.

Grasemann, B., Ratschbacher, L., and Hacker, B., 1998, Exhumation of ultrahigh-pressure rocks: Thermal boundary conditions and cooling history, *in* Hacker, B., and Liou, J. G., eds., When continents collide: Geodynamics and geochemistry of ultrahigh-pressure rocks: Dordrecht, Kluwer Acad. Publ., p. 117–140.

Green, H. W., and Dobrzhinetskaya, L., 1997, Response to determining the origin of ultrahigh-pressure lherzolites, by Hacker et al. 1997: Science, v. 278, p. 704–707.

Green, H. W., Dobrzhinetskaya, L., Riggs, E. M., and Jin, Z. M., 1997, Alpe Arami: A peridotite massif from the mantle transition zone?: Tectonophys., v. 279, p. 1–21.

Hacker, B. R., and Peacock, S. M., 1995, Creation, preservation, and exhumation of UHPM rocks, *in* Coleman R. G., and Wang, X., eds., Ultrahigh-pressure metamorphism: New York, Cambridge University Press, p. 159–181.

Hacker, B. R., Sharp, T., Zhang, R. Y., Liou, J. G., and Hervig, R. L., 1997, Determining the origin of ultrahigh-pressure lherzolites: Science, v. 278, p. 702–704.

Harley, S. L., and Carswell, D. A., 1995, Ultradeep crustal metamorphism: A prospective view: Jour. Geophys. Res., v. 100, p. 8367–8380.

Harlow, G. E., and Veblen, D. R., 1991, Potassium in clinopyroxene inclusions from diamond: Science, v. 251, 652–655.

Henry, C., Michard, A., and Chopin, C., 1993, Geometry and structural evolution of high-pressure rocks from the Dora-Maira massif, western Alps, Italy: Jour. Struct, Geol., v. 15, p. 965–981.

Hirajima, T., 1996, Effect of jadeite content on the garnet-clinopyroxene geothermometer for an ultrahigh-pressure eclogite complex: Proc. Japan Acad., v. 72, p. 208–213.

Hsü, K. J., 1991, Exhumation of high pressure metamorphic rocks: Geology, v. 19, p. 107–110.

Hynes, A., Arkani-Hamed, J., and Greiling, R., 1996, Subduction of continental margins and the uplift of high-pressure metamorphic rocks: Earth Planet. Sci. Lett., v. 140, p. 13–25.

Iizuka, Y., and Nakamura, E., 1995, Experimental study of the slab-mantle interaction and implications for the formation of titanoclinohurnite at deep subduction zone: Proc. Japan Acad., v. 71, p. 159–164.

Liou, J. G., Banno, S., and Ernst, W. G., 1995, Ultrahigh-pressure metamorphism and tectonics: The Island Arc, v. 4, p. 233–239.

Liou, J. G., Maruyama, S., and Cong, B., 1998, Introduction to geodynamics for high- and ultrahigh-pressure metamorphism: The Island Arc, v. 7, p. 1–5.

Liou, J. G., Maruyama, S., and Ernst, W. G., 1997, Seeing a mountain in a grain of garnet: Science, v. 274, p. 48–49.

Liou, J. G., Zhang, R., and Ernst, W. G., 1994, An introduction to ultrahigh-P metamorphism: The Island Arc, v. 3, p. 1–24.

Liou, J. G., Zhang, R. Y., Ernst, W. G., Rumble, D., and Maruyama, S., 1998, High-pressure minerals from deeply subducted metamorphic rocks, *in* Heraley, R. J., ed., Ultrahigh-pressure mineralogy: Physics and chemistry of the Earth's deep interior: Rev. Mineral., v. 37, p. 33–96.

Luth, R. W., 1993, Diamonds, eclogites, and the oxidation state of the Earth's mantle: Science, v. 261, p. 66–68.

Maruyama, S., and Liou, J. G., 1998, Initiation of UHP metamorphism and its significance on the Proterozoic/Phanerozoic boundary: The Island Arc, v. 7, p. 6–35.

Maruyama, S., Liou, J. G., and Terabayashi, M., 1996, Bleuschists/eclogites of the world and their exhumation: INT. GEOL. REV., v. 38, p. 490–596.

Massonne, H., Bernhardt, H., Dettmar, D., Kessler, E., Medenbach, O., and Westphal, T., 1998, Simple identification and quantification of microdiamonds in rock thin-sections: Eur. Jour. Mineral., v. 10, p. 497–504.

Matte, P., 1998, Continental subduction and exhumation of HP rocks in Paleozoic orogenic belts: Uralides and Variscides: Geol. Foren. Stockholm Forhandl., v. 120, p. 209–222.

Medaris, L. G., and Carswell, D. A., 1990, Petrogenesis of Mg-Cr garnet-peridotites in European metamorphic belts, *in* Carswell, D. A., ed., Eclogite facies rocks: Glasgow, Blackie and Sons, p. 260–290.

Medaris, L. G., Wang, H. F., Misar, Z., and Jelinek, E., 1990, Thermobarometry, diffusion modeling and cooling rates of crustal garnet peridotites: Two examples from the Moldanubian zone of the Bohemian Massif: Lithos, v. 25, p. 189–202.

Michard, A., Chopin, C., and Henry, C., 1993, Compression versus extension in the exhumation of the Dora-Maira coesite-bearing unit, Western Alps, Italy: Tectonophys., v. 221, p. 173–193.

Nakamura, D., and Banno, S., 1997, Thermodyamic modeling of sodic pyroxene solid-solution and its application in a garnet-omphacite-kyanite-coesite geothermobarometer for UHP metamorphic rocks: Contrib. Mineral. Petrol., v. 130, p. 93–102.

Pearson, D. G., Davies, G. R., and Nixon, P. H., 1995, Orogenic ultramafic rocks of UHP diamond facies origin, *in* Coleman, R. G., and Wang, X., eds., Ultrahigh pressure metamorphism: New York, Cambridge University Press, p. 456–510.

Perchuk, L. L., Yapaskurt, V. O., and Okay, A., 1995, Comparative petrology of diamond-bearing complexes: Petrology, v. 3, p. 238–277.

Philippot, P., and van Roermund, H. L. M., 1992, Deformation processes in eclogitic rocks: Evidence for the rheological delamination of the oceanic crust in deeper levels of subduction zones: Jour. Struct. Geol., v. 14, p. 1059–1077.

Platt, J. P., 1986, Dynamics of orogenic wedges and the uplift of high-pressure metamorphic rocks: Geol. Soc. Amer. Bull., v. 97, p. 1037–1053.

_______, 1993, Exhumation of high-pressure rocks: A review of concepts and processes: Terra Nova, v. 5, p. 119–133.

Renner, J., Zerbian, A., and Stockhert, B., 1997, Microstructures of synthetic polycrystalline coesite aggregates. The effect of pressure, temperature and time: Lithos, v. 41, p. 169–184.

Rumble, D., 1998, Stable isotope geochemistry of ultrahigh-pressure rocks, *in* Hacker, B., and Liou, J. G., eds., When continents collide: Geodynamics and geochemistry of ultrahigh-pressure rocks: Dordrecht, Kluwer Acad. Publ., p. 241–260.

Scaillet, S., 1998, K-Ar, $^{40}Ar/^{39}Ar$ geochronology of ultrahigh-pressure rocks, *in* Hacker, B., and Liou, J. G., eds,, When continents collide: Geodynamics and geochemistry of ultrahigh-pressure rocks: Dordrecht, Kluwer Acad. Publ., p. 161–202.

Schreyer, W., 1995, Ultradeep metamorphic rocks: The retrospective viewpoint: Jour. Geophys. Res., v. 100, p. 8353–8366.

Schreyer, W., and Werding, G., 1997, High-pressure behaviour of selected boron minerals and the question of boron distribution between fluids and rocks: Lithos, v. 41, p. 251–266.

Schreyer, W., Massonne, H.-J., and Chopin, C., 1987, Continental crust subducted to mantle depths near 100 km: Implications for magma and fluid genesis in collision zones. Magmatic processes: Physicochemical principles: Geochem. Soc. Spec. Publ., v. 1, p. 155–163.

Sorenson, S. S., Grossman, J. N., and Perfit, M. R., 1997, Phengite-hosted LILE enrichment in eclogite and related rocks: Implications for fluid-mediated mass transfer in subduction zones and arc magma genesis: Jour. Petrol., v. 38, p. 3–34.

Stöckhert, B., and Renner, J., 1998, Rheology of crustal rocks at ultrahigh pressure, *in* Hacker, B., and Liou, J. G., eds., When continents collide: Geodynamics and geochemistry of ultrahigh-pressure rocks: Dordrecht, Kluwer Acad. Publ., p. 57–96.

Ton, S., and Wortel, M. J. R., 1997, Slab detachment in continental collision zones: An analysis of controlling parameters: Geophys. Res. Lett., v. 24, p. 2095–2098.

van der Molen, I., and van Roermund, H. L. M., 1986, The pressure path of solid inclusions in minerals: The retention of coesite inclusions during uplift: Lithos, v. 19, p. 317–324.

van Roermund, H. L. M., 1992, Thermal and deformation induced omphacite microstructures from eclogites: Implications for formation and uplift of (U)HP metamorphic terrains, *in* Jacob, A., ed., Trends in mineralogy, v. I. Research trends: Trivandrum, India, p. 117–152.

von Blankenburg, F., and Davies, J. H., 1995, Slab breakoff. A model for syncollisional magmatism and tectonics in the Alps: Tectonics, v. 14, p. 120–131.

EXPERIMENTAL STUDIES

Akaogi, M., Yusa, H., Shiraishi, K., and Suzuki, T., 1995, Thermodynamic properties of α-quartz, coesite, and stishovite and equilibrium phase relations at high pressure and high temperature: Jour. Geophys. Res., v. 100, p. 22,337–22,347.

Bose, K., and Ganguly, J., 1995, Quartz-coesite transition revisited: Reversed experimental determination at 500–1200°C and retrieved thermodynamic properties: Amer. Mineral., v. 80, p. 231–238.

Bose, K., and Navrotsky, A., 1998, Thermochemistry and phase equilibria of hydrous phases in the system $MgO-SiO_2-H_2O$: Implications for volatile transport to the mantle: Jour. Geophys. Res., v. 103, p. 9713–9719.

Chopin, C., 1986, Phase relationships of ellenbergerite, a new high-pressure Mg-Al-Ti-silicate in pyrope-coesite-quartzite from the Western Alps: Geol. Soc. Amer. Mem., v. 164, p. 31–42.

Chopin, C., and Schreyer, W., 1983, Magnesiocarpholite and magnesiochloritoid: Two index minerals of pelitic blueschists and their preliminary phase relations in the model system $MgO-Al_2O_3-SiO_2-H_2O$: Amer. Jour. Sci., v. 283-A, p. 72–96.

Domanik, K. J., and Holloway, J. R., 1996, The stability and composition of phengitic muscovite and associated phases from 5.5 to 11 GPa: Implications for deeply subducted sediments: Geochim. et. Cosmochim. Acta, v. 60, p. 4133–4150.

Doroshev, A. M., Sobolev, N. V., and Brey, G., 1992, Experimental evidence of high-pressure origin of the potassium-bearing pyroxenes [abs.]: Proc. 29th Int. Geol. Cong., v. 2, p. 602.

Fockenberg, T., 1995, Synthesis and chemical variability of Mg-staurolite in the system $MgO-Al_2O_3-SiO_2-H_2O$ as a function of water pressure: Eur. Jour. Mineral., v. 7, p. 1373–1380.

______, 1998, An experimental investigation on the P-T stability of Mg-staurolite in the system $MgO-Al_2O_3-SiO_2-H_2O$: Contrib. Mineral. Petrol., v. 130, p. 187–198.

______, 1998, An experimental study of the pressure-temperature stability of MgMgAl-pumpellyite in the system $MgO-Al_2O_3-SiO_2-H_2O$: Amer. Mineral., v. 83, p. 220–227.

Gillet, P., 1993, Stability of magnesite $MgCO_3$ at mantle pressure and temperature conditions: A Raman spectroscopic study: Amer. Mineral., v. 78, p. 1328–1331.

Harlow, G. E., 1997, K in clinopyroxene at high pressure and temperature: An experimental study: Amer. Mineral., v. 82, 259–269.

Harlow, G. E., and Veblen, D. R., 1991, Potassium in clinopyroxene inclusion from diamonds: Science, v. 251, p. 652–655.

Hemingway, B. S., Bohlen, S. R., Hankins, W. B., Westrum, E. F., Jr., and Kuskov, O. L., 1998, Heat capacity and thermodynamic properties for coesite and jadeite, reexamination of the quartz-coesite equilibrium boundary: Amer. Mineral., v. 83, p. 409–418.

Irifune, T., et al., 1986, The eclogite-garnetite transformation at high pressure and some geophysical implications: Earth Planet. Sci. Lett., v. 77, p. 245–256.

______, 1989, Eclogite-garnetite transformation at high pressure and its bearing on the occurrence of garnet inclusions in diamond, *in* Ross, J., et al., eds., Spec. Publ., Geol. Soc. Austral., v. 14, p. 877–882.

Kerr, R. A., 1997, Deep-sinking slabs stir the mantle: Science, v. 275, p. 613–615.

Konzett, J., and Ulmer, P., 1998, The stability of hydrous potassic phases in lherzolitic mantle—an experimental study to 9.5 GPa in simplified and natural bulk composition: Jour. Petrol., v. 40, p. 629–652.

Luth, R. W., 1993, Diamonds, eclogites, and the oxidation state of the Earth's mantle: Science, v. 261, p. 66–68.

______, 1997, Experimental study of the system phologopite-diopside from 3.5 to 17 GPa: Amer. Mineral., v. 62, p. 1198–1209.

Massonne, H.-J., 1995, Experimental and petrogenetic study of UHPM, *in* Coleman, R. G., and Wang, X., eds., Ultrahigh pressure metamorphism: New York, Cambridge University Press, p. 33–95.

Massonne, H.-J., and Schreyer, W., 1987, Phengite geobarometry based on the limiting assemblage with K-feldspar, phlogopite, and quartz: Contrib. Mineral. Petrol., v. 96, p. 212–224.

______, 1989, Stability field of the high-pressure assemblage talc + phengite and two new phengite barometers: Eur. Jour. Mineral., v. 1, p. 391–410.

Massonne, H.-J., and Szpurka, Z., 1997, Thermodynamic properties of white micas on the basis of high-pressure experiments in the systems K_2O-MgO-Al_2O_3-SiO_2-H_2O and K_2O-FeO-Al_2O_3-SiO_2-H_2O: Lithos, v. 41, p. 229–250.

Mirwald, P. W., and Massonne, H.-J., 1980, The low-high quartz and quartz-coesite transition to 40 kbar between 600° and 1600°C and some reconnaissance data on the effect of $NaAlO_2$ component on the low quartz-coesite transition: Jour. Geophys. Res., v. 85, p. 6983–6990.

Mosenfelder, J. L., and Bohlen, S. R., 1997, Kinetics of the coesite to quartz transformation: Earth Planet. Sci. Lett., v. 153, p. 133–147.

Mysen, B. O., Ulmer, P., Konzett, J., and Schmidt, M. W., 1998, The upper mantle near convergent plate boundaries, *in* Hemley, R. J., ed., Ultrahigh-pressure mineralogy: Physics and chemistry of the Earth's deep interior: Rev. Mineral., v. 37, p. 97–138.

Nishiyama, T., 1998, Kinetic modeling of the coesite-quartz transition in an elastic field and its implication for the exhumation of ultrahigh-pressure metamorphic rocks: The Island Arc, v. 7, p. 70–81.

Ogasawara, Y., Liou, J. G., and Zhang, R. Y., 1995, Petrogenetic grid for ultrahigh-P metamorphism in the model system CaO-MgO-SiO_2-CO_2-H_2O: The Island Arc, v. 4, p. 240–253.

______, 1997, Thermochemical calculation of log fO_2-T-P stability relations of diamond-bearing assemblages in the model system CaO-MgO-SiO_2-CO_2-H_2O: Russ. Geol. Geophys., v. 38, p. 587–598.

Okamoto, K., and Maruyama, S., 1998, Multi-anvil re-equilibration experiments of a Dabie Shan ultrahigh-pressure eclogite within the diamond-stability fields: The Island Arc, v. 7, p. 52–69.

______, in press, The high pressure stability limits of lawsonite in the MORB + H_2O system: Amer. Mineral.

Ono, S., 1998, Stability limits of hydrous minerals in sediment and mid-ocean ridge basalt compositions: Implication for water transport in subduction zones: Jour. Geophys. Res., v. 103, p. 18,253–18,267.

Patiño Douce, A. E., and McCarthy, T. C., 1998, Melting of crustal rocks during continental collision and subduction, *in* Hacker, B., and Liou, J. G., eds., When continents collide: Geodynamics and geochemistry of ultrahigh-pressure rocks: Dordrecht, Kluwer Acad. Publ., p. 27–56.

Pawley, A. R., 1992, Experimental study of the compositions and stabilities of synthetic nyböite and nyböite-glaucophane amphiboles: Eur. Jour. Mineral., v. 4, p. 171–192.

______, 1994, The pressure and temperature stability limits of lawsonite: Implications for H_2O recycling in subduction zones: Contrib. Mineral. Petrol., v. 118, p. 99–108.

______, 1998, The reaction talc + forsterite = enstatite + H_2O: New experimental results and petrological implications: Amer. Mineral., v. 83, p. 51–57.

Poli, S., and Schmidt, M. W., 1995, H_2O transport and release in subduction zones: Experimental constraints on basaltic and andesitic systems: Jour. Geophys. Res. v. 100, p. 22,299–22,314.

______, 1997, The high-pressure stability of hydrous phases in orogenic belts: An experimental approach on eclogite-forming processes: Tectonophys., v. 273, p. 169–184.

Sato, K., Katsura, T., and Ito, E., 1997, Phase relations of natural phlogopite with and without enstatite up to 8 GPa: Implication for mantle metasomatism: Earth Planet Sci. Lett., v. 146, p. 511–526.

Scambelluri, M., Muentener, O., Hermann, J., Piccardo, G. B., and Trommsdorff, V., 1995, Subduction of water into the mantle: History of an alpine peridotite: Geology, v. 23, p. 459–462.

Schmidt, M. W., 1996, Experimental constraints on recycling of potassium from subducted oceanic crust: Science, v. 272, p. 1227–1230.

Schmidt, M. W., and Poli, S., 1994, The stability of lawsonite and zoisite at high pressure: Experiment in CASH to 92 kbar and implications for the presence of hydrous phases in subducted lithosphere: Earth Planet. Sci. Lett., v. 124, p. 105–118.

______, 1998, What causes the position of the volcanic front? Experimentally based water budget for dehydrating slabs and consequences for arc magma generation: Earth Planet Sci. Lett., v. 163, p. 361–379.

Schreyer, W., 1988, Subduction of continental crust to mantle depths: Petrological evidence: Episodes, v. 11, p. 97–104.

______, 1988, Experimental studies on metamorphism of crustal rocks under mantle pressures: Mineral. Mag,, v. 51, p. 1–26.

Ulmer, P., and Trommsdorff, V., 1999, Phase relations of hydrous mantle subducting to 300 km, *in* Fei, Y. W., Bertka, C., and Mysen, B. O., eds., Mantle petrology: Field observations and high-pressure experimentation: Geochemical Society, in press.

Wunder, B., 1998, Equilibrium experiments in the system MgO-SiO_2-H_2O (MSH): Stability fields of clinohumite [$Mg_9Si_4O_{16}(OH)_{12}$], chondrodite-OH [$Mg_5Si_2O_8(OH)_2$], and phase A [$Mg_7Si_2O_8(OH)_6$]: Contrib. Mineral. Petrol., v. 132, p. 111–120.

Wunder, B., and Schreyer, W., 1997, Antigorite: High-pressure stability in the system MgO-SiO_2-H_2O (MSH): Lithos, v. 41, p. 213–227.

REGIONAL STUDIES

Africa

Caby, R., 1994, Precambrian coesite from northern Mali: First record and implications for plate tectonics in the trans-Saharan segment of the Pan-African belt: Eur. Jour. Mineral., v. 6, p. 235–244.

Caby, R., Andreopoulos-Renaud, U., and Pin, C., 1989, Late Proterozoic arc-continent and continent-continent collision in the Pan-African trans-Saharan belt of Mali: Can. Jour. Earth Sci., v. 26, p. 1136–1146.

Frimmel, H. E., and Harmady, C. J. H., 1992, Blue amphiboles and their significance for the metamorphic history of the Pan-African Gariep belt, Namibia: Jour. Metamor. Geol., v. 10, p. 651–669.

Pearson, D. G., Davies, G. R., and Nixon, P. H., 1993, Geochemical constraints on the petrogenesis of diamond facies pyroxenites from the Beni Bousera peridotite massif, N. Morocco: Jour. Petrol., v. 34, p. 125–172.

Pearson, D. G., Davies, G. R., Nixon, P. H., Greenwood, P. B., and Mattey, D. P., 1991, Oxygen isotope evidence for the origin of pyroxenites in the Beni Bousera peridotite massif, North Morocco: Derivation from subducted oceanic lithosphere: Earth Planet. Sci. Lett., v. 102, p. 289–301.

Pearson, D. G., Davies, G. R., Nixon, P. H., and Mattey, D. P., 1991, A carbon isotope study of diamond facies pyroxenites and associated rocks from the Beni Bousera. peridotite massif, North Morocco: Jour. Petrol., v. 32, p. 175–189.

Pearson, D. G., Davies, G. R., Nixon, P. H., and Milledge, H. J., 1989, Graphitized diamonds from a peridotite massif in Morocco and implications for anomalous diamond occurrences: Nature, v. 338, p. 60–62.

Europe

Dora-Maira massif, Western Alps

Arnaud, N. O., and Kelley, S. P., 1995, Evidence for excess argon during high pressure metamorphism in the Dora Maira massif, western Alps, Italy, using an ultra-violet laser ablation microprobe ^{40}Ar-^{39}Ar technique: Contrib. Mineral. Petrol., v. 121, p. 1–11.

Avigad, D., 1992, Exhumation of coesite-bearing rocks in the Dora-Maira Massif, western Alps, Italy: Geology, v. 20, p. 947–950.

Avigad, D., Chopin, C., Goffé, B., and Michard, A., 1993, Tectonic model for the evolution of the western Alps: Geology, v. 21, p. 659–662.

Barnicoat, A. C., and Frey, N., 1986, High-pressure metamorphism in the Zermatt-Saas Fee ophiolite zone, Switzerland: Jour. Geol. Soc. Lond., v. 143, p. 607–618.

Becker, H., 1993, Garnet peridotite and eclogite Sm-Nd mineral ages from the Lepontine dome, Swiss Alps: New evidence for Eocene high-pressure metamorphism in the central Alps: Geology, v. 21, p. 599–602.

Biino, G. G., 1995, Pre-Variscan evolution of the ecologitized mafic rocks from the Helvetic basement of the central Alps: Eur. Jour. Mineral., v. 7, p. 57–70.

Biino, G., and Compagnoni, R., 1992, Very high-pressure metamorphism of the Brossasco coronite metagranite, southern Dora Maira Massif, western Alps: Schweiz. Mineral. Petrol. Mitteil., v. 72, p. 347–363.

Blake, M. C., Jr., and Jayko, A., 1990, Uplift of very high pressure rocks in the western Alps: Evidence for structural attenuation along low-angle faults: Mém. Soc. Géol. France, v. 156, p. 237–246.

Blake, M. C. R., Moore, D. E., and Jayko, A. S., 1996, The role of serpentinite mélanges in the unroofing of UHPM rocks: An example from the Western Alps of Italy, *in* Coleman, R. G., and Wang, X., eds., Ultrahigh-pressure metamorphism: New York, Cambridge University Press, p. 182–205.

Bouffette, J., et al., 1993, Le Passage des granulites aux eclogites dans les metapelites de l'unite de la Punta Muret, Massif Dora-Maira, Alpes occidentales: Comptes Rendus Acad. Sci. Paris, Serie 2, v. 317, p. 1617–1624.

Bowtell, S. A., Cliff, R. A., and Barnicoat, A. C., 1994, Sm-Nd isotopic evidence on the age of eclogitization in the Zermatt-Saas ophiolite: Jour. Metamor. Geol., v. 12, p. 187–196.

Burchard, M., and Schreyer, W., 1995, Melting experiments on the granitic country rock of the Dora Maira pyrope quartzites: Bochumer Geol. Geotech. Ar., v. 44, p. 23–27.

Chopin, C., 1984, Coesite and pure pyrope in high grade pelitic blueschists of the Western Alps: A first record and some consequences: Contrib. Mineral. Petrol., v. 86, p. 107–118.

______, 1987, Very high-pressure metamorphism in the Western Alps: Implications for subduction of continental crust: Phil. Trans. Roy. Soc. London, v. A321, p. 183–197.

Chopin, C., Henry, C., and Michard, A., 1991, Geology and petrology of the coesite-bearing terrain, Dora Maria Massif, Western Alps: Eur. Jour. Mineral., v. 3, p. 263–291.

Chopin, C., and Monié, P., 1984, A unique magnesiochloritoid-bearing, high-pressure assemblage from the Monte Rosa: A petrologic and ^{40}Ar-^{39}Ar study: Contrib. Mineral. Petrol., v. 87, p. 388–398.

Chopin, C., Brunet, F., Gebert, W., Medenbach, O., and Tillmanns, E., 1993, Bearthite, $Ca_2Al[PO_4]_2(OH)$, a new mineral from high-pressure terranes of the western Alps: Schweiz. Mineral. Petrogr. Mitteil., v. 73, p. 1–9.

Chopin, C., Ferraris, G., Ivaldi, G, Schertl, H.-P., Schreyer, W., Compagnoni, R., Davidson, C., and Davis, A.M., 1995, Magnesiodumortierite, a new mineral from very-high-pressure rocks, western Alps. II. Crystal chemistry and petrological significance: Eur. Jour. Mineral., v. 7, p. 525–535.

Compagnoni, R., et al., 1991, High-pressure metamorphism in the Western Alps, introduction: Jour. Metamor. Geol., v. 9, p. 34.

Compagnoni, R., Hirajima, T., and Chopin, C., 1995, Ultrahigh-pressure metamorphic rocks in the Wetern Alps, *in* Coleman, R. G., and Wang, X., eds., Ultrahigh pressure metamorphism: New York, Cambridge University Press, p. 206–243.

Compagnoni, R., Hirajima, T., Turello, R., and Castelli, D., 1994, The Brosasco-Isaca unit of the Dora Maira massif, *in* Compagnoni, R., and Messiga, B., eds., High-pressure metamorphism in the Western Alps. Guidebook to the field excursion B1: 16th Gen. Meet. Int. Mineral. Assoc., Pisa, p. 87–105.

D'amico, C., Campana, R., Felice, G., and Ghedini, M., 1995, Eclogites and jades as prehistoric implements in Europe. A case of petrology applied to cultural heritage: Eur. Jour. Mineral., v. 7, p. 29–41.

Davies, G. R., Nixon, P. H., Pearson, D. G., and Obata, M., 1992, The tectonic implications of graphitised diamonds from the Ronda peridotite massif, S. Spain: Geology, v. 21, p. 471–474.

Dobrzhinetskaya, L., Green, H. W., and Wang, S., 1996, Alpe Arami: A peridotite massif from depths of more than 300 kilometers: Science, v. 271, p. 1841–1845.

Droop, G. T. R., Lombardo, B., and Pognante, U., 1990, Formation and distribution of eclogite facies rocks in the Alps, *in* Carswell, D. A., ed., Eclogite facies rocks: Glasgow, Blackie, p. 225–259.

Escher, A., and Beaumont, C., 1997, Formation, burial and exhumation of basement nappes at crustal scale: A geometric model based on the western Swiss-Italian Alps: Jour. Struct. Geol., v. 19, p. 955–974.

Gebauer, D., 1996, A P-T-t path for an ultra-high-pressure ultramafic/mafic association and their felsic country-rocks based on SHRIMP-dating of magmatic and metamorphic zircon domains. Example: Alpe Arami, Central Swiss Alps, *in* Basu, A., and Hart, S., eds., Earth processes: Reading the isotopic code: Amer. Geophys. Union, Geophys. Monogr., v. 95, p. 307–329.

Gebauer, D., Grunenfelder, M., Tilton, G. R., Trommsdorff, V., and Schmid, S., 1992, The geodynamic evolution of garnet-peridotites, garnet-pyroxenites, and eclogites of Alpe Arami and Cima di Gagnone, Central Alps, from Early Proterozoic to Oligocene: Schweiz. Mineral. Petrog. Mitteil., v. 72, p. 107–111.

Gebauer, D., Schertl, H.-P, Brix, M., and Schreyer, W., 1997, 35 Ma old ultrahigh-pressure metamorphism and evidence for very rapid exhumation in the Dora Maira Massif, Western Alps: Lithos, v. 41, p. 5–24.

Gillet, P., Ingrin, J., and Chopin, C., 1984, Coesite in subducted continental crust: P-T history deduced from an elastic model: Earth Planet. Sci. Lett., v. 70, p. 426–436.

Goffé, B., and Chopin, C., 1986, High-pressure metamorphism in the Western Alps: Petrography of metapelites, chronology and consequences: Schweiz. Mineral. Mitt., v. 66, p. 41–52.

Green, H. W., Dobrzhinetskaya, L., Riggs, E. M., and Zhen-Ming, J., 1997, Alpe Arami: A peridotitie massif from the mantle transition zone?: Tectonophys., v. 279, p. 1–21.

Henry, C., Michard, A., and Chopin, C., 1993, Geometry and structural evolution of ultra-high-pressure and high-pressure rocks from the Dora-Maira massif, Western Alps, Italy: Jour. Struct. Geol., v. 15, p. 965–981.

Hirajima, T., and Compagnoni, R., 1993, Petrology of a jadeite-quartz-coesite-almandine-phengite fels with retrograde ferro-nyböite from the Dora-Maira massif, western Alps: Eur. Jour. Mineral., v. 5, p. 943–955.

Kienast, J. R., Lombardo, B., Biino, G., and Pinardon, J. L., 1991, Petrology of very-high-pressure eclogitic rocks from the Brossasco-Isasca Complex, Dora-Maira Massif, Italian western Alps: Jour. Metamor. Geol., v. 9, p. 19–34.

Klemd, R., Matthes, S., and Schüssler, U., 1994, Reaction textures and fluid behaviour in very high-pressure calc-silicate rocks of the Münchberg gneiss complex, Bavaria, Germany: Jour. Metamor. Geol., v. 12, p. 735–745.

Marchant, R. H., and Stampli, G. M., 1997, Subduction of continental crust in the western Alps: Tectonophys., v. 269, p. 217–235.

Matte, P., 1998, Continental subduction and exhumation of HP rocks in Paleozoic orogenic belts: Uralides and Variscides: GFF, v. 120, p. 209–222.

Michard, A., Chopin, C., and Henry, C., 1993, Compression versus extension in the exhumation of the Dora-Maira coesite-bearing unit, West Alps, Italy: Tectonophys., v. 221, p. 173–193.

Michard, A., Henry, C., and Chopin, C., 1995, Structures in ultra-high-pressure metamorphic rocks, *in* Coleman, R., and Wang, X.., eds., Ultrahigh pressure metamorphism: New York, Cambridge University Press, p. 132–158.

Monié, P. J., and Chopin, C., 1991, $^{40}Ar/^{39}Ar$ dating in coesite-bearing and associated units of the Dora-Maira massif, Western Alps: Eur. Jour. Mineral., v. 3, p. 239–262.

Monié, P. J., Galindo-Zaldivar, J., Gonzalez-Lodeiro, F., Goffé, B., and Jabaloy, A., 1991, $^{40}Ar/^{39}Ar$ geochronology of Alpine tectonism in the Betic Cordilleras, southern Spain: Jour. Geol. Soc. Lond., v. 148, p. 289–297.

Paquette, J.-L., Chopin, C., and Peucat, J. J., 1989, U-Pb zircon, Rb-Sr and Sm-Nd geochronology of high to very high pressure meta-acidic rocks from the Western Alps: Contrib. Mineral. Petrol., v. 101, p. 280–289.

Paquette, J.-L., Montel, J.-M., and Chopin, C., 1999, U-Th-Pb dating of the Brossasco ultrahigh-pressure metagranite, Dora-Maira massif, western Alps: Eur. Jour. Mineral., v. 11, p. 69–77.

Philippot, P., 1993, Fluid–melt-rock interaction in mafic eclogites and coesite-bearing metasediments, constraints on volatile recycling during subduction: Chem. Geol., v. 108, p. 93–112.

Philippot, P., Chopin, C., and Dubessy, J., 1993, Fluids in the pyrope-coesite rocks of the Dora-Maira massif: Evidence for early fluids and for fluid partitioning during decompression: Archiwum Mineral., v. XLIX, p. 161–164.

Philippot, P., Chevallier, P., Chopin, C., and Dubessy, J., 1995, Fluid composition and evolution in coesite-bearing rocks, Dora-Maira massif, Western Alps: Implications for element recycling during subduction: Contrib. Mineral. Petrol., v. 121, p. 29–44.

Pognante, U., 1991, Petrological constraints on the eclogite- and blueschist-facies metamorphism and P-T-t paths in the Western Alps: Jour. Metamor. Geol., v. 9, p. 5–17.

Pognante, U., et al., 1987, High pressure metamorphism in the nappes of the Valle dell'Orco traverse, Western Alps collisional belt: Jour. Metamor. Geol., v. 5, p. 397–414.

______, 1989, Eclogites in the northern Dora-Maira nappe, Western Alps, Italy: Contrib. Mineral. Petrol., v. 40, p. 57–71.

Reinecke, T., 1991, Very-high-pressure metamorphism and uplift of coesite-bearing metasediments, from the Zermatt-Saas zone, Western Alps: Eur. Jour. Mineral, v. 3, p. 7–17.

Reinecke, T., Schreyer, W., and Chopin, C., 1991, The pyrope-coesite rocks and their country rocks at Parigi, Dora Maira massif, western Alps: Detailed petrography, mineral chemistry, and PT-path: Contrib. Mineral. Petrol., v. 108, p. 1–21.

Reinecke, T., 1998, Prograde high- to ultrahigh-pressure metamorphism and exhumation of oceanic sediments at Lago di Cignana, Zermatt-Saas Zone, western Alps: Lithos, v. 42, p. 147–190.

Scaillet, S., 1996, Excess ^{40}Ar transport scale and mechanism in high-pressure phengites: A case study from an eclogitized metabasite of the Dora Maira nappe, western Alps: Geochim. et Cosmochim. Acta, v. 60, p. 1075–1090.

Scaillet, S., Feraud, G., Ballevre, M., and Amouric, M., 1992, Mg/Fe and [(Mg,Fe)Si-Al_2] compositional control on argon behavior in high-pressure white micas: A $^{40}Ar/^{39}Ar$ continuous laser-probe study from the Dora Maira nappe of the internal Western Alps, Italy: Geochim et Cosmochim. Acta, v. 56, p. 2851–2872.

Scambelluri, M., Hoogerduuijin Strating, E. H., Piccardo, G. B., Vissers, L. M., and Rampone, E., 1991, Alpine olivine– and titanian clinohumite–bearing assemblages in the Effo-Tobbio peridotite, NW Italy: Jour. Metamor. Geol., v. 9, p. 79–91.

Scambelluri, M., Müntener, O., Hermann, J., Piccardo, G. B., and Trommsdroff, V., 1995, Subduction of water into the mantle: History of an Alpine peridotite: Geology, v. 23, p. 459–462.

Schertl, H. J., and Schreyer, W., 1996, Mineral inclusions in heavy minerals of the ultrahigh-pressure metamorphic rocks of the Dora-Maira Massif and their bearing in the relative timing of the petrological events, *in* Basu, A., and Hart, S., eds., Earth processes: Reading the isotopic code: Amer. Geophys. Union, Geophys. Monog. 95, p. 331–342.

Schertl, H. J., Schreyer, W., and Chopin, C., 1991, The pyrope-coesite rocks and their country rocks at Parigi, Dora Maira massif, western Alps: Detailed petrography, mineral chemistry, and P-T path: Contrib. Mineral. Petrol., v. 108, p. 1–21.

Sharp, Z. D., Essene, E. T., and Hunziker, J. C., 1993, Stable isotope geochemistry and phase equilibria of coesite-bearing whiteschists. Dora Maira Massif, Western Alps: Contrib. Mineral. Petrol., v. 114, p. 1–12.

Simon, G., Chopin, C., and Schenk, V., 1997, Near-end-member magnesiochloritoid in prograde-zoned pyrope, Dora-Maira massif, western Alps: Lithos, v. 41, p. 37–57.

Tilton, G. R., Ames, L., Schertl, H.-P., and Schreyer, W., 1997, Reconnaissance isotopic investigations on rocks of an undeformed granite contact within the coesite-bearing unit of the Dora Maira Massif: Lithos, v. 41, p. 25–36.

Tilton, G. R., Schreyer, W., and Schertl, H. P., 1989, Pb-Sr-Nd isotopic behavior of deeply subducted crustal rocks from the Dora Maria massif, Western Alps, Italy: Geochim. et Cosmochim. Acta., v. 53, p. 1391–1400.

______, 1991, Pb-Sr-Nd isotopic behavior of deeply subducted crustal rocks from the Dora-Maira Massif, Western Alps-II: What is the age of the ultrahigh-pressure metamorphism?: Contrib. Mineral. Petrol., v. 108, p. 22–33.

Vallis, F., and Scambelluri, M., 1996, Redistribution of high-pressure fluids during retrograde metamorphism of eclogite-facies rocks, Voltri Massif, Italian Western Alps: Lithos, v. 39, p. 81–96.

van der Klauw, S. N. G. C., Reinecke, T., and Stöckhert, B., 1997, Exhumation of ultrahigh-pressure metamorphic oceanic crust from Lago di Cignana, Piemontese zone, western Alps: The structural record in metabasites: Lithos, v. 41, p. 79–102.

van der Molen, I., et al., 1986, The pressure path of solid inclusions in minerals, the retention of coesite inclusions during uplift: Lithos, v. 19, p. 317–324.

Western Gneiss Region, Norway

Andersen, T., et al., 1989, Nitrogen-bearing, aqueous fluid inclusions in some eclogites from the Western Gneiss Region of the Norwegian Caledonides: Contrib. Mineral. Petrol., v. 103, p. 153–165.

______, 1990, Uplift of deep crust during orogenic extensional collapse, a model based on field studies in the Sogn-Sunnfjord region of western Norway: Tectonics, v. 9, p. 1097–1111.

______, 1991, Mineral-fluid-melt interactions in high-pressure shear zones in the Bergen Arc nappe complex, Caledonides of W. Norway: Implications for the fluid regime in Caledonian eclogite-facies metamorphism: Lithos, v. 27, p. 187–204.

______, 1993, N_2 and CO_2 in deep crustal fluids, evidence from the Caledonides of Norway: Chem. Geol., v. 108, p. 113–132.

______, 1994, Deep crustal fabrics and a model for the extensional collapse of Southwest Norwegian Caledonides: Jour. Struct. Geol., v. 16, p. 1191–1203.

Andersen, T. B., Jamtveit, B., Dewey, J. F., and Swensson, E., 1991, Subduction and eduction of continental crust: Major mechanisms during continent-continent collision and orogenic extensional collapse, a model based on the south Norwegian Caledonides: Terra Nova, v. 3, p. 303–310.

Austrheim, H., 1987, Eclogitization of lower crustal granulites by fluid migration through shear zones: Earth Planet. Sci. Lett., v. 81, p. 221–232.

______, 1990, The granulite-eclogite facies transition, a comparison of experimental work and a natural occurrence in the Bergen arcs, western Norway: Lithos, v. 25, p. 163–169.

______, 1991, Eclogite formation and dynamics of crustal roots under continental collision zones: Terra Nova, v. 3, p. 492–499.

Austrheim, H., et al., 1994, Pseudotachylytes generated during seismic faulting and eclogitization of the deep crust: Science, v. 265, p. 82–83.

Boundy, T. M., Mezger, K., and. Essene, E. J., 1997, Temporal and tectonic evolution of the granulite-eclogite association from the Bergen Arcs, western Norway: Lithos, v. 39, p. 159–178.

Bowtell, S. A., et al., 1994, Sm-Nd isotopic evidence on the age of eclogitization in the Zermatt-Saas Ophiolite: Jour. Metamor. Geol., v. 12, p. 187–196.

Brueckner, H. K., Gilotti, J. A., and Nutman, A., 1997, Caledonian eclogite facies metamorphism of early Proterozoic protoliths from the north-east Greenland Eclogite Province: Contrib. Mineral. Petrol., v. 130, p. 103–120.

Brueckner, H. K., and Medaris, G. L., 1998, A tale of two orogens: The contrasting TPt history and geochemical evolution of mantle in high- and ultrahigh-pressure metamorphic terranes of the Norwegian Caledonides and the Czech Variscides: Schweiz. Mineral. Petrog. Mitteil., v. 78, p. 293–307.

Carswell, D. A., 1986, The metamorphic evolution of Mg-Cr type Norwegian garnet peridotites: Lithos, v. 19, p. 279–297.

Dobrzhinetskaya, L. F., Eide, E. A., Larsen, R. B., Sturt, B. A., Tronnes, R. G., Smith, D. C., Taylor, W. R., and Posukhova, T. V., 1995, Microdiamond in high-grade metamorphic rocks of the Western Gneiss region, Norway: Geology, v. 23, p. 597–600.

Jamtveit, B., et al., 1990, Fluid controlled eclogitization of granulites in deep crustal shear zones, Bergen arcs, Western Norway: Contrib. Mineral. Petrol., v. 104, p. 184–193.

Jamtveit, B., Carswell, D. A., and Mearns, E. W., 1991, Chronology of the high-pressure metamorphism of Norwegian garnet peridotites/pyroxenites: Jour. Metamor. Geol., v. 9, p. 125–139.

Gilotti, J. A., 1993, Discovery of a medium-temperature eclogite province in the Caledonides of North-East Greenland: Geology, v. 21, p. 523–526.

Krabbendam, M., and Dewey, J. F., 1998, Exhumation of UHP rocks by transtension in the Western Gneiss Region, Scandinavian Caledonides, *in* Continental transpressional and transtensional tectonics: Geol. Soc. Lond., Spec. Publ., v. 135, p. 159–181.

Krabbendam, M., and Wain, A., 1997, Late-orogenic structures, differential retrogression and structural position of ultrahigh-pressure rocks from the Nordfjord-Stadtlandet area of the Western Gneiss region: Norges Geol. Underssok. Bull., v. 432, p. 127–139.

Krogh, E. J., and Carwsell, D. A., 1995, HP and UHP eclogites and garnet peridotites in the Scandinavian Caledonides, *in* Coleman, R. G., and Wang, X., eds., Ultrahigh pressure metamorphism: New York, Cambridge University Press, p. 244–298.

Larsen, R. B., Eide, E. A., et al., in press, Evolution of metamorphic volatiles during exhumation of microdiamond-bearing granulites in the Western Gneiss region, Norway: Contrib. Mineral. Petrol.

Smith, D. C., 1984, Coesite in clinopyroxene in the Caledonides and its implication for geodynamics: Nature, v. 310, p. 641 –644.

______, 1988, A review of the peculiar mineralogy of the "Norwegian coesite-eclogite province" with crystal-chemical, petrological, geochemical, and geodynamical notes and an extensive bibliography," *in* Smith, D. C., ed., Eclogites and eclogite-facies rocks: Amsterdam, Elsevier, p. 1–206.

______, 1995, Microcoesites and microdiamonds in Norway: An overview, *in* Coleman, R. G., and Wang, X., eds., Ultrahigh pressure metamorphism: New York, Cambridge University Press, p. 299–355.

Smith, D. C., and Lappin, M. A., 1989, Coesite in the Straumen kyanite-eclogite pod, Norway: Terra Nova, v. 1, p. 47–56.

Smith, D. C., et al., 1985, Petrochemistry of opaque minerals in eclogites from the Western Gneiss Region, Norway, 2. Chemistry of the ilmenite mineral group: Chem. Geol., v. 50, p. 251–266.

Wain, A., 1997, New evidence for coesite in eclogite and gneisses: Defining an ultrahigh-pressure province in the Western Gneiss region of Norway: Geology, v. 25, p. 927–930.

van Roermund, H. L. M., 1989, High-pressure ultramafic rocks from the allochthonous nappes of the Swedish Caledonides, *in* Gayer, R. D., ed., The Caledonide geology of Scandinavia: London, Graham and Trotman, p. 205–220.

The Bohemian massif and the Saxonian Erzgebirge

Bakun-Czubarow, N., 1991, Geodynamic significance of the Variscan HP eclogite-granulite series of the Zlote Mountains in the Sudetes: Publ: Inst. Geophys., ser. A: Phys. Earth Inter., v. 19, p. 215–244.

______, 1991, On the possibility of occurrence of quartz pseudomorphs after coesite in the eclogite-granulite rock series of the Zlote Mountains in the Sudetes, SW Poland: Archiwurn Mineral., v. 48, p. 3–25.

Beard, B. L., Medaris, L. G., Johnson, C. M., Jelinek, E., Tonika, J., and Riciputi, L. R., 1995, Geochronology and geochemistry of eclogites from the Marianske Lazne Complex, Czech Republic: Implications for Variscan orogenesis: Geol. Rundsch., v. 84, p. 552–567.

Becker, H., 1996, Geochemistry of garnet peridotite massifs from lower Austria and the composition of deep lithosphere beneath a Paleozoic convergent plate margin: Chem. Geol., v. 134, p. 49–65.

______, 1997, Sm-Nd garnet ages and cooling history of high-temperature garnet peridotites and high-pressure granulites from lower Austria: Contrib. Mineral. Petrol., v. 127, p. 224–236.

Brocker, M., and Klemd, R., 1996, Ultrahigh-pressure metamorphism in the Snieznik Mountains (Sudets, Poland): P-T constraints and geological implications: Jour. Geol., v. 104, p. 417–433.

Brueckner, H. K., Blusztain, J., and Bakun-Czubarow, N., 1996, Trace element and Sm-Nd "age" zoning in garnets from peridotites of the Caledonian and Variscan Mountains and tectonic implications: Jour. Metamor. Geol., v. 14, p. 61–73.

Massonne, H.-J., in press, A new occurrence of microdiamonds in quartzofeldspathic rocks of the Saxonian Erzgebirge, Germany, and their metamorphic evolution: Proc. 7th Int. Kimberlite Conf., Cape Town.

Medaris, L. G., Jr., Beard, B. L., Johnson, C. M., Valley, J. W., Spicuzza, M. J., Jelinek, E., and Misar, Z., 1995, Garnet pyroxenite and eclogite in the Bohemian massif: Geochemical evidence for Variscan recycling of subducted lithosphere: Geol. Rundsch., v. 84, p. 489–505.

Medaris, L. G., Jr., Jelinek, E., and Misar, Z., 1995, Czech eclogites: Terrane setting and implications for Variscan tectonic evolution of the Bohemian massif: Eur. Jour. Mineral., v. 7, p. 7–28.

O'Brien, J., 1993, Partially retrograded eclogites of the Muenchberg massif, Germany: Records of a multistage Variscan uplift history in the Bohemian Massif: Jour. Metamor. Geol., v. 11, p. 241–260.

______, 1997, Garnet zoning and reaction textures in overprinted eclogites, Bohemian massif, European Variscides: A record of their thermal history during exhumation: Lithos, v. 41, p. 119–133.

Schmadicke, E., 1991, Quartz pseudomorphs after coesite in eclogites from the Saxonian Erzgebirge: Eur. Jour. Mineral., v. 1, p. 231–238.

Schmadicke, E., and Evans, B. W., 1997, Garnet-bearing ultramafic rocks from the Erzgebirge, and their relation to other settings in the Bohemian massif: Contrib. Mineral. Petrol., v. 127, p. 57–74.

Schmadicke, E., Mezger, K., Cosca, M. A., and Okrusch, M., 1995, Variscan Sm-Nd and Ar-Ar ages of eclogite facies rocks from the Erzgebirge, Bohemian massif: Jour. Metamor. Geol., v. 13, p. 537–552.

Schmadicke, E., Okrusch, M., and Schmidt, W., 1992, Eclogite-facies rocks in the Saxonian Erzgebirge, Germany: High-pressure metamorphism under contrasting P-T conditions: Contrib. Mineral. Petrol., v. 110, p. 226–241.

Steltenpohl, M. G., Cymerman, Z., Krogh, E. J., and Kunk, A. J., 1993, Exhumation of eclogitized continental basement during Variscan lithospheric delamination and gravitational collapse, Sudety Mountains, Poland: Geology, v. 21, p. 1111–1114.

Other European regions

Altheff, R., and Kalt, A., 1996, Metamorphic evolution of ultrahigh-pressure garnet peridotites from the Variscan Vosges Mts., France: Chem. Geol., v. 134, p. 27–47.

Davis, G. R., Nixon, P. H., Pearson, D. G., and Obata, M., 1992, The tectonic implications of graphitized diamonds from the Ronda peridotite massif, S. Spain: Geology, v. 21, p. 471–474.

Trommsdorff, V., Lopez Sanchez-Vizcaino, V., Gomez-Pugnaire, A. T., and Muntener, O., 1998, High-pressure breakdown of antigorite to spinifex-texture olivine and orthopyroxene, SE Spain: Contrib. Mineral. Petrol., v. 132, p. 139–148.

Asia

East-Central China

Ames, L., Tilton, G. R., and Zhou, G., 1993, Timing of collision of the Sino-Korean and Yangtze cratons: U-Pb zircon dating of coesite-bearing eclogites: Geology, v. 21, p. 339–342.

Ames, L., Zhou, G., and Xiong, B., 1996, Geochronology and geochemistry of ultrahigh-pressure metamorphism with implications for collision of the Sino-Korean and Yangtze cratons, central China: Tectonics, v. 15, p. 472–489.

Baker, J., Matthews, A., Mattey, D., Rowley, D. B., and Xue, F., 1997, Fluid-rock interaction during ultra-high pressure metamorphism, Dabie Shan, China: Geochim. et Cosmochim. Acta, v. 61, p. 1685–1696.

Carswell, D. A., O'Brien, P. J., Wilson, R. N., and Zhai, M., 1997, Thermobarometry of phengite-bearing eclogites in the Dabie Mountains of central China: Jour. Metamor. Geol., v. 15, p. 239–252.

Carwsell, D. A., Wilson, R. N., and Zhai, M., 1996, Ultrahigh pressure aluminous titanites in carbonate-bearing eclogites at Shuanghe in Dabieshan, central China: Mineral. Mag., v. 60, p. 461–471.

Chavagnac, V., and Jahn, B. M., 1996, Coesite-bearing eclogites from the Bixiling Complex, Dabie Mountains, China: Sm-Nd ages, geochemical characteristics, and tectonic implications: Chem. Geol., v. 133, p. 29–51.

Chen, J., Xie, Z., and Liu, S., 1995, Cooling age of Dabie orogen, China: Determined by Ar/Ar and fission track techniques: Sci. China, ser. B., v. 38, p. 749–755.

Chen, J., Wang, Q., Zhai, M., and Ye, K., 1995, Ductile deformation of garnet in eclogite: Sci. China, ser. B, v. 25, p. 1115–1120 (in Chinese).

Chen, W. J., Harrison, T. M., Li, Q., Heizler, M. T., Liu, R. X., Ma, B., and Li, J., 1992, The cooling history of mélange zone in north Jiangsu–south Shandong region: Evidence from multiple diffusion domain $^{40}Ar/^{39}Ar$ thermal geochronology: Acta Petrol. Sinica, v. 8, p. 1–17 (in Chinese).

Chen, Y., Li, S., and Cong. B., 1992, Time and genesis of eclogites from the Jiaonan area—evidence from Sr, Nd isotope geochemistry and chronology: Chinese Sci. Bull., v. 37, p. 2109 (in Chinese).

Cheng, Z. X., 1990, Basic characteristics of the Donghai eclogite-type rutile deposit in Jiangsu: Mineral Deposits, v. 9, p. 86–90.

Cong, B., and Wang, Q., 1994, Review of research on ultrahigh-pressure metamorphic rocks in China: Chinese Sci. Bull., v. 39, p. 2068–2075.

______, 1995, Ultrahigh pressure metamorphic rocks in China: Episodes, v. 18, p. 91–94.

Cong, B., Wang, Q., Zhai, M., Zhang, R., Zhao, Z., and Ye, K., 1994, UHP metamorphic rocks in the Dabie-Su-Lu region, China: Their formation and exhumation: The Island Arc, v. 3, p. 135–150.

Cong, B., Zhai, M., Carswell, D. A., Wilson, R. N., Wang, Q., Zhao, Z., and Windley, B. F., 1995, Petrogenesis of the ultrahigh-pressure rocks and their country rocks at Shuanghe in Dabieshan, central China: Eur. Jour. Mineral., v. 7, p. 119–138.

Cong, B., Zhang, R., Li, S., and Zhai, M., 1992, Preliminary study of isotope chronology of eclogites in northern Jiangsu and eastern Shandong Provinces, China, *in* Exploration of volcanoes and rocks in Japan, China, and Antarctica: Commemorative papers for Prof. Yukio Matsumoto: Yamaguchi, Japan, Yamaguchi University, p. 411–417.

Cong, B., Zhang, R. Y., Liou, J. G., Ye, K., and Wang, Q., 1996, Metamorphic evolution of UHPM rocks, *in* Cong, B. ed., Ultrahigh-pressure metamorphic rocks in the Dabieshan-Sulu Region of China: Beijing, Sci. Press China, p. 128–160.

Dong, S. W., Chen, J., and Huang, D., 1998. Differential exhumation of tectonic units and ultrahigh-pressure metamorphic rocks in the Dabie Mountains, China: The Island Arc, v. 7, p. 174–183.

Dong, S. W., Wang, X. F., and Huang, D. Z., 1997, Discovery of low-grade metamorphic volcanic rock sheets within UHP in Dabie Mts., and its implications: Chinese Sci. Bull., v. 42, p. 1199–1203.

Dong, S. W., Wu, X., Gao, R., Lu, D., Li, Y., He, Y., Tang, J., Cao, F., Hou, M., and Huang, D., 1996, Preliminary study on deep geology of Dabie orogenic belt: Continental Dynamics, v. 1, p. 103–108.

Eide, E. A., 1995, A model for the tectonic history of HP and UHPM regions in east central China, *in* Coleman, R. G., and Wang, X., eds., Ultrahigh-pressure metamorphism: New York, Cambridge University Press, p. 391–426.

Eide, E., McWilliams, M. O., and Liou, J. G., 1994, $^{40}Ar/^{39}Ar$ geochronology and exhumation of high-pressure to ultrahigh-pressure metamorphic rocks in east-central China: Geology, v. 22, p. 601–604.

Enami, M., Cong, B., Yoshida, T., and Kawabe, I., 1995, A mechanism for Na incorporation in garnet: An example from garnet in orthogneiss from the Su-Lu terrane, eastern China: Amer. Mineral., v. 80, p. 475–482.

Enami, M., and Nagasaki, A., 1999, Prograde P-T path of kyanite eclogites from Junan in the Su-lu ultrahigh-pressure province, eastern China: The Island Arc, v. 8, in press.

Enami, M., Suzuki, K., Zhai, M., and Zheng, X., 1993, The chemical Th-U-total Pb isochron ages of Jiandong and Jiaonan metamorphic rocks in the Shandong Peninsula, eastern China: The Island Arc, v. 2, p. 104–113.

Enami, M., and Zang, Q., 1988, Magnesian staurolite in garnet-corundum rocks and eclogite from the Donghai district, Jiangsu province, east China: Amer. Mineral., v. 73, p. 48–56.

______, 1990, Quartz pseudomorphs after coesite in eclogites from Shandong province, east China: Amer. Mineral., v. 75, p. 381–386.

Enami, M., Zang, Q., and Yin, Y., 1993, High-pressure eclogites in northern Jiangsu–southern Shandong province, eastern China: Jour. Metamor. Geol., v. 11, p. 589–603.

Enami, M., Zhao, Z., and Wang, Q., 1993, A calderitic garnet paragenesis in granitic gneisses in the Su-Lu ultrahigh-pressure terrane, eastern China: Mineral. Jour., v. 16, p. 268–277.

Ernst, W. G., and Liou, J. G., 1995, Contrasting plate-tectonic styles of the Quiling-Dabie-Sulu, Alpine-type, and Franciscan, Pacific-type, metamorphic belts: Geology, v. 23, p. 353–356.

Ernst, W. G., Zhou, G., Liou, J. G., Eide, E., and Wang, X., 1991, High-pressure and superhigh-pressure metamorphic terranes in the Qinling-Dabie mountain belt, central China: Early- to mid-Phanerozoic accretion of the western paleo-Pacific Rim: Pacific Sci. Assoc. Infor. Bull., v. 43, p. 6–15.

Fan, Q., Liu, R., and Ma, B., 1992, The corundum garnetite of Zhimafang, Donghai district: Acta Petrol. Sinica, v. 9, p. 291–295 (in Chinese).

Gao, S., Jin, Z., Jin, S., Kern, H., and Popp, T., 1998, Seismic velocity structure and composition of continental crust in the Dabie-Sulu area: Continental Dynamics, v. 3, p. 108–112.

Ge, Ningjie, et al., 1993, Mineral composition and metamorphic P-T conditions of eclogitic rocks in the eastern part of the Dabie Mountains and their genetic implications: Acta Geol. Sinica, v. 67, p. 109–122.

Hacker, B. R., 1996, Eclogite formation and the rheology, buoyancy, seismicity and H_2O content of oceanic crust, *in* Bebout, G. E. et al.., eds., Subduction: Top to bottom: Amer. Geophys. Union Monog., v. 96, p. 337–346.

Hacker, B. R., Ratschbacher, L., Webb, L., and Dong S., 1995, What brought them up? Exhumation of the Dabie Shan ultrahigh-pressure rocks: Geology, v. 23, p. 743–746.

Hacker, B.R., Ratschbacher, L., Webb, L. Ireland, T., Walker, D., and Dong, S., 1998, U/Pb zircon ages constrain the architecture of the ultrahigh-pressure Qinling-Dabie Orogen, China: Earth Planet. Sci. Lett., v. 161, p. 215–230.

Hacker, B. R., and Wang, Q., 1995, Ar/Ar geochronology of ultrahigh-pressure metamorphism: Tectonics, v. 14, p. 994–1006.

Hacker, B. R., Wang, X., Eide, E. A., and Ratschbacher, L., 1996, The Qinling-Dabie ultrahigh-pressure collisional orogen, *in* Harrison, M. T., and Yin, A., eds., The tectonic development of Asia: New York, Cambridge University Press, Rubey vol. VIII, p. 345–370.

Hacker, B. R., Zhang, R. Y., Liou, J. G., and Hervig, R. L., 1997, Low-Ti olivines contradict a transition-zone origin for ultrahigh-pressure lherzolites: Science, v. 278, p. 702–704.

Han, Z., 1992, Petrology, mineral chemistry and genesis of eclogite and garnet peridotite from Rongcheng, Shandong Province: Jour. Ocean University Qingdao, v. 22, p. 61–70.

______, 1992, Study on crystallization experiment for eclogite at high-temperature and high-pressure: Jour. Ocean University Qingdao, v. 22, p. 71–79.

______, 1993, Eclogite suite types and their geological significance: Jour. Ocean University Qingdao, v. 23, p. 69–74.

______, 1994, Geochemistry of eclogites from the Su-Lu district: Jour. Ocean University Qingdao, v. 24, p. 100–112 (in Chinese).

Han, Z., et al., 1992, Characteristics of eclogite-facies metamorphic paleo-ocean crust from Qingdao and its genetic significance: Jour. Ocean University Qingdao, v. 22, p. 81–89.

Han, Z., Yuan, Y., and Sheng, X., 1989, Petrology and geochemistry of Bixiling eclogite and associated garnet peridotite, Yuexi County, Anhui Province: Acta Petrol. Sinica, v. 4, p. 22–73 (in Chinese).

Han, Z., Zhang, G., Guo, H., Zhang, Z., and Wang, L., 1993, Sm-Nd isochron age of eclogite from southern Shandong Province and its tectonic implications: Jour. Ocean University Qingdao, v. 23, p. 119–124 (in Chinese).

Hirajima, T., Ishiwatari, A., Cong, B., Zhang, R., Banno, S., and Nozaka, T., 1990, Coesite from Mengzhong eclogite at Donghai county, northeastern Jiangsu province, China: Mineral. Mag., v. 54, p. 579–583.

Hirajima, T., Wallis, S. R., Zhai, M., and Ye, K., 1993, Eclogitized metagranitoid from the Sulu ultrahigh-pressure province, eastern China: Proc. Japan Acad., v. 69, p. 249–254.

Hirajima, T., Zhang, R. Y., Li, J., and Cong, B., 1992, Nyböite from the Donghai area, Jiangsu Province, eastern China: Mineral. Mag., v. 56, p. 37–46.

Hiramatsu, N., Banno, S., Hirajima, T., and Cong, B., 1995, Ultrahigh-pressure garnet lherzolite from Chijiadian, Rongcheng County, in the Su-Lu region of eastern China: The Island Arc, v. 4, p. 324–333.

Hiramatsu, N., and Hirajima, T., 1995, Petrology of the Hujialin garnet clinopyroxenite in the Su-Lu ultrahigh-pressure province, eastern China: The Island Arc, v. 4, p. 310–323.

Hu, K., 1989, Geology of eclogite belt in North Hubei Province: Jour. Changchu Geol. Inst. (Spec. iss. on the geology of blueschist in Hubei and Anhui provinces), p. 84–95 (in Chinese).

Ishizaka, K., Hirajima, T., and Zheng, X., 1994, Rb-Sr dating for the Jiaodong gneiss of the Su-Lu ultrahigh pressure province, eastern China: The Island Arc, v. 3, p. 232–241.

Jahn, B-M., 1998, Geochemical and isotopic characteristics of UHP eclogites of the Dabie orogen: Implications for continental subduction and collisional tectonics, *in* Hacker, B., and Liou, J. G., eds., When continents collide: Geodynamics and geochemistry of ultrahigh-pressure rocks: Dordrecht, Kluwer Acad. Publ., p. 203–240.

Jahn, B.-M., Cornichet, J., Cong, B., and Yui, T.-F., 1996, Ultrahigh-Nd eclogites from an ultrahigh-pressure metamorphic terrane of China: Chem. Geol., v. 127, p. 61–79.

Jian, Y., Li, Z., Xiong, C., and Zhang, Z., 1994, Isotopic chronology and path of eclogite from the Dabie orogenic belt and implications for tectonics, *in* Chen, H., ed., Research on isotope geochemistry: Hangzhou, Zhejiang University Press, p. 205–213 (in Chinese).

Kang, W. G., et al., 1993, The discovery and geological features of kyanite-topazite in high-pressure belt from central China: Scientia Sinica, ser. B, v. 36, p. 856–862.

Kato, T., Enami, A., and Zhai, M., 1997, Ultrahigh-pressure marble and eclogite in the Su-Lu UHP terrane, eastern China: Jour. Metamor. Geol., v. 15, p. 169–182.

Kern, H., Popp, T., Gao, S., and Jin, Z., 1998, P- and S-wave velocities and densities of rocks from the Dabie ultrahighpressure (UHP) metamorphic belt to pressure of 600 MPa and to temperatures of 600°C: Continental Dynamics, v. 3, p. 96–102.

Li, S., Chen, Y., Cong, B., Zhang, Z., Zhang, R., Liou, D., Hart, S. R., and Ge, N., 1993, Collision of the North China and Yangtze blocks and formation of coesite-bearing eclogites: Timing and processes: Chem. Geol., v. 109, p. 70–89.

Li, S., Chen, Y., Ge, N., Liou, D., Zhang, Z., Zhang, Q., and Zhou, D., 1993, Zircon U-Pb ages of eclogite and gneiss from Qingdao area—evidence for Jinning magma episode in Jiaonan Group: Chinese Sci. Bull., v. 38, p. 1773–1777 (in Chinese).

Li, S., Chen, Y., Song, M., Zhang, Z., Yang, C., and Zhao, D., 1994, U-Pb zircon ages of amphibolite from the Haiyangsuo area, eastern Shandong province—an example of the influence of multimetamorphism to lower and upper intercepts of zircon discordia line at the concordia curve: Acta Geosci. Sinica, v. 1, p. 37–42.

Li, S., Ge, N., Liou, D., Zhang, Z., Ye, X., Zheng, S., and Peng, C., 1989, The Sm-Nd isotopic age of C-type eclogite from the Dabie Group in the Northern Dabie Mountains and its tectonic implication: Chinese Sci. Bull., v. 34, p. 1623–1628.

Li, S., Hart, S. R., and Wu, T., 1990, Rb-Sf and Sm-Nd isotopic dating of an early Precambrian spilite-keratophyre sequence in the Wutaishan area, North China: Preliminary evidence for Nd-isotopic homogenization in the mafic and felsic lavas during low-grade metamorphism: Precambrian Res., v. 47, p. 191–203.

Li, S., Hart, S. R., Zhang, X., Guo, A., Liu, D., and Zhang, G., 1989, Sm-Nd age for collision between the Sino-Korean and Yangtze cratons, China: Acad. Sinica, v. 8, p. 312–319.

Li, S., Hart, S. R., Zheng, S., Liou, D., Zhang, G., and Guo, A., 1989, Timing of collision between the North and South China blocks—Sm-Nd isotopic age evidence: Sci. China, ser. B, v. 32, p. 1391–1400.

Li, S., Jagoutz, E., Zhang, Z., Lo, C., Chen, W., and Chen, Y., 1998, Geochemical and geochronological constraints on the tectonic outline of the Dabie Mountains, central China: A continent-microcontinent-continent collision model: Continental Geodynamics, v. 3, p. 14–31.

Li, S., Liou, D., Chen, Y., and Ge, N., 1992, The Sm-Nd isotopic age of coesite-bearing eclogite from the southern Dabie Mountains: Chinese Sci. Bull., v. 37, p. 1638–1641.

Li, S., Liou, D., Chen, Y., Zhang, G., and Zhang, Z., 1991, A chronological table of the major tectonic events for the Qinling-Dabie orogenic belt and its implications: Xian, Northwest University Publ. House, p. 229–237 (in Chinese).

Li, S., Wang, S., Chen, Y., Zhou, H., Zhang, Z., Liou, D., and Qiou, J., 1994, Excess argon in phengite of eclogite: Evidence from comparing dating of eclogite by Sm-Nd, Rb-Sr, and $^{40}Ar/^{39}Ar$ isotope methods: Chem. Geol., v. 112, p. 343–350.

Li, S., Xiao, Y., Chen, Y., Liou, D., and Ge, N., 1992, A four-stage tectonic model for evaluation of the Qinling orogenic belt, China: Beijing, Sci. Tech. Publ. House, p. 121–125 (in Chinese).

Li, X., et al., 1993, Discovery of coesite in eclogite and kyanite quartzite from the Dabie Mountains, Henan Province: Jour. Changchun Coll. Geol., v. 23, p. 256–261.

Liang, W., and Jing, Y., 1990, Geology of coesite-bearing eclogite in the Dabie Mountains, Anhui Province: Jour. Changchun Geol. Inst., v. 4, p. 393–398.

Liang W., et al., 1990, The geological characteristics of coesite-bearing eclogites in Dabie Mountains, Anhui: Jour. Changchun Coll. Geol., v. 20, p. 393–398.

Lin, C. H., and Roecker, S. W., 1998, Active crustal subduction and exhumation in Taiwan, *in* Hacker, B., and Liou, J. G., eds., When continents collide: Geodynamics and geochemistry of ultrahigh-pressure rocks: Dordrecht, Kluwer Acad. Publ., p. 1–26.

Liou, J. G., Wang, Q., Zhai, M., Zhang, R. Y., and Cong, B., 1995, Ultrahigh-P metamorphic rocks and their associated lithologies from the Dabie mountains, central

China: A field trip guide to the 3rd Int. Eclogite Field Symp.: Chinese Sci. Bull., v. 40 (suppl.), p. 1–71.

Liou, J. G., and Zhang, R. Y., 1995, Significance of ultrahigh-P talc-bearing eclogitic assemblages: Mineral. Mag., v. 59, p. 93–102.

______, 1996, Occurrence of intergranular coesite in Sulu ultrahigh-P rocks from China: Implications for fluid activity during exhumation: Amer. Mineral., v. 81, p. 1217–1221.

______, 1998, Petrogenesis of ultrahigh-P garnet-bearing ultramafic body from Maowu, the Dabie Mountains, Central China: The Island Arc, v. 7, p. 115–134.

Liou, J. G., Zhang, R. Y., and Ernst, W. G., 1995, Occurrence of some hydrous and carbonate phases in ultrahigh-P rocks from east-central China—implications for the role of fluids in deep subduction zones: The Island Arc, v. 4, p. 362–375.

______, 1997, Lack of fluid during ultrahigh-P metamorphism in the Dabie-Sulu region, eastern China: Proc. 30th Int. Geol. Cong., v. 17, p. 141–156.

Liou, J. G., Zhang, R. Y., and Jahn, B. M., 1997, Petrogenesis of ultrahigh-pressure jadeite quartzite from the Dabie region, east-central China: Lithos., v. 41 (spec. iss. on High-Pressure Metamorphism in Nature and Experiment), p. 59–78.

Liou, J. G., Zhang, R. Y., Wang, X. M., Eide, E. A., Ernst, W. G., and Maruyama, S., 1996, Metamorphism and tectonics of high-P and ultrahigh-P belts in the Dabie-Sulu region, eastern central China, *in* Harrison, M. T., and Yin, A., eds., The tectonic development of Asia: New York, Cambridge University Press, Rubey vol. VIII, p. 300–344.

Liu, J., and Liou, J. G., 1995, Kyanite-anthophyllite schist and southwest extension of the Dabie Mountains ultrahigh- to high-pressure belt: The Island Arc, v. 4, p. 334–346.

Liu, R., 1994, Zircon U-Pb age of the Bixiling eclogite from Dabie Mountains, *in* Developments in research on the mineralogy, petrology, and geochemistry of China: Lanzhou, Lanzhou University Press, p. 169–170 (in Chinese).

Liu, R., Ma, B., Zhang, Z., Yang, H., and Hu, L., 1989, Geological features of the tectonic mélange in northern Jiangsu, southern Shandong region and lithological column: Seism. Geol., v. 11, p. 47–54 (in Chinese).

Liu, X, 1991, The paired high-pressure metamorphic belts in Eastern China: Geol. Rev., v. 37, p. 300–309.

Liu, X., et al., 1993, Study of several ultrahigh pressure metamorphic minerals in eclogites in Dabie Mountains area: Jour. Mineral. Petrol., v. 13, p. 21–26.

Liu, X., and Hao, J., 1989, Structure and tectonic evolution of the Tonbai-Dabie range in the east Qinling collisional belt: Tectonics, v. 8, p. 637–645.

Liu, X., Zhou, H., Ma, Z., and Chang, L., 1998, Chrome-rich clinopyroxene in orthopyroxenite from Maowu, Dabie Mountains, central China: A second record and its implications for petrogenesis: The Island Arc, v. 7, p. 135–141.

Ma, W., 1989, Tectonics of the Tongbai-Dabie fold belt: Jour. Southeast. Asian Sci., v. 3, p. 77–85.

Maruyama, S., Liou, J. G., and Zhang, R., 1994, Tectonic evolution of the ultrahigh-pressure and high-pressure metamorphic belts from central China: The Island Arc, v. 3, p. 112–121.

Maruyama, S., Nutman, A. P., Morikawa, T., and Liou, J. G., 1998, SHRIMP U-Pb geochronology of ultrahigh-pressure metamorphic rocks of the Dabie Mountains, central China: Continental Dynamics, v. 3, p. 72–85.

Mattauer, M., Matte, P., Malavieille, J., Tapponnier, P., Maluski, H., Xu, Z., Lun, L. Y., and Tan, Y., 1985, Tectonics of the Qingling belt: Buildup and evolution of eastern Asia: Nature, v. 317, p. 496–500.

Nagasaki, A., and Enami, M., 1998, Sr-bearing zoisite and epidote in ultrahigh-pressure metamorphic rocks from the Sulu province, eastern China: An important Sr reservoir under UHP conditions: Amer. Mineral., v. 83, p. 240–247.

Nie, S., Yin, A., Rowley, D. B., and Jin, Y., 1994, Exhumation of the Dabie Shan ultrahigh-pressure rocks and accumulation of the Songpan-Ganzi flysch sequence, central China: Geology, v. 11, p. 999–1002.

Ogasawara, Y., Zhang, R. Y., and Liou, J. G., 1998, Petrogenesis of dolomitic marbles from Rongcheng in the Sulu UHP metamorphic terrane, eastern China: The Island Arc, v. 7, p. 82–97.

Okay, A. I., 1993, Petrology of a diamond and coesite-bearing metamorphic terrain, Dabie Shan, China: Eur. Jour. Mineral., v. 5, p. 659–675.

______, 1994, Sapphirine and Ti-clinohumite in ultrahigh-pressure garnet pyroxenite and eclogite from Dabie Shan, China: Contrib. Mineral. Petrol., v. 116, p. 145–155.

______, 1995, Paragonite eclogites from Dabie Shan, China: Re-equilibration during exhumation?: Jour. Metamor. Geol., v. 13, p. 449–460.

Okay, A. L., and Sengör, A. M. C., 1992, Evidence for intracontinental thrust-related exhumation of the ultra-high-pressure rocks in China: Geology, v. 20, p. 411–414.

Okay, A. I., Xu, S., and Sengör, A. M. C., 1989, Coesite from the Dabie Shan eclogites, central China: Eur. Jour. Mineral., v. 1, p. 595–598.

Okay, A. I., Sengör, A. M. C., and Saw, M., 1993, Tectonics of an ultrahigh-pressure metamorphic terrane: The Dabie Shan/Tongbai Shan orogen, China: Tectonics, v. 12, p. 1320–1334.

Omori, S., Liou, J. G., Zhang, R. Y., and Ogasawara, Y., 1998, Petrogenesis of impure dolomitic marble from the Dabie Mountains, central China: The Island Arc, v. 7, p. 98–114.

Pan, G., Jing, Y., Xia, M., Liang, W., Wang, X., Liou, J. G., Maruyama, S., and Mao, H., 1990, Discovery of coesite-bearing eclogite from the Dabie Mountains: Geol. Rev., v. 4, p. 136–139 (in Chinese).

Rowley, D. B., Xue, F., Tucker, R. D., Peng, Z. X., Baker, J., and Davis, A,, 1997, Ages of ultrahigh pressure metamorphism and protolith orthogneisses from the eastern Dabie Shan: U/Pb zircon geochronology: Earth Planet. Sci. Lett., v. 151, p. 21–36.

Rumble, D., 1998, Stable isotope geochemistry of ultrahigh pressure rocks, *in*: Hacker, B., and Liou, J. G., eds., When continents collide: Geodynamics and geochemistry of ultrahigh-pressure rocks: Dordrecht, Kluwer Acad. Publ., p. 241–260.

Rumble, D., and Yui, T. F., 1998, The Qinglongshan oxygen and hydrogen isotope anomaly near Doughai in Jiangsu Province, China: Geochim. et Cosmochim. Acta, v. 62, p. 3307–3321.

Sang, L., 1991, The petrochemistry of the Lower Proterozoic metamorphic rocks from the Dabieshan-Lianyungang area, the southeastern margin of the North China Platform: Mineral. Mag., v. 55, p. 263–276.

Schertl, H. P., and Okay, A. I., 1994, A coesite inclusion in dolomite in Dabie Shan, China: Petrological and rheological significance: Eur. Jour. Mineral., v. 6, p. 995–1000.

Su, S., Wang, R., Gu, D., and Lai, X., 1994, High-pressure metamorphism of eclogitic country rocks, Jiaonan Group, Zhucheng District, Shandong province: Geo. Sci., v, 8, p. 65–72.

Su, W., Xu, S., Jiang, L., and Liu, Y., 1996, Coesite from quartz-jadeitite in the Dabie Mountains, eastern China: Mineral. Mag., v. 60, p. 659–662.

Sun, J., Pan, M., Chen, H., Xue, L., Xu, Q., and Jing, Y., 1992, Macrosopic deformation characteristics of Donghai eclogites in pre-Sinian metamorphic complex in northeastern Jiangsu Province: Acta Petrol. Sinica, v. 8, p. 161–167 (in Chinese).

Suo, S., Sang, L., Han, Y., and You, Z., 1993, The petrology and tectonics of Dabie Precambrian metamorphic terranes, Central China: China University Geosci. Press, Beijing, 259 p. (in Chinese).

Tabata, H., Maruyama, S., and Shi, Z., 1998, Metamorphic zoning and thermal structure of the Dabie UHP-HP terrane, central China: The Island Arc, v. 7, p. 142–158.

Tabata, H., Yamauchi, K., Maruyama, S., and Liou, J. G., 1998, Tracing the extent of a UHP metamorphic terrane: Mineral-inclusion study of zircons in gneisses from the Dabie Shan, *in* Hacker, B., and Liou, J. G., eds., When continents collide: Geodynamics and geochemistry of ultrahigh-pressure rocks: Kluwer Acad. Publ., p. 261–274.

Wallis, S., Ishiwatari, A., Hirajima, T., Ye, K., Guo, J., Nakamura, D., Takao, T., Zhai, M., Enami, M., Cong, B., and Banno, S., 1997, Occurrence and field relationships of ultrahigh-pressure metagranitoid and coesite eclogite in the Su-Lu terrane, eastern China: Jour. Geol. Soc. Lond., v. 154, p. 45–54.

Wang, C. Y., Zeng, R. S., Mooney, W. D., and Hacker, R., in press, Ultrahigh-pressure Dabie Shan orogenic belt: Support for an Alpine subduction model from deep seismic-refraction profiling: Jour. Geophys. Res.

Wang, Q., 1996, Tectonic implication of UHP rocks from the Dabie Mountains: Science in China, ser. D, v. 39, p. 311–318.

Wang, Q., and Cong, B., 1998, Supracrustal sleeve experienced ultrahigh-pressure metamorphism: Continental Dynamics, v. 3, p. 54–62.

Wang, Q., Liu, X., Maruyama, S., and Cong, B., 1994, Top boundary of the Dabie UHPM rocks, central China: Jour. Southeast Asia Earth Sci., v. 11, p. 295–300.

Wang, Q., Zhang, R.., Cong, B., and Zhao, Z., 1992, Eclogites in eastern Shandong–northern Jiangsu region: Their structure and exhumation mechanism: Acta Petrol. Sinica, v. 8, p. 153–160 (in Chinese).

Wang, Q., Ishiwatari, A., Zhao, Z., Hirajima, T., Enami, M., Zhai, M., Li, J., and Cong, B., 1993, Coesite-bearing granulite retrograded from eclogite in Weihai, eastern China: A preliminary study: Eur. Jour. Mineral., v. 5, p. 141–152.

Wang, X., and Liou, J. G., 1991, Regional ultrahigh-pressure metamorphic terrane in central China: Evidence from coesite-bearing eclogite, marble, and metapelite: Geology, v. 19, p. 933–936.

______, 1993, Ultrahigh-pressure metamorphism of carbonate rocks in the Dabie Mountains, central China: Jour. Metamor. Geol., v. 11, p. 575–588.

Wang, X., Liou, J., and Mao, H. K., 1989, Coesite-bearing eclogite from the Dabie Mountains in central China: Geology, v. 17, p. 1085–1088.

Wang, X., Liou, J. G., and Maruyama, S., 1992, Coesite-bearing eclogites from the Dabie Mountains, Central China: Petrology and P-T path: Jour. Geol., v. 100, p. 231–250.

Wang, X., Jing, Y., Liou, J. G., Pan, G., Liang, W., Xia, M., and Maruyama, S., 1990, Field occurrences and petrology of eclogite from the Dabie Mountain, Anhui, Central China: Lithos, v. 25, p. 119–131.

Wang, X., Zhang, R., and Liou, J. G., 1995, Ultrahigh pressure metamorphic terrane in eastern central China, *in* Coleman, R. G., and Wang, X., eds., Ultrahigh-pressure metamorphism: New York, Cambridge University Press, p. 356–390.

Wang, X. Y., Cui, W., Liu, S., Zhang, L., and Wei, C., 1993, Discovery of coesite-bearing eclogites from Southern Henan Province: Acta Petrol. Sinica, v. 9, p. 181–184 (in Chinese).

Webb, L. E., Hacker, B. R., Ratschbacher, L., and Dong, S., in press, Tectonics of high and ultrahigh-pressure rocks in the Qinling-Dabie orogen: $^{40}Ar/^{39}Ar$ thermochronologic constraints on deformation and cooling history: Tectonics.

______, in press, Mesozoic tectonics in the Hong'an and Tongbai Shan of the Qinling-Dabie collisional orogen, China: Implications for the exhumation of ultrahigh-pressure rocks, *in* Hendrix, M. S., and Davis, G. A., eds., Paleozoic and Mesozoic tectonic evolution of central Asia —from continental assembly to intracontinental deformation: Geol. Soc. Amer. Spec. Pap.

Wu, W., Xu, S., Jiang, L., Liu, Y., and Su, W., 1998, Quartz-jadeitite in ultrahigh pressure metamorphic belt in the Dabie Mountains, eastern China: Acta Petrol. Sinica, v. 14, p. 60–70 (in Chinese).

Xiao, Y., and Li, S., 1993, P-T-t paths and its tectonic significance of eclogites from Shima, Dabie Mountains: Geotect. Metallogen., v. 17, p. 239–250 (in Chinese).

Xie, Douke, et al., 1984, A study on the paleosuture in the Dabie Mountain: Bull. Chinese Acad. Geol. Sci., v. 10, p. 167–178.

Xu, G., and Hao, J., 1988, The characteristics of Foziling Group and the geotectonic environment of its formation in the northern foot of the Dabie Mountains: Scientia Geol. Sinica, no, 1, p. 97–109.

Xu, G., and Wang, Y., 1992, The geological evidence for the trench–island arc–backarc basin system in the Tongbai-Dabieshan orogenic belt before the collision time, *in* Xu, G., and Chang, C., eds., Structure of continental lithosphere and mineral resources: Beijing, Oceanic Press, p. 140–160.

Xu, S., Chen, G., Dong, S., Zhou, H., and Zhang, W., 1987, Deformation and tectonic evolution of the major structural elements in Anhui Province, China: Beijing, Ocean Press, 145 p.

Xu, S., Jiang, L., Liu, Y., and Zhang, Y., 1992, Tectonic framework and evolution of the Dabie Mountains in Anhui, eastern China: Acta Geol. Sinica, v. 66, p. 1–14.

Xu, S., Jiang, L., and Zhang, Y., 1992, Tectonic framework and evolution of the Dabie Mountains in Anhui, eastern China: Acta Petrol. Sinica, v. 7, p. 1–13 (in Chinese).

Xu, S., Liu, Y., Jiang, L., Su, W., and Ji, S., 1994, Tectonic regime and evolution of Dabie Mountains: Beijing, Science Press, p. 175 (in Chinese).

Xu, S., Okay, A.I., Ji, S., Sengör, A. M. C., Su, W., Liu, Y., and Jiang, L., 1992, Diamond from the Dabie Shan metamorphic rocks and its implication for tectonic setting: Science, v. 256, p. 80–82.

Xu, Z., Yang, W., Zhang, Z., and Yang, T., 1998, Scientific significance and site-selection research of the first Chinese continental scientific deep drillhole: Continental Dynamics, v. 3, p. 1–13.

Xue, F., Rowley, D. B., and Baker, J., 1996, Refolded synultrahigh pressure thrust sheets in the south Dabie complex, China: Field evidence and tectonic significance: Geology, v. 24, p. 455–458.

Xue, F., Rowley, D. B., Tucker, R. D., and Peng, Z. X., 1997, U-Pb geochronology of granitoid rocks in the north Dabie complex and its tectonic significance: Jour. Geol., v. 105, p. 744–753.

Yang, J., 1991, Eclogite, garnet pryoxenites and related ultrabasics in Shangdong and North Jiangsu of east China: Beijing, Geol. Publ. House, 99 p. (in Chinese).

Yang, J., Godard, G., and Smith, C., 1998, K-feldspar-bearing coesite pseudomorphs in an eclogite from Lanshantou (Eastern China): Eur. Jour. Mineral., v. 10, p. 969–985.

Yang, J., Godard, G., Kienast, J. R., Lu, Y., and Sun, J., 1993, Ultrahigh-pressure, 60 kbar, magnesite-bearing garnet peridotites from northeastern Jiangsu, China: Jour. Geol., v. 101, p. 541–554.

Yang, W., 1997, Crustal structure and development of Sulu UHPM terrane in east-central China: Episodes, v. 20, p. 100–103.

Yang, W., and Chen, Z., 1998, Crustal structure and development of the Dabie UHPM terrane in east-central China: Continental Dynamics, v. 3, p. 86–95.

Yang, W., Wang, L., Han, Y., Zhang, Z., and Liou, R., 1994, Chronology of blueschist and eclogite from the Dabie area, *in* Chen, H., ed., Research on isotope geochemistry: Hangzhou, Zhejiang University Press, p. 175–186 (in Chinese).

Ye, K., and Cong, B., 1995, Petrology of ultrahigh-pressure meta-calc-silicate rock from Tengjiaji, Rongcheng County, Shangdong Province, China: Chinese Sci. Bull., v. 40, p. 730-733 (in Chinese).

Ye, K., Cong, B., and Ye, D., 1997, Ultrahigh-pressure P-bearing titanite from eclogite at Houshuichegou, Shandong Province, eastern China: Scientia Geol. Sinica, v. 6, p. 7–18.

Ye, K., and Hirajima, T., 1996, High-pressure marble at Yangguantun, Rongcheng County, Shandong Province, eastern China: Mineral. Petrol., v. 57, p. 151–165.

Ye, K., and Xu, P., 1992, Petrogenesis of eclogites and related rocks at Datuan area, Rongcheng County, Shandong Province, China: Acta Petrol. Sinica, v. 8, p. 27–40 (in Chinese).

Ye, K., and Ye, D., 1996, Significance of phosphorous, P- and Mg-bearing high-Al titanite in high-pressure marble from Yangguantun, Rongcheng County, Shandong province: Chinese Sci. Bull., v. 41, p. 1194–1197.

Yin, A., and Nie, X., 1993, An indentation model for the north and south China collision and the development of the Tan-Lu and Honam fault systems, eastern Asia: Tectonics, v. 12, p. 801–813.

Yin, Y., 1991, Characteristics and petrogenesis of the eclogites in Shandong and Jiangsu: Acta Petrol. Mineral., v. 10, p. 11–20.

You, Z., Han, Y., Suo, S., Chen, N., and Zhong, Z., 1993, Metamorphic history and tectonic evolution of the Qinling complex, eastern Qinling Mountains, China: Jour. Metamor. Geol., v. 11, p. 549–560.

You, Z., Suo, S., Zhong, Z., Zhang, Z., and Wei, B., 1998, The extensional tectonics and retrometamorphic processes subsequent to the high-pressure and ultrahigh-pressure metamorphic events in Dabieshan, China: Continental Dynamics, v. 3, p. 63–71.

Yuan, X. C., Xu, M. C., Tang, W. B., and Wang, Q. H., 1994, Seismic reflection profile on eastern Qinling continental crust: Acta Geophys. Sinica, v. 37, p. 749–758 (in Chinese).

Yui, T. F., Rumble, D., Chen, C. H., and Lo, C. H., 1997, Stable isotope characteristics of eclogites from the ultrahigh-pressure metamorphic terrain, east-central China: Chem. Geol., v. 137, p. 135–147.

Yui, T. F., Rumble, D., and Lo, C. H., 1995, Unusually low $\delta^{18}O$ ultrahigh-pressure metamorphic rocks from Su-Lu terrane China: Geochim. et Cosmochim. Acta, v. 59, p. 2859–2864.

Zhai, M., and Cong, B., 1996, Major and trace element geochemistry of eclogites and related rocks, *in* Cong, B., ed., Ultrahigh-pressure metamorphic rocks in the Dabieshan-Sulu region of China: Beijing, Science Press, p. 69–89.

Zhai, M., Cong, B., Zhao, Z., Wang, Q., and Li, J., 1992, High pressure jadeitic quartzite from the Dabie eclogite belt and its geological implication: Chinese Sci. Bull., v. 37, p. 1013–1015 (in Chinese).

Zhai, M., Cong, B., Zhao, Z., Wang, Q., Wang, G., and Jiang, L., 1995, Petrological-tectonic units in the coesite-bearing metamorphic terrain of the Dabie Mountains, central China and their geotectonic implications: Jour. Southeast Asian Earth Sci., v. 11, p. 1–13.

Zhang, G., Zhou, D., and Yu, Z., 1991, The Dabie orogenic belt and the Zhoukou fault-downwarp basin: *in* Ma, X., Yang, S., and Zhu, Z., eds., Memoirs of continental tectonics in China: Beijing, China University Geosciences Press, Beijing, 14–24 (in Chinese).

Zhang, H., 1989, The typomorphic features of some metamorphic minerals in eclogites at Taihu, Anhui: Earth Sci. Jour. Wuhan College Geol., v. 14, p. 599–606.

Zhang, Q., Ma, B., Liou, R., Zhao, D., Fan, Q., Li, Q., and Li, X., 1995, Geochemistry of ultramafic rocks from Raobazai, Anhui Province: A remnant of continental lithosphere over subduction zone: Science in China, v. 38, p. 1522–1529.

Zhang, Q., Zhao, D., Liu, R., Ma, B., Fan, Q., and Li, Q., 1993, Protoliths of garnet peridotite–eclogite complex in Bixiling Wuhe, Yuexi County, Anhui Province: Chinese Sci. Bull., v. 38, p. 1694–1697.

Zhang, R. Y., Hirajima, T., Banno, S., Cong, B., and Liou, J. G., 1995, Petrology of ultrahigh-pressure rocks from the southern Sulu region, eastern China: Jour. Metamor. Geol., v. 13, p, 659–675.

Zhang, R. Y., and Liou, J. G., 1994, Coesite-bearing eclogite in Henan Province, central China: Detailed petrography, glaucophane stability, and PT path: Eur. Jour. Mineral., v. 6, p. 217–233.

______, 1994, Significance of magnesite paragenesis in ultrahigh-P metamorphic rocks: Amer. Mineral., v. 79, p. 397–400.

______, 1996, Significance of coesite inclusions in dolomite from eclogite in the southern Dabie mountains, China: Amer. Mineral., v. 80, p. 181–186.

______, 1997, Partial transformation of gabbro to coesite-bearing eclogites from Yangkou, the Sulu Terrane, eastern China: Jour. Metamor. Geol., v. 15, p. 183–202.

______, 1998, Ultrahigh-pressure metamorphism of the Sulu terrane, eastern China: A prospective view: Continental Dynamics, v. 3, p. 32–53.

Zhang, R. Y., Liou, J. G., and Cong, B., 1994, Petrogenesis of garnet-bearing ultramafic rocks and associated eclogites in the Su-Lu ultrahigh-P metamorphic terrane, eastern China: Jour. Metamor. Geol., v. 12, p. 169–186.

______, 1995, Ultrahigh-pressure metamorphosed talc-, magnesite-, and Ti-clinohumite–bearing mafic-ultramafic complex, Dabie Mountains, east central China: Jour. Petrol., v. 36, p. 1011–1038.

Zhang, R. Y., Liou, J. G., and Ernst, W. G., 1995, Ultrahigh-pressure metamorphism and decompressional P-T path of eclogites and country rocks from Weihai, eastern China: The Island Arc, v. 4, p. 293–309.

Zhang, R. Y., Liou, J. G., and Tsai, C. H., 1996, Petrogeneses of a high-temperature metamorphic belt: A new tectonic interpretation for the north Dabieshan, Central China: Jour. Metamor. Geol., v. 14, p. 319–333.

Zhang, R. Y., Liou, J. G., and Ye, K., 1996, Petrography of UHPM rocks and their country rock gneisses: *in* Cong, B., ed., Ultrahigh-pressure metamorphic rocks in the Dabieshan-Sulu Region of China: Beijing, Science Press, China, p. 49–68.

______, 1996, Mineralogy of UHPM rocks, *in* Cong, B., ed., Ultrahigh-pressure metamorphic rocks in the Dabieshan-Sulu Region of China: Beijing, Science Press, p. 106–127.

Zhang, R. Y., Rumble, D., Liou, J. G., and Wang, Q. C., 1998, Low $\delta^{18}O$ ultrahigh-P garnet mafic-ultramafic rocks from Dabieshan, China by in-situ UV Laser Probe: Chem. Geol., v. 150, p. 161–170.

Zhang, R. Y., Shu, J. F., Mao, H. K., and Liou, J. G., 1999, Magnetite lamellae in olivine and clinohumite from the Dabie UHP ultramafic rocks, central China: Amer. Mineral., v. 84, p. 564–569.

Zhang, W., 1998, The proposed technical plan for the drilling of Dabie-Sulu scientific borehole: Continental Dynamics, v. 3, p. 103–107.

Zhang, X., Chen, Y., and Wang, G., 1992, Characteristics of eclogites in Junan country, southeastern Shandong: Acta Petrol. Sinica, v. 8, p. 40–49 (in Chinese).

Zhang, Y., Jiang, L., and Liu, Y., 1991, Characteristics of ultra-high-pressure coesite eclogite and metamorphism in Southern Dabie orogenic belt, Anhui Province: Acta Petrol. Sinica, v. 7, p. 1–13.

Zhang, Z., 1990, Significance of the study of eclogites and its new achievements: Geol. Sci. Tech. Information, v. 9, p. 25–30.

Zhang, Z., et al., 1992, Metamorphic P-T-t path of the coesite eclogites from the Dabieshan, central China, and some geodynamic implications: Earth Sci. Jour. China Univ Geosci., v., 17, p. 141–149.

Zhang, Z., Liu, D., and Fu, G., 1994, Isotopic dating of metamorphic strata in northern Qinling: Beijing, Geol. Publ. House, 191 p.

Zhang, Z., and You, Z., 1993, Development in research on the Dabieshan eclogite belt: Advances in Earth Science, v. 8, p. 38–43.

Zhang, Z., You, Z., and Han, Y., 1994, Geochemistry and chronology of the eclogite belt in central China: Acta Mineral. Sinica, v. 14, p. 215–222 (in Chinese).

Zhang, Z., You, Z., Han, Y., and Sang, L., 1996, Petrology, metamorphic process and genesis of the Dabie-Sulu eclogite belt, eastern-central China: Acta Geol. Sinica, v. 9, p. 134–156.

Zhang, Z., You, Z., Han, Y., Yang, W., and Wei, B., 1995, High-pressure and ultrahigh-pressure metamorphic belt in Tongbai–Dabie Mountains, central China: Jour. China University Geosci., v. 6, p. 139–145.

Zhao, D., Zhang, Q., Liou, R., Fan, Q., and Ma, B., 1993, Preliminary study of transformation from spinel peridotite to garnet-peridotite at Suoluoshu in Shandong province: Chinese Sci. Bull., v. 38, p. 1581–1584.

Zhao, G., et al., 1992, The micro-deformation and its paleo-tectonic setting of the mylonitic rocks from Jiaonan tectonic complex belt, information from the eclogite country rocks: Jour. Ocean University Qingdao, v. 22, p. 71–80.

Zhao, X., and Coe, R. S., 1987, Palaeomagnetic constraints on the collision and rotation of North and South China: Nature, v. 327, p. 141–144.

Zhao, Z., Wang, Q., and Cong, B., 1992, Coesite-bearing ultramafic pressure metamorphic rocks from Donghai, northern Jiangsu province, eastern China: "Foreign" or "in situ"?: Scientia Geol. Sinica, v. 1, p. 43–58.

Zheng, Y. F., 1997, Oxygen and carbon isotope anomalies in the ultrahigh pressure metamorphic rocks of the Dabie-Sulu teranes: Implications for geodynamics: Episodes, v. 20, p. 104–109.

Zheng, Y. F., Fu, B., Cong, B., and Li, S. G., 1966, Extreme ^{18}O depletion in eclogite from the Su-Lu terrane in East China: Eur. Jour. MineraL, v. 8, p. 317–323.

Zheng, Y. F., Fu, B., Cong, B., and Wang, Z. R., 1998, Carbon isotope anomaly in marbles associated with eclogites from the Dabie Mountains in China: Jour. Geol., v. 106, p. 97–104.

Zheng, Y. F., Fu, B., Li, Y., Ziao, Y., and Li, S., 1998, Oxygen and hydrogen isotope geochemistry of ultrahigh-pressure eclogites from the Dabie Mountains and the Sulu terrane: Earth Planet. Sci. Lett., v. 155, p. 113–129.

Zhou, D., and Graham, S. A. 1996, Songpan-Ganzi complex of the west Qinling Shan as a Triassic remnant ocean basin, *in* Harrison, T. M., and Yin, A., eds., The tectonic development of Asia: New York, Cambridge University Press, Rubey vol. VIII, p. 281–299.

Zhou, G., Liou, J. G., and Liu, Y., 1996, High- and ultra-high-P metamorphic terranes of northern Hubei Province, China: China Science Press, 242 p. (in Chinese).

Zhou, G., Liu, Y. J., Eide, E. A., Liou, J. G., and Ernst, W. G., 1993, Progressive, high pressure–low temperature metamorphism in northern Hubei Province, central China: Jour. Metamor. Geol., v. 11, p. 561–574.

The Maksyutov Complex

Beane, R., Liou, J. G., Coleman, R. G., and Leech, M. L., 1995, Petrology and retrograde P-T path for eclogites of the Maksyutov Complex, Southern Ural Mountains, Russia: The Island Arc, v. 4, p. 254–266.

Dobretsov, N. L., and Dobretsova, L. V., 1988, New mineralogic data on the Maksyutov eclogite-glaucophane schist complex, southern Urals: Doklady. Akad. Nauk SSSR, v. 300, p. 111–116.

Dobretsov, N. L., Shatsky, V. S., Coleman, R. G., Lennykh, V. L, Valizer, P. M., Liou, J. G., Zhang, R. Y., and Beane, R. J., 1996, Tectonic setting and petrology of ultrahigh-pressure metamorphic rocks in the Maksyutov Complex, Ural Mountains, Russia: INT. GEOL. REV., v. 38, p. 136–160.

Leech, M. L., and Ernst, W. G., 1998, Graphite pseudomorphs after diamond? A carbon isotope and spectroscopic study of graphite cuboids from the Maksyutov Complex, south Ural Mountains, Russia: Geochim. et Cosmochim. Acta, v. 62, p. 2143–2154.

Leech, M. L., and Stockli, D. F., 1999, Exhumation of the high- to ultrahigh-pressure Maksyutov Complex, south Ural Mountains, Russia: Tectonics, in review.

Lennykh, V., Valizer, P. M., Beane, R., Leech, M., and Ernst, W. G., 1995, Petrotectonic evolution of the Maksyutov Complex, south Urals, Russia: Implications for ultrahigh-pressure metamorphism: INT. GEOL. REV., v. 37, p. 584–600.

Matte, P., and Chemenda, A., 1996, A mechanism for exhumation of high-pressure metamorphic rocks during continental subduction in Southern Urals. Geodynamics: Comptes Rendus, Acad. Sci. Paris, ser. IIa, v. 323, p. 525–530.

Shatsky, V. S., Jagoutz, E., and Koz'menko, O. A., 1997, Sm-Nd dating of the high-pressure metamorphism of the Maksyutov Complex, Southern Urals: Trans. Russian Acad. Sci., Earth Sci. Sect. v. 353, p. 285–287.

The Kokchetav Massif

Claoué-Long, J. C., Sobolev, N. V., Shatsky, V. S., and Sobolev, A. V., 1991, Zircon response to diamond-pressure metamorphism in the Kokchetav massif, USSR: Geology, v., 19, p. 710–713.

De Corte, K., Carigny, P., Shatsky, V. S., Sobolev, N. V., and Javoy, M., in press, First evidence of inclusions in metamorphic microdiamonds from the Kokchetav massif, northern Kazakhstan: Geochim. et Cosmochim. Acta.

Dobretsov, N. L., Sobolev, N. V., Shatsky, V. S., Coleman, R. G., and Ernst, W. G., 1995, Geotectonic evolution of diamondiferous paragneisses, Kokchetav Complex, northern Kazakhstan—the geologic enigma of ultrahigh-pressure crustal rocks within a Phanerozoic foldbelt: The Island Arc, v. 4, p. 267–279.

Korsakov, A. V., Shatskiy, V. S., and Sobolev, N. V., 1998, First occurrence of coesite in eclogites from the Kokchetav massif: Doklady Akad. Nauk, Earth Sci. Sect., v. 359, p. 77–81.

Shatsky, V. S., Jagoutz, E., Koz'menko, O. A., Blinchik, T. M., and Sobolev, N. V., 1993, Age and genesis of eclogites from the Kokchetav massif, Northern Kazakhstan: Russ. Geol. Geophys., v. 34, p. 40–50.

Shatsky, V. S., Jagoutz, E., Sobolev, N. V., Kozmenko, O. A., Parkhomenko, V. S., and Troesch, M., in press, Geochemistry and age of ultra-high pressure rocks from the Kokchetav massif, northern Kazakhstan: Contrib. Mineral. Petrol.

Shatsky, V. S., Theynissen, K., Dobretsov, N. L., and Sobolev, N. V., 1998, New indication of ultrahigh-pressure metamorphism in the mica schists of the Kulet site of the Kokchetav massif: Geol. Geofiz., v. 9, p. 1039–1044 (in Russian).

Sobolev, N. V., Dobretsov, N. L., Bakirov, A. B., and Shatsky, V. S., 1986, Eclogites from various types of metamorphic complexes in the USSR and the problems of their origin: *in* Evans, B. W., and Brown, E. H., eds., Blueschists and eclogites: Geol. Soc. Amer. Memoir, v. 164, p. 349–363.

Sobolev, N. V., and Shatsky, V. S., 1990, Diamond inclusions in garnets from metamorphic rocks: A new environment for diamond formation: Nature, v. 343, p. 742–746.

Sobolev, N. V., Shatsky, V. S., Vavilov, M. A., and Goryainov, S. V., 1991, Coesite inclusion in zircon from diamondiferous gneiss of Kokchetav massif—first finding of coesite in metamorphic rocks in the territory of the USSR: Doklady Akad. Nauk SSSR, v. 321, p. 184–188 (in Russian).

Yaroslavtseva, N. S., 1994, Genesis of polymetamorphites of the Kokshetau massif in Paleozoic structures (formation, development, mineralogy): Geology of Kazakhstan, no. 6, p. 69–81 (in Russian).

Zhang, R. Y., Liou, J. G., Coleman, R. G., Ernst, W. G., Sobolev, N. V., and Shatsky, V. S., 1997, Metamorphic evolution of diamond-bearing and associated rocks from the Kokchetav Massif, northern Kazakhstan: Jour. Metamor. Geol., v. 15, p. 479–496.

Other Asian regions

Bai, W. J., Zhou, M. F., and Robinson, P. T., 1993, Diamond-bearing mantle peridotites and podiform chromitites in the Luobusa and Donqiao ophiolite massifs, Tibet: Can. Jour. Earth Sci., v. 30, p. 1650–1659.

Bai, W., Robinson, P. T., and Zhou, M., 1993, Diamond-bearing peridotites from Tibetan ophiolites: Implications for a subduction-related origin of diamond: Proc. Mid-Continent Diamonds Symp., p. 77–92.

Parkinson, C. D., Miyazaki, K., Wakita, K., Barber, A. J., and Carswell, D. A., 1998, An overview and tectonic synthesis of the pre-Tertiary very-high-pressure metamorphic and associated rocks of Java, Sulawesi, and Kalimantan, Indonesia: The Island Arc, v. 7, p. 184–200.

Parkinson, C. D., in review, Coesite-bearing quartzites from Sulawesi, Indonesia: A first record of UHP metamorphism in SE Asia, *in* Maruyama, S., and Parkinson, D. D., eds, Current concepts of orogeny and future perspectives: Tectonophys., spec. vol.

Tagiri, M., and Bakirov, A., 1990, Quartz pseudomorph after coesite in garnet from a garnet-chloritoid-talc schist, northern Tien-Shan, Kirghiz, USSR: Proc. Japan Acad., v. 66, p. 135–139.

Tagiri, M., Yano, T., Bakirov, A. Nakajima, T., and Uchiumi, S., 1995, Mineral parageneses and metamorphic P-T paths of ultrahigh-pressure eclogites from Kyrgyzstan Tien-Shan: The Island Arc, v. 4, p. 280–292.

Yan, B., Liang, R., Fan, Q., Yang, F., and Wan, C., 1986, Characteristics of diamond and diamond-bearing ultramafic rocks in Qiaoxi and Hongqu, Xizang: Bull. Inst. Geol., Chinese Acad. Geol. Sci., v. 14, p. 61–124.